Inference for Diffusion Processes

Christiane Fuchs

Inference for Diffusion Processes

With Applications in Life Sciences

Springer

Christiane Fuchs
Institute for Bioinformatics and Systems Biology
Helmholtz Zentrum München
Neuherberg
Germany

ISBN 978-3-642-25968-5 ISBN 978-3-642-25969-2 (eBook)
DOI 10.1007/978-3-642-25969-2
Springer Heidelberg New York Dordrecht London

Library of Congress Control Number: 2012950004

Printed on acid-free paper

Springer is part of Springer Science+Business Media (www.springer.com)

To Florian

Foreword

Beginning with Brownian motion and its modifications, diffusion processes and stochastic differential equations have a long tradition as mathematical models for stochastic phenomena evolving in continuous time, with applications in diverse substantive fields such as engineering, finance and life sciences. While many textbooks, in particular in financial mathematics, include concise and rigorous introductions to diffusion processes and stochastic calculus, they often provide only limited information on modern statistical inference and usually neglect the question of how to adequately approximate original phenomena through a diffusion process at all. These two issues are, however, of fundamental importance in a number of applied fields, in particular in modern life sciences: First, the original processes typically have a large but discrete state space, and approximations through diffusion processes can be challenging. Second, data are usually observed with low frequency and often at non-equidistant time points, involve measurement error etc. This requires modern tools such as simulation-based Bayesian inference. Motivated through applications in epidemiology and molecular biology, this book closes the existing gap. For the first time, it provides a unified presentation of the approximation techniques, previously often developed for special cases only, and a thorough account of modern statistical inference, including a powerful Bayesian approach. Together with the two application chapters, this book will be of high value for theoretical and applied work. For me, it was a pleasure to watch it grow, to be able to give some advice, to read the final version, and, at last, to see it on my bookshelf!

Munich, Germany Ludwig Fahrmeir

Preface

This book originates from my doctoral thesis. One of the first problems of my work was to describe the spread of an infectious disease by a diffusion process and to statistically estimate the involved model parameters. At that time, I did not expect such a seemingly straightforward task to surface so many diverse open problems to fill an entire book. As a mathematician by training, I knew about stochastic calculus, but I did not anticipate the troubles of deriving diffusion approximations and inferring their parameters from real data.

When delving into the diffusion approximation literature, I noticed that there were several, at first sight, contradicting approaches, some of them formulated in generality, others being carried out for particular problems. Their appropriateness, differences and conformities, however, were unclear as well as their extension to more complex, e.g. multidimensional, processes. Furthermore, parameter estimation for diffusions is a challenging problem, in particular if the application of interest involves multi-dimensional processes, few observation times, latent variables and considerable measurement error. Under these circumstances, probably the only applicable technique is a popular Bayesian approach which is used in a number of scientific papers. I was astonished that I could not find any textbook which comprehensively explained it. Moreover, the method has a well-known but hard-to-grasp convergence problem, which has not been detailed in any book or review so far. Since I am convinced that these are subjects of wide-spread interest and importance, I dedicated to them the major chapter in each of the first two parts of this book. The third part finally addresses the initial project which triggered the theoretical questions: to estimate a diffusion model for the spread of diseases.

In contrast to existing literature, this book treats modelling and inference for diffusions under one umbrella. It thus covers both steps that necessarily arise in a real application. Importance is attached to presenting the methods both comprehensibly and mathematically well-founded. As such, the book addresses both theoreticians, like mathematicians and statisticians, as well as practitioners, like bioinformaticians and biologists. The reader is required to have basic knowledge about deterministic differential equations, probability theory and statistics. An introduction to stochastic calculus, in particular to diffusions, is provided in this book.

Everybody who supported me during the writing of my thesis "Bayesian Inference for Diffusion Processes with Applications in Life Sciences", submitted in 2010 at Ludwig-Maximilians-Universität Munich under my maiden name Christiane Dargatz, also supported the making of this book. My sincere gratitude is due to my supervisors Ludwig Fahrmeir and Gareth Roberts. They enriched my work through their advice, ideas and encouragement. I deeply appreciate the careful proof-reading and helpful comments by Michael Höhle. Katrin Schneider and Lothar Schermelleh deserve my thanks for having initiated the collaboration on the FRAP project and having collected all the data. I thank my former and present colleagues at Ludwig-Maximilians-Universität Munich and Helmholtz Centre Munich for their interest and helpful discussions. Furthermore, I thank Niels Thomas and Alice Blanck from Springer for the friendly and constructive cooperation. My family has been a constant source of support, which I greatly acknowledge. My heartfelt gratitude is due to my husband Florian, who caringly accompanied me and my work all chapters long.

Munich, Germany Christiane Fuchs

Contents

Notation

Symbols

$\mathbb{N} = \{1, 2, 3, \ldots\}$	the natural numbers
$\mathbb{N}_0 = \mathbb{N} \cup \{0\}$	the non-negative whole numbers
$\mathbb{Z}$	the whole numbers
$\mathbb{Q}$	the rational numbers
$\mathbb{R}$	the real numbers
$\mathbb{R}_+$	the strictly positive real numbers
$\mathbb{R}_0 = \mathbb{R}_+ \cup \{0\}$	the non-negative real numbers

$\boldsymbol{x}$	a column vector
$\boldsymbol{x}'$	the transpose of $\boldsymbol{x}$
$\boldsymbol{0} = (0, \ldots, 0)'$	the null vector
$\boldsymbol{e}_i = (0, \ldots, 1, \ldots, 0)'$	the ith unit vector
$\boldsymbol{I}$	the identity matrix
$\operatorname{diag}(\boldsymbol{A})$	the main diagonal of the quadratic matrix $\boldsymbol{A}$
$\operatorname{diag}(\boldsymbol{a})$	quadratic matrix with main diagonal $\boldsymbol{a}$ and zero entries otherwise

$\lvert a \rvert$	the absolute value of $a \in \mathbb{R}$
$\lVert \boldsymbol{A} \rVert$	Euclidean distance, i. e. $\lVert \boldsymbol{A} \rVert^2 = \operatorname{tr}(\boldsymbol{A}'\boldsymbol{A})$ for a vector or matrix $\boldsymbol{A}$

$\lvert \boldsymbol{k} \rvert = \sum_{i=1}^{n} k_i$, where $\boldsymbol{k} = (k_1, \ldots, k_n)'$ (in the context of Sect. B.1)

$\boldsymbol{u} \diamond \boldsymbol{v} = (u_1 v_1, \ldots, u_n v_n)'$, where $\boldsymbol{u} = (u_1, \ldots, u_n)'$ and $\boldsymbol{v} = (v_1, \ldots, v_n)'$

$(\Omega, \mathcal{F}^*, \mathcal{F}, \mathbb{P})$	a filtered probability space with sample space Ω, σ-algebra $\mathcal{F}^*$, natural filtration $\mathcal{F} = (\mathcal{F}_t)_{t \geq 0}$ and probability measure $\mathbb{P}$
$\mathcal{L}$	the σ-algebra of Lebesgue subsets of $\mathbb{R}$

$\mathbb{L}$	Lebesgue measure
$\mathbb{W}$	Wiener measure
$\mathbb{P}_1 \ll \mathbb{P}_2$	the measure $\mathbb{P}_1$ is absolutely continuous with respect to $\mathbb{P}_2$
$\mathbb{P}_1 \perp \mathbb{P}_2$	the measures $\mathbb{P}_1$ and $\mathbb{P}_2$ are mutually singular, i. e. they have disjoint support
$\mathbb{P}_1 \otimes \mathbb{P}_2$	factorisation of measures
$\mathfrak{L}(X)$	the distribution of the random variable X
$\mathbb{1}(A)$	indicator function; equal to one if A is true and zero otherwise
$\delta(x-y)$	Dirac delta function; equal to ∞ if $x=y$ and zero othrwise
$a = o(h)$	$\Leftrightarrow \lim_{h\to 0}(a/h) = 0$
$\biguplus_{i=1}^{n} A_i$	disjoint union of sets $A_1, \ldots, A_n$
Γ	Gamma function, defined as $\Gamma(x) = \int_0^\infty t^{x-1}\exp(-t)\mathrm{d}t$
$\mathrm{d}/\mathrm{d}t$	total derivative, i. e.

$$\frac{\mathrm{d}}{\mathrm{d}t} f(t,\xi_1,\ldots,\xi_m) = \frac{\partial f(t,\xi_1,\ldots,\xi_m)}{\partial t} + \sum_{j=1}^{n} \frac{\partial f(t,\xi_1,\ldots,\xi_m)}{\partial \xi_j}\frac{\mathrm{d}\xi_j}{\mathrm{d}t}$$

$\partial/\partial t$	partial derivative, where all arguments but t remain constant

Abbreviations

SDE	stochastic differential equation
ODE	ordinary differential equation
i.i.d.	independent and identically distributed
a.s.	almost surely

Distributions

Normal distribution	$\boldsymbol{X} \sim \mathcal{N}(\boldsymbol{\mu}, \boldsymbol{\Sigma})$ with $\boldsymbol{\mu} \in \mathbb{R}^n$, $\boldsymbol{\Sigma} \in \mathbb{R}^{n\times n}$ symmetric and positive definite
Log-normal distribution	$\boldsymbol{X} \sim \mathcal{LN}(\boldsymbol{\mu}, \boldsymbol{\Sigma}) \Leftrightarrow \log(\boldsymbol{X}) \sim \mathcal{N}(\boldsymbol{\mu}, \boldsymbol{\Sigma})$
Truncated normal distribution	$X \sim \mathcal{N}_{\text{trunc}}(\mu, \sigma^2)$; generates random numbers from $\mathcal{N}(\mu, \sigma^2)$ restricted to the positive real line, i. e. the density $f(x)$ of this distribution is proportional to the density of $\mathcal{N}(\mu, \sigma^2)$ for $x > 0$ and zero otherwise
Multivariate t distribution	$\boldsymbol{X} \sim t_\nu(\boldsymbol{\mu}, \boldsymbol{\Sigma})$ with $\nu \in \mathbb{R}_+$ degrees of freedom, $\boldsymbol{\mu} \in \mathbb{R}^n$, $\boldsymbol{\Sigma} \in \mathbb{R}^{n\times n}$ symmetric and positive definite and density

$$f(\boldsymbol{x}) = \frac{\Gamma\left(\frac{\nu+n}{2}\right)}{\Gamma\left(\frac{\nu}{2}\right)(\nu\pi)^{n/2}} |\boldsymbol{\Sigma}|^{-\frac{1}{2}}$$

$$\left(1 + \frac{1}{\nu}(\boldsymbol{x} - \boldsymbol{\mu})'\boldsymbol{\Sigma}^{-1}(\boldsymbol{x} - \boldsymbol{\mu})\right)^{-\frac{\nu+n}{2}} \quad \text{for } x \in \mathbb{R}$$

Uniform distribution	$X \sim U(A)$ for a discrete or continuous set A
Poisson distribution	$X \sim \mathrm{Po}(\lambda)$ with $\lambda \in \mathbb{R}_+$
Exponential distribution	$X \sim \mathrm{Exp}(\lambda)$ with $\lambda \in \mathbb{R}_+$
Gamma distribution	$X \sim \mathrm{Ga}(a, b)$ with $a, b \in \mathbb{R}_+$ and density

$$f(x) = \frac{b^a}{\Gamma(a)} x^{a-1} \exp(-bx) \quad \text{for } x \in \mathbb{R}_+$$

Inverse gamma distribution	$X \sim \mathrm{IG}(a, b) \;\Leftrightarrow\; 1/X \sim \mathrm{Ga}(a, b)$

Chapter 1
Introduction

Life sciences cover a diverse spectrum of scientific studies of life, ranging from intracellular processes at molecular level up to the worldwide spread of infectious diseases in humans. Mathematical models are an indispensable tool for the understanding of such complex natural phenomena.

In order to describe the time-continuous evolution of a given system, deterministic models are often favoured as they allow comparatively simple simulation and estimation techniques. Such models, however, do not capture the randomness of the underlying dynamics and therefore turn out to be inadequate in many applications. The utilisation of exact individual-based stochastic models, on the other hand, typically proves to be infeasible in practice when the considered organism involves large numbers of objects. A natural and powerful compromise is the application of stochastic differential equations (SDEs) whose solutions are given by diffusion processes. Hence, diffusions have become an increasingly important tool for the statistical analysis of real world phenomena.

However, approximation of a given dynamic system is often done heuristically in the literature, leading to diffusions that do not correctly mirror the true dynamics of the original process. Furthermore, the statistical inference for diffusions typically turns out to be demanding in real data situations as described below. Hence, the statistical estimation of complex diffusion models as applied to real datasets is not widely spread. These issues are addressed in the present book.

1.1 Aims of This Book

The main objectives of this book are threefold: First of all, given a dynamical system of interest with the aim to describe its temporal evolution by means of a diffusion process, one needs to construct this process such that it appropriately mirrors the characteristics of the considered real phenomenon. In applications in life sciences, the original process typically concerns whole numbers of objects such as the numbers of infectious individuals in a population or the numbers of proteins in a

C. Fuchs, *Inference for Diffusion Processes: With Applications in Life Sciences*,
DOI 10.1007/978-3-642-25969-2_1, © Springer-Verlag Berlin Heidelberg 2013

cell nucleus. Since the paths of diffusion processes are almost surely continuous, a representation in terms of diffusions automatically involves an approximation of the exact dynamics. The transition from discrete to continuous state space causes internal fluctuations which appear as a noise term in the characterising SDEs. These disturbances are small when the system is large. Depending on the underlying problem, their correct specification may be a challenging task. As authors typically work through specific examples, there is no universal standard procedure. One objective of this book is to investigate the systematic derivation of diffusion approximations with the aim to provide a general framework which is both mathematically well-founded and attainable for practitioners. Moreover, according procedures are required also for the case when the underlying system is characterised by more than one size parameter. This problem is investigated in this book for the first time as well.

Next, assume that a diffusion model for some problem of interest is given in parametric form as the solution of an SDE. Provided time-discrete observations of the underlying dynamics, one often wishes to statistically infer on the model parameters. In a first step, the present book investigates the state of the art concerning this objective. As a consequence, maximum likelihood estimation would be the first choice as it yields consistent and asymptotically efficient estimates. However, the likelihood function of the time-continuous diffusion process is typically unknown when the process is observed discretely in time, and hence maximum likelihood estimation is not an option. There is comprehensive literature on alternative frequentist methodology concerning statistical inference for diffusions. However, the application of most such methods becomes problematic either when inter-observation times are large or non-equidistant, or for multi-dimensional diffusion processes, or when some components of the state vector are latent or measured with error. Unfortunately, many datasets in life sciences possess at least one of these properties. A powerful technique to overcome this problem is to estimate the model parameters in a Bayesian framework. A well-known approach is based on the idea to introduce auxiliary data points as additional observations. These are estimated by application of Markov chain Monte Carlo (MCMC) techniques which alternately update the auxiliary data and the model parameter. However, there is one notorious convergence problem caused by a close link between the model parameters and the quadratic variation of the diffusion path. A practical solution for this problem has not yet been proven for multi-dimensional diffusion processes. This open question is answered in this book.

Finally, a third aim of the present work is the application of the above theoretical investigations to real datasets from life sciences. In particular, the spatial spread of human influenza and the in vivo binding behaviour of molecules in a cell nucleus shall be statistically analysed. These are of large interest for life scientists. The considered datasets comprise several of the above mentioned properties so that the utilisation of the newly developed Bayesian estimation technique is required.

1.2 Outline of This Book

In accordance with the just formulated aims, the main chapters of this book are structured in three parts as illustrated in Table 1.1: Chaps. 2–5 deal with modelling especially by means of diffusions, Chaps. 6 and 7 concern the statistical inference for such models, and Chaps. 8 and 9 contain the two just mentioned application studies, combining the former theoretical contributions.

Chapter 2 introduces the reader to mathematical modelling in life sciences, with the focus on human epidemiology and molecular biology as two emerging fields. Typical modelling approaches are explained, where emphasis is put on the importance of using stochastic as opposed to deterministic models. Examples from this chapter are recurrently employed throughout the entire book. For instance, the application studies in Chaps. 8 and 9 originate from the above two research areas.

As a basis for the stochastic analysis of diffusions, which will be carried out in the remainder of this book, Chap. 3 provides a compact introduction to diffusion processes and their characterising SDEs. The contents of this overview are oriented towards the needs of this book. The reader who is familiar with stochastic calculus may skip this chapter and refer to it when required.

Chapter 4 addresses the above mentioned approximation of Markov jump processes by diffusions. For the first time, it provides a detailed overview of such techniques in a multi-dimensional context. To that end, established methods from the literature are supplemented by new formulations and extended to multi-dimensional diffusion processes where necessary. Moreover, this chapter extends all approaches to a more advanced framework, where the dimension of a system is characterised through multiple size parameters rather than a single one.

The theoretical investigations from Chap. 4 are illustrated in Chap. 5, where diffusion approximations for distinguished models from epidemiology are derived. More specifically, this chapter considers a standard model for the spread of infectious diseases and proposes an extension which allows for host heterogeneity. The resulting diffusion processes form the basis of Chap. 8.

Table 1.1 Outline of this book

		1.	Introduction
I.	*Stochastic modelling*	2.	Stochastic modelling in life sciences
		3.	Stochastic differential equations and diffusions in a nutshell
		4.	Approximation of Markov jump processes by diffusions
		5.	Diffusion models in life sciences
II.	*Statistical inference*	6.	Parametric inference for discretely-observed diffusions
		7.	Bayesian inference for diffusions with low-frequency Observations
III.	*Applications*	8.	Application I: spread of influenza
		9.	Application II: analysis of molecular binding
		10.	Summary and future work
		A.–C.	Appendix

Statistical inference for discretely-observed diffusion processes is a challenging task. As indicated before, maximum likelihood estimation is possible only in rare cases which usually do not match the complex dynamics of processes in life sciences. Chapter 6 introduces the reader to the theoretical background of parametric inference for discretely-observed diffusions and reviews frequentist methods from this highly developing research area. The techniques of this and the following chapter are of course also applicable to datasets from other scientific areas than life sciences.

Most techniques that are presented in Chap. 6 struggle when inter-observation times of the considered phenomenon are large. Datasets in life sciences, however, may well be of such low-frequency type. For the first time, Chap. 7 reviews in detail MCMC techniques which base on the introduction of missing data such that the union of missing values and observations forms a high-frequency dataset. Such techniques are also suitable for irregularly spaced observation intervals, multivariate diffusions with possibly latent components and for observations that are subject to measurement error. However, as already described in the aims of this book, the considered concept suffers from convergence problems which are due to strong dependence structures between the model parameters and the quadratic variation of the diffusion path. As a consequence, the MCMC algorithm experiences arbitrarily slow mixing. Chapter 7 newly formulates a modified technique for conditioned diffusions on infinite-dimensional state spaces and provides the mathematical proof that the so-constructed MCMC scheme converges. For practical usability, the proposed scheme is also formulated in algorithmic form. All algorithms are implemented in `R`, which is a freely distributed software available at http://www.r-project.org. Simulation studies certify moderate computing times and a sound performance of the proposed scheme.

Finally, with the modelling and estimation tools from Chaps. 2–7 at hand, it is now possible to statistically analyse complex dynamics in life sciences. Applying the diffusion approximations derived in Chap. 5 and the Bayesian estimation techniques developed in Chaps. 7 and 8 investigates the spread of human influenza, which is one of the most common and severe diseases worldwide. More precisely, statistical inference is carried out for a well-known dataset on an influenza outbreak in a British boarding school and for the spatial spread of influenza in Germany during the season 2009/10, in which the 'swine flu' virus was prevalent. Spatial mixing of individuals is derived based on commuter data. This chapter provides a first application of statistical parameter estimation for spatial epidemic models by utilisation of diffusion approximations.

As a second application, Chap. 9 investigates the binding behaviour of the protein Dnmt1 to chromatin. This protein plays a major role in the maintenance of DNA methylation patterns and is hence of great interest. Suitable data is extracted by application of a fluorescence microscopy technique called FRAP. Appropriate kinetic models are derived as diffusion processes by means of the techniques from Chap. 4, and statistical inference is performed by application of the techniques from Chap. 7. This analysis supplies new insight into cell cycle dependent kinetic properties of Dnmt1. It is the first application of diffusion approximations in the FRAP literature, where deterministic models are prevalent.

Chapter 10 briefly concludes this book and gives an outlook on projects which can be based on its contributions. Supplementary material for the main chapters is provided in Appendices A–C.

Part I
Stochastic Modelling

Chapter 2
Stochastic Modelling in Life Sciences

The dynamics of natural phenomena such as the growth of populations of species, the spread of epidemics, changes in gene frequencies or the course of chemical reactions are all subject to random variation. Their evolution is not exactly predictable. However, the application of mathematical models enables insight into such complex processes.

This chapter motivates and reviews representative application fields from life sciences and appropriate mathematical models. These applications and models will recur throughout the entire book. They give rise to the model constructions in Chaps. 3–5 and the investigation and development of estimation procedures in Chaps. 6 and 7. Moreover, they form the basis for the application studies in Chaps. 8 and 9.

The emphasis of this and the following chapters is on the important role of chance. In the literature, there is a vast number of works for modelling the mentioned dynamics where randomness is not taken into account. Such deterministic models provide a convenient and sometimes also appropriate way to represent a situation of interest. For comparison purposes, this deterministic approach is also introduced here. In general, however, deterministic models are not able to capture the natural stochastic behaviour of a real-world phenomenon. For instance, a deterministic model for the spread of an infectious disease may predict a major outbreak in a marginal situation and possibly prove wrong (cf. Sect. 2.2). Deterministic models for the dynamics of chemical reactions typically fail when the number of reactants is small (e.g. McQuarrie 1967). As another example, Lande et al. (2003) invoke harvest strategies, say in fishery, which may do harm to small populations of endangered species when they are developed based on deterministic models. For that reason, this book particularly focuses on the application of stochastic models. These account for random fluctuations of the considered processes and assign probabilities to critical events.

The structure of the present chapter is as follows: Sect. 2.1 introduces the very general class of compartment models. From such a model, both deterministic and stochastic processes can be derived. Sections 2.2 and 2.3 provide introductions to two emerging fields of life sciences, namely to models for the spread of infectious

C. Fuchs, *Inference for Diffusion Processes: With Applications in Life Sciences*,
DOI 10.1007/978-3-642-25969-2_2, © Springer-Verlag Berlin Heidelberg 2013

diseases and to models for processes in molecular biology, biochemistry and genetics. Both sections start from a compartmental representation and then consider three types of models. These are stochastic jump processes, deterministic continuous processes and stochastic diffusion processes. The first type of process mirrors the exact dynamics of the compartmental system, whereas the second and third can be considered as approximations of the first. The development of an exact simulation algorithm for the jump process in 1976 hence meant a considerable advancement in the field of statistical modelling. This algorithm is presented in Sect. 2.4. In many situations, however, its application is computationally costly. Hence, numerical approximation algorithms for the second and third type of process are outlined as well. Section 2.5 concludes this chapter.

2.1 Compartment Models

In a *compartment model*, all objects involved in a system of interest are arranged in a finite number of *compartments*, i.e. in groups of objects that are defined through certain specified properties (Jacquez 1972). The compartments are mutually disjoint, and the assignment of each object to a compartment is unambiguous. The elements of each compartment are assumed to be homogeneous and well-mixed. Interaction between different compartments happens through the exchange of objects which is described by transition equations. Such passages are assigned with some rate that typically depends on the concentrations of objects from the distinct compartments. In this book, the considered compartmental systems are usually *closed*, i.e. there is no flow of objects to and from the environment.

The classification of objects into different compartments may, for example, be due to the location of animals or humans in a geographical region, the kinetic properties of molecules, or the age or physical conditions of individuals that are susceptible to a disease. Figures 2.1 and 2.2 display two compartment models from the fields of applications that are considered in Sects. 2.2 and 2.3.

A compartment model is a convenient fundament for a dynamical system one wishes to represent. From this model, different types of processes can be derived, all of them standing for the same considered phenomenon. This book will consider

Fig. 2.1 Compartmental representation of the susceptible–infectious–removed (SIR) model that will be investigated in Sect. 2.2.2. In this model, a population of interest is classified into susceptible, infectious and removed individuals. Transitions between these three groups are due to infections and recoveries

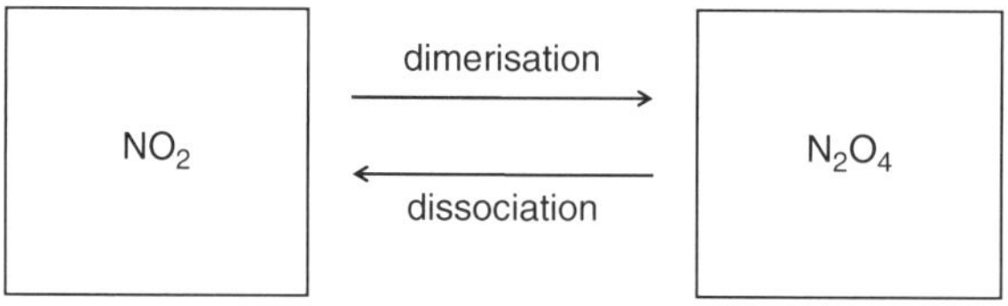

Fig. 2.2 Compartmental representation of the dimerisation of nitrogen dioxide. In this model, all nitrogen dioxide (NO_2) and dinitrogen tetroxide (N_2O_4) molecules in a gas are summarised in two compartments. Depending on the temperature and pressure, two of the NO_2 molecules may dimerise, yielding one N_2O_4 molecule. The other way round, an N_2O_4 molecule may dissociate into two NO_2 molecules

the following three kinds: pure Markov jump process, deterministic processes with continuous sample paths, and diffusion processes. First examples are shown in the next two sections.

2.2 Modelling the Spread of Infectious Diseases

Epidemics of infectious diseases have shaped the history of humankind. They have directly affected economy, politics and demography, the course of wars, social behaviour and religious beliefs (McNeill 1976; Cunha 2004; Smallman-Raynor and Cliff 2004; Sherman 2006; Oldstone 2010).

Devastating historic epidemics and pandemics include the Black Death in 1347–1350 with 25 million deaths in Europe, where there was up to 50% mortality of the urban population in England and Southern Europe; outbreaks of smallpox, measles and typhus in Mexico in 1518–1520 with 2–15 million deads out of a population of 20 million; several cholera epidemics in India during the seventeenth century with more than 20 million deaths; and the Spanish influenza pandemic in 1918–1920 with estimated numbers of worldwide deaths lying between 25 and 50 million (Dobson and Carper 1996; Smallman-Raynor and Cliff 2004; Vasold 2008).

Present-day pandemics comprise for instance the acquired immunodeficiency syndrome (AIDS) caused by the human immunodeficiency virus (HIV) which was identified in the 1980s. It is assumed that in 2008 there were 2.7 million new infections, 2 million AIDS-related deaths and 33.4 million people living with the virus worldwide (UNAIDS 2009). Quite recently, in 2009, an influenza pandemic spread from Mexico over the whole world within a few months. It possibly affected between 11 and 21% of the global population (Kelly et al. 2011) and caused more than 18,000 deaths (WHO 2010). During the early stages of the epidemic, one even feared much higher mortality. Hence, the spread of diseases is still a serious concern in both the developed and developing world.

The elimination of infectious disease epidemics is desirable not only from a humane viewpoint but also regarding economic factors such as manpower and public health costs. Even for diseases with relatively mild courses it is generally favourable to invest in prevention rather than cure. Considerable progress in understanding the propagation of infectious diseases from a medical point of view has been achieved by Louis Pasteur (1822–1895) and Robert Koch (1843–1910), who discovered the cause of infections by microorganisms. Targeted intervention against the spread of diseases, such as vaccination or isolation, however requires an overall comprehension of the typically complex dynamics of an epidemic. This is achieved by application of mathematical modelling (Brauer 2009).

The objectives of this section are the following: First, to introduce basic models for the spread of infectious diseases, and second, to motivate the utilisation of stochastic rather than deterministic models. This presentation is oriented towards the needs of subsequent chapters. For further information, the reader is referred to Bailey (1975), Anderson (1982), Becker (1989), Anderson and May (1991), Daley and Gani (1999), Andersson and Britton (2000), Diekmann and Heesterbeek (2000) and Keeling and Rohani (2008).

2.2.1 History of Epidemic Modelling

Detailed statistics on disease counts go back to John Graunt (1620–1674) who recorded weekly death counts in London together with their causes. The first mathematical model for the spread of infectious diseases, however, is generally accredited to Daniel Bernoulli (1700–1782), but epidemic modelling has not received much attention until the beginning of the twentieth century. Early works include En'ko (1889), Hamer (1906), Ross (1915) and Kermack and McKendrick (1927). Detailed historical accounts on the development of mathematical epidemiology can be found in Bailey (1975), Dietz (1967), Anderson and May (1991) and Daley and Gani (1999).

In the early stages of epidemic modelling, the spread of diseases was generally formulated as a deterministic process. According to Bailey (1975), the first author who included a random component in an epidemic model was McKendrick (1926), but that particular approach was only continued 20 years later. Instead, the class of *chain binomial models*, independently introduced by Lowell Reed and Wade Hampton Frost (see Abbey 1952 or Costa Maia 1952) and Greenwood (1931), established itself. A model of this type considers the evolution of an epidemic at discrete time points. To that end, the number of susceptible and infectious individuals in a population is assumed to be binomially distributed, conditioned on the state of the epidemic at the previous time point. An overview about chain binomial models is given in Becker (1989) and Daley and Gani (1999).

In subsequent years, both stochastic and deterministic models were refined and their mathematical analysis was extended; see e.g. Isham (2004) for a review. The class of *susceptible–infective–removed (SIR) models*, which is introduced in the next section, emerged as the most prominent description of the spread of infectious disease epidemics.

While the comprehension of disease dynamics and the development of mathematical tools progresses, the general framework of modelling the spread of diseases changes as well: First of all, the increased mobility of humans raises the risk of fast spreading pandemics. On the other hand, detailed medical knowledge of infection processes and improved hygienic conditions in many countries help prevent transmission of diseases. Modern epidemiological models take into account travel, social behaviour, the effect of intervention such as vaccination or isolation, and many other aspects.

The following section introduces a standard model from epidemiology which serves as the basis for many extensions as indicated in Sect. 2.2.3 and implemented in Chap. 5. This section concentrates on infectious diseases for humans. The considered diseases are assumed to be directly transmittable rather than vector-borne, i.e. transmitted for example by insects.

2.2.2 *SIR Model*

An *SIR model* (Kermack and McKendrick 1927; Bartlett 1949) classifies a population of fixed size N into susceptible (S), infectious (I) and removed (R) individuals. Transitions between these classes are

$$\text{Ⓢ} + \text{Ⓘ} \xrightarrow{\alpha} 2\,\text{Ⓘ} \quad \text{and} \quad \text{Ⓘ} \xrightarrow{\beta} \text{Ⓡ}. \tag{2.1}$$

The first transition means that each contact between a susceptible and an infectious individual will cause an infection with rate $\alpha \in \mathbb{R}_{+}$, resulting in two infectious individuals. The second transition denotes that each of these infectious individuals will be removed with rate $\beta \in \mathbb{R}_{+}$ due to being recovered and immune, or quarantined, or dead. The parameter α is the *contact rate* of an infectious individual for spreading the disease, and β is the *reciprocal average infectious period*. Some authors also refer to α and β as the *infection rate* and *removal rate*, respectively.

Modifications of the SIR model e.g. disregard recovery (SI), allow a return to the susceptible status (SIS, SIRS), or incorporate a latent/exposed period (SLIR/SEIR). For simplicity, we assume in this section that an individual is infectious as soon as it is infected. The terms *infected*, *infectious* and *infective* are considered interchangeable.

The SIR model is conveniently described as a time-homogeneous Markov process. Unless otherwise stated, we assume the population closed during the time of consideration, ignoring births, non-related deaths, and migration. Furthermore, the population is presumed to mix homogeneously.

Different constructions of the SIR model can be found in the literature, see for example Andersson and Britton (2000) for an overview. The following paragraphs present three of the most common descriptions.

Representation as Pure Markov Jump Process

Denote by S and I the absolute numbers of susceptible and infectious individuals in the population under consideration. Due to the fixed population size N, the current state of an SIR process is completely described by the tuple $(S, I)'$, which is an element of the state space $\mathcal{D} = \{(S, I)' \in [0, N]^2 \cap \mathbb{N}_0^2 \,|\, S + I \leq N\}$; the number of removed individuals can be calculated as $R = N - S - I$.

Hence, let $(S, I)' \in \mathcal{D}$ be the state of the process at time $t \in \mathbb{R}_0$. Assuming that at most one event can occur within a small time interval of length Δt, there are three possibilities for the state of the process at time $t + \Delta t$:

1. $(S - 1, I + 1)'$ in case one infection occurs,
2. $(S, I - 1)'$ in case one recovery occurs,
3. $(S, I)'$ in case nothing happens.

These transitions come up with probabilities

$$p_1 = \alpha\, SI/N\ \Delta t + o(\Delta t), \quad p_2 = \beta I\ \Delta t + o(\Delta t) \quad \text{and} \quad p_3 = 1 - p_1 - p_2, \tag{2.2}$$

respectively, where $o(\Delta t)/\Delta t \to 0$ as $\Delta t \to 0$. See Sect. 5.1.2 for the derivation of (2.2). For $(S, I)' \notin ([0, N-1] \times [1, N-1]) \cap \mathcal{D}$, the above target states may not be an element of $\mathcal{D}$. In those cases, however, the respective transition probabilities leading to them are $o(\Delta t)$. For an initial condition $(S_0, I_0)' \in \mathcal{D}$, the process can therefore never leave the admissible state space.

A Markov process with the above described dynamics is also termed the *general stochastic epidemic*. Section 2.4.1 describes how an according Markov chain can exactly be simulated. Figure 2.3a shows a realisation of such a Markov chain.

A notable insight into the dynamics of the general stochastic epidemic is the following stochastic threshold result: Let $(S_0, I_0)' \in \mathcal{D}$ denote the initial state of the process and define $\mathcal{R}_0 = \alpha/\beta$. Then, in large populations, a major outbreak will occur with probability tending to

$$1 - \left(\min\left\{1, \frac{N}{S_0}\,\mathcal{R}_0^{-1}\right\}\right)^{I_0}$$

as N and $S_0 = N - I_0$ grow to infinity for fixed I_0 (Whittle 1955; Williams 1971; Ball 1983). This probability is positive if and only if the relative removal rate $\mathcal{R}_0^{-1}$ is smaller than the initial fraction of susceptibles S_0/N. In this formulation, the term *major outbreak* means that the fraction S/N of susceptibles will fall below $\mathcal{R}_0^{-1}$ roughly as far as it was above this threshold before, provided that the difference between S_0/N and $\mathcal{R}_0^{-1}$ is not too large. For more details, see for example Daley and Gani (1999, Chap. 3.4). $\mathcal{R}_0$ is called the *basic reproductive ratio* and interpreted as the average number of infections caused by an infectious individual during its entire infectious period, provided that the infective enters a totally susceptible population.

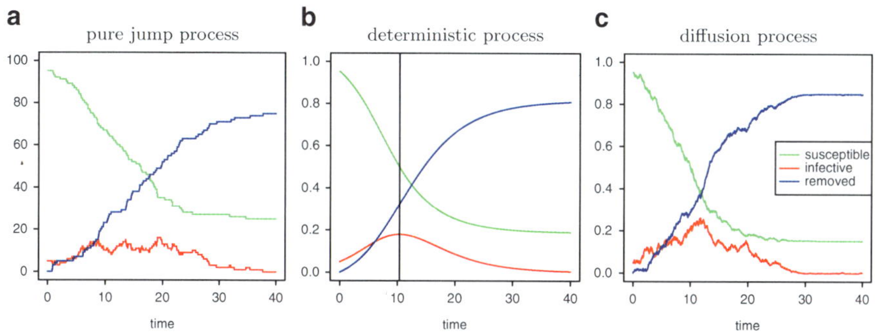

Fig. 2.3 Illustration of SIR model for parameters $\alpha = 0.5$, $\beta = 0.25$ and population size $N = 100$. (**a**) Temporal evolution of numbers of susceptible, infective and removed individuals in the stochastic SIR model with transition probabilities (2.2) for initial value $(S_0, I_0)' = (95, 5)'$. The graphs have been simulated by application of Gillespie's Algorithm, i.e. Algorithm 2.1 on p. 26. (**b**) Temporal evolution of fractions of susceptible, infective and removed individuals in the standard deterministic SIR model (2.3) for initial value $(s_0, i_0)' = (0.95, 0.05)'$. The graphs have been obtained by application of the standard Euler scheme with step length 0.025. The *vertical line* marks the instant at which the fraction of susceptibles falls below $\mathcal{R}_0^{-1} = \beta/\alpha = 0.5$. The fraction of infectives reaches its maximum at this point. (**c**) Temporal evolution of fractions of susceptible, infective and removed individuals in the SIR diffusion model (2.4) for initial value $(s_0, i_0)' = (0.95, 0.05)'$. The graphs have been obtained by application of the Euler-Maruyama scheme from Sect. 6.3.2 with step length 0.025

Representation Through a System of Ordinary Differential Equations

Another possibility to describe the infection dynamics in the SIR model is a deterministic representation via the set of ordinary differential equations (ODEs)

$$\mathrm{d}s/\mathrm{d}t = -\alpha si, \qquad \mathrm{d}i/\mathrm{d}t = \alpha si - \beta i, \tag{2.3}$$

where $s = S/N$ and $i = I/N$ denote the fractions of susceptible and infectious individuals. In this description, the state space $\mathcal{C} = \{(s,i)' \in [0,1]^2 \cap \mathbb{R}_0^2 \,|\, s+i \leq 1\}$ is considered continuous, which is an eligible assumption for large populations. The remaining fraction $r = R/N$ can again be obtained as $r = 1-s-i$. The ODEs (2.3) are subject to an initial condition $(s_0, i_0)' \in \mathcal{C}$. See Sect. 5.1.4 for their formal derivation.

Figure 2.3b shows the typical evolution of an epidemic following the deterministic description (2.3). While recovery follows a linear process, infections occur at high rate only when both the fractions of susceptibles and infectives are sufficiently large. As the ODEs are not explicitly solvable, the trajectories have been obtained numerically by application of the standard Euler scheme (cf. Sect. 2.4.2). Figure 2.4 displays the course of the deterministic SIR process for different values of α and β.

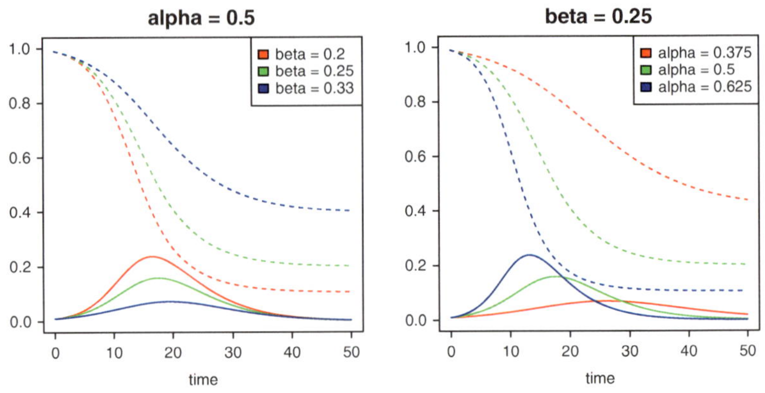

Fig. 2.4 Fractions of susceptibles (*dashed*) and infectives (*solid lines*) in an SIR epidemic following the deterministic model (2.3) for different values of α and β and initial value $(s_0, i_0)'$ equal to $(0.99, 0.01)'$. The graphs have been obtained by application of the standard Euler scheme with step length 0.025 for solving the ODE system. In both graphics, the parameters correspond to $\mathcal{R}_0 = \alpha/\beta \in \{1.5, 2.0, 2.5\}$

The first equation in (2.3) implies that the fraction of susceptibles is strictly decreasing as long as both s and i are nonzero. Solving $\mathrm{d}i/\mathrm{d}t < 0$ leads to $s < \beta/\alpha$. That means, when $\mathcal{R}_0^{-1} := \beta/\alpha$ is greater than the initial fraction of susceptibles s_0, no epidemic will develop. Otherwise, the epidemic will rise first but fall off as soon as the fraction s drops below this threshold. This is the famous *threshold theorem* by Kermack and McKendrick (1927). An obvious strategy to eradicate an epidemic is hence to vaccinate the population until the latter requirement is met. The vertical line in Fig. 2.3b indicates the first time point at which the fraction of susceptibles falls below $\mathcal{R}_0^{-1}$. Apparently, this mark agrees with the time point at which the epidemic reaches its maximum with respect to the number of infected individuals.

Representation Through a System of Stochastic Differential Equations

A third variant to express the SIR dynamics (2.1) as a mathematical process is by a stochastic differential equation (SDE)

$$\begin{pmatrix} \mathrm{d}s \\ \mathrm{d}i \end{pmatrix} = \begin{pmatrix} -\alpha si \\ \alpha si - \beta i \end{pmatrix} dt + \frac{1}{\sqrt{N}} \begin{pmatrix} \sqrt{\alpha si} & 0 \\ -\sqrt{\alpha si} & \sqrt{\beta i} \end{pmatrix} \begin{pmatrix} \mathrm{d}B_1 \\ \mathrm{d}B_2 \end{pmatrix}. \tag{2.4}$$

In this equation, s and i denote again the fractions of susceptible and infectious individuals in the population. The right hand side of the differential equation (2.4) consists of a deterministic and a stochastic component, that is the first and the second summand, respectively. B_1 and B_2 are independent Brownian motions, representing stochasticity in disease transmission and recovery. As for the multivariate ODE (2.3), an appropriate initial condition has to be specified for the SDE (2.4).

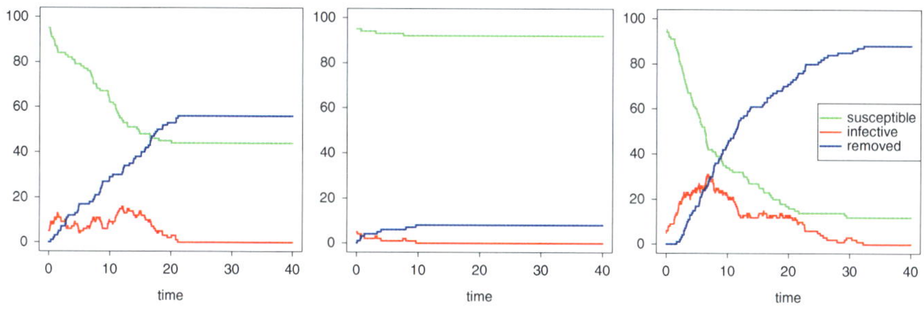

Fig. 2.5 Different courses of stochastic SIR model with transition probabilities (2.2). The simulations base on parameters $\alpha = 0.5$, $\beta = 0.25$, population size $N = 100$ and initial value $(S_0, I_0)' = (95, 5)'$. The graphs have been obtained by application of Gillespie's Algorithm (Algorithm 2.1)

Stochastic differential equations and their solutions, which are typically diffusion processes, will be formally introduced in Chap. 3. Diffusion processes possess extremely wiggly but almost surely continuous trajectories. Figure 2.3c displays the course of an SIR epidemic as described by Eq. (2.4).

Concluding Remarks

This section introduced three different representations of the standard SIR model. There naturally arises the question which type of process is the most appropriate one. The pure Markov jump process, considered first, mirrors the exact dynamics following the transitions (2.1). In many cases, however, this type of process is rather inconvenient for the purpose of simulation and statistical inference. The ODE representation, considered next, has the advantage of a non-individual-based view point. It facilitates interpretation and mathematical analysis, but unfortunately ignores possible variation by chance. In particular, the ODEs (2.3) do not even take into account the population size N and hence unrealistically predict identical fractions of infectives and susceptibles in small and large populations. Finally, the representation of the SIR model in terms of a multivariate SDE consists of both a deterministic and a stochastic component and this way compromises on the former two processes. For this reason, the utilisation of SDEs is favourable in many contexts. Their statistical analysis is the subject of this book. A more elaborate discussion concerning the three above representations is the topic of Chap. 4.

In order to further ellucidate the impact of random events in the SIR model, recall the above deterministic and stochastic threshold results. Both the stochastic model with transition probabilities (2.2) and the deterministic model following the ODEs (2.3) possess the same threshold $\mathcal{R}_0^{-1} = s_0$. The interpretation of this threshold, however, differs substantially in these two models: In the deterministic case, a major epidemic will always occur whenever $\mathcal{R}_0^{-1} < s_0$. In the stochastic case, a major outbreak does not necessarily happen if $\mathcal{R}_0^{-1} < s_0$. The probability for this event lies strictly between zero and one. Figure 2.5 illustrates that different realisations of the course of an epidemic may clearly differ in a stochastic framework. A deterministic simulation for the same model parameters is displayed in Fig. 2.3b. A further investigation of the SDE (2.4) requires its formal definition, which is the subject of Chap. 3. An illustration of this model is for example given in Sect. 5.1.5.

Epidemics will usually terminate due to a lack of infectives, not due to a lack of susceptibles, i.e. at the end of an epidemic outbreak not all individuals will typically have suffered from the disease. According to the above thresholds, major epidemics occur or have positive probability, respectively, when $\mathcal{R}_0 < s$. Suppose that this is the case. Then, there are three general measures to weaken the strength of an epidemic: First, to reduce the number of susceptibles, typically by vaccination, i.e. to decrease the fraction s. Second, to reduce the number of potentially infectious contacts, possibly by closing schools or simply invoking caution, i.e. to decrease α. Third, to reduce the time until an infectious individuals goes over to the removed class, for example by isolation, i.e. to reduce the average infectious period β^{-1}. Each of these three strategies aims at lowering the difference between $\mathcal{R}_0 = \alpha/\beta$ and s, at best accomplishing $\mathcal{R}_0 > s$. The fact that an epidemic does not start or fades out after sufficiently many individuals have left the susceptible state is known as *herd immunity*. The subject of herd immunity, including many examples, is discussed by Anderson and May (1985) and Fine (1993), corresponding control strategies by Morton and Wickwire (1974).

2.2.3 Model Extensions

So far, the SIR model considered in the previous section is fairly simplistic, assuming a homogeneously mixing population, homogeneity of individuals and a time-homogeneous course of an epidemic. In most contexts, some modifications are necessary in order to adapt the mathematical model to a real life situation in which an epidemic develops. Some of these aspects are outlined in the following.

First of all, one very often experiences heterogeneity in contacts among the population. In those cases, individuals typically mix homogeneously in certain subgroups but not with respect to the entire population. It is then meaningful to incorporate patterns into the model such as the age structure of the population e.g. for childhood diseases, a risk structure e.g. for sexually transmitted infections, a geographical structure like an assignment of individuals to different cities or countries, or social structures such as households, schools or circles of friends.

Moreover, there is typically heterogeneity among individuals in the population. For example, susceptible persons may differ in their degree of susceptibility, such as children or elderly people that possibly have a weaker immune system, or individuals that have acquired partial immunity to a disease due to previous epidemics.

In some cases, it is also appropriate to extend an epidemic model such that it accounts for time-varying background conditions. For example, the weather and temperature may well have an effect on the susceptibility of individuals. Furthermore, there may be changes in social behaviour, either independently or dependently on the course of an ongoing epidemic, leading to a variation of contact rates. When observing the spread of a disease over a long period of time, demographic changes such as births and non-related deaths may be included in the model. Other models consider endemic components, i.e. the sustained presence of a certain number of infectious cases in the population, or the presence of carriers that are apparently healthy but infective.

Ample examples and references for the above model extensions are given by Isham (2004) and Keeling and Rohani (2008). In order to mention just a few of them, multipopulation epidemics are for example investigated by Rushton and Mautner (1955), Ball (1986), Sattenspiel (1987), Sattenspiel and Dietz (1995) and Ball et al. (1997). Such models can often be applied to any kind of contact heterogeneity but are in most cases described for the division of a population into several communities in distinct geographical areas. Chapter 5 in this book introduces a multitype SIR model for arbitrary contact heterogeneities as well. Concerning the remaining model modifications mentioned above, Hethcote (2000) takes into account the age of individuals, and Hethcote (1994) gives many references for models which take into account varying population sizes. Neal (2007) analyses an epidemic model where individuals differ with respect to both their susceptibility and infectivity. Ireland et al. (2007) consider seasonality in birth-rates of hosts. Riley (2007) reviews some recent approaches for spatial modelling. Finally, Lloyd-Smith et al. (2005) and Galvani and May (2005) investigate the impact of the presence of superspreaders, that are individuals that communicate a disease in a substantially greater extent than other individuals.

Appropriate modifications of the basic SIR model improve the compatibility between the model assumptions and reality and hence increase the applicability of the model. On the other hand, each extension automatically requires additional information such as community sizes or contact patterns between groups. One should hence balance carefully between complex and oversimplistic models. Stochastic models typically get along with fewer details as minor aspects can be covered by random fluctuations. Chapters 5 and 8 in this book derive and statistically infer on a probabilistic multitype model for the spread of an infectious disease.

2.3 Modelling Processes in Molecular Biology, Biochemistry and Genetics

Understanding the mechanisms of heredity and variation of living organisms, senescence and the emergence of diseases such as cancer has fascinated mankind within living memory. Nowadays one knows that these phenomena are based on chemical processes in living organisms and the structures and functions of living cells.

This section briefly considers mathematical modelling in the overlapping areas of molecular biology, biochemistry and genetics. These fields comprise an enormous variety of different applications and models, the complete review of which would be far beyond the scope of this book. Hence, this section exemplarily addresses one specific branch of the above research areas: That is, applications which utilise the framework of chemical reactions for the modelling of selected key processes. This section hence starts with historical background information and a mathematical review on that subject in Sects. 2.3.1 and 2.3.2, followed by an outline of cross connections to other disciplines in Sect. 2.3.3.

2.3.1 History of Chemical Reaction Modelling

The first landmark in the development of chemical reaction modelling was set in 1850 by Ludwig Wilhelmy, who empirically derived a mathematical expression for the progress of the inversion of cane sugar in the presence of acids (McQuarrie 1967; Arnaut et al. 2007). In several articles published between 1864 and 1879, Cato Maximilian Goldberg and Peter Waage proposed the *law of mass action*, which says that the hazard of an elementary reaction is proportional to the product of the concentrations of all reactants; cf. Sect. 2.3.2 for details. Important further contributions to the understanding of the order and temperature dependence of chemical reactions were made between 1865 and 1889 by Augustus Harcourt, William Esson, Jacobus Henricus van't Hoff, Wilhelm Ostwald and Svante Arrhenius (Laidler 1993). Until 1940, many mathematical models were formulated which described the mechanism of a chemical reaction in a deterministic way. According to McQuarrie (1967), Kramers (1940) was the first author who applied the theory of stochastic processes to chemical reactions models.

Nowadays, detailed knowledge about molecular structures and mechanisms is available, in addition to sophisticated mathematical and statistical modelling tools. This enables the description and analysis of complex chemical networks. A detailed historical review on chemical kinetics modelling is provided by Arnaut et al. (2007). McQuarrie (1967) considers this subject from a statistician's point of view.

2.3.2 Chemical Reaction Kinetics

Chemical reactions are typically specified by reaction equations of the form

$$a_1 A_1 + \ldots + a_k A_k \longrightarrow b_1 B_1 + \ldots + b_l B_l. \tag{2.5}$$

This equation describes a reaction in which k different *reactants* $A_1, \ldots, A_k$ are transformed into l distinct *products* $B_1, \ldots, B_l$. The numbers a_i, $i = 1, \ldots, k$, and b_j, $j = 1, \ldots, l$, are the *stoichiometries* of the reaction and denote the numbers of reactants A_i and products B_j involved. They are assumed to be natural numbers with greatest common divisor equal to one. In this chapter, equations like (2.5) are declared to represent *elementary reactions*, i.e. reactions that do not consist of several intermediate steps. Equation (2.1) on p. 13 was of type (2.5) as well.

As in the context of modelling the spread of infectious diseases in the previous section, there are various approaches to mathematically describe the dynamics of a process in which reactions such as (2.5) occur. In what follows, three possibilities are briefly introduced in the same order as for the SIR model in Sect. 2.2.2. All representations have in common that they assume the underlying system well-stirred and the process to be Markovian and time-homogeneous. In particular, external parameters such as temperature and pressure are presumed to be constant.

Representation as Pure Markov Jump Process

The sets of reactants $\{A_1, \ldots, A_k\}$ and products $\{B_1, \ldots, B_l\}$ are typically non-disjoint subsets of a collection $\{C_1, \ldots, C_m\}$ of particles that are present in the considered system. The reaction equation (2.5) can hence be rewritten as

$$c_1 C_1 + \ldots + c_m C_m \longrightarrow \tilde{c}_1 C_1 + \ldots + \tilde{c}_m C_m, \tag{2.6}$$

where

$$c_i = \begin{cases} a_j & \text{if } C_i = A_j \\ 0 & \text{if } C_i \notin \{A_1, \ldots, A_k\} \end{cases} \quad \text{and} \quad \tilde{c}_i = \begin{cases} b_j & \text{if } C_i = B_j \\ 0 & \text{if } C_i \notin \{B_1, \ldots, B_l\}. \end{cases}$$

For $i \in \{1, \ldots, m\}$, let X_i denote the number of particles C_i in the system and define $(X_1, \ldots, X_m)'$ as the state variable of a stochastic process describing the system dynamics. The chemical reaction (2.6) then causes a state change

$$\begin{pmatrix} X_1 \\ \vdots \\ X_m \end{pmatrix} \longrightarrow \begin{pmatrix} X_1 - (c_1 - \tilde{c}_1) \\ \vdots \\ X_m - (c_m - \tilde{c}_m) \end{pmatrix}. \tag{2.7}$$

In real applications, one typically has several chemical reactions such as (2.6), each causing a transition like (2.7). Every reaction is associated with a *reaction rate* indicating the hazard with which the specific reaction is going to occur within the next infinitesimal time interval. These rates are assumed to depend on the left hand side of (2.6) only. Wilkinson (2006) exemplarily states the following reactions and associated reaction rates, where the current state of the process is $(X_1, \ldots, X_m)'$:

$$C_i \longrightarrow \tilde{c}_1 C_1 + \ldots + \tilde{c}_m C_m \ \textit{(first-order reaction)} \text{ with rate } k_1 X_i \tag{2.8}$$

$$C_i + C_j \longrightarrow \tilde{c}_1 C_1 + \ldots + \tilde{c}_m C_m \text{ (second-order reaction) with rate } k_2 X_i X_j \quad (2.9)$$

$$2C_i \longrightarrow \tilde{c}_1 C_1 + \ldots + \tilde{c}_m C_m \text{ (second-order reaction) with rate } k_3 X_i (X_i - 1)/2. \quad (2.10)$$

In the second equation, one requires $i \neq j$. The variables $k_1, k_2, k_3 \in \mathbb{R}_+$ are called *rate constants*. They are usually unknown and hence the subject of statistical inference based on available experimental data. The remaining parts of the reaction rates result from combinatorial considerations, counting the number of possible collisions between the reactants, and the fact that the hazard of two specific particles colliding is constant (Gillespie 1992).

As a consequence of the above specified reaction rates, the probability that, for example, reaction (2.8) will occur within a time interval of length Δt, provided that the current number of particles C_i is X_i, equals $k_1 X_i \Delta t + o(\Delta t)$, where $o(\Delta t)/\Delta t \to 0$ as $\Delta t \to 0$. Without any other reactions taking place, the expected time until the occurrence of this reaction is exponentially distributed with mean $k_1 X_i$.

Representation Through a System of Ordinary Differential Equations

A different possibility to describe the state of a system which is subject to elementary chemical reactions of type (2.6) is via the rates of change of the concentrations of all reaction participants. To that end, consider the concentrations $x_1, \ldots, x_m$ of the particles $X_1, \ldots, X_m$. These concentrations are considered continuous rather than discrete quantities. The chemical reaction (2.6) induces a change of the current state $(x_1, \ldots, x_m)'$ which is typically described by a set of ordinary differential equations (ODEs): For all $i = 1, \ldots, m$, one has

$$\mathrm{d}x_i/\mathrm{d}t = \bar{k}\,(\tilde{c}_i - c_i)\, x_1^{c_1} \cdot \ldots \cdot x_m^{c_m}$$

for some positive *(stochastic) rate constant* $\bar{k}$. This equation results from the law of mass action, which was already mentioned in Sect. 2.3.1. The sum of exponents $c_1 + \ldots + c_m$ is called the *order* of the reaction (McQuarrie 1967). The right hand side of the ODE is positive if $c_i < \tilde{c}_i$, i.e. if the chemical reaction described by (2.6) increases the amount of particles X_i in the system. It is negative or equal to zero if the reaction decreases the number X_i or leaves it unaltered, respectively. If there is more than one possible reaction, each reaction is assigned a different rate constant, and the ODEs resulting from each reaction equation are added in order to arrive at a description for the whole reaction dynamics. For example, consider the following set of coupled reactions for $m = 2$, which is a special case of Eqs. (2.8)–(2.10):

$$C_1 \longrightarrow \tilde{c}_1^{(1)}\, C_1 + \tilde{c}_2^{(1)}\, C_2 \quad (2.11)$$

$$C_1 + C_2 \longrightarrow \tilde{c}_1^{(2)}\, C_1 + \tilde{c}_2^{(2)}\, C_2 \quad (2.12)$$

$$2C_2 \longrightarrow \tilde{c}_1^{(3)}\, C_1 + \tilde{c}_2^{(3)}\, C_2. \quad (2.13)$$

For these reactions, one obtains the ODEs

$$\mathrm{d}x_1/\mathrm{d}t = \bar{k}_1\left(\tilde{c}_1^{(1)} - 1\right)x_1 + \bar{k}_2\left(\tilde{c}_1^{(2)} - 1\right)x_1x_2 + \bar{k}_3\tilde{c}_1^{(3)}x_2^2 \tag{2.14}$$

$$\mathrm{d}x_2/\mathrm{d}t = \bar{k}_1\tilde{c}_2^{(1)}x_1 + \bar{k}_2\left(\tilde{c}_2^{(2)} - 1\right)x_1x_2 + \bar{k}_3\left(\tilde{c}_2^{(3)} - 2\right)x_2^2 \tag{2.15}$$

for appropriate rate constants $\bar{k}_1, \bar{k}_2, \bar{k}_3 > 0$. Additionally, a suitable initial state of the process needs to be specified. The constants k_1, k_2, k_3 in Eqs. (2.8)–(2.10) and the constants $\bar{k}_1, \bar{k}_2, \bar{k}_3$ in (2.14)–(2.15) depend on the units of X_1, X_2 and x_1, x_2, respectively, and are not necessarily the same. See Wilkinson (2006, Chap. 6.6) for the conversion from k_i to $\bar{k}_i$ in case the concentrations are measured in moles per litre.

Representation Through a System of Stochastic Differential Equations

Finally, a third way to represent the evolution of a system which is subject to chemical reactions utilises stochastic differential equations (SDEs). In case of the reactions (2.11)–(2.13), the multi-dimensional SDE reads

$$\begin{pmatrix}\mathrm{d}x_1\\ \mathrm{d}x_2\end{pmatrix} = \begin{pmatrix}\bar{k}_1\left(\tilde{c}_1^{(1)} - 1\right)x_1 + \bar{k}_2\left(\tilde{c}_1^{(2)} - 1\right)x_1x_2 + \bar{k}_3\tilde{c}_1^{(3)}x_2^2\\ \bar{k}_1\tilde{c}_2^{(1)}x_1 + \bar{k}_2\left(\tilde{c}_2^{(2)} - 1\right)x_1x_2 + \bar{k}_3\left(\tilde{c}_2^{(3)} - 2\right)x_2^2\end{pmatrix}\mathrm{d}t + \begin{pmatrix}\sigma_{11} & \sigma_{12}\\ \sigma_{21} & \sigma_{22}\end{pmatrix}\begin{pmatrix}\mathrm{d}B_1\\ \mathrm{d}B_2\end{pmatrix},$$

where σ_{11}, σ_{12}, σ_{21} and σ_{22} are functions of the state variables, rate constants and stoichiometries not explicitly given here. The first summand on the right hand side represents the deterministic component of the process and agrees with Eqs. (2.14) and (2.15). The second summand stands for the probabilistic component with B_1 and B_2 being two independent Brownian motion processes. SDEs and Brownian motion will formally be defined in Chap. 3.

2.3.3 Reaction Kinetics in the Biological Sciences

Reaction equations and their associated mathematical theory are convenient tools also in the biological sciences. They are particularly used to describe the natural laws which underlie the functioning of cells. This section gives some examples.

Chemical work can be performed by cells only if there is enough energy available. Such energy is gained through cellular catabolism, which is a mechanism consisting of a series of enzymatic reactions like

$$\text{enzyme} + \text{substrate} \quad \longleftrightarrow \quad \text{complex} \quad \longrightarrow \quad \text{enzyme} + \text{product},$$

where the enzyme acts as a catalyst (Keener and Sneyd 1989). Double-sided arrows mean that the reaction can take place in both directions. Kinetic models for metabolic systems are, for example, developed by Demin et al. (2005).

Within each cell, there are several thousand types of interacting proteins. Depending on its environment, a cell determines the required amount of each protein by means of *transcription networks* (Alon 2007). Transcription is one out of several regulatory mechanisms in genetic networks and can be described by a set of coupled elementary reactions (Wilkinson 2006). At a less detailed level, transcription and other key processes can be assembled to construct genetic networks. For example, the following components of a prokaryotic auto-regulatory network are summarised by Wilkinson (2006):

$$\begin{array}{rclr}
g & \longrightarrow & g+r & \text{(transcription)} \\
g+P_2 & \longleftrightarrow & g\cdot P_2 & \text{(repression)} \\
r & \longrightarrow & r+P & \text{(translation)} \\
2P & \longleftrightarrow & P_2 & \text{(dimerisation)} \\
r & \longrightarrow & \emptyset & \text{(mRNA degradation)} \\
P & \longrightarrow & \emptyset & \text{(protein degradation).}
\end{array}$$

In these equations, P stands for a protein, P_2 for the compound of two of these proteins, g for a gene and r for a *transcript* of g. The empty set $\emptyset$ indicates that the product of a reaction is not part of the model, and a dot represents the compound of two components.

The close connection between models for chemical reactions and genetic mechanisms is hardly surprising as genetics is based on the chemistry of nucleid acids. There are, however, also cases of compartmental systems in cellular biology where reaction equations represent transitions other than chemical reactions. In the application in Chap. 9, for example, the location of a diffusing protein between a bleached and an unbleached part of the cell nucleus is observed. This can be written as

$$X^{\text{bleached}} \longleftrightarrow X^{\text{unbleached}}.$$

A molecule that undergoes this transition does not change any of its chemical or kinetic properties but only its location, so the compartments reflect the spatial dimension of the problem here.

Plenty of further applications are, for example, presented in Jacquez (1972) and McQuarrie (1967). Ehrenberg et al. (2003) give a brief overview about current research questions in systems biology. For general reviews on mathematical models in biology, see Goel and Richter-Dyn (1974), Renshaw (1991), Allen (2003) or Lande et al. (2003).

Though representing entirely different natural phenomena, the above mentioned applications have in common that they are intrinsically stochastic. A number of papers is devoted to the importance of the utilisation of probabilistic instead of deterministic models in systems biology, biochemistry and genetics, see for example Kimura (1964), Zheng and Ross (1991), Arkin et al. (1998), Sveiczer et al. (2001), Rao et al. (2002), Bahcall (2005), Tian et al. (2007) and Boys et al. (2008). In agreement with this point of view, the present book motivates, constructs and statistically infers on stochastic models from life sciences.

2.4 Algorithms for Simulation

In Sects. 2.2 and 2.3, different kinds of processes were considered to represent the dynamics of different phenomena in life sciences. For the simulation of these processes, one requires algorithms for the exact or approximate generation of according sample paths. Such algorithms have already been applied for the generation of Figs. 2.3–2.5.

2.4.1 Simulation of Continuous-Time Markov Jump Processes

Continuous-time pure Markov jump processes can always exactly be simulated. An according algorithm is presented in what follows.

Consider a system consisting of n different types of objects such as molecules in a fluid, predator and prey in a specified region or susceptibles and infectives in a population. Assume that the time-continuous evolution of these objects can be described by a time-homogeneous stochastic Markov process with state variable $\boldsymbol{X}(t) = (X_1(t), \ldots, X_n(t))' \in \mathbb{Z}^n$, where $X_i(t)$ is the number of type i objects at time $t \in \mathbb{R}_+$. Suppose that there are m possible events $k \in \{1, \ldots, m\}$ like chemical reactions or interactions within a population, each causing a change $\boldsymbol{\Delta}_k \in \mathbb{Z}^n \setminus \{\boldsymbol{0}\}$ in the state variable. Let $\lambda_k = f_k(\boldsymbol{X})$ denote the hazard for event k, where f_k is an appropriate function depending on the state $\boldsymbol{X}$. That means, the probability that a type k event will occur within the next time interval of length Δt conditioned on the current state $\boldsymbol{X}$ is $\lambda_k \Delta t + o(\Delta t)$, where $o(\Delta t)/\Delta t \to 0$ as $\Delta t \to 0$. The objective is to exactly simulate realisations of the considered process, that means to successively draw pairs $(\tau, k) \in \mathbb{R}_+ \times \{1, \ldots, m\}$, where τ is the waiting time until the occurrence of the next event, and k is the type of event happening at that time.

Denote by $p(\tau, k)$ the joint probability density function of τ and k. Under the assumption that only one event can happen at the same time, Gillespie (1976) shows that

$$p(\tau, k) = \lambda_k \exp\left(-\tau \sum_{j=1}^{m} \lambda_j\right) = \lambda_k \exp(-\lambda\tau) \text{ for } \tau \in \mathbb{R}_+ \text{ and } k \in \{1, \ldots, m\},$$

where $\lambda = \sum_{j=1}^{m} \lambda_j$. This joint density can be expressed as $p(\tau, k) = p(\tau)p(k|\tau)$, where

$$p(\tau) = \sum_{k=1}^{m} p(\tau, k) = \lambda \exp(-\lambda\tau), \qquad \text{i.e.} \quad \tau \sim \text{Exp}(\lambda),$$

and

$$p(k|\tau) = \frac{p(\tau, k)}{p(\tau)} = \frac{\lambda_k}{\lambda} \tag{2.16}$$

are the density of τ and the conditional probability function of k, respectively.

This leads to an exact and efficient method to obtain sample trajectories of the considered process on a time interval $[t_{\min}, t_{\max}]$. The procedure has been called *stochastic simulation algorithm (SSA)* by its originator, but is usually known as *Gillespie's algorithm*:

Algorithm 2.1 (Gillespie's Algorithm, Gillespie 1976).

1. *Set* $t = t_{\min}$ *and initialise* $\boldsymbol{X}(t)$.
2. *While* $t < t_{\max}$:
   i. *Calculate* λ_k *for all* k *and their sum* λ. *Terminate if the system has reached an absorbing state, i.e.* $\lambda = 0$.
   ii. *Draw* $\tau \sim \text{Exp}(\lambda)$. *Set* $\tau^* = \min\{\tau, t_{\max} - t\}$.
   iii. *Draw* k *from* (2.16).
   iv. *Set* $\boldsymbol{X}(s) = \boldsymbol{X}(t)$ *for all* $s \in (t, t+\tau^*)$ *and* $\boldsymbol{X}(t+\tau^*) = \boldsymbol{X}(t) + \boldsymbol{\Delta}_k \mathbb{1}(\tau^* = \tau)$.
   v. *Set* $t = t + \tau$.

Estimates of the average or the variation of the sample paths can be obtained by respective Monte Carlo statistics. For further details and experimental results, see Gillespie (1976, 1977). Extensions, later elaborations and improvements with respect to computing time are contained in Gillespie (2007). Manninen et al. (2006) provide ample references for different implementations of the Gillespie algorithm, such as the *next reaction method* by Gibson and Bruck (2000), and alternative approaches, for example the *StochSim algorithm* by Le Novère and Shimizu (2001). Another good review is Wilkinson (2006, Chap. 8).

2.4.2 *Simulation of Solutions of ODEs and SDEs*

When a system consists of a large number of objects, the just described simulation of a pure Markov jump process becomes expensive in terms of computing time. In contrast, the most convenient process with respect to its simulation is the deterministic process described by a set of ODEs, because this process has no random component. If there is an analytically explicit solution of the ODEs available, one can simply calculate the according multivariate sample path without any approximation error. Otherwise, numerical schemes such as the Euler scheme can be applied to obtain approximate trajectories. Such algorithms can be found in any standard textbook on numerical mathematics.

Similarly, a stochastic process described by a set of SDEs can exactly be simulated if an explicit solution for the differential equations is known. Otherwise, numerical approximation schemes are utilised. The consideration of respective procedures is postponed to Sect. 3.3 in the next chapter, because this subject requires a preliminary introduction to stochastic calculus. The numerical approximation of a solution of an ODE arises as a special case of the algorithm for an SDE.

2.5 Conclusion

Assessment of key mechanisms in life sciences cannot be imagined without the application of mathematical models. Moreover, real situations can particularly be rendered by the consideration of random events. This chapter provided an introduction to established models in life sciences, starting with the general class of compartment models in Sect. 2.1 and then proceeding to applications in mathematical epidemiology and biology in Sects. 2.2 and 2.3. To that end, three types of processes were considered, namely stochastic jump processes, deterministic continuous processes and stochastic diffusion processes, the simulation of which is the subject of Sect. 2.4. The latter type of process emerges as a convenient compromise between the former two, and hence this book focuses on diffusion processes.

However, diffusions have not been defined formally in this book yet. For that reason, Chap. 3 introduces the theory of stochastic calculus to an extent which is oriented towards the needs of subsequent chapters. Chapter 4 discusses the application of the three above process classes and considers the derivation of diffusion processes from the compartmental description of some phenomenon. This methodology is applied in Chap. 5, where a multitype SIR model for heterogeneous contact patterns is developed.

Until that point, this book is mainly concerned with the construction of models, which enables the simulation of a considered mechanism for given sets of model parameters. In practice, however, such parameters are unknown and hence to be estimated statistically based on available observations. Therefore, Chaps. 6 and 7 consider the important subject of statistical inference for diffusion processes.

The methodology of all preceding parts is applied in Chaps. 8 and 9 on the example of modelling the spread of influenza and the binding behaviour of molecules, respectively. These chapters also point out challenges arising from typical data situations such as partial observations or measurement errors.

References

Abbey H (1952) An examination of the Reed-Frost theory of epidemics. Hum Biol 24:201–233
Allen L (2003) An introduction to stochastic processes with applications to biology. Pearson Prentice Hall, Upper Saddle River
Alon U (2007) An Introduction to systems biology. Design principles of biological circuits. Chapman and Hall, Boca Raton
Anderson R (1982) The population dynamics of infectious diseases. Chapman and Hall, London
Andersson H, Britton T (2000) Stochastic epidemic models and their statistical analysis. Lecture Notes in Statistics, vol 151. Springer, New York
Anderson R, May R (1985) Vaccination and herd immunity to infectious diseases. Nature 318: 323–329
Anderson R, May R (1991) Infectious diseases of humans. Oxford University Press, Oxford
Arkin A, Ross J, McAdams H (1998) Stochastic kinetic analysis of developmental pathway bifurcation in phage λ-infected Escherichia coli cells. Genetics 149:1633–1648
Arnaut L, Formosinho S, Burrows H (2007) Chemical kinetics: from molecular structure to chemical reactivity. Elsevier, Amsterdam/Oxford
Bahcall O (2005) Single cell resolution in regulation of gene expression. Mol Syst Biol 1 (article number 2005.0015)
Bailey N (1975) The mathematical theory of infectious diseases, 2nd edn. Charles Griffin, London
Ball F (1983) The threshold behaviour of epidemic models. J Appl Probab 20:227–241
Ball F (1986) A unified approach to the distribution of total size and total area under the trajectory of infectives in epidemic models. Adv in Appl Probab 18:289–310
Ball F, Mollison D, Scalia-Tomba G (1997) Epidemics with two levels of mixing. Ann Appl Probab 7:46–89
Bartlett M (1949) Some evolutionary stochastic processes. J R Stat Soc Ser B 11:211–229
Becker N (1989) Analysis of infectious disease data. Monographs on statistics and applied probability. Chapman and Hall, London
Boys R, Wilkinson D, Kirkwood T (2008) Bayesian inference for a discretely observed stochastic kinetic model. Stat Comput 18:125–135
Brauer F (2009) Mathematical epidemiology is not an oxymoron. BMC Public Health 9:S2
Costa Maia J (1952) Some mathematical developments on the epidemic theory formulated by Reed and Frost. Hum Biol 24:167–200
Cunha B (2004) Historical aspects of infectious diseases, part I. Infect Dis Clin N Am 18(1):xi–xv
Daley D, Gani J (1999) Epidemic modelling: an introduction. Cambridge studies in mathematical biology, vol 15. Cambridge University Press, Cambridge
Demin O, Plyusnina T, Lebedeva G, Zobova E, Metelkin E, Kolupaev A, Goryanin I, Tobin F (2005) Kinetic modelling of the E. coli metabolism. In: Alberghina L, Westerhoff H (eds) Systems biology. Definitions and perspectives. Springer, Berlin/Heidelberg, pp 31–67
Diekmann O, Heesterbeek J (2000) Mathematical epidemiology of infectious diseases: model building, analysis and interpretation. Wiley, Chichester
Dietz K (1967) Epidemics and rumours: a survey. J R Stat Soc Ser A 130:505–528
Dobson A, Carper E (1996) Infectious diseases and human population history. Bioscience 46: 115–126
Ehrenberg M, Elf J, Aurell E, Sandberg R, Tegnér J (2003) Systems biology is taking off. Genome Res 13:2377–2380
En'ko P (1889) On the course of epidemics of some infectious diseases. Vrach St Petersburg 10:1008–1010, 1039–1042, 1061–1063
Fine P (1993) Herd immunity: history, theory, practice. Epidemiol Rev 15:265–302
Galvani A, May R (2005) Dimensions of superspreading. Nature 438:293–295
Gibson M, Bruck J (2000) Efficient exact stochastic simulation of chemical systems with many species and many channels. J Phys Chem A 104:1876–1889

Gillespie D (1976) A general method for numerically simulating the stochastic time evolution of coupled chemical reactions. J Comput Phys 22:403–434
Gillespie D (1977) Exact stochastic simulation of coupled chemical reactions. J Phys Chem 81:2340–2361
Gillespie D (1992) A rigorous derivation of the chemical master equation. Phys A 188:404–425
Gillespie D (2007) Stochastic simulation of chemical kinetics. Annu Rev Phys Chem 58:35–55
Goel N, Richter-Dyn N (1974) Stochastic models in biology. Academic, New York
Greenwood M (1931) On the statistical measure of infectiousness. J Hyg 31:336–351
Hamer W (1906) The Milroy lectures on epidemic disease in England – the evidence of variability and of persistency of type (Lecture I). Lancet 167:569–574
Hethcote H (1994) A thousand and one epidemic models. In: Levin S (ed) Frontiers in mathematical biology, Lecture notes in biomathematics. Springer, Berlin, pp 504–515
Hethcote H (2000) The mathematics of infectious diseases. SIAM Rev 42:599–653
Ireland J, Mestel B, Norman R (2007) The effect of seasonal host birth rates on disease persistence. Math Biosci 206:31–45
Isham V (2004) Stochastic models for epidemics. Research Report No 263, Department of Statistical Science, University College London
Jacquez J (1972) Compartmental analysis in biology and medicine. Elsevier, Amsterdam
Keeling M, Rohani P (2008) Modeling infectious disease in humans and animals. Princeton University Press, Princeton
Keener J, Sneyd J (1989) Mathematical physiology. Springer, New York
Kelly H, Peck H, Laurie K, Wu P, Nishiura H, Cowling B (2011) The age-specific cumulative incidence of infection with pandemic influenza H1N1 2009 was similar in various countries prior to vaccination. PLoS ONE 6:e21 828
Kermack W, McKendrick A (1927) A contribution to the mathematical theory of epidemics. Proc R Soc London Ser A 115:700–721
Kimura M (1964) Diffusion models in population genetics. J Appl Probab 1:177–232
Kramers H (1940) Brownian motion in a field of force and the diffusion model of chemical reactions. Physica 7:284–304
Laidler K (1993) The world of physical chemistry. Oxford University Press, New York
Lande R, Engen S, Sæther B (2003) Stochastic population dynamics in ecology and conservation. Oxford University Press, New York
Le Novère N, Shimizu T (2001) StochSim: modelling of stochastic biomolecular processes. Bioinformatics 17:575–576
Lloyd-Smith J, Schreiber S, Kopp P, Getz W (2005) Superspreading and the effect of individual variation on disease emergence. Nature 438:355–359
Manninen T, Linne ML, Ruohonena K (2006) Developing Itô stochastic differential equation models for neuronal signal transduction pathways. Comput Biol Chem 30:280–291
McKendrick A (1926) Application of mathematics to medical problems. Proc Edinb Math Soc 44:98–130
McNeill W (1976) Plagues and people. Anchor, New York
McQuarrie D (1967) Stochastic approach to chemical kinetics. J Appl Probab 4:413–478
Morton R, Wickwire K (1974) On the optimal control of a deterministic epidemic. Adv Appl Probab 6:622–635
Neal P (2007) Coupling of two SIR epidemic models with variable susceptibilities and infectivities. J Appl Probab 44:41–57
Oldstone M (2010) Viruses, plagues, and history: past, present and future. Oxford University Press, Oxford/New York
Rao C, Wolf D, Arkin A (2002) Control, exploitation and tolerance of intracellular noise. Nature 420:231–237
Renshaw E (1991) Modelling biological populations in space and time. Cambridge University Press, Cambridge
Riley S (2007) Large-scale spatial-transmission models of infectious disease. Science 316: 1298–1301

Ross R (1915) Some a priori pathometric equations. Br Med J 1:546–547
Rushton S, Mautner A (1955) The deterministic model of a simple epidemic for more than one community. Biometrika 42:126–132
Sattenspiel L (1987) Population structure and the spread of disease. Hum Biol 59:411–438
Sattenspiel L, Dietz K (1995) A structured epidemic model incorporating geographic mobility among regions. Math Biosci 128:71–91
Sherman I (2006) The power of plagues. ASM Press, Washington, DC
Smallman-Raynor M, Cliff A (2004) Impact of infectious diseases on war. Infect Dis Clin N Am 18:341–368
Sveiczer A, Tyson J, Novak B (2001) A stochastic, molecular model of the fission yeast cell cycle: role of the nucleocytoplasmic ratio in cycle time regulation. Biophys Chem 92:1–15
Tian T, Xu S, Gao J, Burrage K (2007) Simulated maximum likelihood method for estimating kinetic rates in gene expression. Bioinformatics 23:84–91
UNAIDS (2009) AIDS epidemic update: November 2009. WHO library cataloguing-in-publication data. Available at http://www.unaids.org
Vasold M (2008) Grippe, Pest und Cholera: eine Geschichte der Seuchen in Europa. Franz Steiner Verlag, Stuttgart
Whittle P (1955) The outcome of a stochastic epidemic – a note on Bailey's paper. Biometrika 42:116–122
WHO (2010) Pandemic (H1N1) 2009 – update 112 (from 6 August 2010). Available at http://www.who.int/csr/don/2010_08_06/en/index.html
Wilkinson D (2006) Stochastic modelling for systems biology. Chapman and Hall, Boca Raton
Williams T (1971) An algebraic proof of the threshold theorem for the general stochastic epidemic. Adv Appl Probab 3:223
Zheng Q, Ross J (1991) Comparison of deterministic and stochastic kinetics for nonlinear systems. J Chem Phys 94:3644–3648

Chapter 3
Stochastic Differential Equations and Diffusions in a Nutshell

Stochastic differential equations are a powerful and natural tool for the modelling of complex systems that change roughly in continuous time. Application areas include econometrics and finance (Robinson 1959; Black and Scholes 1973; Merton 1976; Cox et al. 1985; Bibby and Sørensen 2001; Elerian et al. 2001; Eraker 2001; Chiarella et al. 2009), physics (van Kampen 1965, 1981; Ramshaw 1985; Tuckwell 1987; Seifert 2008), biology (Leung 1985; Elf and Ehrenberg 2003; Sjöberg et al. 2009), systems biology (Golightly and Wilkinson 2005, 2006, 2008), medicine (Walsh 1981; Fogelson 1984; Capasso and Morale 2009), epidemiology (Barbour 1974; Clancy and French 2001; Hufnagel et al. 2004; Chen and Bokka 2005; Alonso et al. 2007), population biology (Ferm et al. 2008), genetics (Kimura 1964; Fearnhead 2006; Tian et al. 2007), social sciences (Cobb 1981; de la Lama et al. 2006), geostatistics (Duan et al. 2009) and traffic control (McNeil 1973).

This chapter provides a short introduction to stochastic differential equations and their solutions, which under regularity conditions agree with the class of diffusion processes. The contents of this primer are selected according to the needs of the remaining parts of this book; it by no means claims to cover completely the theory of stochastic calculus. Thorough works include Arnold (1973), Stroock and Varadhan (1979), Gardiner (1983), Karatzas and Shreve (1991), Revuz and Yor (1991), Kloeden and Platen (1999) and Øksendal (2003).

Generally speaking, a *stochastic differential equation (SDE)* is a differential equation—i.e. an equation relating a process to one or several of its derivatives—which involves any kind of randomness. This might be because of random coefficients, a random initial value or some dependence on a stochastic force.

However, a reasonable further classification as in Arnold (1973), Gard (1988) or Kloeden and Platen (1999) distinguishes between the driving force being a regular or irregular process. In the former case, solution processes of such equations have differentiable sample paths and do not differ substantially from ordinary differential equations. Such equations are referred to as *random differential equations* and are of no further interest here. The second class contains stochastic differential equations

C. Fuchs, *Inference for Diffusion Processes: With Applications in Life Sciences*,
DOI 10.1007/978-3-642-25969-2_3,

in the strict sense. These are forced by some irregular noise process—the notion of which will be explained in Sect. 3.1.3—, and the sample paths of corresponding solution processes are almost surely nowhere differentiable.

Section 3.1 conceives Brownian motion and Gaussian white noise as the key processes of stochastic calculus. In Sect. 3.2, the introduction of the stochastic integral allows the formal definition of stochastic differential equations, whose solutions turn out to be essentially diffusion processes. For these, fundamental formulas are stated. Section 3.3 deals with the simulation and numerical approximation of diffusion sample paths; the latter is especially necessary due to the usual absence of explicitly attainable solutions of SDEs.

Throughout this chapter let $(\Omega, \mathcal{F}^*, \mathcal{F}, \mathbb{P})$ be a filtered probability space with sample space Ω, σ-algebra $\mathcal{F}^*$, $\mathcal{F} = (\mathcal{F}_t)_{t\geq 0}$ the natural filtration and $\mathbb{P}$ a probability measure on $(\Omega, \mathcal{F}^*)$. The σ-algebra of Lebesgue subsets of $\mathbb{R}$ will be denoted by $\mathcal{L}$. We will consider continuous jointly $\mathcal{L} \times \mathcal{F}^*$-measurable stochastic processes

$$\boldsymbol{X} : \begin{cases} T \times \Omega \to \mathcal{X} \\ (t, \omega) \mapsto \boldsymbol{X}(t, \omega) \end{cases}$$

with state space $\mathcal{X} \subseteq \mathbb{R}^d$, $d \geq 1$, and non-empty time set $T \subseteq \mathbb{R}_0$, but omit the dependency on ω in the notation $\boldsymbol{X} = (\boldsymbol{X}_t)_{t\in T}$. We will generally assume that for all subsets $\{t_0, \ldots, t_n\} \subseteq T$ and $\{\boldsymbol{x}_{t_0}, \ldots, \boldsymbol{x}_{t_n}\} \subseteq \mathcal{X}$ the joint distribution of $\boldsymbol{X}_{t_0}, \ldots, \boldsymbol{X}_{t_n}$ has a probability density and that conditional probabilities and densities can be defined in the usual way.

3.1 Brownian Motion and Gaussian White Noise

This section defines elementary modules of stochastic calculus on which subsequent considerations are based.

3.1.1 Brownian Motion

A real-valued $\mathcal{F}$-adapted process $B = (B_t)_{t\geq 0}$ is defined to be *Brownian motion*—also called a *Wiener process*[1]—if

1. $B_0 = u$ almost surely for $u \in \mathbb{R}$ fixed,
2. All paths are almost surely continuous,

[1]Some authors denote by a Wiener process the mathematical description given above while Brownian motion stands for the physical movement of a diffusing particle. In this book, both terms are used interchangeably.

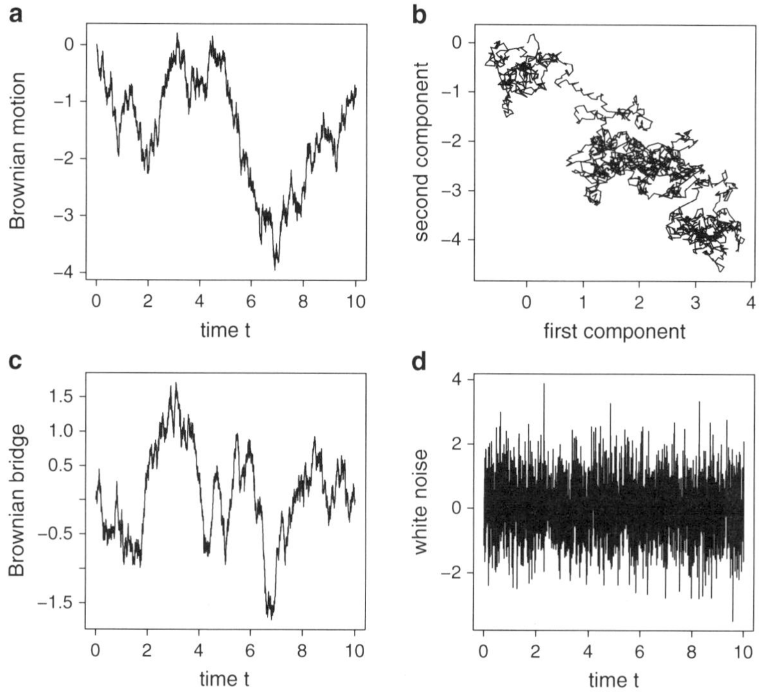

Fig. 3.1 Discrete sample path realisations of (**a**) one-dimensional standard Brownian motion, (**b**) two-dimensional standard Brownian motion, (**c**) a Brownian $(0, 0, 10, 0)$-bridge with volatility parameter $\sigma = 1$ and (**d**) standard Gaussian white noise for equidistant time steps of size 0.005

3. All paths have independent and stationary increments,
4. $B_t \sim \mathcal{N}(0, \sigma^2 t)$ for all $t \geq 0$ and constant *volatility parameter* $\sigma \in \mathbb{R}_+$.

The process is called *standard* for $u = 0$ and $\sigma = 1$. A vector-valued process is said to be *d-dimensional (standard) Brownian motion* if its d components are mutually independent one-dimensional (standard) Brownian motions. The existence of such a process was first proven by Wiener (1923). The probability law induced by standard Brownian motion is thus called *Wiener measure*. Figure 3.1a, b shows typical discrete-time sample path realisations of one- and two-dimensional standard Brownian motion.

Although the paths of one-dimensional Brownian motion are almost surely continuous, almost all paths are nowhere Lipschitz continuous and hence nowhere differentiable. As a consequence, almost all paths are of unbounded *total variation* on any time interval $[s, t]$ on the positive real line, i.e.

$$\sup_{\mathcal{Z}_n} \sum_{i=1}^{h(n)} |B_{t_i}^{(n)} - B_{t_{i-1}}^{(n)}| = \infty,$$

where the supremum is taken over all partitions $\mathcal{Z}_n = (s = t_0^{(n)} < t_1^{(n)} < \ldots < t_{h(n)}^{(n)} = t)$ of $[s, t]$ into $h(n)$ subintervals for arbitrary n and h. However, the sample paths have finite *quadratic variation* $\langle B, B \rangle$; more precisely,

$$\langle B, B\rangle_{[s,t]} = \lim_{\delta(\mathcal{Z}_n)\downarrow 0} \sum_{i=1}^{h(n)} \left(B_{t_i}^{(n)} - B_{t_{i-1}}^{(n)}\right)^2 = (t-s)\,\sigma^2 \quad \text{in } L^2 \tag{3.1}$$

(and hence also in probability), where $\delta(\mathcal{Z}_n)$ denotes the fineness of the partition $\mathcal{Z}_n$ of $[s,t]$. If $\sum_{n=1}^{\infty} \delta(\mathcal{Z}_n) < \infty$, for instance for $t_k^{(n)} = s + 2^{-n}k(t-s)$, $k = 0, \ldots, h(n) = 2^n$, one even obtains almost sure convergence in (3.1) (Arnold 1973). All properties naturally hold for each component of multi-dimensional Brownian motion.

3.1.2 Brownian Bridge

If standard Brownian motion is further conditioned on some end point $\boldsymbol{B}_t = \boldsymbol{v}$, then the conditioned process $(\boldsymbol{B}_\tau)_{\tau\in[0,t]}$ is called a *Brownian bridge*. More generally, Brownian motion $(\boldsymbol{B}_\tau)_{\tau\in[s,t]}$ conditioned on $\boldsymbol{B}_s = \boldsymbol{u}$ and $\boldsymbol{B}_t = \boldsymbol{v}$ will be referred to as a *Brownian* $(s, \boldsymbol{u}, t, \boldsymbol{v})$*-bridge*. Like Brownian motion, Brownian bridges are Gaussian processes, but without independent increments. See Fig. 3.1c for a discrete sample path realisation of a Brownian bridge, obtained with the sampling algorithm introduced in Sect. 3.3.2.

3.1.3 Gaussian White Noise

In many life sciences applications, a system is disturbed by external fluctuations which vary much more rapidly than the system itself; the memory of the environment seems to be short compared to the memory of the system. In the idealised case, the environment is considered memoryless, and the according disturbances are called *white noise*. The paths of such a white noise process are uncorrelated at any two distinct time instants[2] and are extremely irregular.

Formally, white noise is defined as a continuous-time stationary process with mean zero and autocorrelation function proportional to the Dirac delta function. Such a process does not exist in the usual sense but belongs to the class of

[2]This choice of autocorrelation implies a constant nonzero power spectral density of the process, defined as the Fourier transform of its autocorrelation function. That explains the term white noise in analogy to white light, where all visible frequencies occur in equal amounts.

generalised processes (see e.g. Arnold 1973). If the single values of the paths of the white noise process are normally distributed, one speaks of *Gaussian white noise*. In that case uncorrelatedness implies independence. In analogy to the definition of Brownian motion, *d-dimensional white noise* consists of d mutually independent one-dimensional white noise processes.

The choice of white noise as a model for the disturbances—i.e. the choice of a memoryless environment—yields the advantage of retaining the Markov nature of the system. Gaussian white noise can formally be interpreted as the generalised derivative of the (nowhere differentiable) Brownian motion (see e.g. Kloeden and Platen 1999). In the next section, we will hence use the notation $\mathrm{d}\boldsymbol{B}_t = \boldsymbol{\xi}_t \mathrm{d}t$ for standard Brownian motion $\boldsymbol{B}$ and a standard Gaussian white noise process $\boldsymbol{\xi} = (\boldsymbol{\xi}_t)_{t \geq 0}$, i.e. $\mathrm{Var}(\boldsymbol{\xi}_t \mathrm{d}t) = \boldsymbol{I}\mathrm{d}t$ for all t with $\boldsymbol{I}$ being the identity matrix. It is this process that we referred to when we defined stochastic differential equations to be driven by an irregular stochastic process at the beginning of this chapter. Figure 3.1d shows a discrete-time simulation of Gaussian white noise. For a detailed discussion of this process see Horsthemke and Lefever (1984).

3.1.4 Excursus: Lévy Processes

The integration of Brownian motion as a source of noise is generally reasonable for models of perturbed systems in many contexts; the resulting *diffusion processes* (see Sect. 3.2.5) are called *Brownian-driven*.

However, such models turned out to be unsatisfactory in some applications in finance. In these cases more general driving forces are taken, inducing the *Lévy processes*. These are defined as processes whose sample paths almost surely start in zero, are continuous in probability and have independent and stationary increments (Protter 1990). Special cases are the Poisson process and Brownian motion with initial value zero.

Unlike the trajectories of diffusion processes, the paths of such *Lévy-driven* processes may experience jumps. One famous example is the *jump-diffusion model* by Merton (1976). However, the focus of this book lies on diffusion processes as the appropriate model in many applications in life sciences; therefore, Brownian-driven processes are considered exclusively in the following.

3.2 Itô Calculus

Inclusion of white noise as a source of randomness in differential equations leads to difficulties in the application of classical calculus. The need for a new integral definition arises, resulting in stochastic calculus with the Itô calculus as a prominent representative. This is introduced in the following.

3.2.1 Stochastic Integral and Stochastic Differential Equations

Differential equations are eminently appealing for the modelling of phenomena that evolve continuously in time as they express the rates of change. In order to adequately include random fluctuations—which are present in all natural contexts—there is a need to include a stochastic component. As motivated in Sect. 3.1.3, we consider *stochastic differential equations* of the form

$$\frac{\mathrm{d}\boldsymbol{X}_t}{\mathrm{d}t} = \boldsymbol{\mu}(\boldsymbol{X}_t, t) + \boldsymbol{\sigma}(\boldsymbol{X}_t, t)\,\boldsymbol{\xi}_t, \qquad \boldsymbol{X}_{t_0} = \boldsymbol{x}_0, \tag{3.2}$$

with jointly measurable functions $\boldsymbol{\mu} : \mathcal{X} \times T \to \mathbb{R}^d$, $\boldsymbol{\sigma} : \mathcal{X} \times T \to \mathbb{R}^{d \times m}$ and m-dimensional standard Gaussian white noise $\boldsymbol{\xi}$. That means the differential of $\boldsymbol{X} = (\boldsymbol{X}_t)_{t \geq t_0}$ is composed of a systematic component $\boldsymbol{\mu}$ and zero-mean random disturbances $\boldsymbol{\xi}$ intensified by $\boldsymbol{\sigma}$. The noise $\boldsymbol{\xi}$ is called *additive* if $\boldsymbol{\sigma}$ is constant in space and *multiplicative* otherwise. Using the notation $\mathrm{d}\boldsymbol{B}_t = \boldsymbol{\xi}_t \mathrm{d}t$ for standard Brownian motion $\boldsymbol{B}$, we get a differential

$$\mathrm{d}\boldsymbol{X}_t = \boldsymbol{\mu}(\boldsymbol{X}_t, t)\mathrm{d}t + \boldsymbol{\sigma}(\boldsymbol{X}_t, t)\mathrm{d}\boldsymbol{B}_t, \qquad \boldsymbol{X}_{t_0} = \boldsymbol{x}_0, \tag{3.3}$$

or, equivalently, an integral

$$\boldsymbol{X}_t = \boldsymbol{X}_{t_0} + \int_{t_0}^{t} \boldsymbol{\mu}(\boldsymbol{X}_s, s)\mathrm{d}s + \int_{t_0}^{t} \boldsymbol{\sigma}(\boldsymbol{X}_s, s)\mathrm{d}\boldsymbol{B}_s. \tag{3.4}$$

The first integral can be treated as an ordinary Lebesgue integral, but due to the almost surely unbounded total variation of the sample paths of Brownian motion (see Sect. 3.1.1), the second integral cannot be understood as a Lebesgue-Stieltjes integral. Instead, it is defined by taking advantage of the finite quadratic variation of the Brownian motion paths as follows: Let $\mathcal{D}^*_{[t_0,t]}$ be the class of non-anticipating jointly $\mathcal{F}^* \times \mathcal{L}$-measurable functions $\boldsymbol{f} : \Omega \times [t_0, t] \to \mathbb{R}^{k \times l}$ for appropriate k and l with

$$\mathbb{E}\|\boldsymbol{f}(\omega, s)\|^2 < \infty \quad \text{for all } s \in [t_0, t] \text{ and} \quad \int_{t_0}^{t} \mathbb{E}\|\boldsymbol{f}(\omega, s)\|^2 \mathrm{d}s < \infty, \tag{3.5}$$

where $\|\boldsymbol{A}\|^2 = \mathrm{tr}(\boldsymbol{A}'\boldsymbol{A})$ denotes the Euclidean norm for a vector or matrix $\boldsymbol{A}$. This and the following statements shall hold for all $\omega \in \Omega$. For step functions $\boldsymbol{f} \in \mathcal{D}^*_{[t_0,t]}$ with jumps occurring at $t_1 < \ldots < t_{n-1}$, define

$$\int_{t_0}^{t} \boldsymbol{f}(\omega, s)\mathrm{d}\boldsymbol{B}_s := \sum_{i=1}^{n} \boldsymbol{f}(\omega, t_{i-1})\big(\boldsymbol{B}_{t_i} - \boldsymbol{B}_{t_{i-1}}\big) \tag{3.6}$$

for $t_n = t$. For all $\boldsymbol{f} \in \mathcal{D}^*_{[t_0,t]}$, there exists a sequence of step functions $\boldsymbol{f}_n \in \mathcal{D}^*_{[t_0,t]}$ which approximates $\boldsymbol{f}$ in the sense that

$$\lim_{n\to\infty} \int_{t_0}^{t} \mathbb{E}\|\boldsymbol{f}(\omega,s) - \boldsymbol{f}_n(\omega,s)\|^2 \mathrm{d}s = 0. \tag{3.7}$$

With such a sequence $\{\boldsymbol{f}_n\}_{n\in\mathbb{N}}$, the *Itô (stochastic) integral* is now defined for all $\boldsymbol{f} \in \mathcal{D}^*_{[t_0,t]}$ as the mean-square limit of the integrals of the step functions, i.e.

$$\int_{t_0}^{t} \boldsymbol{f}(\omega,s)\mathrm{d}\boldsymbol{B}_s := \lim_{n\to\infty} \int_{t_0}^{t} \boldsymbol{f}_n(\omega,s)\mathrm{d}\boldsymbol{B}_s \quad \text{in } L^2. \tag{3.8}$$

The value of this integral does not depend on the particular choice of the $\boldsymbol{f}_n$. If $\boldsymbol{f}$ covers only

$$\int_{t_0}^{t} \|\boldsymbol{f}(\omega,s)\|^2 \mathrm{d}s < \infty \quad \text{a.s.} \tag{3.9}$$

rather than condition (3.5), then there exists a sequence of functions $\boldsymbol{f}_n$ from this larger class $\mathcal{D}_{[t_0,t]}$ so that

$$\lim_{n\to\infty} \int_{t_0}^{t} \|\boldsymbol{f}(\omega,s) - \boldsymbol{f}_n(\omega,s)\|^2 \mathrm{d}s = 0 \quad \text{a.s.}$$

The Itô integral of such $\boldsymbol{f}$ is defined as in (3.8) with the convergence in L^2 replaced by convergence in probability.

From now on let $\boldsymbol{\mu}, \boldsymbol{\sigma} \in \mathcal{D}_T$ for any $T \subset \mathbb{R}$, where $\boldsymbol{\mu}$ fulfils (3.9) also for non-squared norm and both coefficients depend on $\omega \in \Omega$ through $\boldsymbol{X}(\omega)$, so (3.4) is well-defined.

3.2.2 Different Stochastic Integrals

Like the classical Lebesgue-Stieltjes integral, the general stochastic integral is the limit (now in mean-square) of a sequence of partial sums, but with the crucial difference that the value of the limit depends on the selection of evaluation points within the partition of the time axis. Different choices of evaluation points lead to different stochastic integrals. The most common ones are the above *Itô integral* and the *Stratonovich integral*, introduced by Itô (1944, 1946) and Stratonovich (1966). As in (3.6), the Itô integral takes the left end point of each subinterval as the evaluation point, whereas the Stratonovich integral uses the midpoint of each subinterval. The Itô integral has the most convenient property of being a martingale but does not—unlike the Stratonovich integral—meet the transformation rules of classical calculus (cf. Sect. 3.2.10).

For this book, only the Itô interpretation is of relevance, and we will restrict our attention to that. However, it is possible to switch between the Itô and the Stratonovich calculi whenever it is advantageous (see e.g. Øksendal 2003, Chap. 6, for a respective formula).

3.2.3 Existence and Uniqueness of Solutions

With the definition of the stochastic integral, an $\mathcal{F}$-adapted stochastic process $\boldsymbol{X}$ is now defined to be the solution of the stochastic differential equation (3.3) if and only if $\boldsymbol{X}$ satisfies the stochastic integral equation (3.4) almost surely. Then $\boldsymbol{X}$ is Markovian and called an *Itô process*.

Such a solution exists pathwise uniquely if $\boldsymbol{\mu}$ and $\boldsymbol{\sigma}$ are Lipschitz continuous. Then, by definition, there exists a constant $C > 0$ such that for all $t \in T$ (in case of $T = [t_0, \infty)$ for all $t \in T'$ for all finite subsets $T' \subset T$) and $\boldsymbol{x}, \boldsymbol{y} \in \mathcal{X}$

$$\|\boldsymbol{\mu}(\boldsymbol{x}, t) - \boldsymbol{\mu}(\boldsymbol{y}, t)\| + \|\boldsymbol{\sigma}(\boldsymbol{x}, t) - \boldsymbol{\sigma}(\boldsymbol{y}, t)\| \leq C\|\boldsymbol{x} - \boldsymbol{y}\|. \tag{3.10}$$

The solution will be non-explosive with finite second moments if $\mathbb{E}\|\boldsymbol{X}_{t_0}\|^2 < \infty$ and there is a constant $D > 0$ with

$$\|\boldsymbol{\mu}(\boldsymbol{x}, t)\|^2 + \|\boldsymbol{\sigma}(\boldsymbol{x}, t)\|^2 \leq D\big(1 + \|\boldsymbol{x}\|^2\big) \tag{3.11}$$

for all t and $\boldsymbol{x}$.

Pathwise uniqueness means that if there are two solutions $\boldsymbol{X}$ and $\tilde{\boldsymbol{X}}$ with the same initial value, then

$$\mathbb{P}\left(\sup_{t \in T} \|\boldsymbol{X}_t - \tilde{\boldsymbol{X}}_t\| > 0\right) = 0,$$

which implies equivalence of $\boldsymbol{X}$ and $\tilde{\boldsymbol{X}}$. Such a pathwise unique solution is called a *strong solution*. It can be interpreted as a general functional of the Wiener process. Under weaker assumptions (e.g. Kloeden and Platen 1999, Chap. 4), an SDE may only have a *weak solution* which is obtained for a particular Wiener process. A weak solution is unique if all solutions have the same probability law.

Every strong solution has an almost surely continuous separable version, hence without loss of generality we can assume this property for $\boldsymbol{X}$ in the sequel. However, it is generally not possible to find an explicit solution of an SDE; instead, numerical approximation methods are applied (see Sect. 3.3).

Example 3.1. One of the few cases in which an explicit strong solution is available is given by the SDE

$$\mathrm{d}X_t = \alpha X_t \mathrm{d}t + \sigma X_t \mathrm{d}B_t, \quad X_0 = x_0, \tag{3.12}$$

with parameters $\alpha \in \mathbb{R}$, $\sigma \in \mathbb{R}_+$ and state space $\mathcal{X} = \mathbb{R}_+$ for $x_0 \in \mathbb{R}_+$. This SDE is solved by the *geometric Brownian motion* $X = (X_t)_{t \geq 0}$ with

$$X_t = x_0 \exp\left(\left(\alpha - \frac{1}{2}\sigma^2\right) t + \sigma B_t\right) \tag{3.13}$$

for all $t \geq 0$. This process is described in more detail in Sect. A.1 in the Appendix.

3.2.4 Transition Density and Likelihood Function

The assumptions made on p. 32 regarding the existence of conditional probabilities and densities particularly ensure the existence of the *transition density* $p(s, \boldsymbol{x}, t, \boldsymbol{y})$ defined through

$$\mathbb{P}(\boldsymbol{X}_t \in A | \boldsymbol{X}_s = \boldsymbol{x}) = \int_A p(s, \boldsymbol{x}, t, \boldsymbol{y}) \mathrm{d}\boldsymbol{y} \tag{3.14}$$

for all $\mathcal{F}^*$-measurable sets $A \subseteq \mathcal{X}$. $p(s, \boldsymbol{x}, t, \boldsymbol{y})$ is the density of a Markov process $\boldsymbol{X}$ for going from state $\boldsymbol{x} \in \mathcal{X}$ at time $s \geq 0$ to $\boldsymbol{y} \in \mathcal{X}$ at time $t > s$. For $s = t$, we define

$$p(t, \boldsymbol{x}, t, \boldsymbol{y}) = \delta(\boldsymbol{x} - \boldsymbol{y}),$$

where δ denotes the Dirac delta function. If $\boldsymbol{X}$ is homogeneous in time, i.e. the transition density depends on s and t solely through their difference $t - s$, we also write $p(t-s; \boldsymbol{x}, \boldsymbol{y})$.

In many applications, the transition density further depends on a parameter $\boldsymbol{\theta}$ from a parameter space Θ. For discrete observations $\boldsymbol{x}_{t_1}, \ldots, \boldsymbol{x}_{t_n}$ and given starting value $\boldsymbol{x}_{t_0}$ at time points $t_0 < \ldots < t_n$, the *likelihood function* of $\boldsymbol{\theta}$ reads

$$L(\boldsymbol{\theta}) = \prod_{i=0}^{n-1} p_{\boldsymbol{\theta}}(t_i, \boldsymbol{x}_{t_i}, t_{i+1}, \boldsymbol{x}_{t_{i+1}}) = \prod_{i=0}^{n-1} p_{\boldsymbol{\theta}}(t_{i+1} - t_i; \boldsymbol{x}_{t_i}, \boldsymbol{x}_{t_{i+1}})$$

for a time-homogeneous Markov process and all $\boldsymbol{\theta} \in \Theta$. The case of continuous observations is regarded in Sect. 6.1.1. Estimation of $\boldsymbol{\theta}$ is one focus of this book and will thoroughly be treated in Chaps. 6 and 7.

3.2.5 Itô Diffusion Processes

A *diffusion process* is defined as a Markov process whose transition probability function p meets the following three properties for all $\boldsymbol{x} \in \mathcal{X}$ and $s \geq 0$:

1. For all $\varepsilon > 0$ one has uniformly

$$\lim_{t \downarrow s} \frac{1}{t-s} \int\limits_{\|\boldsymbol{y}-\boldsymbol{x}\|>\varepsilon} p(s,\boldsymbol{x},t,\boldsymbol{y})\mathrm{d}\boldsymbol{y} = 0, \tag{3.15}$$

   i.e. large jumps are improbable over small time intervals, that means the process has almost surely continuous sample paths.
2. For all $\varepsilon > 0$ the uniform limit

$$\boldsymbol{\mu}(\boldsymbol{x},s) = \lim_{t \downarrow s} \frac{1}{t-s} \int\limits_{\|\boldsymbol{y}-\boldsymbol{x}\|\leq\varepsilon} p(s,\boldsymbol{x},t,\boldsymbol{y})(\boldsymbol{y}-\boldsymbol{x})\mathrm{d}\boldsymbol{y} \tag{3.16}$$

   exists. The vector-valued function $\boldsymbol{\mu}$ is called the *drift* and describes the instantaneous rate of change of the conditional expectation of the increments.
3. For all $\varepsilon > 0$ the uniform limit

$$\boldsymbol{\Sigma}(\boldsymbol{x},s) = \lim_{t \downarrow s} \frac{1}{t-s} \int\limits_{\|\boldsymbol{y}-\boldsymbol{x}\|\leq\varepsilon} p(s,\boldsymbol{x},t,\boldsymbol{y})(\boldsymbol{y}-\boldsymbol{x})(\boldsymbol{y}-\boldsymbol{x})'\mathrm{d}\boldsymbol{y} \tag{3.17}$$

   exists. The symmetric and positive semi-definite matrix-valued function $\boldsymbol{\Sigma}$ is called the *diffusion matrix* and reflects the instantaneous rate of change of the conditional covariance of the increments. A matrix $\boldsymbol{\sigma}$ with $\boldsymbol{\Sigma} = \boldsymbol{\sigma}\boldsymbol{\sigma}'$ is called the *diffusion coefficient*.

   Such a decomposition exists due to the positive semi-definiteness of $\boldsymbol{\Sigma}$, but is not necessarily unique, i.e. there might be matrices $\boldsymbol{\sigma} \neq \tilde{\boldsymbol{\sigma}}$ with $\boldsymbol{\sigma}\boldsymbol{\sigma}' = \tilde{\boldsymbol{\sigma}}\tilde{\boldsymbol{\sigma}}' = \boldsymbol{\Sigma}$ which do not even agree in their number of columns. However, as Stroock and Varadhan (1979, Chap. 5.3) show, the particular choice of the diffusion coefficient does not influence the distribution of the process $\boldsymbol{X}$ as long as it is a square root of the diffusion matrix.

An $\mathcal{F}$-adapted process $\boldsymbol{X}$ satisfying the Itô SDE

$$\mathrm{d}\boldsymbol{X}_t = \boldsymbol{\mu}(\boldsymbol{X}_t,t)\mathrm{d}t + \boldsymbol{\sigma}(\boldsymbol{X}_t,t)\mathrm{d}\boldsymbol{B}_t, \qquad \boldsymbol{X}_{t_0} = \boldsymbol{x}_0, \tag{3.18}$$

is an *Itô diffusion* with drift $\boldsymbol{\mu}$ and diffusion matrix $\boldsymbol{\Sigma} = \boldsymbol{\sigma}\boldsymbol{\sigma}'$ if the coefficients $\boldsymbol{\mu}$ and $\boldsymbol{\sigma}$ fulfil the Lipschitz condition (3.10) and growth bound (3.11) and are continuous in time. The other way round, if the transition density of an Itô diffusion with starting value $\boldsymbol{X}_{t_0} = \boldsymbol{x}_0$ is uniquely determined by the drift $\boldsymbol{\mu}$ and positive semi-definite diffusion matrix $\boldsymbol{\Sigma} = \boldsymbol{\sigma}\boldsymbol{\sigma}'$, where again $\boldsymbol{\mu}$ and $\boldsymbol{\sigma}$ satisfy (3.10) and (3.11), it is a solution of the Itô SDE (3.18) (Arnold 1973, Chap. 9.3).

3.2.6 Sample Path Properties

For non-vanishing diffusion coefficient, diffusion processes look like Brownian motion locally in time. Hence many characteristic sample path properties such as the infinite total variation and non-differentiability are inherited from the driving Brownian motion. As the integral exerts a smoothing effect, diffusion processes have almost surely continuous sample paths. Similarly to (3.1), the *quadratic variation* of the above Itô diffusion process on a time interval $[s,t]$ is

$$\langle \boldsymbol{X}, \boldsymbol{X}\rangle_{[s,t]} = \lim_{\delta(\mathcal{Z}_n)\downarrow 0} \sum_{i=1}^{h(n)} \Big(\boldsymbol{X}_{t_i^{(n)}} - \boldsymbol{X}_{t_{i-1}^{(n)}}\Big)\Big(\boldsymbol{X}_{t_i^{(n)}} - \boldsymbol{X}_{t_{i-1}^{(n)}}\Big)' = \int_s^t \boldsymbol{\Sigma}(\boldsymbol{X}_\tau, \tau)\mathrm{d}\tau$$

in probability (and almost surely for sufficiently smooth $\boldsymbol{\Sigma}$), where $\delta(\mathcal{Z}_n)$ is the fineness of a partition $\mathcal{Z}_n$ of $[s,t]$. This turns out to be of great importance in parameter estimation if observations are continuous or on a sufficiently fine time scale (see Sects. 6.1.1 and 7.3).

3.2.7 Ergodicity

For any $\varepsilon > 0$ and $\boldsymbol{x} \in \mathcal{X}$, denote by $\mathcal{U}_\varepsilon(\boldsymbol{x})$ a spherical neighbourhood of radius ε around $\boldsymbol{x}$. Assume that at time t_0, a time-homogeneous diffusion process $\boldsymbol{X}$ is in state $\boldsymbol{x}_0 \in \mathcal{X}$. Let

$$T_\varepsilon(\boldsymbol{x}) = \inf_{t \geq t_0} \big\{\boldsymbol{X}_t \in \mathcal{U}_\varepsilon(\boldsymbol{x})\big\}$$

be the first time at which the process enters $\mathcal{U}_\varepsilon(\boldsymbol{x})$. The process is called *recurrent* if this time is almost surely finite, irrespectively of $\boldsymbol{x}_0$ and $\boldsymbol{x}$, i.e.

$$\forall\, \varepsilon > 0 \quad \forall\, \boldsymbol{x}_0, \boldsymbol{x} \in \mathcal{X} \quad \mathbb{P}\big(T_\varepsilon(\boldsymbol{x}) < \infty \,\big|\, \boldsymbol{X}_{t_0} = \boldsymbol{x}_0\big) = 1.$$

Furthermore, the process is called *positive recurrent* or *ergodic* if the expected value of this time point is finite, i.e.

$$\forall\, \varepsilon > 0 \quad \forall\, \boldsymbol{x}_0, \boldsymbol{x} \in \mathcal{X} \quad \mathbb{E}\big(T_\varepsilon(\boldsymbol{x}) \,\big|\, \boldsymbol{X}_{t_0} = \boldsymbol{x}_0\big) < \infty.$$

The diffusion process then possesses a stationary distribution (Klebaner 2005).

For one-dimensional diffusion processes, the following statements hold: The *scale function* and *speed measure* of a time-homogeneous diffusion process

$$\mathrm{d}X_t = \mu(X_t)\mathrm{d}t + \sigma(X_t)\mathrm{d}B_t, \qquad X_{t_0} = x_0,$$

with positive diffusion $\Sigma = \sigma^2$ are defined as

$$s(x) = \exp\left(-2\int_{x_0}^{x} \mu(z)\Sigma^{-1}(z)\mathrm{d}z\right) \quad \text{and} \quad m(x) = \frac{1}{s(x)}\,\Sigma^{-1}(x)$$

for all $x \in \mathcal{X}$. The process is recurrent if and only if

$$\lim_{x\to-\infty}\int_{x_0}^{x} s(z)\mathrm{d}z = -\infty \quad \text{and} \quad \lim_{x\to\infty}\int_{x_0}^{x} s(z)\mathrm{d}z = \infty.$$

It is positive recurrent if and only if

$$\int_{\mathcal{X}} m(z)\mathrm{d}z < \infty.$$

If these two conditions are fulfilled, the diffusion process is ergodic, i.e. there exists a *stationary* (or *invariant*) *density* $\pi : \mathcal{X} \to \mathbb{R}$ such that for a random variable $\xi \sim \pi$ and any measurable function h with $\mathbb{E}|h(\xi)| < \infty$ one has almost surely

$$\lim_{t\to\infty}\frac{1}{t-t_0}\int_{t_0}^{t} h(X_s)\mathrm{d}s = \mathbb{E}\,h(\xi). \tag{3.19}$$

Equation (3.19) relates the long-term time average of the paths to the spatial average with respect to π. The stationary density results as

$$\pi(x) = m(x)\Big/\int_{\mathcal{X}} m(x)\mathrm{d}x$$

(Kutoyants 2004, Chap. 1.2). Several of the inference techniques in Chap. 6 require the existence of a stationary distribution. For ergodicity conditions for multi-dimensional diffusion processes, see Klebaner (2005, Chap. 6).

3.2.8 Kolmogorov Forward and Backward Equations

Suppose the transition density p of an Itô diffusion process

$$\mathrm{d}\boldsymbol{X}_t = \boldsymbol{\mu}(\boldsymbol{X}_t, t)\mathrm{d}t + \boldsymbol{\sigma}(\boldsymbol{X}_t, t)\mathrm{d}\boldsymbol{B}_t, \qquad \boldsymbol{X}_{t_0} = \boldsymbol{x}_0,$$

is smooth enough such that the derivatives in the following partial differential equations exist and are continuous. Then p satisfies the *Kolmogorov forward equation*

$$\begin{aligned} &\frac{\partial p(s,\boldsymbol{x},t,\boldsymbol{y})}{\partial t} \\ &= -\sum_{i=1}^{d} \frac{\partial\left[\mu_i(\boldsymbol{y},t)p(s,\boldsymbol{x},t,\boldsymbol{y})\right]}{\partial y^{(i)}} + \frac{1}{2}\sum_{i,j=1}^{d} \frac{\partial^2\left[\Sigma_{ij}(\boldsymbol{y},t)p(s,\boldsymbol{x},t,\boldsymbol{y})\right]}{\partial y^{(i)}\partial y^{(j)}} \end{aligned} \tag{3.20}$$

for fixed $\boldsymbol{x}$ and s and the *Kolmogorov backward equation*

$$\begin{aligned} &-\frac{\partial p(s,\boldsymbol{x},t,\boldsymbol{y})}{\partial s} \\ &= \sum_{i=1}^{d} \mu_i(\boldsymbol{x},s)\frac{\partial p(s,\boldsymbol{x},t,\boldsymbol{y})}{\partial x^{(i)}} + \frac{1}{2}\sum_{i,j=1}^{d} \Sigma_{ij}(\boldsymbol{x},s)\frac{\partial^2 p(s,\boldsymbol{x},t,\boldsymbol{y})}{\partial x^{(i)}\partial x^{(j)}} \end{aligned} \tag{3.21}$$

for fixed $\boldsymbol{y}$ and t, where $\boldsymbol{x},\boldsymbol{y} \in \mathcal{X}$ and $t > s \geq 0$, and i, j denote the respective components of $\boldsymbol{x}$, $\boldsymbol{y}$, $\boldsymbol{\mu}$ and $\boldsymbol{\Sigma} = \boldsymbol{\sigma}\boldsymbol{\sigma}'$. Remarkably, each of these equations uniquely determines the transition density p (subject to an appropriate initial condition), and hence diffusion processes are, like Gaussian processes, already completely defined by their instantaneous mean and variance $\boldsymbol{\mu}$ and $\boldsymbol{\Sigma}$. Furthermore, if the transition density of a stochastic process fulfils the Kolmogorov forward or backward equation, then it is an Itô diffusion process.

Equations (3.20) and (3.21) are sometimes also called the *forward* and *backward diffusion equations*. The terms forward and backward arise from the equations describing the evolution of the process with respect to a later and former state, respectively. The Kolmogorov forward equation is additionally known as the *Fokker-Planck equation*. For shorter notation, introduce the two operators $\mathcal{L}^F_{\boldsymbol{\mu},\boldsymbol{\Sigma}}$ and $\mathcal{L}^B_{\boldsymbol{\mu},\boldsymbol{\Sigma}}$ such that

$$\frac{\partial p}{\partial t} = \mathcal{L}^F_{\boldsymbol{\mu},\boldsymbol{\Sigma}}\, p \qquad \text{and} \qquad -\frac{\partial p}{\partial s} = \mathcal{L}^B_{\boldsymbol{\mu},\boldsymbol{\Sigma}}\, p. \tag{3.22}$$

The Kolmogorov forward and backward equations are important tools in the approximation of pure Markov jump processes by diffusions as considered in Chap. 4. The above equations correspond to diffusion equations of the Itô type. Stratonovich (1989) deals with counterparts of (3.20) in other stochastic calculi.

3.2.9 Infinitesimal Generator

The *infinitesimal generator* $\mathcal{G}$ of a Markov process $\boldsymbol{X}$ is defined by

$$\mathcal{G}f(\boldsymbol{x},t) = \lim_{\Delta t \downarrow 0} \frac{1}{\Delta t}\,\mathbb{E}\big(f\big(\boldsymbol{X}_{t+\Delta t}, t+\Delta t\big) - f\big(\boldsymbol{x},t\big)\big|\boldsymbol{X}_t = \boldsymbol{x}\big)$$

for all measurable bounded functions $f : \mathcal{X} \times T \to \mathbb{R}$ for which the uniform limit exists. That is the expected infinitesimal rate of change of $f(\boldsymbol{X}_t, t)$ given $\boldsymbol{X}_t = \boldsymbol{x}$. Like the Kolmogorov forward and backward equations, $\mathcal{G}$ uniquely determines a diffusion process for a given initial value. The infinitesimal generator of a diffusion $\boldsymbol{X}$ with drift vector $\boldsymbol{\mu}$ and diffusion matrix $\boldsymbol{\Sigma}$ is related to the Kolmogorov backward operator (3.22) through

$$\mathcal{G}\tilde{p} = \left(\frac{\partial}{\partial s} + \mathcal{L}^B_{\boldsymbol{\mu},\boldsymbol{\Sigma}}\right)\tilde{p}$$

with $\tilde{p}(\boldsymbol{x}, s) = p(s, \boldsymbol{x}, t, \boldsymbol{y})$ and t and $\boldsymbol{y}$ fixed. Hence, if one is able to derive the infinitesimal generator of a diffusion process, one can directly read out its drift vector and diffusion matrix. We will make use of this in Chap. 4 when deriving diffusion approximations from Markov jump processes.

3.2.10 Itô Formula

Let $\boldsymbol{X} = (\boldsymbol{X}_t)_{t\in T}$ be an Itô process with state space $\mathcal{X} \subseteq \mathbb{R}^d$ and $\boldsymbol{g} : \mathcal{X} \times T \to \mathbb{R}^l$ a jointly measurable function that is twice continuously differentiable in space and once in time. Then $\boldsymbol{Y} = (\boldsymbol{Y}_t)_{t\in T}$ with $\boldsymbol{Y}_t = \boldsymbol{g}(\boldsymbol{X}_t, t)$ is again an Itô process, and for its kth component we get the *Itô formula*

$$\begin{aligned} \mathrm{d}Y_t^{(k)} &= \frac{\partial g^{(k)}(\boldsymbol{X}_t, t)}{\partial t}\,\mathrm{d}t + \sum_{i=1}^{d} \frac{\partial g^{(k)}(\boldsymbol{X}_t, t)}{\partial x^{(i)}}\,\mathrm{d}X_t^{(i)} \\ &\quad + \frac{1}{2}\sum_{i,j=1}^{d} \frac{\partial^2 g^{(k)}(\boldsymbol{X}_t, t)}{\partial x^{(i)}\partial x^{(j)}}\,\mathrm{d}X_t^{(i)}\mathrm{d}X_t^{(j)} \end{aligned} \tag{3.23}$$

for $k = 1, \ldots, l$, where the upper indices denote the respective component numbers. The terms $\mathrm{d}X_t^{(i)}\mathrm{d}X_t^{(j)}$ are to be calculated according to the mean-square rules

$$(\mathrm{d}t)^2 = \mathrm{d}t \cdot \mathrm{d}B_t^{(i)} = \mathrm{d}B_t^{(i)} \cdot \mathrm{d}t = 0 \qquad \text{and} \qquad \mathrm{d}B_t^{(i)}\mathrm{d}B_t^{(j)} = \delta_{ij}\mathrm{d}t, \tag{3.24}$$

where δ_{ij} is the Kronecker delta, in combination with the SDE defining $\boldsymbol{X}_t$.

Formula (3.23) is the Itô stochastic counterpart of the deterministic chain rule in classical calculus (and also in Stratonovich calculus), where the second sum is absent.

3.2.11 Lamperti Transformation

An application of Itô's formula is the following: Consider a one-dimensional Itô diffusion $(X_t)_{t\geq 0}$ with time-homogeneous diffusion coefficient, i.e.

$$\mathrm{d}X_t = \mu(X_t, t)\mathrm{d}t + \sigma(X_t)\mathrm{d}B_t,\ X_{t_0} = x_0,$$

and its *Lamperti transform* $Y = (Y_t)_{t\geq 0}$, where

$$Y_t = g(X_t) = \int_a^{X_t} \frac{\mathrm{d}u}{\sigma(u)}$$

for any a in the state space. Then Y fulfils the Itô SDE

$$\mathrm{d}Y_t = \left(\frac{\mu(g^{-1}(Y_t), t)}{\sigma(g^{-1}(Y_t))} - \frac{1}{2}\frac{\partial\sigma}{\partial x}\left(g^{-1}(Y_t)\right)\right)\mathrm{d}t + \mathrm{d}B_t, \qquad Y_{t_0} = g(x_0),$$

i.e. it has unit diffusion. This is a convenient property in the context of parameter estimation, and hence this transformation will frequently be used in the methods covered in Chaps. 6 and 7. Unfortunately, there is no such transform for general multi-dimensional diffusion processes $\boldsymbol{X}$ with diffusion matrix $\boldsymbol{\Sigma}$. A transformation $\boldsymbol{Y} = \boldsymbol{g}(\boldsymbol{X})$ with unit diffusion requires $\boldsymbol{g} : \mathcal{X} \to \mathbb{R}^l$ to be an invertible function fulfilling the conditions of the Itô formula and

$$\nabla\boldsymbol{g}(\boldsymbol{X}_t)\boldsymbol{\Sigma}(\boldsymbol{X}_t)\nabla\boldsymbol{g}(\boldsymbol{X}_t)' = \boldsymbol{I},$$

where

$$\nabla\boldsymbol{g} = \begin{pmatrix} \frac{\partial g^{(1)}}{\partial x^{(1)}} & \cdots & \frac{\partial g^{(1)}}{\partial x^{(d)}} \\ \vdots & \ddots & \vdots \\ \frac{\partial g^{(l)}}{\partial x^{(1)}} & \cdots & \frac{\partial g^{(l)}}{\partial x^{(d)}} \end{pmatrix}$$

(Papaspiliopoulos et al. 2003). Such $\boldsymbol{g}$ cannot be found in general. Aït-Sahalia (2008), however, provides a necessary and sufficient condition for the availability of an appropriate transform.

3.2.12 Girsanov Formula

Let $\mathbb{P}_{\boldsymbol{\sigma}}$ be the probability measure induced by the solution of the Itô SDE

$$\mathrm{d}\boldsymbol{X}_\tau = \boldsymbol{\mu}(\boldsymbol{X}_\tau, \tau)\mathrm{d}\tau + \boldsymbol{\sigma}(\boldsymbol{X}_\tau, \tau)\mathrm{d}\boldsymbol{B}_\tau$$

for $\tau \in [t_0, t]$ and a fixed starting value at time t_0, and let $\mathbb{W}_{\boldsymbol{\sigma}}$ be the law of the respective driftless process. Suppose that $\boldsymbol{\Sigma} = \boldsymbol{\sigma}\boldsymbol{\sigma}'$ is invertible and $\boldsymbol{\mu}$ fulfils the *Novikov condition*

$$\mathbb{E}_{\mathbb{P}_{\boldsymbol{\sigma}}}\left[\exp\left(\frac{1}{2}\int_{t_0}^{t} \|\boldsymbol{\mu}(\boldsymbol{X}_\tau, \tau)\|^2 \mathrm{d}\tau\right)\right] < \infty.$$

Then $\mathbb{P}_{\boldsymbol{\sigma}}$ and $\mathbb{W}_{\boldsymbol{\sigma}}$ are equivalent measures with Radon-Nikodym derivative given by *Girsanov's formula*

$$\frac{\mathrm{d}\mathbb{P}_{\boldsymbol{\sigma}}}{\mathrm{d}\mathbb{W}_{\boldsymbol{\sigma}}}(\boldsymbol{X}_{[t_0,s]}) = \exp\left(\int_{t_0}^{s} \boldsymbol{\mu}'\,\boldsymbol{\Sigma}^{-1}\mathrm{d}\boldsymbol{X}_\tau - \frac{1}{2}\int_{t_0}^{s} \boldsymbol{\mu}'\,\boldsymbol{\Sigma}^{-1}\boldsymbol{\mu}\,\mathrm{d}\tau\right) \tag{3.25}$$

for all $s \in [t_0, t]$ and $\boldsymbol{X}_{[t_0,s]} = (\boldsymbol{X}_\tau)_{\tau \in [t_0,s]}$. The coefficients $\boldsymbol{\mu}$ and $\boldsymbol{\Sigma}$ in the integrals are evaluated at $\boldsymbol{X}_\tau$ and τ.

The right-hand side of (3.25) is a $\mathbb{W}_{\boldsymbol{\sigma}}$-martingale and states the density of the law of $\boldsymbol{X}$ with respect to $\mathbb{W}_{\boldsymbol{\sigma}}$. For continuous observation of $\boldsymbol{X}$ and known diffusion coefficient, (3.25) serves as the likelihood of the parameters entering the drift function (cf. Sect. 6.1.1).

3.3 Approximation and Simulation

If the solution of a diffusion process is explicitly known, it is straightforward to sample from its distribution at discrete time instants since the increments of the driving Brownian motion process are just Gaussian random variables. However, as mentioned before, solutions of SDEs are usually unattainable in closed form. Sections 3.3.1 and 3.3.2 hence deal with numerical methods to approximate the Itô diffusion process

$$\mathrm{d}\boldsymbol{X}_t = \boldsymbol{\mu}(\boldsymbol{X}_t, t)\mathrm{d}t + \boldsymbol{\sigma}(\boldsymbol{X}_t, t)\mathrm{d}\boldsymbol{B}_t, \qquad \boldsymbol{X}_{t_0} = \boldsymbol{x}_0, \tag{3.26}$$

on a discrete time grid $t_0 < t_1 < \ldots < t_n$, where the Lipschitz and growth bound conditions (3.10) and (3.11) are assumed to be fulfilled and $\mathbb{E}\|\boldsymbol{X}_{t_0}\| < \infty$. These approximation techniques immediately yield approximate sampling algorithms. Section 3.3.2 also covers the exact simulation of Brownian bridges, whose probability distributions are explicitly known.

In the sequel, an approximation or simulation of $\boldsymbol{X}_{t_k}$ will be denoted by $\boldsymbol{Y}_k$, $k = 0, \ldots, n$. Moreover, introduce the increments $\Delta t_k = t_{k+1} - t_k$ and $\Delta \boldsymbol{B}_k = \boldsymbol{B}_{t_{k+1}} - \boldsymbol{B}_{t_k} \sim \mathcal{N}(\boldsymbol{0}, \Delta t_k \boldsymbol{I})$ for $k = 0, \ldots, n-1$ and the maximum time step $\Delta = \max_k \Delta t_k$. The resulting approximation or exact realisation $\boldsymbol{Y}$ is considered a time-continuous process although the according sampling schemes naturally yield values only for a collection of discrete time instants. Intermediate data is usually obtained by linear interpolation. However, as mentioned in Sect. 3.2.6, the paths of a diffusion process are extremely irregular, which cannot be reproduced this way.

3.3.1 Convergence and Consistency

As for the definition of the stochastic integral, one has to take care when deriving a stochastic approximation method from its deterministic counterpart. As e.g. shown in Fahrmeir (1976), such generalisations might for instance result in wrong drift coefficients. It is hence crucial to evaluate approximation schemes through their convergence and consistency properties. For stochastic differential equations, these exist in a weak and a strong sense, where the first concerns distributional and the second pathwise approximations. In this book, we only consider the latter.

With the above notation, an approximation $\boldsymbol{Y}$ of a process $\boldsymbol{X}$ on a time interval $[t_0, t_n]$ is said to *converge strongly* of order $p > 0$ if there exist positive constants C and Δ_0 such that

$$\mathbb{E}\|\boldsymbol{X}_{t_n} - \boldsymbol{Y}_n\| \leq C\Delta^p$$

for all $\Delta \in (0, \Delta_0)$. It is called *strongly consistent* if there exists a non-negative function $\gamma(\Delta)$ which tends to zero as $\Delta \to 0$ such that for all $k = 0, \ldots, n-1$

$$\mathbb{E}\left\|\Delta_k^{-1}\mathbb{E}\left(\boldsymbol{Y}_{k+1} - \boldsymbol{Y}_k|\mathcal{F}_{t_k}\right) - \boldsymbol{\mu}(\boldsymbol{Y}_k, t_k)\right\|^2 \leq \gamma(\Delta)$$

and

$$\Delta_k^{-1}\mathbb{E}\left\|\boldsymbol{Y}_{k+1} - \boldsymbol{Y}_k - \mathbb{E}\left(\boldsymbol{Y}_{k+1} - \boldsymbol{Y}_k|\mathcal{F}_{t_k}\right) - \boldsymbol{\sigma}(\boldsymbol{Y}_k, t_k)\Delta\boldsymbol{B}_k\right\|^2 \leq \gamma(\Delta).$$

Note that the order of strong convergence can be higher in special cases, e.g. for constant drift and diffusion coefficients.

3.3.2 Numerical Approximation

An obvious way to obtain numerical approximation methods is to employ a truncated version of the *Itô-Taylor expansion* (Kloeden and Platen 1991)

$$\boldsymbol{X}_t = \boldsymbol{X}_{t_0} + \boldsymbol{\mu}(\boldsymbol{X}_{t_0}, t_0)\int_{t_0}^{t} \mathrm{d}s + \boldsymbol{\sigma}(\boldsymbol{X}_{t_0}, t_0)\int_{t_0}^{t} \mathrm{d}\boldsymbol{B}_s + \boldsymbol{R}_3$$

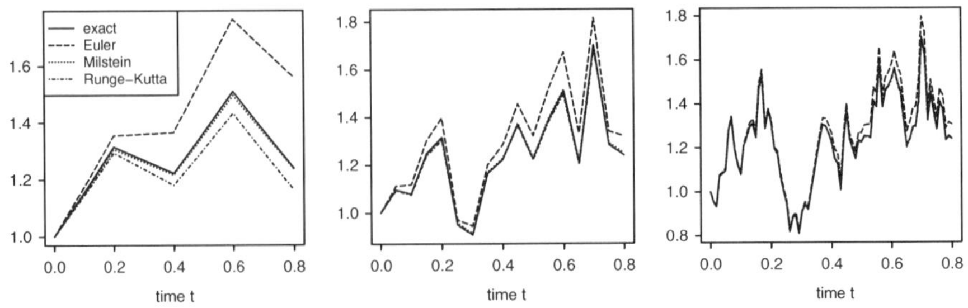

Fig. 3.2 Discrete realisations of geometric Brownian motion, defined through SDE (3.12) or its explicit solution (3.13) on p. 39, for $x_0 = 1$, $\alpha = 0.5$ and $\sigma = 0.9$: Exact sampling (*solid lines*), Euler approximation (*dashed*), Milstein approximation (*dotted*) and Runge-Kutta approximation (*dash-dotted*) for equidistant time steps 0.2 (*left*), 0.05 (*middle*) and 0.01 (*right*), each with respect to the same driving Brownian motion

with a remainder term $\boldsymbol{R}_3$ with ith component

$$R_3^{(i)} = \sum_{j=1}^{m}\sum_{l=1}^{m}\sum_{r=1}^{d} \sigma_{rj}(\boldsymbol{X}_{t_0}, t_0)\left(\frac{\partial \sigma_{il}}{\partial x^{(r)}}(\boldsymbol{X}_{t_0}, t_0)\right)\int_{t_0}^{t}\int_{t_0}^{s} \mathrm{d}B_u^{(j)}\mathrm{d}B_s^{(l)} + \ldots,$$

where the sub- and superscripts denote the components of $\boldsymbol{\sigma} \in \mathbb{R}^{d\times m}$, $\boldsymbol{x} \in \mathbb{R}^d$ and $\boldsymbol{B} \in \mathbb{R}^m$. The following Euler and Milstein methods are applications of this. The Milstein scheme is of higher order of strong convergence than the Euler approximation but involves the more elaborate computation and evaluation of derivatives of the diffusion coefficient. The latter is avoided by the Runge-Kutta scheme which is introduced thereafter. All approximations converge to the solution of (3.26) in the Itô sense.

Figure 3.2 shows discrete approximations of geometric Brownian motion—represented by the explicitly solvable SDE (3.12) on p. 39—obtained with the Euler, Milstein and Runge-Kutta schemes in comparison with exact simulation for different step sizes.

There are several more numerical approximation methods (Fahrmeir and Beeck 1974; Rümelin 1982; Chang 1987; Newton 1991; Kloeden and Platen 1999, and the references therein), but the selection made here covers the needs of this book.

Euler Scheme

The *Euler approximation* (also called *Euler-Maruyama approximation*) of $\boldsymbol{X}$ is obtained by setting $\boldsymbol{Y}_0 = \boldsymbol{x}_0$ and then successively[3]

[3]Contrarily to common matrix notation, but consistently with the differential equation representation, the scalar Δt_k is multiplied with the vector $\boldsymbol{\mu}$ from the right—a consetude that will be kept throughout this book.

$$\boldsymbol{Y}_{k+1} = \boldsymbol{Y}_k + \boldsymbol{\mu}(\boldsymbol{Y}_k, t_k)\Delta t_k + \boldsymbol{\sigma}(\boldsymbol{Y}_k, t_k)\Delta \boldsymbol{B}_k \tag{3.27}$$

for $k = 0, \ldots, n-1$. It is strongly consistent and has strong order of convergence $p = 0.5$.

Milstein Scheme

The *Milstein method* yields approximate values by setting $\boldsymbol{Y}_0 = \boldsymbol{x}_0$ and then successively for the ith component

$$\begin{aligned} \boldsymbol{Y}_{k+1}^{(i)} &= \boldsymbol{Y}_k^{(i)} + \mu_i(\boldsymbol{Y}_k, t_k)\Delta t_k + \sum_{j=1}^{m} \sigma_{ij}(\boldsymbol{Y}_k, t_k)\Delta B_k^{(j)} \\ &+ \sum_{j=1}^{m}\sum_{l=1}^{m}\sum_{r=1}^{d} \sigma_{rj}(\boldsymbol{Y}_k, t_k)\left(\frac{\partial \sigma_{il}}{\partial x^{(r)}}(\boldsymbol{Y}_k, t_k)\right) \int_{t_k}^{t_{k+1}}\int_{t_k}^{s} \mathrm{d}B_u^{(j)}\mathrm{d}B_s^{(l)} \end{aligned}$$

for $k = 0, \ldots, n-1$. For $j = l$ (and thus especially for one-dimensional Brownian motion) the double integral simplifies to

$$\int_{t_k}^{t_{k+1}}\int_{t_k}^{s} \mathrm{d}B_u^{(j)}\mathrm{d}B_s^{(l)} = \frac{1}{2}\left(\left(\Delta B_k^{(j)}\right)^2 - \Delta t_k\right),$$

but otherwise its computation is generally more demanding (cf. e.g. Kloeden and Platen 1999). Suppose that $\boldsymbol{\mu}$ and $\boldsymbol{\sigma}$ are twice continuously differentiable with uniformly Lipschitz continuous derivatives. Then the Milstein scheme is strongly consistent and strongly convergent of order $p = 1$.

Runge-Kutta Scheme

One possible alternative to the computation of the derivative of the diffusion coefficient in the Milstein scheme is the application of finite differences as in the *Runge-Kutta method*

$$\begin{aligned} \widetilde{\boldsymbol{Y}}_{k,j} &= \boldsymbol{Y}_k + \boldsymbol{\mu}(\boldsymbol{Y}_k, t_k)\Delta t_k + \boldsymbol{\sigma}_{\bullet j}(\boldsymbol{Y}_k, t_k)\sqrt{\Delta t_k} \\ \boldsymbol{Y}_{k+1}^{(i)} &= \boldsymbol{Y}_k^{(i)} + \mu_i(\boldsymbol{Y}_k, t_k)\Delta t_k + \sum_{j=1}^{m} \sigma_{ij}(\boldsymbol{Y}_k, t_k)\Delta B_k^{(j)} \\ &+ \frac{1}{\sqrt{\Delta t_k}} \sum_{j=1}^{m}\sum_{l=1}^{m} \left(\sigma_{il}(\widetilde{\boldsymbol{Y}}_{k,j}, t_k) - \sigma_{il}(\boldsymbol{Y}_k, t_k)\right) \int_{t_k}^{t_{k+1}}\int_{t_k}^{s} \mathrm{d}B_u^{(j)}\mathrm{d}B_s^{(l)} \end{aligned}$$

for $k = 0, \dots, n-1$, $j = 1, \dots, m$ and $i = 1, \dots, d$, where $\boldsymbol{\sigma}_{\bullet j}$ denotes the jth column of $\boldsymbol{\sigma}$. This derivative-free approximation is strongly consistent and of strong order $p = 1$ if the coefficients are twice continuously differentiable with uniformly bounded derivatives.

3.3.3 *Simulation of Brownian Bridge*

In Sect. 3.1.2 a Brownian $(s, \boldsymbol{u}, t, \boldsymbol{v})$-bridge $\tilde{\boldsymbol{B}} = (\tilde{\boldsymbol{B}}_\tau)_{\tau \in [s,t]}$ was defined as Brownian motion conditioned on $\tilde{\boldsymbol{B}}_s = \boldsymbol{u}$ and $\tilde{\boldsymbol{B}}_t = \boldsymbol{v}$. This process can exactly be sampled at discrete time instants as follows (Beskos et al. 2006):

1. Simulate Brownian motion at times $s = t_0 < t_1 < \ldots < t_n = t$. This is done by setting $\boldsymbol{B}_0 = \boldsymbol{0}$ and then successively drawing for fixed $\sigma \in \mathbb{R}_+$

$$\boldsymbol{B}_{k+1} \sim \mathcal{N}(\boldsymbol{B}_k, \sigma^2 \Delta t_k \boldsymbol{I}) \quad \text{for} \quad k = 0, \dots, n-1.$$

2. Construct a Brownian $(s, \boldsymbol{0}, t, \boldsymbol{0})$-bridge $\bar{\boldsymbol{B}}$ from the Brownian motion seeds via

$$\bar{\boldsymbol{B}}_k = \boldsymbol{B}_k - \frac{t_k - s}{t - s} \boldsymbol{B}_n \quad \text{for} \quad k = 0, \dots, n.$$

3. Transform this to a Brownian $(s, \boldsymbol{u}, t, \boldsymbol{v})$-bridge $\tilde{\boldsymbol{B}}$ through

$$\tilde{\boldsymbol{B}}_k = \bar{\boldsymbol{B}}_k + \frac{t - t_k}{t - s} \boldsymbol{u} + \frac{t_k - s}{t - s} \boldsymbol{v} \quad \text{for} \quad k = 0, \dots, n.$$

3.4 Concluding Remarks

This chapter gives an overview of stochastic differential equations and diffusion processes to the extent which is required as a basis for the remaining parts of this book. It covers the motivation and introduction of stochastic integrals as opposed to the classical Lebesgue-Stieltjes integral, the definition of diffusion processes, material properties and formulas from stochastic calculus and finally numerical approximation and exact sampling methods. References to monographs on these subjects were provided at the beginning of this chapter.

Regularity conditions were stated whenever necessary. For the purposes of this book, let the following assumptions from now on hold unless otherwise stated: $\boldsymbol{\mu}$ and $\boldsymbol{\sigma}$ denote jointly $\mathcal{F}^* \times \mathcal{L}$-measurable drift and diffusion coefficients of a diffusion process. Dependence on a parameter $\boldsymbol{\theta}$ will be included in the notation later in this book. Both $\boldsymbol{\mu}$ and $\boldsymbol{\sigma}$ are supposed to be such that the stochastic integral is well-defined (cf. Sect. 3.2.1), to fulfil the Lipschitz condition (3.10) and growth bound (3.11), and to be twice continuously differentiable with respect to all arguments. The diffusion matrix $\Sigma = \sigma\sigma'$ is assumed positive definite and invertible. These regularity conditions are usually fulfilled in applications in life sciences.

The following Chaps. 4–7 show how to utilise diffusion processes for modelling phenomena in life sciences and how to perform inference on the model parameters. This is implemented in two applications in Chaps. 8 and 9.

References

Aït-Sahalia Y (2008) Closed-form likelihood expansions for multivariate diffusions. Ann Stat 36:906–937

Alonso D, McKane A, Pascual M (2007) Stochastic amplification in epidemics. J R Soc Interface 4:575–582

Arnold L (1973) Stochastische Differentialgleichungen. Oldenbourg, München

Barbour A (1974) On a functional central limit theorem for Markov population processes. Adv Appl Probab 6:21–39

Beskos A, Papaspiliopoulos O, Roberts G, Fearnhead P (2006) Exact and computationally efficient likelihood-based estimation for discretely observed diffusion processes (with comments). J R Stat Soc Ser B 68:333–382

Bibby B, Sørensen M (2001) Simplified estimating functions for diffusion models with a high-dimensional parameter. Scand J Stat 28:99–112

Black F, Scholes M (1973) The pricing of options and corporate liabilities. J Polit Econ 81:637–654

Capasso V, Morale D (2009) Stochastic modelling of tumour-induced angiogenesis. J Math Biol 58:219–233

Chang CC (1987) Numerical solution of stochastic differential equations with constant diffusion coefficients. Math Comp 49:523–542

Chen WY, Bokka S (2005) Stochastic modeling of nonlinear epidemiology. J Theor Biol 234: 455–470

Chiarella C, Hung H, Tô TD (2009) The volatility structure of the fixed income market under the HJM framework: a nonlinear filtering approach. Comput Stat Data Anal 53:2075–2088

Clancy D, French N (2001) A stochastic model for disease transmission in a managed herd, motivated by Neospora caninum amongst dairy cattle. Math Biosci 170:113–132

Cobb L (1981) Stochastic differential equations for the social sciences. In: Cobb L, Thrall R (eds) Mathematical frontiers of the social and policy sciences. Westview Press, Boulder

Cox J, Ingersoll J, Ross S (1985) An intertemporal general equilibrium model of asset prices. Econometrica 53:363–384

de la Lama M, Szendro I, Iglesias J, Wio H (2006) Van Kampen's expansion approach in an opinion formation model. Eur Phys J B 51:435–442

Duan J, Gelfand A, Sirmans C (2009) Modeling space-time data using stochastic differential equations. Bayesian Anal 4:413–437

Elerian O, Chib S, Shephard N (2001) Likelihood inference for discretely observed nonlinear diffusions. Econometrica 69:959–993

Elf J, Ehrenberg M (2003) Fast evaluation of fluctuations in biochemical networks with the linear noise approximation. Genome Res 13:2475–2484

Eraker B (2001) MCMC analysis of diffusion models with application to finance. J Bus Econom Stat 19:177–191

Fahrmeir L (1976) Approximation von Stochastischen Differentialgleichungen auf Digital- und Hybridrechnern. Computing 16:359–371

Fahrmeir L, Beeck H (1974) Zur Simulation stetiger stochastischer Wirtschaftsmodelle. In: Transactions of the seventh Prague conference and of the European meeting of statisticians, Prague, pp 113–122
Fearnhead P (2006) The stationary distribution of allele frequencies when selection acts at unlinked loci. Theor Popul Biol 70:376–386
Ferm L, Lötstedt P, Hellander A (2008) A hierarchy of approximations of the master equation scaled by a size parameter. J Sci Comput 34:127–151
Fogelson A (1984) A mathematical model and numerical method for studying platelet adhesion and aggregation during blood clotting. J Comput Phys 56:111–134
Gard T (1988) Introduction to stochastic differential equations. Monographs and textbooks in pure and applied mathematics, vol 114. Dekker, New York
Gardiner C (1983) Handbook of stochastic methods. Springer, Berlin/Heidelberg
Golightly A, Wilkinson D (2005) Bayesian inference for stochastic kinetic models using a diffusion approximation. Biometrics 61:781–788
Golightly A, Wilkinson D (2006) Bayesian sequential inference for stochastic kinetic biochemical network models. J Comput Biol 13:838–851
Golightly A, Wilkinson D (2008) Bayesian inference for nonlinear multivariate diffusion models observed with error. Comput Stat Data Anal 52:1674–1693
Horsthemke W, Lefever R (1984) Noise-induced transitions: theory and applications in physics, chemistry, and biology. Springer, Berlin
Hufnagel L, Brockmann D, Geisel T (2004) Forecast and control of epidemics in a globalized world. Proc Natl Acad Sci U S A 101:15124–15129
Itô K (1944) Stochastic integral. Proc Jpn Acad 20:519–524
Itô K (1946) On a stochastic integral equation. Proc Jpn Acad 22:32–35
Karatzas I, Shreve S (1991) Brownian motion and stochastic calculus, 2nd edn. Graduate texts in mathematics. Springer, New York
Kimura M (1964) Diffusion models in population genetics. J Appl Probab 1:177–232
Klebaner F (2005) Introduction to stochastic calculus with applications, 2nd edn. Imperial College Press, London
Kloeden P, Platen E (1991) Stratonovich and Itô stochastic Taylor expansions. Math Nachr 151: 33–50
Kloeden P, Platen E (1999) Numerical solution of stochastic differential equations, 3rd edn. Springer, Berlin/Heidelberg/New York
Kutoyants Y (2004) Statistical inference for ergodic diffusion processes. Springer series in statistics. Springer, London
Leung H (1985) Expansion of the master equation for a biomolecular selection model. Bull Math Biol 47:231–238
McNeil D (1973) Diffusion limits for congestion models. J Appl Probab 10:368–376
Merton R (1976) Option pricing when underlying stock returns are discontinuous. J Finan Econ 3:125–144
Newton N (1991) Asymptotically efficient Runge-Kutta methods for a class of Itô and Stratonovich equations. SIAM J Appl Math 51:542–567
Øksendal B (2003) Stochastic differential equations. An introduction with applications, 6th edn. Springer, Berlin/Heidelberg
Papaspiliopoulos O, Roberts G, Sköld M (2003) Non-centered parameterisations for hierarchical models and data augmentation (with discussion). In: Bernardo J, Bayarri M, Berger J, Dawid A, Heckerman D, Smith A, West M (eds) Bayesian statistics 7. Lecture notes in computer science, vol 4699. Oxford University Press, Oxford, pp 307–326
Protter P (1990) Stochastic integration and differential equations. Applications of mathematics, vol 21. Springer, Berlin/Heidelberg
Ramshaw J (1985) Augmented Langevin approach to fluctuations in nonlinear irreversible processes. J Statist Phys 38:669–680
Revuz D, Yor M (1991) Continuous martingales and Brownian motion. A series of comprehensive studies in mathematics, vol 293. Springer, Berlin/Heidelberg

Robinson E (1959) A stochastic diffusion theory of price. Econometrica 27:679–684

Rümelin W (1982) Numerical treatment of stochastic differential equations. SIAM J Numer Anal 19:604–613

Seifert U (2008) Stochastic thermodynamics: principles and perspectives. Eur Phys J B 64: 423–431

Sjöberg P, Lötstedt P, Elf J (2009) Fokker-Planck approximation of the master equation in molecular biology. Comput Vis Sci 12:37–50

Stratonovich R (1966) A new representation for stochastic integrals and equations. SIAM J Control Optim 4:362–371

Stratonovich R (1989) Some Markov methods in the theory of stochastic processes in nonlinear dynamical systems. In: Moss F, McClintock P (eds) Noise in nonlinear dynamical systems. Theory of continuous Fokker-Planck systems, vol 1. Cambridge University Press, Cambridge, pp 16–71

Stroock D, Varadhan S (1979) Multidimensional diffusion processes. A series of comprehensive studies in mathematics, vol 233. Springer, New York

Tian T, Burrage K, Burrage P, Carletti M (2007) Stochastic delay differential equations for genetic regulatory networks. J Comput Appl Math 205:696–707

Tuckwell H (1987) Diffusion approximations to channel noise. J Theor Biol 127:427–438

van Kampen N (1965) Fluctuations in nonlinear systems. In: Burgess R (ed) Fluctuation phenomena in solids. Academic, New York, pp 139–177

van Kampen N (1981) The validity of nonlinear Langevin equations. J Stat Phys 25:431–442

Walsh J (1981) A stochastic model of neural response. Adv Appl Probab 13:231–281

Wiener N (1923) Differential space. J Math Phys 2:131–174

Chapter 4
Approximation of Markov Jump Processes by Diffusions

In many applications in life sciences one is concerned with the time-continuous evolution of numbers of individual objects such as the number of molecules in a gas, the number of infectives in a population, the number of animals in some region, the number of bacteria in a microscopic field etc. These numbers are stochastic quantities, and the state space of an according stochastic process is a subset of the set of integers or a multi-dimensional equivalent. If the process possesses the Markov property, transitions from one state to another are most adequately described by a so-called *master equation*. That is a differential-difference equation, i.e. a first order differential equation in the continuous time variable and difference equation in the discrete space variable. The discrete state space naturally implies discontinuity of the trajectories. The considered processes are *Markov jump processes*.

However, the sizes of the jumps are often infinitesimally small compared to the total size of the system. An approximation of the discontinuous paths by continuous curves is then justified. For example, consider a large number of different types of molecules which move around randomly, and assume that there is at most one collision possible within an infinitesimally small time interval. If the collision causes a reaction, the numbers of the involved types of molecules will change. On a macroscopic view of the according trajectories, however, these individual jumps will hardly be noticeable.

This chapter deals with the approximation of such Markov jump processes by Markov processes with continuous state space and almost surely continuous sample paths. The reward is the replacement of the master equation by a partial differential equation that is more convenient to deal with in a sense that will be elaborated soon. In order to maintain the strong Markov character of the original jump processes, we employ as approximations the only class of stochastic processes that are both strongly Markovian and have almost surely continuous sample paths. These are the *diffusion processes* introduced in the previous chapter. The counterparts of the master equations are the *Kolmogorov equations* discussed in Sect. 3.2.8.

The crucial point in the approximating procedure is that the sources of both the systematic and fluctuating part of the resulting stochastic differential equation are bound to agree with the initial description of the jump process. This is a complicated

C. Fuchs, *Inference for Diffusion Processes: With Applications in Life Sciences*,
DOI 10.1007/978-3-642-25969-2_4,

matter which has been the subject of numerous modelling attempts and source of considerable confusion (see van Kampen 1965, Chap. 1.C, for an overview). While it is often straightforward to derive the drift of the approximating diffusion process correctly, it is difficult to determine the strengths of the noise terms that arise from internal fluctuations. Hence, several authors avoid a rigorous mathematical derivation and set up the noise terms using heuristic arguments. The present chapter reviews and further develops proper approximation techniques in order to take remedial action.

This chapter is organised as follows: Sect. 4.1 categorises the key processes of this survey and important properties of their transition densities. The necessity of diffusion approximations and the intention of this chapter are emphasised in Sect. 4.2. Different techniques for the transition from a Markov jump process to a diffusion process are presented in Sect. 4.3 in detail. To that end, established methods from the literature are supplemented by new formulations and constructive algorithms in this book. Furthermore, results are presented in a multi-dimensional framework in this chapter—in contrast to the existing literature, where formulas are usually derived for the one-dimensional case. As a novelty, Sect. 4.4 extends the approaches from Sect. 4.3 to a more general framework, where the size of the considered system is characterised through multiple rather than a single size parameter. Section 4.5 discusses the appropriateness of different stochastic integrals for the considered modelling purposes. The outcomes of the entire chapter are summarised and collated in Sect. 4.6. For a reader who is primarily interested in explicit formulas for diffusion approximations rather than in the ideas of the underlying approximating procedures, it might suffice to work through this conclusion.

The methods of this chapter establish an indispensable part of this book as they are utilised in Chaps. 5 and 9 for the approximation of jump models in life sciences. The resulting diffusion approximations, in turn, form the basis of Chaps. 8 and 9, where the spread of influenza and the molecular binding behaviour of proteins are analysed.

For demonstration purposes, the epidemic *susceptible–infected (SI) model* is employed as a running example to which the various approximation techniques are applied after their derivation. This model is briefly introduced in Example 4.1. It arises as a special case of the *susceptible–infected–removed (SIR) model*, which is extensively considered in Sect. 5.1, if one sets the recovery rate equal to zero. Owing to the detailed and illustrative application of all approximation techniques in Chap. 5, the examples in the current chapter are restrained to the statement of intermediate results. For full calculations, the reader is referred to the next chapter.

4.1 Characterisation of Processes

We start with a brief note on the three classes of stochastic models which we will deal with in the course of this chapter: Markov jump processes, deterministic Markov processes with continuous sample paths, and diffusions. These are characterised through the transition density $p(t_0, \boldsymbol{x}_0, t, \boldsymbol{x})$ for the process arriving at state $\boldsymbol{x}$ at time t conditioned on an initial state $\boldsymbol{x}_0$ at time t_0. In the following we will fix t_0 and $\boldsymbol{x}_0$ and also use the notation $p(t, \boldsymbol{x})$ for $p(t_0, \boldsymbol{x}_0, t, \boldsymbol{x})$. p is assumed to fulfil the initial condition $p(t_0, \boldsymbol{x}_0, t_0, \boldsymbol{x}) = \delta(\boldsymbol{x} - \boldsymbol{x}_0)$ and to be smooth enough such that the derivatives in this chapter exist.

Gardiner (1983, Chap. 3.4) derives a differential equation for p, which he calls the *(forward) differential Chapman-Kolmogorov equation*. It is valid for all Markov processes in the interior of the state space $\mathcal{X} \subseteq \mathbb{R}^n$ and reads

$$\begin{aligned} \frac{\partial p(t, \boldsymbol{x})}{\partial t} &= \int\limits_{\mathcal{X}} \Big[W(t, \boldsymbol{y}, \boldsymbol{x} - \boldsymbol{y}) p(t, \boldsymbol{y}) - W(t, \boldsymbol{x}, \boldsymbol{y} - \boldsymbol{x}) p(t, \boldsymbol{x}) \Big] \mathrm{d}\boldsymbol{y} \\ &\quad - \sum_{i=1}^{n} \frac{\partial \big[\mu_i(\boldsymbol{x}, t) p(t, \boldsymbol{x}) \big]}{\partial x_i} + \frac{1}{2} \sum_{i,j=1}^{n} \frac{\partial^2 \big[\Sigma_{ij}(\boldsymbol{x}, t) p(t, \boldsymbol{x}) \big]}{\partial x_i \partial x_j}, \end{aligned} \tag{4.1}$$

where W is the *transition rate*, $\boldsymbol{\mu} = (\mu_i)_{i=1,\ldots,n}$ the *drift vector* and $\boldsymbol{\Sigma} = (\Sigma_{ij})_{i,j=1,\ldots,n}$ the *diffusion matrix*. These are defined for all $\varepsilon > 0$ and $i, j = 1, \ldots, n$ as

$$\begin{aligned} W(t, \boldsymbol{x}, \boldsymbol{y} - \boldsymbol{x}) &= \lim_{\Delta t \downarrow 0} \frac{1}{\Delta t} p(t, \boldsymbol{x}, t + \Delta t, \boldsymbol{y}) \\ \mu_i(\boldsymbol{x}, t) &= \lim_{\Delta t \downarrow 0} \frac{1}{\Delta t} \int\limits_{\|\boldsymbol{y} - \boldsymbol{x}\| \le \varepsilon} (y_i - x_i) p(t, \boldsymbol{x}, t + \Delta t, \boldsymbol{y}) \mathrm{d}\boldsymbol{y} \\ \Sigma_{ij}(\boldsymbol{x}, t) &= \lim_{\Delta t \downarrow 0} \frac{1}{\Delta t} \int\limits_{\|\boldsymbol{y} - \boldsymbol{x}\| \le \varepsilon} (y_i - x_i)(y_j - x_j) p(t, \boldsymbol{x}, t + \Delta t, \boldsymbol{y}) \mathrm{d}\boldsymbol{y} \end{aligned} \tag{4.2}$$

(compare with the definitions in Sect. 3.2.5). Higher order terms such as

$$\lim_{\Delta t \downarrow 0} \frac{1}{\Delta t} \int\limits_{\|\boldsymbol{y} - \boldsymbol{x}\| \le \varepsilon} (y_i - x_i)(y_j - x_j)(y_k - x_k) p(t, \boldsymbol{x}, t + \Delta t, \boldsymbol{y}) \mathrm{d}\boldsymbol{y}$$

vanish. An analogous *backward differential Chapman-Kolmogorov equation* is

$$\begin{aligned} \frac{\partial p(\tau, \boldsymbol{u}, t, \boldsymbol{x})}{\partial \tau} &= \int\limits_{\mathcal{X}} W(\tau, \boldsymbol{u}, \boldsymbol{y} - \boldsymbol{u}) \Big[p(\tau, \boldsymbol{u}, t, \boldsymbol{x}) - p(\tau, \boldsymbol{y}, t, \boldsymbol{x}) \Big] \mathrm{d}\boldsymbol{y} \\ &\quad - \sum_{i=1}^{n} \mu_i(\boldsymbol{u}, \tau) \frac{\partial p(\tau, \boldsymbol{u}, t, \boldsymbol{x})}{\partial u_i} - \frac{1}{2} \sum_{i,j=1}^{n} \Sigma_{ij}(\boldsymbol{u}, \tau) \frac{\partial^2 p(\tau, \boldsymbol{u}, t, \boldsymbol{x})}{\partial u_i \partial u_j}. \end{aligned} \tag{4.3}$$

Gardiner highlights three classes of Markov processes. These are

- Pure *jump processes*, where $\boldsymbol{\mu} = \mathbf{0}$ and $\boldsymbol{\Sigma} = \mathbf{0}$. In this case, Eq. (4.1) reduces to the *(forward) master equation*

$$\frac{\partial p(t,\boldsymbol{x})}{\partial t} = \int\limits_{\mathcal{X}} \Big[W(t,\boldsymbol{y},\boldsymbol{x}-\boldsymbol{y})p(t,\boldsymbol{y}) - W(t,\boldsymbol{x},\boldsymbol{y}-\boldsymbol{x})p(t,\boldsymbol{x})\Big]\mathrm{d}\boldsymbol{y}, \tag{4.4}$$

  Eq. (4.3) to the *backward master equation*

$$\frac{\partial p(\tau,\boldsymbol{u},t,\boldsymbol{x})}{\partial \tau} = \int\limits_{\mathcal{X}} W(\tau,\boldsymbol{u},\boldsymbol{y}-\boldsymbol{u})\Big[p(\tau,\boldsymbol{u},t,\boldsymbol{x}) - p(\tau,\boldsymbol{y},t,\boldsymbol{x})\Big]\mathrm{d}\boldsymbol{y}. \tag{4.5}$$

  Processes of this type have piecewise constant sample paths with finite jumps at discrete time points. The paths can only be continuous if $W(\cdot,\cdot,\boldsymbol{z})$ disappears for $\boldsymbol{z} \neq \mathbf{0}$.
- *Deterministic processes*, where $W(\cdot,\cdot,\boldsymbol{z}) = 0$ for all $\boldsymbol{z} \neq \mathbf{0}$, $\boldsymbol{\Sigma} = \mathbf{0}$ and $\boldsymbol{\mu} \neq \mathbf{0}$. Equations (4.1) and (4.3) become the first order partial differential equations

$$\frac{\partial p(t,\boldsymbol{x})}{\partial t} = -\sum_{i=1}^{n} \frac{\partial\big[\mu_i(\boldsymbol{x},t)p(t,\boldsymbol{x})\big]}{\partial x_i} \tag{4.6}$$

  and

$$\frac{\partial p(\tau,\boldsymbol{u},t,\boldsymbol{x})}{\partial \tau} = -\sum_{i=1}^{n} \mu_i(\boldsymbol{u},\tau)\,\frac{\partial p(\tau,\boldsymbol{u},t,\boldsymbol{x})}{\partial u_i},$$

  respectively. Formula (4.6) is called *Liouville's equation*. These deterministic processes are the only Markov processes with continuous and differentiable sample paths.
- *Diffusion processes*, where $W(\cdot,\cdot,\boldsymbol{z}) = 0$ for all $\boldsymbol{z} \neq \mathbf{0}$, $\boldsymbol{\Sigma}$ is nonzero and $\boldsymbol{\mu}$ may be zero or nonzero. Equations (4.1) and (4.3) then equal the *Kolmogorov (forward) equation* (or *Fokker-Planck* or *forward diffusion equation*)

$$\frac{\partial p(t,\boldsymbol{x})}{\partial t} = -\sum_{i=1}^{n} \frac{\partial\big[\mu_i(\boldsymbol{x},t)p(t,\boldsymbol{x})\big]}{\partial x_i} + \frac{1}{2}\sum_{i,j=1}^{n} \frac{\partial^2\big[\Sigma_{ij}(\boldsymbol{x},t)p(t,\boldsymbol{x})\big]}{\partial x_i \partial x_j} \tag{4.7}$$

  and *Kolmogorov backward equation* (or *backward diffusion equation*)

$$\frac{\partial p(\tau,\boldsymbol{u},t,\boldsymbol{x})}{\partial \tau} = -\sum_{i=1}^{n} \mu_i(\boldsymbol{u},\tau)\,\frac{\partial p(\tau,\boldsymbol{u},t,\boldsymbol{x})}{\partial u_i} - \frac{1}{2}\sum_{i,j=1}^{n} \Sigma_{ij}(\boldsymbol{u},\tau)\,\frac{\partial^2 p(\tau,\boldsymbol{u},t,\boldsymbol{x})}{\partial u_i \partial u_j}$$

  from Sect. 3.2.8.

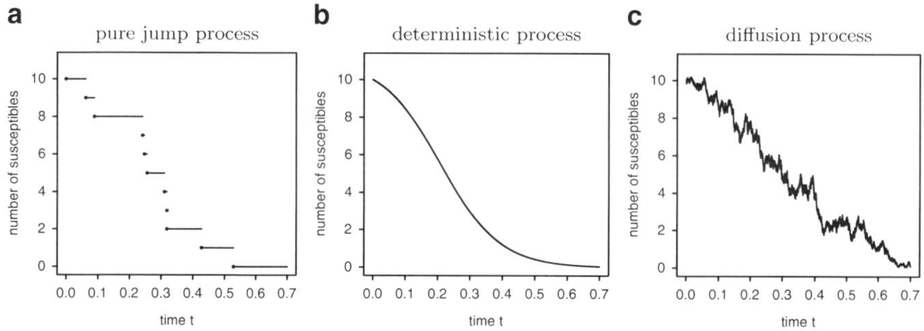

Fig. 4.1 Simulation of the course of an epidemic according to the SI model from Example 4.1 with parameters $N = 11$ and $\lambda = 1$. The model is described by three different classes of Markov processes: (**a**) Representation as a pure Markov jump process according to the master equations (4.9) or (4.10). The realisation is obtained using Gillespie's algorithm (cf. Sect. 2.4.1). (**b**) Representation as a deterministic Markov process with continuous sample paths according to Liouville's equation (4.11) or its backward version (4.12). Equation (4.30) on p. 73 provides an explicit formula for the course of the process whose transition density fulfils these equations. (**c**) Representation as a diffusion process according to the Kolmogorov equations (4.13) or (4.14). The simulated path approximation is obtained by transferring these equations to an SDE and then using the Euler scheme from Sect. 3.3.2 with constant time step 10^{-3}

Figure 4.1 illustrates these three classes of processes on the example of the epidemic SI model, which is investigated in the following Example 4.1. If not further specified, the terms master equation, Kolmogorov equation and diffusion equation usually refer to the forward version. Not included in the above selection are for example general Lévy processes like the jump-diffusion process from Sect. 3.1.4.

If p is a probability instead of a density, it is more convenient to write the forward and backward master equations (4.4) and (4.5) as the sums over all possible jumps $\boldsymbol{\Delta}$, i.e.

$$\frac{\partial p(t,\boldsymbol{x})}{\partial t} = \sum_{\boldsymbol{\Delta}} \Big[W(t, \boldsymbol{x}-\boldsymbol{\Delta}, \boldsymbol{\Delta}) p(t, \boldsymbol{x}-\boldsymbol{\Delta}) - W(t, \boldsymbol{x}, \boldsymbol{\Delta}) p(t, \boldsymbol{x}) \Big] \tag{4.8}$$

and

$$\frac{\partial p(\tau, \boldsymbol{u}, t, \boldsymbol{x})}{\partial \tau} = \sum_{\boldsymbol{\Delta}} W(\tau, \boldsymbol{u}, \boldsymbol{\Delta}) \Big[p(\tau, \boldsymbol{u}, t, \boldsymbol{x}) - p(\tau, \boldsymbol{u}+\boldsymbol{\Delta}, t, \boldsymbol{x}) \Big],$$

respectively. See Sect. 4.3.1 for the derivation and interpretation of these formulas.

Example 4.1. Consider a population of fixed size N within which an epidemic spreads according to the *susceptible–infected (SI) model* as follows: Assume that all individuals can be classified as either susceptible or infected. Denote by $S(t)$ the number of susceptible individuals at time $t \in \mathbb{R}_0$; the number of infecteds then results as $N - S(t)$. The only possible transition in the SI model is an infection, which reduces the number of susceptibles by one and accordingly increments the number of infecteds. Assume that at time zero the population consists of $N-1$ susceptibles and one infected. Furthermore, suppose that all individuals mix homogeneously and that the number of new infections within a short time Δt is approximately proportional to the product of numbers of susceptibles and infecteds. The discrete state space of the according time-homogeneous Markov jump process is $\{0, 1, \ldots, N-1\}$. The only possible jump is $\Delta_1 = -1$ with rate $W(t, S, -1) = \lambda S(N-S)$, where $\lambda \in \mathbb{R}_+$ stands for the *infection rate*. Let $p(\tau, S^*, t, S)$ denote the probability that the process is in state S at time t conditioned on the state S^* at time $\tau \leq t$. The shorter form $p(t, S) = p(0, S_0, t, S)$ refers to the initial state $S_0 = N-1$. The probability p is assumed zero outside the considered state space. Then the Markov jump process is fully described by its forward master equation

$$\frac{\partial p(t,S)}{\partial t} = \lambda(S+1)(N-S-1)p(t,S+1) - \lambda S(N-S)p(t,S) \tag{4.9}$$

or backward master equation

$$\frac{\partial p(\tau,S^*,t,S)}{\partial \tau} = \lambda S^*(N-S^*)\big[p(\tau,S^*,t,S) - p(\tau,S^*-1,t,S)\big] \tag{4.10}$$

for $S^*, S \in \{0, 1, \ldots, N-1\}$. For an approximate description in terms of differential equations one assumes a continuous state space $[0, N)$. The deterministic behaviour of the according process can then be described by Liouville's equation

$$\frac{\partial p(t,S)}{\partial t} = \frac{\partial \lambda S(N-S)p(t,S)}{\partial S} \tag{4.11}$$

or its backward analogue

$$\frac{\partial p(\tau,S^*,t,S)}{\partial \tau} = \lambda S^*(N-S^*)\frac{\partial p(\tau,S^*,t,S)}{\partial S^*}. \tag{4.12}$$

The stochastic dynamics is given by the Kolmogorov forward equation

$$\frac{\partial p(t,S)}{\partial t} = \frac{\partial \lambda S(N-S)p(t,S)}{\partial S} + \frac{1}{2}\frac{\partial^2 \lambda S(N-S)p(t,S)}{\partial S^2} \tag{4.13}$$

or Kolmogorov backward equation

$$\begin{aligned}\frac{\partial p(\tau,S^*,t,S)}{\partial \tau} &= \lambda S^*(N-S^*)\frac{\partial p(\tau,S^*,t,S)}{\partial S^*} \\ &\quad - \frac{1}{2}\lambda S^*(N-S^*)\frac{\partial^2 p(\tau,S^*,t,S)}{\partial (S^*)^2}.\end{aligned} \tag{4.14}$$

In these equations, $p(\tau, S^*, t, S)$ denotes the transition density of the process and fulfils the initial condition $p(0, S) = \delta(S-N+1)$. The remaining chapter explains how to obtain the differential equation descriptions (4.11)–(4.14). Figure 4.1 shows realisations of the SI model according to the three different representations.

4.2 Motivation and Purpose

The just introduced types of Markov processes—pure jump processes, deterministic processes and diffusions—represent three essential types of models that are used to describe the dynamics of natural phenomena in life sciences. Reflecting a system by a master equation, Liouville's equation or a diffusion equation is also referred to as modelling on a *microscopic*, *macroscopic* and *mesoscopic* level, respectively (e.g. Gillespie 1980). Table 4.1 gives a schematic overview over these three types of models.

Jump processes are the most reliable models when numbers of discrete objects are counted as described at the beginning of this chapter. The according master equations are exact; they contain full information on both the macroscopic and microscopic behaviour of the system. For small systems, one would hence stick to the master equation description (see e.g. the references in Ferm et al. 2008). For large systems, however, both simulation (using *Gillespie's algorithm*, Sect. 2.4.1) and parameter estimation (using Monte Carlo methods) from the master equation turn out to be computationally costly (Rao et al. 2002; Sjöberg et al. 2009). The master equation is usually not analytically solvable, but even if a solution was known, it would generally still not provide a decomposition into a systematic and a fluctuating part (cf. the discussion at the beginning of Sect. 4.3.3).

Many authors hence go over to the second class of processes, the deterministic ones, which are included in e.g. Pielou (1969), Eigen (1971), Bailey (1975), Anderson and May (1991), Busenberg and Martelli (1990) and Keeling and Rohani (2008). For this passage, the discontinuous sample paths of the original process are approximated by continuous smooth curves, and the macroscopic behaviour of the process is described by ordinary differential equations. This representation facilitates both simulation and statistical inference substantially. It also contributes to the comprehension of complex systems. However, as many phenomena in life sciences are intrinsically stochastic, such deterministic processes do not provide entirely realistic models. As Gillespie (1976, 1977) emphasises, their formulation may be invalid in the neighbourhood of instabilities of the system. See Rao et al. (2002) for a review article on the urgent need for stochastic models in molecular biology or the ample references in the introduction of McQuarrie (1967) on the same subject in chemical kinetics.

Table 4.1 Scheme of considered modelling levels with according equations fulfilled by the transition density and resulting types of processes

Level	Description	Process
Microscopic	Master equations	Pure Markov jump process
Mesoscopic	Diffusion equations, Kolmogorov equations	Diffusion process
Macroscopic	Liouville's equation and backward analogue	Deterministic Markov process with continuous sample paths

The reconciliation between the desire to seize a convenient model and the demand to maintain the stochastic properties results in the third class of processes, the diffusions. Although the characterising diffusion equations are again intractable, they provide broader possibilities for simulation and interpretation. They immediately reveal the composition of the stochastic process of a deterministic and a stochastic component and enable modest calculation of other interesting quantities. Sensitivity analysis and bifurcation theory become applicable (Rao et al. 2002). In the infectious diseases literature, for example, diffusion approximations are further utilised for the analysis of the duration (Barbour 1975b) or maximum size (Daniels 1974; Barbour 1975c) of an epidemic. Consequently, diffusion models are more and more applied in life sciences as also shown by the references at the beginning of Chap. 3.

However, to set up an approximating diffusion process to an underlying Markov jump process is a demanding task. There actually seems to be no standard procedure; authors usually work through the specific examples which they cover in their works. Unfortunately, such derivations are not always performed very carefully; models are not seldomly motivated by convenience rather than by probabilistic considerations.

The purpose of this chapter is to provide a detailed but compact overview of multi-dimensional diffusion approximation techniques on a level that is both mathematically well-founded and amenable for practitioners. To this end, methods are kept general, and assumptions and full derivations are provided. On the other hand, the design is informal where too much mathematical detail would make the matter incomprehensible. For example, the existence of certain partial derivatives is assumed rather than proved, even for probability functions which are only defined on a discrete state space. Convergence properties of some series and the interchange of certain limits are treated similarly. Indications are given where procedures are heuristic.

A precise mathematical treatment of the approximation of pure Markov jump processes by diffusions involves operator semigroup convergence theorems, martingale characterisations of Markov processes or the convergence of solutions of stochastic equations. Such techniques are explored in Barbour (1972), Kurtz (1981) and Ethier and Kurtz (1986). Further references on weak convergence theory are e.g. Billingsley (1968) and Pollard (1984). More general limit theorems, including higher order approximations, convergence to deterministic models, discrete time models and convergence of non-Markovian processes, are treated in Barbour (1974), Kurtz (1970, 1971) and Norman (1974, 1975). Furthermore, there are some papers in which a rigorous derivation of diffusion approximations concentrates on specific models, for example Feller (1951) on Markov branching processes with an application in genetics, Daley and Kendall (1965); Daley and Gani (1999) on rumours, McNeil (1973) on traffic control, Barbour (1975a) on birth and death processes, Guess and Gillespie (1977) on population growth, Pollett (1990) on a biological model and Andersson and Britton (2000, Chap. 5), Clancy and French (2001), Clancy et al. (2001) and Nåsell (2002) on epidemic models.

The monographs of Gardiner (1983) and van Kampen (1997) contain several diffusion approximation methods but dispense with general multi-dimensional formulas, which are derived in this chapter. Kepler and Elston (2001) review the diffusion approximation of gene regulation models but with the emphasis on different model specifications rather than on distinct approximation approaches. Adressing scientists from biology, Gibson and Mjolsness (2001) briefly sketch the different ideas of the transition from ordinary to stochastic differential equations in an informal way without detailed formulas or derivations. This chapter aims to be complete and compact. New techniques and multi-dimensional formulas are developed, heuristic transitions are evaluated critically, and the framework is extended to settings with multiple system size parameters. References for applications of the various approximation techniques are provided in the respective sections.

4.3 Approximation Methods

This section introduces methods for the systematic derivation of an approximating diffusion process to a Markov jump process $\boldsymbol{X}$. The central assumption in all approaches is that occurring jumps of the approximated process are somehow small. However, the state space of $\boldsymbol{X}$ is usually a subset of the multi-dimensional integer lattice, i.e. the lengths of the jumps are bounded below by one. Hence, a constant parameter N is introduced that appropriately measures the size of the system in the sense that the jump sizes of the *extensive variable* $\boldsymbol{X}$ do not depend on N, but the jumps of the *intensive variable* $\boldsymbol{x} = \boldsymbol{X}/N$ become smaller as N grows larger. This might for example be the number of molecules in a fluid or the carrying capacity of a population. In the limit $N \to \infty$, the sample paths of $\boldsymbol{x}$ become smooth continuous curves. In real applications, however, the jumps of $\boldsymbol{x}$ are of some finite size. The scaled process $\boldsymbol{x}$ is approximated by a diffusion process, where the system size N still enters the diffusion coefficient. The resulting process is hence called a *diffusion approximation* rather than a *diffusion limit*, which would correspond to a deterministic idealisation of the original jump process.

When considering a stochastic system, one distinguishes between two different kinds of noise (Horsthemke and Lefever 1984; Sancho and San Miguel 1984): *External fluctuations* have their sources in the environment of the system and can in some cases be controlled by the experimenter. Such disturbances are not investigated here. *Internal fluctuations* are caused by the discrete nature of particles in the system. They come up when the system is approximated by a continuous process and are thus treated in this chapter. Internal forces are expected to be small when the system size N is large. They usually have an effect of order $\mathcal{O}(N^{-1/2})$ on the macroscopic behaviour of the system and hence vanish as the system size tends to infinity. External forces, on the other hand, do not scale with the system size.

Every diffusion equation can be approximated by a master equation, but the reverse is not true (e.g. Gardiner 1983, Chap. 7.2.1). Roughly speaking, the approximation is only possible if there is some scale parameter δ such that both the average step size and the variance of the step size are proportional to δ, and

Fig. 4.2 Realisations of a scaled Poisson process (*black*) with jump size $\varepsilon = 10^{-5}$ and intensity parameter $\lambda = 0.1$ compared to realisations of the diffusion approximation (*grey*) $\mathrm{d}x_t = \varepsilon\lambda \mathrm{d}t + \varepsilon\sqrt{\lambda}\mathrm{d}B_t$, obtained by the methods in Sects. 4.3.1–4.3.5. The SDE has the exact solution $x_t = \varepsilon\lambda t + \varepsilon\sqrt{\lambda}B_t$ with $x_0 = 0$. The Poisson process does not fulfil the criteria that ensure a satisfactory diffusion approximation

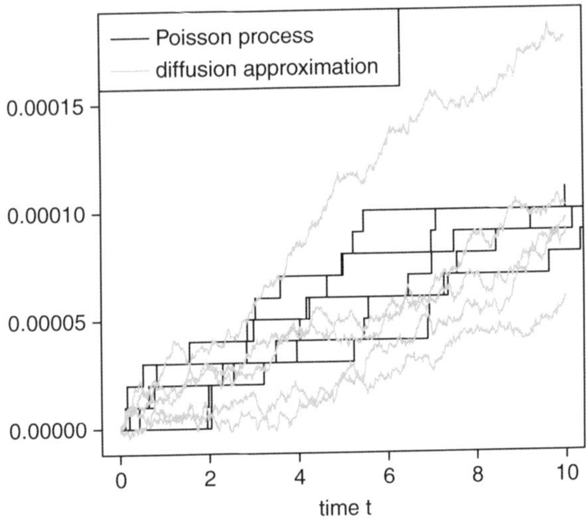

such that the jump probabilities increase as δ decreases. These conditions reflect the defining properties (3.15)–(3.17) of a diffusion process with $t - s$ set to δ and hence ensure consistency. The prototype for a jump process that cannot be approximated by a diffusion is the univariate Poisson process with step size ε, where the jump probability is constant, the average step size is proportional to ε and the variance of the step size is proportional to ε^2. Figure 4.2 compares sample path realisations of such a Poisson process with those of a diffusion approximation as obtained by the techniques in Sects. 4.3.1–4.3.5.

In the following, let $\boldsymbol{X} = (\boldsymbol{X}_t)_{t \geq t_0}$ be a stochastic jump process with state space $\mathcal{D}_N \subseteq \mathbb{Z}^n$ whose memory is so small that a Markov model is appropriate. In most applications in life sciences one has $\mathcal{D}_N \subseteq \mathbb{N}_0^n$, but the above more general state space also allows for e.g. modelling the decrease of a concentration by defining the initial state as state zero. The sample paths of $\boldsymbol{X}$ are assumed to be right-continuous and to have left hand limits. Division by the constant system size N yields the scaled process $\boldsymbol{x} = (\boldsymbol{x}_t)_{t \geq t_0} = \boldsymbol{X}/N$. At time t_0, the two processes are in states $\boldsymbol{X}_0$ and $\boldsymbol{x}_0 = \boldsymbol{X}_0/N$, respectively. Depending on the context, $\boldsymbol{X}$ and $\boldsymbol{x}$ may also just denote a state of the extensive or intensive process. For fixed N, denote by P_N and p_N the transition probabilities of $\boldsymbol{X}$ and $\boldsymbol{x}$, respectively. Similarly, W_N and w_N are the respective transition rates (compare with (4.2)), and $\mathcal{D}_N$ and $\mathcal{C}_N = N^{-1}\mathcal{D}_N$ are the state spaces of $\boldsymbol{X}$ and $\boldsymbol{x}$. It is essential to require that $w_N(t, \boldsymbol{x}, \boldsymbol{\Delta})$ is peaked around $\boldsymbol{x}$, i.e. there exists a bound δ such that $w_N(t, \boldsymbol{x}, \boldsymbol{\Delta}) \approx 0$ for all $\|\boldsymbol{\Delta}/N\| > \delta$, i.e. large jumps $\boldsymbol{\Delta}/N$ of the intensive process are improbable within a small time interval. Furthermore, we assume that $w_N(t, \boldsymbol{x}, \boldsymbol{\Delta})$ varies slowly with $\boldsymbol{x}$ such that Taylor expansions with respect to $\boldsymbol{x}$ are justified.

Different diffusion approximation techniques are now introduced as follows: The first method (Sect. 4.3.1) starts with the setup of the forward master equation of the jump process, expresses it as a sum of difference quotients and considers its approximation by the forward diffusion equation of a corresponding diffusion approximation. In the second approach (Sect. 4.3.2), convergence of the infinitesimal generator of the jump process is investigated. The *Langevin approach* (Sect. 4.3.3) establishes a diffusion approximation as the sum of the deterministic process and a fluctuating term. In the *Kramers-Moyal expansion* (Sect. 4.3.4), the master equation is expanded in a Taylor series with successive terms corresponding to their order of nonlinearity. The diffusion approximation is then chosen such that the corresponding forward diffusion equation equals the terms up to order 2 of the Taylor series. *Van Kampen's expansion* (Sect. 4.3.5) criticises this procedure and suggests a different Taylor series in powers of $N^{-1/2}$ to ensure that the neglected coefficients are small. Under certain regularity conditions, all methods yield the same approximating diffusion process. A detailed comparison of the different outcomes follows in Sect. 4.6. A deterministic model is again an approximation of the stochastic one for $N \to \infty$.

4.3.1 Convergence of the Master Equation

The line of this procedure is as follows: We start by setting up the transition probabilities P_N of the stochastic process in which we count the numbers of individual objects for fixed system size N. The state space of this process is discrete, and the evolution of the transition probabilities is described by the master equation. We then consider a sequence of discrete state space processes in which the state variables denote the intensive variables, i.e. the fractions of different classes of objects. For the system size tending to infinity, this sequence converges to a process with state variables changing continuously in space. The limit of the according sequence of master equations is approximated by a forward diffusion equation which is taken as a description for the limiting process. The limit is obtained by replacing difference quotients by the respective derivatives.

This technique has been used e.g. by Goel and Richter-Dyn (1974) for the approximation of the univariate birth and death process. Gillespie (1980) in a way reverses the method by considering an approximation legitimate only if its discretised version reduces to the master equation; to that end, derivatives are replaced by difference quotients.

The approximation method introduced in this section may seem obvious; however, it apparently has not been formulated in generality in the literature before. Due to space constraints, only main results are shown here. The proofs of some statements which enable this proceeding have been moved to Sect. B.1 in the Appendix.

Assuming that at most one event can occur during a small time interval of length Δt, we can establish an equation for $P_N(t+\Delta t, \boldsymbol{X})$ by summing over all possible nonzero jumps $\Delta \boldsymbol{X} \neq \boldsymbol{0}$ to arrive at state $\boldsymbol{X} \in \mathcal{D}_N$ at time $t+\Delta t$:

$$
\begin{aligned}
P_N(t+\Delta t, \boldsymbol{X}) = & \sum_{\Delta \boldsymbol{X}} P_N(t, \boldsymbol{X}-\Delta \boldsymbol{X}, t+\Delta t, \boldsymbol{X}) P_N(t, \boldsymbol{X}-\Delta \boldsymbol{X}) \\
& + \Big(1 - \sum_{\Delta \boldsymbol{X}} P_N(t, \boldsymbol{X}, t+\Delta t, \boldsymbol{X}+\Delta \boldsymbol{X})\Big) P_N(t, \boldsymbol{X}).
\end{aligned}
$$

The probability $P_N(\cdot, \boldsymbol{X}_1, \cdot, \boldsymbol{X}_2)$ is assumed zero here for all $\boldsymbol{X}_1, \boldsymbol{X}_2 \notin \mathcal{D}_N$. The first line collects all possibilities for transitions to the desired state at the desired time. The second line is the probability that the process has already been in state $\boldsymbol{X}$ at time t and remained there during the considered time interval. That is why the master equation, which results out of this equation, is also called a *gain-loss equation*. Subtract $P_N(t, \boldsymbol{X})$ on both sides, divide by Δt and let $\Delta t \to 0$. We then obtain

$$
\begin{aligned}
& \frac{\partial P_N(t, \boldsymbol{X})}{\partial t} \\
& = \sum_{\Delta \boldsymbol{X}} \Big(W_N(t, \boldsymbol{X}-\Delta \boldsymbol{X}, \Delta \boldsymbol{X}) P_N(t, \boldsymbol{X}-\Delta \boldsymbol{X}) - W_N(t, \boldsymbol{X}, \Delta \boldsymbol{X}) P_N(t, \boldsymbol{X}) \Big)
\end{aligned}
$$

with transition rates

$$
W_N(t, \boldsymbol{X}, \Delta \boldsymbol{X}) = \lim_{\Delta t \downarrow 0} \frac{1}{\Delta t} P_N(t, \boldsymbol{X}, t+\Delta t, \boldsymbol{X}+\Delta \boldsymbol{X})
$$

as a description for the continuous time process with discrete state space. This is the forward master equation (4.8). For an uncountable set of possible jumps, the sum could easily be replaced by an integral. The functional form of W_N is determined by the jump $\Delta \boldsymbol{X}$. For an alternative notation, one can assign to each possible jump an index i from a set I and write $W_{N,i}(t, \boldsymbol{X}) = W_N(t, \boldsymbol{X}, \boldsymbol{\Delta}_i)$ for the corresponding jump $\boldsymbol{\Delta}_i$, resulting in

$$
\begin{aligned}
& \frac{\partial P_N(t, \boldsymbol{X})}{\partial t} \\
& = \sum_{i \in I} \Big(W_{N,i}(t, \boldsymbol{X}-\boldsymbol{\Delta}_i) P_N(t, \boldsymbol{X}-\boldsymbol{\Delta}_i) - W_{N,i}(t, \boldsymbol{X}) P_N(t, \boldsymbol{X}) \Big).
\end{aligned}
\tag{4.15}
$$

Instead of the extensive variable $\boldsymbol{X}$ we now regard the intensive variable $\boldsymbol{x} = \boldsymbol{X}/N$. Consider a sequence of processes with (still discrete) state spaces $\mathcal{C}_N = N^{-1}\mathcal{D}_N$ corresponding to a sequence of numbers N which tends to infinity. The master equation for each process is

$$
\frac{\partial p_N(t, \boldsymbol{x})}{\partial t} = \sum_{i \in I} \Big(w_{N,i}(t, \boldsymbol{x}-\varepsilon \boldsymbol{\Delta}_i) p_N(t, \boldsymbol{x}-\varepsilon \boldsymbol{\Delta}_i) - w_{N,i}(t, \boldsymbol{x}) p_N(t, \boldsymbol{x}) \Big) \tag{4.16}
$$

with $p_N(\tau, \boldsymbol{x}, t, \boldsymbol{y}) = P_N(\tau, N\boldsymbol{x}, t, N\boldsymbol{y})$, $w_{N,i}(t, \boldsymbol{x}) = W_{N,i}(t, N\boldsymbol{x})$ and $\varepsilon = N^{-1}$. In order to approximate the jump process by a diffusion process, this master equation should be approximated by a Kolmogorov equation. That again means that the difference terms in (4.16) should be replaced by derivatives with respect to the components of $\boldsymbol{x}$. The single summands in (4.16) are not of the form of difference quotients though, so this step is not immediately admissible. However, it is always possible to express each of these summands by a collection of difference quotients of some order. This is proven in Lemma B.3 in Sect. B.1 in the Appendix. Then, the master equation becomes

$$\frac{\partial p_N(t, \boldsymbol{x})}{\partial t} = \sum_{i \in I} \sum_{\boldsymbol{k} \in I_i} D_{\boldsymbol{k}}^{|\boldsymbol{k}|}(w_{N,i} \cdot p_N)(t, \boldsymbol{x}) = \sum_{i \in I} \sum_{\boldsymbol{k} \in I_i} \varepsilon^{|\boldsymbol{k}|} \frac{D_{\boldsymbol{k}}^{|\boldsymbol{k}|}(w_{N,i} \cdot p_N)(t, \boldsymbol{x})}{\varepsilon^{|\boldsymbol{k}|}}, \quad (4.17)$$

where the notation $D_{\boldsymbol{k}}^{|\boldsymbol{k}|}$ stands for difference operators as introduced in Definition B.1, the I_i are appropriate sets of vectors $\boldsymbol{k} = (0, k_1, \ldots, k_n)'$ as used in Lemma B.3, and $|\boldsymbol{k}| = \sum_{j=1}^{n} k_j$. The first component of $\boldsymbol{k}$ is zero because t is fixed on the right hand side of Eq. (4.17), i.e. there is no differentiation with respect to the time variable.

It seems feasible now to approximate the difference quotients $D_{\boldsymbol{k}}^{|\boldsymbol{k}|}/\varepsilon^{|\boldsymbol{k}|}$ by proper derivatives as ε goes to zero. However, the consideration of ε tending to zero, i.e. N tending to infinity, involves *two* limiting procedures: First, convergence of the difference quotients, and second, convergence of the functions p_N and $w_{N,i}$. Accurate mathematical treatment of this limit is elaborate and beyond the purpose of this chapter. However, in many examples the scaled function $w_i = N^{-1}w_{N,i}$ does not depend on N anymore. We assume that this is the case here (at least asymptotically), so that Eq. (4.16) equals (if necessary, asymptotically)

$$\frac{\partial p_N(t, \boldsymbol{x})}{\partial t} = \sum_{i \in I} \frac{w_i(t, \boldsymbol{x} - \varepsilon \boldsymbol{\Delta}_i) p_N(t, \boldsymbol{x} - \varepsilon \boldsymbol{\Delta}_i) - w_i(t, \boldsymbol{x}) p_N(t, \boldsymbol{x})}{\varepsilon}. \quad (4.18)$$

Furthermore, it seems plausible that Eq. (4.17) remains true if p_N is replaced by its limit function p (which is assumed to exist), so Eq. (4.17) turns into

$$\frac{\partial p(t, \boldsymbol{x})}{\partial t} = \sum_{i \in I} \sum_{\boldsymbol{k} \in I_i} \varepsilon^{|\boldsymbol{k}|-1} \frac{D_{\boldsymbol{k}}^{|\boldsymbol{k}|}(w_i \cdot p)(t, \boldsymbol{x})}{\varepsilon^{|\boldsymbol{k}|}}.$$

Provided that p and the w_i are sufficiently often differentiable, it follows that—regarding the limits of the difference quotients as ε tends to zero—the master equation becomes

$$\frac{\partial p(t, \boldsymbol{x})}{\partial t} = \sum_{\boldsymbol{k}=(k_1, \ldots, k_n)'} \varepsilon^{|\boldsymbol{k}|-1} \left(\frac{\partial^{|\boldsymbol{k}|} f_{\boldsymbol{k}}(t, \boldsymbol{x}) p(t, \boldsymbol{x})}{\partial x_1^{k_1} \cdots \partial x_n^{k_n}} \right) \quad (4.19)$$

for some finite set of differentiable functions $f_{\boldsymbol{k}}$, $\boldsymbol{k} \in \mathbb{N}_0^n$. Assume that the derivatives are bounded. After restriction to terms up to order $\mathcal{O}(\varepsilon)$, i.e. ignoring smaller terms with $|\boldsymbol{k}| \geq 3$, Eq. (4.19) can then be rewritten as

$$\frac{\partial p(t,\boldsymbol{x})}{\partial t} = -\sum_{j=1}^{n} \frac{\partial\left[\mu_j(\boldsymbol{x},t)p(t,\boldsymbol{x})\right]}{\partial x_j} + \frac{1}{2N}\sum_{j,k=1}^{n} \frac{\partial^2\left[\Sigma_{jk}(\boldsymbol{x},t)p(t,\boldsymbol{x})\right]}{\partial x_j \partial x_k}, \tag{4.20}$$

where μ_j and Σ_{jk} with $j,k = 1,\ldots,n$ are the components of a vector $\boldsymbol{\mu}$ and a matrix $\boldsymbol{\Sigma}$. These can be determined according to Algorithm B.1 and Example B.1 in the Appendix. In some special cases, there are also explicit formulas for $\boldsymbol{\mu}$ and $\boldsymbol{\Sigma}$—see for instance Example B.2.

Heuristically, the space-continuous limit of the initial jump process is described by Eq. (4.20). That is the forward diffusion Eq. (4.7) if $\boldsymbol{\Sigma}$ is positive definite. This equation corresponds to a diffusion process with drift vector $\boldsymbol{\mu}$ and diffusion matrix $\boldsymbol{\Sigma}/N$, i. e. the intensive Markov jump process can be approximated by a diffusion satisfying the SDE

$$\mathrm{d}\boldsymbol{x}_t = \boldsymbol{\mu}(\boldsymbol{x}_t,t)\mathrm{d}t + \frac{1}{\sqrt{N}}\boldsymbol{\sigma}(\boldsymbol{x}_t,t)\mathrm{d}\boldsymbol{B}_t, \quad \boldsymbol{x}_{t_0} = \boldsymbol{x}_0,$$

where $\boldsymbol{\sigma}$ is a square root of $\boldsymbol{\Sigma}$, i.e. $\boldsymbol{\Sigma} = \boldsymbol{\sigma}\boldsymbol{\sigma}'$. The matrix $\boldsymbol{\sigma}$ is not necessarily unique as already discussed on p. 40.

Strictly speaking, since the Kolmogorov equation has been obtained using heuristic arguments, the Lipschitz continuity of $\boldsymbol{\mu}$ and $\boldsymbol{\sigma}$ needs to be checked at this point in order to ensure the existence of a solution to the above SDE (cf. Sect. 3.2.3). Note that such a solution is an approximation and not a limit as the system size parameter N is still part of the diffusion matrix.

The expansion of the backward master equation can be performed similarly and is a special case of the approximation of the infinitesimal generator considered in the next section.

Example 4.2. Recall the SI model from Example 4.1 on p. 59. The stochastic process counting the absolute number S of susceptibles in a population of size N is described by the master equation (4.9). Now consider the fraction $s = S/N$ of susceptible individuals and define $\alpha = \lambda N$. Let $p_N(t,s)$ denote the transition probability of the according intensive process, $w_{N,1}(t,s) = W_N(t,Ns,-1) = N\alpha s(1-s)$ the transition rate and $w_1(t,s) = \alpha s(1-s)$ the scaled transition rate, i.e. $w_N = Nw_1$. Then the master equation of the intensive jump process reads

$$\frac{\partial p_N(t,s)}{\partial t} = \frac{w_1(t,s+\varepsilon)p_N(t,s+\varepsilon) - w_1(t,s)p_N(t,s)}{\varepsilon},$$

where $\varepsilon = N^{-1}$. This corresponds to Eq. (4.18) above. The right hand side of the master equation is already of the form of a difference quotient with respect to a fixed vector $(\cdot,\varepsilon)'$,

$$\frac{\partial p_N(t,s)}{\partial t} = \frac{D^1_{(0,1)',(\cdot,\varepsilon)'}(w_1 \cdot p_N)(t,s)}{\varepsilon}.$$

The dot in the vector $(\cdot, \varepsilon)'$ of small parameters means that it is needless to fix its first component as no derivative with respect to the first argument of $w_1 \cdot p_N$ is considered. According to Example B.1, the above quotient should not be approximated by $(\partial/\partial s)(w_1 \cdot p)(t,s)$ but by Formula (B.7), i.e.

$$\frac{\partial p(t,s)}{\partial t} = \frac{\partial(w_1 \cdot p)(t,s)}{\partial s} + \frac{\varepsilon}{2}\frac{\partial^2(w_1 \cdot p)(t,s)}{\partial s^2}, \tag{4.21}$$

where p_N has been replaced by its limit function p. This is the Kolmogorov forward equation that has already been stated by Eq. (4.13) for the extensive process. The Kolmogorov equation (4.21) corresponds to a diffusion process with drift $\mu(s,t) = -w_1(t,s) = -\alpha s(1-s)$ and diffusion $N^{-1}\Sigma(s,t) = \varepsilon w_1(t,s) = \alpha s(1-s)/N$, i.e. to the solution of the SDE

$$\mathrm{d}s_t = -\alpha s_t(1-s_t)\mathrm{d}t + \frac{1}{\sqrt{N}}\sqrt{\alpha s_t(1-s_t)}\mathrm{d}B_t \tag{4.22}$$

with an appropriate initial value. For $N \to \infty$, (4.21) and (4.22) reduce to

$$\frac{\partial p(t,s)}{\partial t} = \frac{\partial(w_1 \cdot p)(t,s)}{\partial s}$$

and

$$\mathrm{d}s_t = -\alpha s_t(1-s_t)\mathrm{d}t.$$

4.3.2 Convergence of the Infinitesimal Generator

In this method, we aim to approximate the infinitesimal generator $\mathcal{G}$ of the diffusion approximation by the limit of the infinitesimal generator $\mathcal{G}_N$ of the intensive jump process $\boldsymbol{x} = (\boldsymbol{x}_t)_{t \geq t_0}$. The generator $\mathcal{G}$ allows us to directly read out the drift vector and diffusion matrix of a corresponding diffusion approximation as seen in Sect. 3.2.9. This idea follows the line of the very theoretical work of Kurtz (1981), where the weak convergence of sequences of processes is related to the convergence of corresponding generators as characterising semigroups. For better applicability, this subsection presents it in a constructive form.

Let $f : \mathcal{C} \times T \to \mathbb{R}$ be a measurable twice continuously differentiable function, where $\mathcal{C}$ is the state space of the approximating diffusion process and T is the time set. Note that the state space $\mathcal{C}_N$ of $\boldsymbol{x}$ is a subset of $\mathcal{C}$. The infinitesimal generator of $\boldsymbol{x}$ is defined as

$$\mathcal{G}_N f(\boldsymbol{u},t) = \lim_{\Delta t \downarrow 0} \frac{1}{\Delta t}\mathbb{E}_N\big(f(\boldsymbol{x}_{t+\Delta t}, t+\Delta t) - f(\boldsymbol{u},t)\big|\boldsymbol{x}_t = \boldsymbol{u}\big) \tag{4.23}$$

for $\boldsymbol{u} \in \mathcal{C}_N$ and $t \in T$, where $\mathbb{E}_N$ denotes the expectation with respect to the transition probability p_N of $\boldsymbol{x}$. Now proceed as for the approximation of the master equation in Sect. 4.3.1: Label each possible jump (now including the jump of length zero) with an index i from an eligible set I and define

$$w_i(t,\boldsymbol{u}) = w(t,\boldsymbol{u},\boldsymbol{\Delta}_i) = N^{-1} w_N(t,\boldsymbol{u},\boldsymbol{\Delta}_i) = \lim_{\Delta t \downarrow 0} N^{-1} \frac{p_N\big(t,\boldsymbol{u},t+\Delta t,\boldsymbol{u}+\varepsilon\boldsymbol{\Delta}_i\big)}{\Delta t},$$

where it is again assumed that the transition rate w does not depend on N. For $\varepsilon = N^{-1}$, the generator (4.23) agrees with

$$\begin{aligned}
&\lim_{\Delta t \downarrow 0} \frac{1}{\Delta t} \sum_{i \in I} \big(f\big(\boldsymbol{u}+\varepsilon\boldsymbol{\Delta}_i, t+\Delta t\big) - f\big(\boldsymbol{u},t\big)\big) p_N\big(t,\boldsymbol{u},t+\Delta t,\boldsymbol{u}+\varepsilon\boldsymbol{\Delta}_i\big) \\
&= \lim_{\Delta t \downarrow 0} \frac{1}{\Delta t} \sum_{i \in I} \big(f\big(\boldsymbol{u}+\varepsilon\boldsymbol{\Delta}_i, t+\Delta t\big) - f\big(\boldsymbol{u}+\varepsilon\boldsymbol{\Delta}_i, t\big) \\
&\quad + f\big(\boldsymbol{u}+\varepsilon\boldsymbol{\Delta}_i, t\big) - f\big(\boldsymbol{u},t\big)\big) p_N\big(t,\boldsymbol{u},t+\Delta t,\boldsymbol{u}+\varepsilon\boldsymbol{\Delta}_i\big) \\
&= \sum_{i \in I} \frac{\partial f(\boldsymbol{u}+\varepsilon\boldsymbol{\Delta}_i, t)}{\partial t} \lim_{\Delta t \downarrow 0} p_N\big(t,\boldsymbol{u},t+\Delta t,\boldsymbol{u}+\varepsilon\boldsymbol{\Delta}_i\big) \\
&\quad + \sum_{i \in I} w_i(t,\boldsymbol{u}) \frac{f\big(\boldsymbol{u}+\varepsilon\boldsymbol{\Delta}_i, t\big) - f\big(\boldsymbol{u},t\big)}{\varepsilon}.
\end{aligned}$$

Note that $\lim_{\Delta t \downarrow 0} p_N(t,\boldsymbol{u},t+\Delta t,\boldsymbol{u}+\varepsilon\boldsymbol{\Delta}_i)$ equals 1 if $\varepsilon\boldsymbol{\Delta}_i = \boldsymbol{0}$ and zero otherwise, i.e.

$$\mathcal{G}_N f(\boldsymbol{u},t) = \frac{\partial f(\boldsymbol{u},t)}{\partial t} + \sum_{i \in I} w_i(t,\boldsymbol{u}) \frac{f\big(\boldsymbol{u}+\varepsilon\boldsymbol{\Delta}_i, t\big) - f\big(\boldsymbol{u},t\big)}{\varepsilon}. \tag{4.24}$$

For each i, expand $f(\boldsymbol{u}+\varepsilon\boldsymbol{\Delta}_i, t) - f(\boldsymbol{u},t)$ as in Sect. 4.3.1 (or Sect. B.1, respectively) and consider the resulting difference quotients as ε tends to zero. Then

$$\mathcal{G}_N f(\boldsymbol{u},t) = \frac{\partial f(\boldsymbol{u},t)}{\partial t} + \sum_{\boldsymbol{k}=(k_1,\dots,k_n)'} \varepsilon^{|\boldsymbol{k}|-1} \left(\sum_{i \in I_{\boldsymbol{k}}} w_i(t,\boldsymbol{u}) \right) \frac{\partial^{|\boldsymbol{k}|} f(\boldsymbol{u},t)}{\partial u_1^{k_1} \cdots \partial u_n^{k_n}}$$

for a finite number of vectors $\boldsymbol{k} \in \mathbb{N}_0^n$ and appropriate sets $I_{\boldsymbol{k}} \subseteq I$. Assume that these derivatives are bounded and neglect all terms of order higher than $\mathcal{O}(\varepsilon)$. The result can be taken as an approximation of the infinitesimal generator $\mathcal{G}$ and in that case attains the form

$$\mathcal{G}f(\boldsymbol{u},t) = \frac{\partial f(\boldsymbol{u},t)}{\partial t} + \sum_{j=1}^{n} \mu_j(\boldsymbol{u},t) \frac{\partial f(\boldsymbol{u},t)}{\partial u_j} + \frac{1}{2N} \sum_{j,k=1}^{n} \Sigma_{jk}(\boldsymbol{u},t) \frac{\partial^2 f(\boldsymbol{u},t)}{\partial u_j \partial u_k}$$

i.e.

$$\mathcal{G} = \frac{\partial}{\partial t} + \mathcal{L}^B_{\boldsymbol{\mu},\boldsymbol{\Sigma}/N}.$$

As in the preceding Sect. 4.3.1, there are no explicit formulas for $\boldsymbol{\mu} = (\mu_j)_{j=1,\ldots,n}$ and $\boldsymbol{\Sigma} = (\Sigma_{jk})_{j,k=1,\ldots,n}$, but this book provides a constructive algorithm for their derivation by Algorithm B.1 in Sect. B.1 in the Appendix. If $\boldsymbol{\Sigma}$ is positive definite, $\mathcal{L}^B_{\boldsymbol{\mu},\boldsymbol{\Sigma}/N}$ is the Kolmogorov backward operator from Sect. 3.2.8 with drift $\boldsymbol{\mu}$ and diffusion matrix $\boldsymbol{\Sigma}/N$. This generator can be associated to the Itô diffusion approximation

$$\mathrm{d}\boldsymbol{x}_t = \boldsymbol{\mu}(\boldsymbol{x}_t, t)\mathrm{d}t + \frac{1}{\sqrt{N}}\boldsymbol{\sigma}(\boldsymbol{x}_t, t)\mathrm{d}\boldsymbol{B}_t, \quad \boldsymbol{x}_{t_0} = \boldsymbol{x}_0,$$

where $\boldsymbol{\sigma}$ is a square root of $\boldsymbol{\Sigma}$.

Under regularity conditions, the obtained $\boldsymbol{\mu}$ and $\boldsymbol{\Sigma}$ agree with those from Sect. 4.3.1. The same is true for the results from the following sections. Differences and similarities are discussed in the conclusion in Sect. 4.6.

Example 4.3. Consider again the process describing the fractions of susceptibles during an epidemic which evolves according to the SI model. With the notation from Example 4.2, the infinitesimal generator $\mathcal{G}_N$ of this process fulfils

$$\mathcal{G}_N f(s,t) = \frac{\partial f(s,t)}{\partial t} + w_1(t,s)\frac{f(s-\varepsilon,t) - f(s,t)}{\varepsilon}$$

for a measurable twice continuously differentiable function $f : [0,1] \times \mathbb{R}_0 \to \mathbb{R}$. The difference $f(s-\varepsilon,t) - f(s,t)$ can be written as $D^1_{(1,0)',(-\varepsilon,\cdot)'} f(s,t)$ with the notation from Definition B.1. As before, the dot in the subscript indicates that the respective argument does not have to be specified here. With the approximation rule (B.7), one obtains

$$\mathcal{G}f(s,t) = \frac{\partial f(s,t)}{\partial t} - w_1(t,s)\,\frac{\partial f(s,t)}{\partial s} + \frac{\varepsilon}{2}\,w_1(t,s)\,\frac{\partial^2 f(s,t)}{\partial s^2},$$

where $\mathcal{G}$ denotes the infinitesimal generator of the limiting space-continuous process. This generator agrees with the Kolmogorov backward operator with drift $\mu(s,t) = -w_1(t,s)$ and diffusion $N^{-1}\Sigma(s,t) = \varepsilon w_1(t,s)$. This results in the same diffusion approximation as in Example 4.2.

4.3.3 Langevin Approach

In the *Langevin approach* we postulate rather than derive that the process $\boldsymbol{X}$ can be represented by a diffusion approximation. It should however be ensured that this postulation is justifiable in the sense that occurring jumps of the sample paths are sufficiently small. Therefore consider again the scaled process $\boldsymbol{x} = \boldsymbol{X}/N$ and require that it fulfils the Itô SDE

$$\mathrm{d}\boldsymbol{x}_t = \boldsymbol{\mu}(\boldsymbol{x}_t, t)\mathrm{d}t + \frac{1}{\sqrt{N}}\boldsymbol{\sigma}(\boldsymbol{x}_t, t)\mathrm{d}\boldsymbol{B}_t\,, \quad \boldsymbol{x}_{t_0} = \boldsymbol{x}_0. \tag{4.25}$$

Equation (4.25) is also referred to as *Langevin equation*, and $\mathrm{d}\boldsymbol{B}_t$ is called *Gaussian Langevin force* in this context. The original way to obtain the coefficients $\boldsymbol{\mu}$ and $\boldsymbol{\sigma}$ is as follows (see e.g. van Kampen 1981b, 1997, Chap. 9): The deterministic behaviour of $\boldsymbol{x}$ is often known by the *macroscopic equation* (or *phenomenological law*)

$$\mathrm{d}\mathbb{E}\boldsymbol{x}_t = \widetilde{\boldsymbol{\mu}}\big(\mathbb{E}\boldsymbol{x}_t, t\big)\mathrm{d}t \tag{4.26}$$

for some function $\widetilde{\boldsymbol{\mu}}$. The drift function of (4.25) is then set to be identical with $\widetilde{\boldsymbol{\mu}}$, and the diffusion coefficient is chosen such that it appropriately represents the fluctuations of the trajectories around the deterministic course. On the other hand, the master equation yields the exact cohesion

$$\mathrm{d}\mathbb{E}\boldsymbol{x}_t = \mathbb{E}\boldsymbol{\mu}\big(\boldsymbol{x}_t, t\big)\mathrm{d}t \tag{4.27}$$

with

$$\boldsymbol{\mu}(\boldsymbol{u}, t) = \lim_{\Delta t \downarrow 0} \frac{1}{\Delta t}\,\mathbb{E}\left(\boldsymbol{x}_{t+\Delta t} - \boldsymbol{x}_t \,\middle|\, \boldsymbol{x}_t = \boldsymbol{u}\right)$$

for $\boldsymbol{u} \in \mathcal{C}_N$ (van Kampen, 1997, Chap. 5.8). For nonlinear $\boldsymbol{\mu}$, the terms $\boldsymbol{\mu}(\mathbb{E}\boldsymbol{x}_t, t)$ and $\mathbb{E}\boldsymbol{\mu}(\boldsymbol{x}_t, t)$ do not coincide. To be more precise, expanding $\boldsymbol{\mu}(\boldsymbol{x}_t, t)$ around $\mathbb{E}\boldsymbol{x}_t$ in a Taylor series and then taking expectations on both sides yields

$$\mathbb{E}\boldsymbol{\mu}\big(\boldsymbol{x}_t, t\big) = \boldsymbol{\mu}\big(\mathbb{E}\boldsymbol{x}_t, t\big) + \frac{1}{2}\,\boldsymbol{\mu}''\big(\mathbb{E}\boldsymbol{x}_t, t\big)\cdot\mathbb{E}\big((\boldsymbol{x}_t - \mathbb{E}\boldsymbol{x}_t)(\boldsymbol{x}_t - \mathbb{E}\boldsymbol{x}_t)'\big) + \ldots, \tag{4.28}$$

where the prime denotes differentiation with respect to the state variable. This means that $\mathbb{E}\boldsymbol{\mu}(\boldsymbol{x}_t, t)$ and $\boldsymbol{\mu}(\mathbb{E}\boldsymbol{x}_t, t)$ might differ by a term which is of the same order as the fluctuations. If one is only interested in the macroscopic behaviour of $\boldsymbol{x}$, these additional terms are neglected anyway. If one however takes fluctuations into account, identification of $\boldsymbol{\mu}$ with $\widetilde{\boldsymbol{\mu}}$ might result in a wrong diffusion coefficient. Figure 4.3 exemplarily displays the deviation between the determinstic course and the expectation of the stochastic course of a susceptible–infected epidemic. A detailed overview of difficulties arising from the above inconsistency and the attempts of different authors to correct for this has been given by van Kampen (1965); see also Hänggi (1982) or the example in Sect. 4.6.

Example 4.4. Once more, turn to the epidemic SI model and consider the process that counts the absolute numbers of susceptible individuals. The following formulas allow a direct comparison of the deterministic process $\bar{S}$ and the expectation of the stochastic process $\widetilde{S}$. These are taken from Renshaw (1991, Chap. 10), who again refers to Haskey (1954).

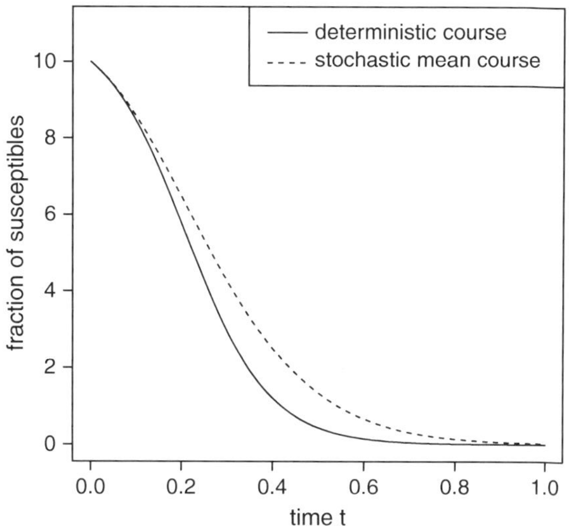

Fig. 4.3 Comparison of the deterministic course $\bar{S}(t)$ (*solid line*) and stochastic mean course $\mathbb{E}\widetilde{S}(t)$ (*dashed line*) of an epidemic following the SI model. Displayed are the numbers of susceptibles plotted against time. The population size equals $N = 11$ with one infected individual at time 0, and the infection rate is $\lambda = 1$. The explicit formulas for $\bar{S}(t)$ and $\mathbb{E}\widetilde{S}(t)$ are shown in Example 4.4. As suspected from Eq. (4.28), the two curves do not coincide

In the deterministic setting, the number $\bar{S}(t)$ of susceptibles at time t can be modelled through the ODE

$$\mathrm{d}\bar{S}(t) = -\lambda\,\bar{S}(t)\,(N - \bar{S}(t))\mathrm{d}t \quad , \bar{S}(0) = N - 1, \tag{4.29}$$

assuming a continuous state space. This representation corresponds to Eq. (4.26) with

$$\widetilde{\mu}(\bar{S}(t), t) = -\lambda\,\bar{S}(t)\,(N - \bar{S}(t)).$$

The explicit solution of the differential equation (4.29) is given by

$$\bar{S}(t) = \frac{N(N-1)}{N - 1 + \exp(N\lambda t)} \tag{4.30}$$

for all $t \in \mathbb{R}_0$. The stochastic course of the epidemic, on the other hand, is best expressed via the master equation

$$\frac{\partial P_N(t,S)}{\partial t} = \lambda(S+1)(N-S-1)P_N(t,S+1) - \lambda S(N-S)P_N(t,S)$$

for $S = 0, 1, \ldots, N-1$, where $P_N(t,S)$ denotes the probability that there are S susceptibles at time $t \in \mathbb{R}_0$ given that there were $N-1$ susceptibles at time 0. This has already been stated in Eq. (4.9) on p. 60. For even $N-1$, the expected number of susceptible individuals at time t is then explicitly given by

$$\mathbb{E}\widetilde{S}(t)$$
$$= \lambda \sum_{j=1}^{\frac{N-1}{2}} \frac{(N-1)!\exp\big(-\lambda j(N-j)t\big)}{(N-j-1)!(j-1)!}\left((N-2j)^2\lambda t + 2 - (N-2j)\sum_{k=j}^{N-j-1}\frac{1}{k}\right). \tag{4.31}$$

An explicit representation of the function μ in Eq. (4.27) is not directly available for general $t \in \mathbb{R}_0$. However, graphical analysis is possible: Fig. 4.3 compares the graph of the deterministic variable $\bar{S}(t)$ in Eq. (4.30) with the expectation $\mathbb{E}\widetilde{S}(t)$ in Formula (4.31) for $N = 11$ and $\lambda = 1$. The discrepancy is considerable. This is due to the deviation shown in Eq. (4.28).

An improved approach is the following one (see e.g. Walsh 1981; Allen 2003; Lande et al. 2003): From the definition of diffusion processes (see Sect. 3.2.5) one has

$$\boldsymbol{\mu}(\boldsymbol{u}, t) = \lim_{\Delta t \downarrow 0} \frac{1}{\Delta t} \mathbb{E}\left(\boldsymbol{x}_{t+\Delta t} - \boldsymbol{x}_t \,\middle|\, \boldsymbol{x}_t = \boldsymbol{u}\right), \tag{4.32}$$

$$N^{-1}\boldsymbol{\Sigma}(\boldsymbol{u}, t) = \lim_{\Delta t \downarrow 0} \frac{1}{\Delta t} \mathbb{E}\left((\boldsymbol{x}_{t+\Delta t} - \boldsymbol{x}_t)(\boldsymbol{x}_{t+\Delta t} - \boldsymbol{x}_t)' \,\middle|\, \boldsymbol{x}_t = \boldsymbol{u}\right) \tag{4.33}$$

$$= \lim_{\Delta t \downarrow 0} \frac{1}{\Delta t} \mathrm{Cov}\left(\boldsymbol{x}_{t+\Delta t} - \boldsymbol{x}_t \,\middle|\, \boldsymbol{x}_t = \boldsymbol{u}\right),$$

where $\boldsymbol{\Sigma} = \boldsymbol{\sigma}\boldsymbol{\sigma}'$. As in Sect. 4.3.1, assume that there is a countable set of possible transitions for the process $\boldsymbol{X}$ causing jumps $\boldsymbol{\Delta}_i \in \mathbb{Z}^n$—i.e. jumps of sizes $\boldsymbol{\Delta}_i/N$ for the process $\boldsymbol{x}$—, where $i \in I$ for some index set I. Then, consulting again the transition rates $w_{N,i}(t, \boldsymbol{x}) = W_N(t, N\boldsymbol{x}, \boldsymbol{\Delta}_i)$ from the previous sections, (4.32) and (4.33) arise as

$$\boldsymbol{\mu}(\boldsymbol{u}, t) = N^{-1} \sum_{i \in I} w_{N,i}(t, \boldsymbol{u}) \boldsymbol{\Delta}_i, \tag{4.34}$$

$$N^{-1}\boldsymbol{\Sigma}(\boldsymbol{u}, t) = N^{-2} \sum_{i \in I} w_{N,i}(t, \boldsymbol{u}) \boldsymbol{\Delta}_i \boldsymbol{\Delta}_i'. \tag{4.35}$$

This result can also be illustrated as follows (see e.g. Golightly and Wilkinson 2005, 2006, 2010): The transition rates $w_{N,i}$ represent the hazards of the respective events to occur. Hence, if the current state of $\boldsymbol{x}$ at time t is $\boldsymbol{u}$ and all transitions happen independently of each other, the waiting time until the occurrence of the next event of type i is exponentially distributed with rate $w_{N,i}(t, \boldsymbol{u})$. As a consequence, the number Z_i of type i events within the small time interval $(t, t + \Delta t]$ is Poisson distributed with rate $w_{N,i}(t, \boldsymbol{u})\Delta t$. Hence,

$$\boldsymbol{\mu}(\boldsymbol{u}, t) = \lim_{\Delta t \downarrow 0} \frac{1}{\Delta t} \sum_{i \in I} \mathbb{E}(Z_i N^{-1} \boldsymbol{\Delta}_i | \boldsymbol{x}_t = \boldsymbol{u}) = N^{-1} \sum_{i \in I} w_{N,i}(t, \boldsymbol{u}) \boldsymbol{\Delta}_i,$$

$$N^{-1}\boldsymbol{\Sigma}(\boldsymbol{u}, t) = \lim_{\Delta t \downarrow 0} \frac{1}{\Delta t} \sum_{i \in I} \mathrm{Cov}\,(Z_i N^{-1} \boldsymbol{\Delta}_i | \boldsymbol{x}_t {=} \boldsymbol{u}) = N^{-2} \sum_{i \in I} w_{N,i}(t, \boldsymbol{u}) \boldsymbol{\Delta}_i \boldsymbol{\Delta}_i'.$$

Note that in this approach the consideration of N tending to infinity entered only in the assumption that jumps are sufficiently small. In most cases, the functions

$N^{-1}w_{N,i}$ converge to bounded functions w_i which do not depend on N. In the limit, we have then established a drift vector and diffusion matrix of order $\mathcal{O}(1)$ and $\mathcal{O}(N^{-1})$, respectively.

The Langevin approach has been applied in numerous fields such as finance (Bouchaud and Cont 1998), genetics (Tian et al. 2007), systems biology (Golightly and Wilkinson 2005, 2006, 2008), physics (Ramshaw 1985; Kleinhans et al. 2005; Pierobon et al. 2005; Seifert 2008; Song et al. 2008) and medicine (Capasso and Morale 2009). Be aware that some of these authors also use *general* instead of *Gaussian* Langevin forces.

Example 4.5. In the SI model, the only possible jump is $\Delta_1 = -1$ with according transition rate $w_{N,1}(t,s) = N\alpha s(1-s)$; this has been determined in Example 4.1 on p. 59. The drift and diffusion of an according diffusion approximation follow immediately with the above formulas (4.34) and (4.35) as

$$\begin{aligned}\mu(s,t) &= N^{-1}w_{N,1}(t,s)\Delta_1 = -\alpha s(1-s),\\ N^{-1}\Sigma(s,t) &= N^{-2}w_{N,1}(t,s)\Delta_1^2 = \alpha s(1-s)/N.\end{aligned}$$

This approximation agrees again with those from the previous examples.

4.3.4 Kramers-Moyal Expansion

This section introduces another, widely used approach to approximate the forward master equation

$$\frac{\partial p_N(t,\boldsymbol{x})}{\partial t} = \sum_{\boldsymbol{\Delta}}\Big(w_N(t,\boldsymbol{x}-\varepsilon\boldsymbol{\Delta},\boldsymbol{\Delta})p_N(t,\boldsymbol{x}-\varepsilon\boldsymbol{\Delta}) - w_N(t,\boldsymbol{x},\boldsymbol{\Delta})p_N(t,\boldsymbol{x})\Big)$$

for the transition probability p_N of the process $\boldsymbol{x} = \boldsymbol{X}/N$, where the sum is again taken over all possible jumps $\boldsymbol{\Delta} = (\Delta_1,\dots,\Delta_n)'$ of the trajectories of $\boldsymbol{X}$, and $\varepsilon = N^{-1}$. Expansion of $w_N(\cdot,\boldsymbol{x}-\varepsilon\boldsymbol{\Delta},\cdot)p_N(\cdot,\boldsymbol{x}-\varepsilon\boldsymbol{\Delta})$ in a Taylor series around $\boldsymbol{x}$ yields

$$\begin{aligned}&\frac{\partial p_N(t,\boldsymbol{x})}{\partial t}\\ &= -\sum_{m=1}^{\infty}(-\varepsilon)^{m-1}\sum_{\boldsymbol{k}\in\mathcal{K}_m}\left(\prod_{j=1}^{n}\frac{1}{k_j!}\right)\frac{\partial^m}{\partial x_1^{k_1}\cdots\partial x_n^{k_n}}a_{m,\boldsymbol{k}}(t,\boldsymbol{x})p_N(t,\boldsymbol{x})\end{aligned} \tag{4.36}$$

with

$$\mathcal{K}_m = \big\{\boldsymbol{k} = (k_1,\dots,k_n)' \in \mathbb{N}_0^n \,\big|\, |\boldsymbol{k}| = m\big\} \tag{4.37}$$

and the *Kramers-Moyal moments*

$$a_{m,\boldsymbol{k}}(t,\boldsymbol{x}) = \sum_{\boldsymbol{\Delta}} \left(\prod_{j=1}^{n} \Delta_j^{k_j} \right) N^{-1} w_N(t,\boldsymbol{x},\boldsymbol{\Delta}) \tag{4.38}$$

for all $m \in \mathbb{N}_0$. Equation (4.36) is called *Kramers-Moyal expansion* as it has first been derived by Kramers (1940) and further been developed by Moyal (1949, Chap. 8). Note that the derivatives of p_N with respect to state variables are formally not defined as p_N is a discrete probability measure. However, as already elucidated in Sect. 4.2, the above expansion is to be seen as a heuristic approximation of the master equation as the system size tends to infinity, and hence the notation is left imprecise in this respect.

Terminating the right hand side of (4.36) after $m = 2$ and letting p_N tend to p (assuming that this limit exists) results in

$$\begin{aligned} &\frac{\partial p(t,\boldsymbol{x})}{\partial t} \\ &= -\sum_{i=1}^{n} \frac{\partial}{\partial x_i} a_{1,i}(t,\boldsymbol{x}) p(t,\boldsymbol{x}) + \frac{1}{2N} \sum_{i,j=1}^{n} \frac{\partial^2}{\partial x_i \partial x_j} a_{2,(i,j)}(t,\boldsymbol{x}) p(t,\boldsymbol{x}) \end{aligned} \tag{4.39}$$

with $a_{1,i} = a_{1,\boldsymbol{e}_i}$ and $a_{2,(i,j)} = a_{2,\boldsymbol{e}_i+\boldsymbol{e}_j}$. In the heuristic framework of this chapter, this is a description of the continuous state space process, i.e. for

$$\boldsymbol{\mu}(\boldsymbol{x},t) = (a_{1,i}(t,\boldsymbol{x}))_{i=1,\ldots,n} = N^{-1} \sum_{\boldsymbol{\Delta}} w_N(t,\boldsymbol{x},\boldsymbol{\Delta})\boldsymbol{\Delta} \tag{4.40}$$

and positive definite

$$\boldsymbol{\Sigma}(\boldsymbol{x},t) = (a_{2,(i,j)}(t,\boldsymbol{x}))_{i,j=1,\ldots,n} = N^{-1} \sum_{\boldsymbol{\Delta}} w_N(t,\boldsymbol{x},\boldsymbol{\Delta})\boldsymbol{\Delta}\boldsymbol{\Delta}', \tag{4.41}$$

Eq. (4.39) is equivalent to representing the process $\boldsymbol{x}$ by the diffusion approximation

$$\mathrm{d}\boldsymbol{x}_t = \boldsymbol{\mu}(\boldsymbol{x}_t,t)\mathrm{d}t + \frac{1}{\sqrt{N}} \boldsymbol{\sigma}(\boldsymbol{x}_t,t)\mathrm{d}\boldsymbol{B}_t\,, \quad \boldsymbol{x}_{t_0} = \boldsymbol{x}_0,$$

where $\boldsymbol{\sigma}$ is a square root of $\boldsymbol{\Sigma}$.

Pawula's theorem (Pawula 1967a,b) states that the Kramers-Moyal expansion either terminates after $m = 1$ or $m = 2$, or it contains infinitely many terms. That leads to the fact that the accuracy of the approximation of the master equation does not necessarily improve when the truncation is done after some $m \geq 3$, although this might be true in special cases: For example, Risken and Vollmer (1987) show that the careful inclusion of more than only the first two terms yields better approximation results for the Poisson process than the truncation after the second term (compare also with the discussion of the Poisson process on p. 64).

The Kramers-Moyal expansion has been applied in various areas such as physics (Kishida et al. 1976; Robertson et al. 1996; Naert et al. 1997), geophysics (Strumik and Macek 2008), finance (Karth and Peinke 2003) and infectious disease epidemiology (Hufnagel et al. 2004). Risken (1984, Chap. 4.2) also contains an analogous *Kramers-Moyal backward expansion.*

Example 4.6. The formulas (4.40) and (4.41) for the drift and diffusion of an approximation through the Kramers-Moyal expansion are exactly the same as Eqs. (4.34) and (4.35) in the Langevin approach. Hence, see the previous Example 4.5 for the approximation of the SI epidemic process using the Kramers-Moyal expansion.

4.3.5 Van Kampen Expansion

Equation (4.39) is obtained under the assumption that $a_{m,\boldsymbol{k}}$ is sufficiently small for $m \geq 3$ and large N, which has explicitly been demanded by Moyal (1949). However, the smallness of these coefficients is not generally guaranteed. This lack has been criticised by van Kampen and gave rise to the method in this section—now known as *van Kampen expansion*—, where the master equation is systematically expanded in powers of a small parameter. It has been introduced in van Kampen (1961), but is more comprehensively described in van Kampen (1965, Chap. 3.D, 1997).

Although the expansion has been applied to multi-dimensional settings (e.g. Gardiner 1983, Chap. 7.6; van Kampen 1997, Chap. 10.5; Chen and Bokka 2005; Alonso et al. 2007), it seems that it has not been derived in a general multi-dimensional framework in the literature. This is considerably more elaborate than a univariate analysis—as described in Gardiner (1983, p. 266): "This is so complicated that it will not be explicitly derived here." The following paragraphs develop the van Kampen expansion for multi-dimensional processes.

As before, let P_N be the transition probability function of the extensive variable $\boldsymbol{X}$. Define $\boldsymbol{\phi}(t) = (\phi_1(t), \ldots, \phi_n(t))'$ with $\boldsymbol{\phi}(t_0) = \boldsymbol{x}_0$ as the solution of an ordinary differential equation describing the dynamics of the intensive variable $\boldsymbol{x} = \boldsymbol{X}/N$ deterministically in a sense that is specified more precisely in Eq. (4.53) below. The probability function $P_N(t, \boldsymbol{X})$ is peaked around $N\boldsymbol{\phi}(t)$ with width proportional to $N^{\frac{1}{2}}$. In order to ease this dependence on N, introduce $\boldsymbol{z} = (z_1, \ldots, z_n)'$ as the time-dependent transformation

$$\boldsymbol{z} = \frac{\boldsymbol{X} - N\boldsymbol{\phi}(t)}{N^{\frac{1}{2}}} = N^{\frac{1}{2}}\big(\boldsymbol{x} - \boldsymbol{\phi}(t)\big)$$

with probability function π_N satisfying

$$P_N(t, \boldsymbol{X}) = P_N(t, N\boldsymbol{\phi}(t) + N^{\frac{1}{2}}\boldsymbol{z}) = \pi_N(t, \boldsymbol{z}). \tag{4.42}$$

Equating the total differentials with respect to time, i.e. $\mathrm{d}P_N(t,\boldsymbol{X})/\mathrm{d}t{=}\mathrm{d}\pi_N(t,\boldsymbol{z})/\mathrm{d}t$, yields

$$\frac{\partial P_N(t,\boldsymbol{X})}{\partial t} = \frac{\partial \pi_N(t,\boldsymbol{z})}{\partial t} - N^{\frac{1}{2}} \sum_{j=1}^{n} \frac{\mathrm{d}\phi_j(t)}{\mathrm{d}t} \frac{\partial \pi_N(t,\boldsymbol{z})}{\partial z_j} \tag{4.43}$$

as

$$\frac{\mathrm{d}z_j}{\mathrm{d}t} = -N^{\frac{1}{2}} \frac{\mathrm{d}\phi_j(t)}{\mathrm{d}t} \tag{4.44}$$

for all j. Note that the symbol d is used for the total differential and ∂ for partial derivatives (cf. the notation tables on pp. xvii). The comments from Sect. 4.3.4 on derivatives of discrete probability functions apply here as well. Assume that there are functions Φ_l for $l \in \mathbb{N}_0$, which do not depend on N, and a positive function f such that

$$W_N(t,\boldsymbol{X},\boldsymbol{\Delta}) = f(N) \sum_{l=0}^{\infty} N^{-l}\, \Phi_l\left(t, N^{-1}\boldsymbol{X}, \boldsymbol{\Delta}\right). \tag{4.45}$$

The factor $f(N)$ represents the fact that large systems evolve slowlier than small systems. For most cases such as the examples in Chap. 5 it would actually suffice to consider a function Φ with $W_N(t,\boldsymbol{X},\boldsymbol{\Delta}) = N\Phi(t, N^{-1}\boldsymbol{X},\boldsymbol{\Delta})$ (as has been done in the methods in Sects. 4.3.1–4.3.4), but the above setting leaves the method more general. As before, the transition probability P_N fulfils the master equation

$$\frac{\partial P_N(t,\boldsymbol{X})}{\partial t} = \sum_{\boldsymbol{\Delta}} \big(W_N(t,\boldsymbol{X}-\boldsymbol{\Delta},\boldsymbol{\Delta})P_N(t,\boldsymbol{X}-\boldsymbol{\Delta}) - W_N(t,\boldsymbol{X},\boldsymbol{\Delta})P_N(t,\boldsymbol{X})\big),$$

where the sum is taken over all possible jumps of size $\boldsymbol{\Delta} = (\Delta_1,\ldots,\Delta_n)'$. With (4.45), this equation now becomes

$$\begin{aligned} \frac{\partial P_N(t,\boldsymbol{X})}{\partial t} = f(N) \sum_{\boldsymbol{\Delta}} \sum_{l=0}^{\infty} N^{-l} \Big[\Phi_l\Big(t, \frac{\boldsymbol{X}-\boldsymbol{\Delta}}{N}, \boldsymbol{\Delta}\Big) P_N(t,\boldsymbol{X}-\boldsymbol{\Delta}) \\ - \Phi_l\Big(t, \frac{\boldsymbol{X}}{N}, \boldsymbol{\Delta}\Big) P_N(t,\boldsymbol{X})\Big]. \end{aligned} \tag{4.46}$$

Using (4.42) and (4.43), this expression can be written in terms of π_N as

$$\begin{aligned} &\frac{\partial \pi_N(t,\boldsymbol{z})}{\partial t} - N^{\frac{1}{2}} \sum_{j=1}^{n} \frac{\mathrm{d}\phi_j(t)}{\mathrm{d}t} \frac{\partial \pi_N(t,\boldsymbol{z})}{\partial z_j} \\ &\quad = f(N) \sum_{\boldsymbol{\Delta}} \sum_{l=0}^{\infty} N^{-l} \Big[\Phi_l\left(t, \boldsymbol{\phi}(t){+}N^{-\frac{1}{2}}(\boldsymbol{z}{-}N^{-\frac{1}{2}}\boldsymbol{\Delta}), \boldsymbol{\Delta}\right) \pi_N(t, \boldsymbol{z}{-}N^{-\frac{1}{2}}\boldsymbol{\Delta}) \end{aligned}$$

$$-\Phi_l\left(t, \boldsymbol{\phi}(t) + N^{-\frac{1}{2}}\boldsymbol{z}, \boldsymbol{\Delta}\right)\pi_N(t, \boldsymbol{z})\Big]$$

$$= f(N)\sum_{\boldsymbol{\Delta}}\sum_{l=0}^{\infty} N^{-l}\left[\Psi_l\left(\boldsymbol{z} - N^{-\frac{1}{2}}\boldsymbol{\Delta}\right) - \Psi_l(\boldsymbol{z})\right] \tag{4.47}$$

with $\Psi_l(\boldsymbol{z}) = \Phi_l(t, \boldsymbol{\phi}(t) + N^{-\frac{1}{2}}\boldsymbol{z}, \boldsymbol{\Delta})\pi_N(t, \boldsymbol{z})$ for all $l \in \mathbb{N}_0$ and fixed $\boldsymbol{\Delta}$ and t. Taylor expansion of $\Psi_l(\boldsymbol{z} - N^{-\frac{1}{2}}\boldsymbol{\Delta})$ around $\boldsymbol{z}$ yields

$$\Psi_l(\boldsymbol{z} - N^{-\frac{1}{2}}\boldsymbol{\Delta}) = \sum_{m=0}^{\infty}(-1)^m N^{-\frac{m}{2}} \sum_{\boldsymbol{k}\in\mathcal{K}_m}\left(\prod_{j=1}^{n}\frac{\Delta_j^{k_j}}{k_j!}\right)\frac{\partial^m \Psi_l(\boldsymbol{z})}{\partial z_1^{k_1}\cdots\partial z_n^{k_n}} \tag{4.48}$$

with $\mathcal{K}_m$ defined as in (4.37). The combination of (4.47) and (4.48) yields

$$\frac{\partial \pi_N(t,\boldsymbol{z})}{\partial t} - N^{\frac{1}{2}}\sum_{j=1}^{n}\frac{\mathrm{d}\phi_j(t)}{\mathrm{d}t}\frac{\partial \pi_N(t,\boldsymbol{z})}{\partial z_j}$$

$$= f(N)\sum_{\boldsymbol{\Delta}}\Big(\left[\Psi_0\left(\boldsymbol{z} - N^{-\frac{1}{2}}\boldsymbol{\Delta}\right) - \Psi_0(\boldsymbol{z})\right]$$

$$+N^{-1}\left[\Psi_1\left(\boldsymbol{z} - N^{-\frac{1}{2}}\boldsymbol{\Delta}\right) - \Psi_1(\boldsymbol{z})\right] + \mathcal{O}(N^{-2})\Big)$$

$$= f(N)\sum_{\boldsymbol{\Delta}}\Big(\Big[-N^{-\frac{1}{2}}\sum_{j=1}^{n}\Delta_j\frac{\partial \Psi_0(\boldsymbol{z})}{\partial z_j} + \frac{1}{2}N^{-1}\sum_{j=1}^{n}\sum_{k=1}^{n}\Delta_j\Delta_k\frac{\partial^2\Psi_0(\boldsymbol{z})}{\partial z_j\partial z_k}$$

$$+\mathcal{O}(N^{-\frac{3}{2}})\Big] + N^{-1}\Big[\mathcal{O}(N^{-\frac{1}{2}})\Big] + \mathcal{O}(N^{-2})\Big).$$

This measurement in powers of N is possible due to (4.45) and because π_N and the Φ_l are assumed not to be too irregular. Similarly to (4.38), define

$$\tilde{a}_{1,j}(t,\boldsymbol{z}) = \sum_{\boldsymbol{\Delta}}\Delta_j\Phi_0(t,\boldsymbol{z},\boldsymbol{\Delta}), \quad \tilde{a}_{2,(j,k)}(t,\boldsymbol{z}) = \sum_{\boldsymbol{\Delta}}\Delta_j\Delta_k\Phi_0(t,\boldsymbol{z},\boldsymbol{\Delta}) \tag{4.49}$$

(often called first and second *jump moments*) for all $j, k = 1, \ldots, n$ and resubstitute $\Psi_l(\boldsymbol{z})$ for $\Phi_l(t, \boldsymbol{\phi}(t) + N^{-1/2}\boldsymbol{z}, \boldsymbol{\Delta})\pi_N(t,\boldsymbol{z})$. We then obtain

$$\frac{\partial \pi_N(t,\boldsymbol{z})}{\partial t} - N^{\frac{1}{2}}\sum_{j=1}^{n}\frac{\mathrm{d}\phi_j(t)}{\mathrm{d}t}\frac{\partial \pi_N(t,\boldsymbol{z})}{\partial z_j}$$

$$= f(N)\Big(-N^{-\frac{1}{2}}\sum_{j=1}^{n}\frac{\partial}{\partial z_j}\tilde{a}_{1,j}\big(t,\boldsymbol{\phi}(t) + N^{-\frac{1}{2}}\boldsymbol{z}\big)\pi_N(t,\boldsymbol{z})$$

$$+\frac{1}{2}N^{-1}\sum_{j=1}^{n}\sum_{k=1}^{n}\frac{\partial^2}{\partial z_j\partial z_k}\tilde{a}_{2,(j,k)}\big(t,\boldsymbol{\phi}(t) + N^{-\frac{1}{2}}\boldsymbol{z}\big)\pi_N(t,\boldsymbol{z}) + \mathcal{O}(N^{-\frac{3}{2}})\Big).$$

Rescale the time such that $N^{-1}f(N)t = s$, i.e. apply $\mathrm{d}\pi_N/\mathrm{d}s = (\mathrm{d}\pi_N/\mathrm{d}t)\cdot(\mathrm{d}t/\mathrm{d}s)$. Then

$$\frac{\partial \pi_N(s,\boldsymbol{z})}{\partial s} - N^{\frac{1}{2}} \sum_{j=1}^{n} \frac{\mathrm{d}\phi_j(s)}{\mathrm{d}s} \frac{\partial \pi_N(s,\boldsymbol{z})}{\partial z_j}$$

$$= -N^{\frac{1}{2}} \sum_{j=1}^{n} \frac{\partial}{\partial z_j} \tilde{a}_{1,j}\big(s, \boldsymbol{\phi}(s) + N^{-\frac{1}{2}}\boldsymbol{z}\big)\pi_N(s,\boldsymbol{z})$$

$$+ \frac{1}{2} \sum_{j=1}^{n}\sum_{k=1}^{n} \frac{\partial^2}{\partial z_j \partial z_k} \tilde{a}_{2,(j,k)}\big(s, \boldsymbol{\phi}(s) + N^{-\frac{1}{2}}\boldsymbol{z}\big)\pi_N(s,\boldsymbol{z}) + \mathcal{O}\big(N^{-\frac{1}{2}}\big).$$

Taylor expansion of $\tilde{a}_{1,j}(s, \boldsymbol{\phi}(s) + N^{-1/2}\boldsymbol{z})$ and $\tilde{a}_{2,(j,k)}(s, \boldsymbol{\phi}(s) + N^{-1/2}\boldsymbol{z})$ around $\boldsymbol{\phi}(s)$ yields

$$\frac{\partial \pi_N(s,\boldsymbol{z})}{\partial s} - N^{\frac{1}{2}} \sum_{j=1}^{n} \frac{\mathrm{d}\phi_j(s)}{\mathrm{d}s} \frac{\partial \pi_N(s,\boldsymbol{z})}{\partial z_j} \tag{4.50}$$

$$= -N^{\frac{1}{2}} \sum_{j=1}^{n} \frac{\partial}{\partial z_j} \Big[\tilde{a}_{1,j}\,(s, \boldsymbol{\phi}(s)) + N^{-\frac{1}{2}} \sum_{i=1}^{n} z_i \tilde{a}_{1,j}^{(i)}(s, \boldsymbol{\phi}(s)) + \mathcal{O}\big(N^{-1}\big)\Big]\pi_N(s,\boldsymbol{z}) \tag{4.51}$$

$$+ \frac{1}{2} \sum_{j=1}^{n}\sum_{k=1}^{n} \frac{\partial^2}{\partial z_j \partial z_k} \Big[\tilde{a}_{2,(j,k)}(s, \boldsymbol{\phi}(s)) + \mathcal{O}\big(N^{-\frac{1}{2}}\big)\Big]\pi_N(s,\boldsymbol{z}) + \mathcal{O}\big(N^{-\frac{1}{2}}\big), \tag{4.52}$$

where $\tilde{a}_{1,j}^{(i)}$ denotes the first derivative of $\tilde{a}_{1,j}$ with respect to the ith component of the state variable. The terms of order $N^{\frac{1}{2}}$ cancel if

$$\sum_{j=1}^{n} \frac{\mathrm{d}\phi_j(s)}{\mathrm{d}s} \frac{\partial \pi_N(s,\boldsymbol{z})}{\partial z_j} = \sum_{j=1}^{n} \tilde{a}_{1,j}(s, \boldsymbol{\phi}(s)) \frac{\partial \pi_N(s,\boldsymbol{z})}{\partial z_j},$$

i.e.

$$\frac{\mathrm{d}\phi_j(s)}{\mathrm{d}s} = \tilde{a}_{1,j}(s, \boldsymbol{\phi}(s)) \tag{4.53}$$

for all $j = 1, \ldots, n$. This is assumed to be fulfilled by definition of $\boldsymbol{\phi}$. Furthermore, $\boldsymbol{\phi}$ is supposed to be the unique stable solution of (4.53), and $\tilde{a}_{1,j}$ shall fulfil certain regularity conditions such that all solutions of (4.53) converge to $\boldsymbol{\phi}$ fast enough (van Kampen 1997, Chap. 10.3). As N tends to infinity, only the terms of order $\mathcal{O}(1)$ in (4.50) to (4.52) remain. These are

$$\frac{\partial \pi_N(s,\boldsymbol{z})}{\partial s} = -\sum_{i,j=1}^{n} \tilde{a}_{1,j}^{(i)}(s, \boldsymbol{\phi}(s)) \frac{\partial z_i\, \pi_N(s,\boldsymbol{z})}{\partial z_j} + \frac{1}{2} \sum_{j,k=1}^{n} \tilde{a}_{2,(j,k)}(s, \boldsymbol{\phi}(s)) \frac{\partial^2 \pi_N(s,\boldsymbol{z})}{\partial z_j \partial z_k}, \tag{4.54}$$

which is a linear forward diffusion equation for π_N, i.e. the drift term is linear and the diffusion term constant. The solution of such an equation is a Gaussian density (van Kampen 1997, Chap. 8.6). Presumably, Formula (4.54) remains true when the probability π_N is replaced by the density π of the *continuous* variable $\boldsymbol{z}$. In the following we will show that (4.54) is then equivalent to a diffusion equation for $p(t,\boldsymbol{x})$, the transition density of $\boldsymbol{x} = \boldsymbol{\phi}(t) + N^{-1/2}\boldsymbol{z}$, where both $\boldsymbol{z}$ and $\boldsymbol{x}$ are treated as continuous variables. Equating the total differentials $\mathrm{d}p(s,\boldsymbol{x})/\mathrm{d}s = N^{\frac{1}{2}}\mathrm{d}\pi(s,\boldsymbol{z})/\mathrm{d}s$, one obtains

$$\frac{\partial p(s,\boldsymbol{x})}{\partial s} = N^{\frac{1}{2}}\left(\frac{\partial \pi(s,\boldsymbol{z})}{\partial s} + \sum_{j=1}^{n}\frac{\partial \pi(s,\boldsymbol{z})}{\partial z_j}\frac{\mathrm{d}z_j}{\mathrm{d}s}\right).$$

Plugging in (4.44), (4.53) and (4.54) with π_N replaced by π yields

$$\begin{aligned}
&\frac{\partial p(s,\boldsymbol{x})}{\partial s}\\
&= N^{\frac{1}{2}}\left(-\sum_{j=1}^{n}\sum_{i=1}^{n}\tilde{a}_{1,j}^{(i)}(s,\boldsymbol{\phi}(s))\frac{\partial z_i\,\pi(s,\boldsymbol{z})}{\partial z_j}\right.\\
&\quad\left.+\frac{1}{2}\sum_{j=1}^{n}\sum_{k=1}^{n}\tilde{a}_{2,(j,k)}(s,\boldsymbol{\phi}(s))\frac{\partial^2\pi(s,\boldsymbol{z})}{\partial z_j\partial z_k}\right) - N\sum_{j=1}^{n}\tilde{a}_{1,j}(s,\boldsymbol{\phi}(s))\frac{\partial \pi(s,\boldsymbol{z})}{\partial z_j}\\
&= -N\sum_{j=1}^{n}\frac{\partial}{\partial z_j}\left[\tilde{a}_{1,j}(s,\boldsymbol{\phi}(s)) + N^{-\frac{1}{2}}\sum_{i=1}^{n}z_i\tilde{a}_{1,j}^{(i)}(s,\boldsymbol{\phi}(s))\right]\pi(s,\boldsymbol{z})\\
&\quad+\frac{1}{2}N^{\frac{1}{2}}\sum_{j=1}^{n}\sum_{k=1}^{n}\frac{\partial^2}{\partial z_j\partial z_k}\left[\tilde{a}_{2,(j,k)}(s,\boldsymbol{\phi}(s))\right]\pi(s,\boldsymbol{z})\\
&= -N^{\frac{1}{2}}\sum_{j=1}^{n}\frac{\partial}{\partial z_j}\left[\tilde{a}_{1,j}(s,\boldsymbol{\phi}(s)) + N^{-\frac{1}{2}}\sum_{i=1}^{n}z_i\tilde{a}_{1,j}^{(i)}(s,\boldsymbol{\phi}(s))\right]p(s,\boldsymbol{\phi}(s)+N^{-\frac{1}{2}}\boldsymbol{z}) \qquad (4.55)\\
&\quad+\frac{1}{2}\sum_{j=1}^{n}\sum_{k=1}^{n}\frac{\partial^2}{\partial z_j\partial z_k}\left[\tilde{a}_{2,(j,k)}(s,\boldsymbol{\phi}(s))\right]p(s,\boldsymbol{\phi}(s)+N^{-\frac{1}{2}}\boldsymbol{z}). \qquad (4.56)
\end{aligned}$$

The expressions in the square brackets in (4.55) and (4.56) are the first terms of a Taylor expansion of $\tilde{a}_{1,j}(s,\boldsymbol{\phi}(s) + N^{-1/2}\boldsymbol{z})$ and $\tilde{a}_{2,(j,k)}(s,\boldsymbol{\phi}(s) + N^{-1/2}\boldsymbol{z})$ around $\boldsymbol{\phi}(s)$. The missing terms can be added since they are of order $\mathcal{O}(N^{-1})$ and $\mathcal{O}(N^{-1/2})$, respectively, and will vanish anyway in the limit $N \to \infty$. We can hence regard

$$\begin{aligned}
\frac{\partial p(s,\boldsymbol{x})}{\partial s} &= -N\sum_{j=1}^{n}\frac{\partial\tilde{a}_{1,j}\left(s,\boldsymbol{\phi}(s)+N^{-\frac{1}{2}}\boldsymbol{z}\right)\pi(s,\boldsymbol{z})}{\partial z_j}\\
&\quad+\frac{1}{2}N^{\frac{1}{2}}\sum_{j,k=1}^{n}\frac{\partial^2\tilde{a}_{2,(j,k)}\left(s,\boldsymbol{\phi}(s)+N^{-\frac{1}{2}}\boldsymbol{z}\right)\pi(s,\boldsymbol{z})}{\partial z_j\partial z_k}.
\end{aligned}$$

Changing the time scale from s back to t and differentiation with respect to $\boldsymbol{z}$ to differentiation with respect to $\boldsymbol{x}$ yields

$$\frac{\partial p(t,\boldsymbol{x})}{\partial t} = -\sum_{j=1}^{n} \frac{\partial}{\partial x_j} \tilde{a}_{1,j}(t,\boldsymbol{x}) p(t,\boldsymbol{x}) + \frac{1}{2N} \sum_{j=1}^{n}\sum_{k=1}^{n} \frac{\partial^2}{\partial x_j \partial x_k} \tilde{a}_{2,(j,k)}(t,\boldsymbol{x}) p(t,\boldsymbol{x}). \tag{4.57}$$

This is a forward diffusion equation for $p(t,\boldsymbol{x})$ with drift

$$\boldsymbol{\mu}(\boldsymbol{x},t) = (\tilde{a}_{1,j}(t,\boldsymbol{x}))_{j=1,\dots,n} = \sum_{\boldsymbol{\Delta}} \Phi_0(t,\boldsymbol{x},\boldsymbol{\Delta})\boldsymbol{\Delta} \tag{4.58}$$

and diffusion matrix

$$N^{-1}\boldsymbol{\Sigma}(\boldsymbol{x},t) = N^{-1}(\tilde{a}_{2,(j,k)}(t,\boldsymbol{x}))_{j,k=1,\dots,n} = N^{-1}\sum_{\boldsymbol{\Delta}} \Phi_0(t,\boldsymbol{x},\boldsymbol{\Delta})\boldsymbol{\Delta}\boldsymbol{\Delta}' \tag{4.59}$$

if $\boldsymbol{\Sigma}$ is positive definite. Thus, heuristically, the process $\boldsymbol{x}$ approximately follows a stochastic differential equation

$$\mathrm{d}\boldsymbol{x}_t = \boldsymbol{\mu}(\boldsymbol{x}_t,t)\mathrm{d}t + \frac{1}{\sqrt{N}}\boldsymbol{\sigma}(\boldsymbol{x}_t,t)\mathrm{d}\boldsymbol{B}_t\,, \quad \boldsymbol{x}_{t_0} = \boldsymbol{x}_0,$$

with $\boldsymbol{\sigma}\boldsymbol{\sigma}' = \boldsymbol{\Sigma}$.

Van Kampen's expansion is frequently applied, especially in life sciences, for instance in molecular and cell biology (Leung 1985; Elf and Ehrenberg 2003; Paulsson 2004; Sjöberg et al. 2009), microbiology (Hsu and Wang 1987), social sciences (de la Lama et al. 2006), physics (van Kampen 1961), infectious disease epidemiology (Chen and Bokka 2005; Alonso et al. 2007) or more generally in population biology (McKane and Newman 2004; Ferm et al. 2008).

Example 4.7. In the SI model, we have observed that the only possible jump is $\Delta_1 = -1$ with transition rate $W_N(t,Ns,-1) = w_{N,1}(t,s) = N\alpha s(1-s)$. Van Kampen's expansion requires that W_N can be written in the canonical form (4.45). This is fulfilled for $f(N) = N$, $\Phi_0(t,s,-1) = N^{-1}w_{N,1}(t,s) = w_1(t,s)$ and $\Phi_l \equiv 0$ for all $l \geq 1$. The drift and diffusion of an approximation to the intensive process are given by (4.58) and (4.59). With the above choices, these are

$$\mu(s,t) = \Phi_0(t,s,-1)\Delta_1 = w_1(t,s)\Delta_1 = -\alpha s(1-s),$$
$$N^{-1}\Sigma(s,t) = N^{-1}\Phi_0(t,s,-1)\Delta_1^2 = N^{-1}w_1(t,s)\Delta_1^2 = \alpha s(1-s)/N.$$

Once more, this resembles the results from the previous examples.

4.3.6 Other Approaches

A number of further approximation techniques is proposed by different authors, especially in the physics literature and most often in the context of physical processes:

Drummond et al. (1981) review *quasi-probability methods* for transforming master equations into generalised diffusion equations. In these approaches, positivity of the probability or positive-definiteness of the diffusion matrices cannot always be guaranteed. A prominent example for such a quasiprobability method is the *Poisson representation* (Gardiner and Chaturvedi 1977; Chaturvedi and Gardiner 1978) which expands the probability distribution of the process in Poisson distributions.

Under the assumption of the asymptotic form $p(\boldsymbol{x}, t) \propto \exp(N\psi(\boldsymbol{x}, t))$ for some function ψ, Kubo et al. (1973) approximate the transition density p by a Gaussian distribution whose parameters are expressed through the cumulants of $\boldsymbol{x}$. (In statistical thermodynamics, if the system is in equilibrium, ψ equals the standardised negative free energy for a unit of the system whose size is determined by N.)

Walsh (1981) suggests a *well-timed diffusion approximation* in the sense that the approximating process $\tilde{\boldsymbol{X}} = (\tilde{\boldsymbol{X}}_t)_{t \geq t_0}$ is a diffusion in which the jump process $\boldsymbol{X} = (\boldsymbol{X}_t)_{t \geq t_0}$ can be embedded, and there are stopping times $(T_t)_{t \geq 0}$ with $\mathbb{E}T_t = t$ for all $t \geq 0$ such that $(\boldsymbol{X}_t)_{t \geq t_0}$ and $(\tilde{\boldsymbol{X}}_{T_t})_{t \geq t_0}$ have the same distribution. In other words, the sample paths of both processes cover the same space (in distribution) at the same speed.

A variety of papers is devoted to the problem of processes with special properties like irreversibility or nonstability. Grabert et al. (1983) suggest a technique for the derivation of the forward diffusion equation for models which take into account pressure and temperature fluctuations. The so-obtained drift coefficient differs from that of the Kramers-Moyal expansion only by $\mathcal{O}(N^{-1})$, but the difference between the two diffusion matrices is proportional to the deviation from steady states as measured by a thermodynamic force. Hänggi et al. (1984) show that for bistable systems the Kramers-Moyal expansion overestimates the transition rates between deterministically stable states, while the approach by Grabert et al. (1983) estimates them correctly. Further specialised approximation methods are developed or reviewed in e.g. Green (1952), Grabert and Green (1979); Grabert et al. (1980), Hänggi and Jung (1988), Shizgal and Barrett (1989) and Muñoz and Garrido (1994).

4.4 Extensions to Systems with Multiple Size Parameters

Section 4.3 has introduced diffusion approximation methods for systems whose size is sufficiently described by a single parameter $N \in \mathbb{N}$. In some applications, however, a more reasonable characterisation is given by an entire set $\{N_1, \ldots, N_d\} \subset \mathbb{N}^d$, $d \in \mathbb{N}$, of system size parameters. An example for such an application is the multitype SIR model which will be presented in Sect. 5.2, where N_i may stand for the population size of a geographical region labelled i. In these cases, the considered approximation techniques need to be adjusted as investigated in this section.

As before, consider a pure Markov jump process which in its extensive form is denoted by $\boldsymbol{X}$. Let $\boldsymbol{X} = (\boldsymbol{X}_1', \ldots, \boldsymbol{X}_d')'$ be a partition of the state variable such that the vector $\boldsymbol{X}_i$ is characterised by the size variable N_i, $i = 1, \ldots, d$. Then $\boldsymbol{x} = (N_1^{-1}\boldsymbol{X}_1', \ldots, N_d^{-1}\boldsymbol{X}_d')'$ specifies the respective intensive jump process where occurring jumps are small if the system sizes are large. Define the (invertible) diagonal matrix $\boldsymbol{M}$ such that $\boldsymbol{X} = \boldsymbol{M}\boldsymbol{x}$, and let $N = \sum_{i=1}^d N_i$. If $\mathcal{D}_N^{(d)} \subseteq \mathbb{Z}^n$ denotes the state space of the extensive process, $\mathcal{C}_N^{(d)} = \boldsymbol{M}^{-1}\mathcal{D}_N^{(d)} = \{\boldsymbol{M}^{-1}\boldsymbol{X} \,|\, \boldsymbol{X} \in \mathcal{D}_N^{(d)}\}$ is the state space of the intensive one. (This notation is used although the sum N naturally does not determine the individual sizes $N_1, \ldots, N_d$ unless their ratios are fixed.) Once more, depending on the context, $\boldsymbol{X}$ and $\boldsymbol{x}$ may interchangeably stand for the whole process or a single state.

For appropriate I, let $\{\boldsymbol{\Delta}_i \,|\, i \in I\}$ and $\{\tilde{\boldsymbol{\Delta}}_i \,|\, i \in I\} = \{\boldsymbol{M}^{-1}\boldsymbol{\Delta}_i \,|\, i \in I\}$ denote the sets of nonzero jumps of the extensive and intensive process, respectively. Adopt the notation for the transition probabilities P_N, p_N and transition rates $W_{N,i}$, $w_{N,i}$ from Sect. 4.3. Sections 4.4.1–4.4.5 present how the techniques from Sects. 4.3.1–4.3.5 can be modified. For examples of these approximation procedures, the reader is referred to Chap. 5.

4.4.1 Convergence of the Master Equation

According to Eq. (4.15), the forward master equation for $\boldsymbol{X}$ reads

$$\frac{\partial P_N(t,\boldsymbol{X})}{\partial t} = \sum_{i\in I}\Big(W_{N,i}(t,\boldsymbol{X}-\boldsymbol{\Delta}_i)P_N(t,\boldsymbol{X}-\boldsymbol{\Delta}_i) - W_{N,i}(t,\boldsymbol{X})P_N(t,\boldsymbol{X})\Big).$$

The forward master equation for $\boldsymbol{x}$ is then

$$\frac{\partial p_N(t,\boldsymbol{x})}{\partial t} = \sum_{i\in I}\Big(w_{N,i}(t,\boldsymbol{x}-\tilde{\boldsymbol{\Delta}}_i)p_N(t,\boldsymbol{x}-\tilde{\boldsymbol{\Delta}}_i) - w_{N,i}(t,\boldsymbol{x})p_N(t,\boldsymbol{x})\Big).$$

Replace p_N by its limit function p and assume that there are functions w_i and small but positive δ_i such that $w_{N,i} = \delta_i^{-1}w_i$ for all $i \in I$. Suppose that the w_i depend on $\{N_1, \ldots, N_d\}$ only through some statistic T for which $T(\{N_1, \ldots, N_d\}) = T(\{cN_1, \ldots, cN_d\})$ holds for all $c \in \mathbb{N}$. In other words, w_i does not change for $N \to \infty$ as long as the proportions between the single population sizes remain constant. This facilitates the following limiting procedure between lines (4.61) and (4.62) (compare with the explanation on p. 67). The master equation becomes

$$\frac{\partial p(t,\boldsymbol{x})}{\partial t} = \sum_{i\in I} \delta_i^{-1}\Big(w_i(t,\boldsymbol{x}-\tilde{\boldsymbol{\Delta}}_i)p(t,\boldsymbol{x}-\tilde{\boldsymbol{\Delta}}_i) - w_i(t,\boldsymbol{x})p(t,\boldsymbol{x})\Big). \tag{4.60}$$

As in Sect. 4.3.1, the bracketed terms can be rewritten by sums of difference operators, resulting in

$$\frac{\partial p(t,\boldsymbol{x})}{\partial t} = \sum_{i\in I} \delta_i^{-1}\left(\sum_{\boldsymbol{k}_i=(0,k_{i,1},\ldots,k_{i,n})'} D^{|\boldsymbol{k}_i|}_{\boldsymbol{k}_i,\boldsymbol{\varepsilon}_i} w_i(t,\boldsymbol{x})p(t,\boldsymbol{x})\right) \tag{4.61}$$

$$\approx \sum_{i\in I} \delta_i^{-1}\left(\sum_{\boldsymbol{k}_i=(k_{i,1},\ldots,k_{i,n})'} \varepsilon_{i,1}^{k_{i,1}}\cdots\varepsilon_{i,n}^{k_{i,n}} \frac{\partial^{|\boldsymbol{k}_i|}w_i(t,\boldsymbol{x})p(t,\boldsymbol{x})}{\partial x_1^{k_{i,1}}\cdots\partial x_n^{k_{i,n}}}\right), \tag{4.62}$$

where $\boldsymbol{\varepsilon}_i = (0,\varepsilon_{i,1},\ldots,\varepsilon_{i,n})'$ with $|\varepsilon_{i,k}|^{-1} \in \{N_1,\ldots,N_d\}$ for all $i \in I$ and $k = 1,\ldots,n$. Typically, δ_i^{-1} cancels with one of the $\varepsilon_{i,k}$ (at least up to a finite constant) for all i. Once more, restrict the master equation to terms up to order $\mathcal{O}(\max\{N_1^{-1},\ldots,N_d^{-1}\})$. Then, one again arrives at

$$\frac{\partial p(t,\boldsymbol{x})}{\partial t} = -\sum_{j=1}^{n}\frac{\partial\big[\mu_j(\boldsymbol{x},t)p(t,\boldsymbol{x})\big]}{\partial x_j} + \frac{1}{2}\sum_{j,k=1}^{n}\frac{\partial^2\big[\widetilde{\Sigma}_{jk}(\boldsymbol{x},t)p(t,\boldsymbol{x})\big]}{\partial x_j\partial x_k}$$

for some vector $\boldsymbol{\mu} = (\mu_j)_{j=1,\ldots,n}$ and matrix $\widetilde{\boldsymbol{\Sigma}} = (\widetilde{\Sigma}_{jk})_{j,k=1,\ldots,n}$. These can be derived by application of Algorithm B.1 in Sect. B.1 in the Appendix. If $\widetilde{\boldsymbol{\Sigma}}$ is positive definite, these are the drift vector and diffusion matrix of the diffusion process approximating the jump process $\boldsymbol{x}$.

4.4.2 Convergence of the Infinitesimal Generator

In analogy to Eq. (4.24), for the infinitesimal generator of the intensive jump process $\boldsymbol{x}$ in the new setting one obtains

$$\mathcal{G}_N f(\boldsymbol{x},t) = \frac{\partial f(\boldsymbol{x},t)}{\partial t} + \sum_{i\in I} w_{N,i}(t,\boldsymbol{x})\left(f(\boldsymbol{x}+\tilde{\boldsymbol{\Delta}}_i,t) - f(\boldsymbol{x},t)\right),$$

where $f : \mathcal{C}\times T \to \mathbb{R}$ is a measurable twice continuously differentiable function, $\mathcal{C}$ the continuous state space of the diffusion approximation, $\boldsymbol{x} \in \mathcal{C}$ and $t \in T$. In a similar manner as above,

$$
\begin{aligned}
\mathcal{G}_N f(\boldsymbol{x},t) &= \frac{\partial f(\boldsymbol{x},t)}{\partial t} + \sum_{i\in I} \delta_i^{-1} w_i(t,\boldsymbol{x}) \Big(f(\boldsymbol{x}+\tilde{\boldsymbol{\Delta}}_i,t) - f(\boldsymbol{x},t) \Big) \\
&= \frac{\partial f(\boldsymbol{x},t)}{\partial t} + \sum_{i\in I} \delta_i^{-1} w_i(t,\boldsymbol{x}) \left(\sum_{\boldsymbol{k}_i=(k_{i,1},\dots,k_{i,n},0)'} D_{\boldsymbol{k}_i,\boldsymbol{\varepsilon}_i}^{|\boldsymbol{k}_i|} f(\boldsymbol{x},t) \right)
\end{aligned} \tag{4.63}
$$

with $\boldsymbol{\varepsilon}_i = (\varepsilon_{i,1},\dots,\varepsilon_{i,n},0)'$ and $|\varepsilon_{i,k}|^{-1} \in \{N_1,\dots,N_d\}$. Hence,

$$
\mathcal{G}_N \approx \frac{\partial}{\partial t} + \sum_{i\in I} \delta_i^{-1} w_i(t,\boldsymbol{x}) \left(\sum_{\boldsymbol{k}_i=(k_{i,1},\dots,k_{i,n})'} \varepsilon_{i,1}^{k_{i,1}} \cdots \varepsilon_{i,n}^{k_{i,n}} \frac{\partial^{|\boldsymbol{k}_i|}}{\partial x_1^{k_{i,1}} \cdots \partial x_n^{k_{i,n}}} \right).
$$

Again, take out δ_i^{-1} for one of the $\varepsilon_{i,k}$, $k = 1,\dots,n$, and neglect all terms of order higher than $\mathcal{O}(\max\{N_1^{-1},\dots,N_d^{-1}\})$. Then, $\mathcal{G}_N$ can be approximated by

$$
\mathcal{G} = \frac{\partial}{\partial t} + \sum_{j=1}^{n} \mu_j(\boldsymbol{x},t) \frac{\partial}{\partial x_j} + \frac{1}{2} \sum_{j,k=1}^{n} \widetilde{\Sigma}_{jk}(\boldsymbol{x},t) \frac{\partial^2}{\partial x_j \partial x_k}
$$

for a vector $\boldsymbol{\mu} = (\mu_j)_{j=1,\dots,n}$ and matrix $\widetilde{\boldsymbol{\Sigma}} = (\widetilde{\Sigma}_{jk})_{j,k=1,\dots,n}$. Once again, these can be determined using Algorithm B.1. For positive definite $\widetilde{\boldsymbol{\Sigma}}$, this operator can be seen as the generator of the diffusion approximation, i.e. the drift vector and diffusion matrix are given by $\boldsymbol{\mu}$ and $\widetilde{\boldsymbol{\Sigma}}$, respectively.

4.4.3 Langevin Approach

Suppose the intensive jump process $\boldsymbol{x}$ can be approximated by the solution of an SDE with drift vector $\boldsymbol{\mu}$ and diffusion matrix $\widetilde{\boldsymbol{\Sigma}}$. In the Langevin approach, the formulas for $\boldsymbol{\mu}$ and $\widetilde{\boldsymbol{\Sigma}}$ immediately follow from their definitions

$$
\begin{aligned}
\boldsymbol{\mu}(\boldsymbol{u},t) &= \lim_{\Delta t \downarrow 0} \frac{1}{\Delta t} \mathbb{E}\left(\boldsymbol{x}_{t+\Delta t} - \boldsymbol{x}_t \,\middle|\, \boldsymbol{x}_t = \boldsymbol{u} \right), \\
\widetilde{\boldsymbol{\Sigma}}(\boldsymbol{u},t) &= \lim_{\Delta t \downarrow 0} \frac{1}{\Delta t} \mathbb{E}\left((\boldsymbol{x}_{t+\Delta t} - \boldsymbol{x}_t)(\boldsymbol{x}_{t+\Delta t} - \boldsymbol{x}_t)' \,\middle|\, \boldsymbol{x}_t = \boldsymbol{u} \right)
\end{aligned}
$$

for $\boldsymbol{u} \in \mathcal{C}$ (compare with Eqs. (4.32) and (4.33)), i.e.

$$
\begin{aligned}
\boldsymbol{\mu}(\boldsymbol{u},t) &= \sum_{i\in I} w_{N,i}(t,\boldsymbol{u}) \tilde{\boldsymbol{\Delta}}_i, \\
\widetilde{\boldsymbol{\Sigma}}(\boldsymbol{u},t) &= \sum_{i\in I} w_{N,i}(t,\boldsymbol{u}) \tilde{\boldsymbol{\Delta}}_i \tilde{\boldsymbol{\Delta}}_i'.
\end{aligned}
$$

Note that in this formula the $\tilde{\boldsymbol{\Delta}}_i$, $i \in I$, denote the jumps of the *intensive* jump process.

4.4.4 Kramers-Moyal Expansion

In the Kramers-Moyal expansion, the master equation

$$\frac{\partial p_N(t,\boldsymbol{x})}{\partial t} = \sum_{i\in I}\Big(w_{N,i}(t,\boldsymbol{x}-\tilde{\boldsymbol{\Delta}}_i)p_N(t,\boldsymbol{x}-\tilde{\boldsymbol{\Delta}}_i) - w_{N,i}(t,\boldsymbol{x})p_N(t,\boldsymbol{x})\Big)$$

of the intensive jump process $\boldsymbol{x}$ is approximated by a Taylor expansion in orders of nonlinearity. As before, let $\tilde{\boldsymbol{\Delta}}_i = (\tilde{\Delta}_{i,1},\ldots,\tilde{\Delta}_{i,n})'$, $i \in I$, denote the jumps of $\boldsymbol{x}$. By expansion of $w_{N,i}(\cdot,\boldsymbol{x}-\tilde{\boldsymbol{\Delta}}_i)p_N(\cdot,\tilde{\boldsymbol{\Delta}}_i)$ around $\boldsymbol{x}$ one obtains

$$\frac{\partial p_N(t,\boldsymbol{x})}{\partial t} = \sum_{i\in I}\sum_{m=1}^{\infty}\sum_{\boldsymbol{k}\in\mathcal{K}_m}\left(\prod_{j=1}^{n}\frac{(-\tilde{\Delta}_{i,j})^{k_j}}{k_j!}\right)\frac{\partial^m}{\partial x_1^{k_1}\cdots\partial x_n^{k_n}}\,w_{N,i}(t,\boldsymbol{x})p_N(t,\boldsymbol{x}),$$

where for all $m \in \mathbb{N}_0$

$$\mathcal{K}_m = \big\{\boldsymbol{k} = (k_1,\ldots,k_n)' \in \mathbb{N}_0^n \,\big|\, |\boldsymbol{k}| = m\big\}.$$

Replace p_N by its limit p and terminate the above expansion after $m = 2$. Then

$$\begin{aligned}\frac{\partial p(t,\boldsymbol{x})}{\partial t} &= -\sum_{j=1}^{n}\sum_{i\in I}\tilde{\Delta}_{i,j}\frac{\partial}{\partial x_j}\,w_{N,i}(t,\boldsymbol{x})p(t,\boldsymbol{x})\\ &\quad+\frac{1}{2}\sum_{j,k=1}^{n}\sum_{i\in I}\tilde{\Delta}_{i,j}\tilde{\Delta}_{i,k}\frac{\partial^2}{\partial x_j\partial x_k}\,w_{N,i}(t,\boldsymbol{x})p(t,\boldsymbol{x}).\end{aligned}$$

That means, the forward master equation can be approximated by

$$\frac{\partial p(t,\boldsymbol{x})}{\partial t} = -\sum_{j=1}^{n}\frac{\partial\big[\mu_j(\boldsymbol{x},t)p(t,\boldsymbol{x})\big]}{\partial x_j} + \frac{1}{2}\sum_{j,k=1}^{n}\frac{\partial^2\big[\widetilde{\Sigma}_{jk}(\boldsymbol{x},t)p(t,\boldsymbol{x})\big]}{\partial x_j\partial x_k}$$

with

$$\boldsymbol{\mu}(\boldsymbol{x},t) = (\mu_j)_{j=1,\ldots,n} = \sum_{i\in I} w_{N,i}(t,\boldsymbol{x})\tilde{\boldsymbol{\Delta}}_i$$

and

$$\widetilde{\boldsymbol{\Sigma}}(\boldsymbol{x},t) = (\widetilde{\Sigma}_{jk})_{j,k=1,\ldots,n} = \sum_{i\in I} w_{N,i}(t,\boldsymbol{x})\tilde{\boldsymbol{\Delta}}_i\tilde{\boldsymbol{\Delta}}_i'.$$

For positive definite $\widetilde{\boldsymbol{\Sigma}}$, this is a forward diffusion equation leading to a diffusion approximation with drift vector $\boldsymbol{\mu}$ and diffusion matrix $\widetilde{\boldsymbol{\Sigma}}$.

4.4.5 Van Kampen Expansion

Like van Kampen's expansion in the context of one system size parameter N, its extension to a set $\{N_1, \ldots, N_d\}$ of system sizes considers the fluctuations of the process $\boldsymbol{x}$ around a deterministic process $\boldsymbol{\phi}(t) = (\phi_1(t), \ldots, \phi_n(t))'$ describing the macroscopic behaviour of $\boldsymbol{x}$.

Let M_j stand for the jth main diagonal element of the diagonal matrix $\boldsymbol{M}$, that is $X_j = M_j x_j$ for $j = 1, \ldots, n$. In the multiple size parameter setting, the probability function $P_N(t, \boldsymbol{X})$ is peaked around $\boldsymbol{M}\boldsymbol{\phi}(t)$ with width proportional to $M_j^{1/2}$ in the jth component. Hence, consider the time-dependent transformation

$$\boldsymbol{z} = (z_1, \ldots, z_n)' = \boldsymbol{M}^{\frac{1}{2}}\big(\boldsymbol{x} - \boldsymbol{\phi}(t)\big)$$

and the probability function π_N of $\boldsymbol{z}$, which fulfils

$$P_N(t, \boldsymbol{X}) = P_N(t, \boldsymbol{M}\boldsymbol{\phi}(t) + \boldsymbol{M}^{\frac{1}{2}}\boldsymbol{z}) = \pi_N(t, \boldsymbol{z}).$$

Analogously to the procedure in Sect. 4.3.5, equate the total differentials of P_N and π_N to obtain

$$\frac{\partial P_N(t, \boldsymbol{X})}{\partial t} = \frac{\partial \pi_N(t, \boldsymbol{z})}{\partial t} - \sum_{j=1}^{n} M_j^{\frac{1}{2}} \frac{\mathrm{d}\phi_j(t)}{\mathrm{d}t} \frac{\partial \pi_N(t, \boldsymbol{z})}{\partial z_j}.$$

For the main diagonal elements of $\boldsymbol{M}$ one has $M_j \in \{N_1, \ldots, N_d\}$ for all j. For $v = 1, \ldots, d$, define

$$J_v = \big\{u \in \{1, \ldots, n\} \,\big|\, M_u = N_v\big\}. \tag{4.64}$$

Without loss of generality, let $N_j \neq N_k$ for $j \neq k$. Then $J_1, \ldots, J_d$ is a partition of $\{1, \ldots, n\}$, i.e. a division into pairwise disjoint sets. Hence

$$\frac{\partial P_N(t, \boldsymbol{X})}{\partial t} = \frac{\partial \pi_N(t, \boldsymbol{z})}{\partial t} - \sum_{v=1}^{d} N_v^{\frac{1}{2}} \sum_{u \in J_v} \frac{\mathrm{d}\phi_u(t)}{\mathrm{d}t} \frac{\partial \pi_N(t, \boldsymbol{z})}{\partial z_u},$$

corresponding to Eq. (4.43) in the previous considerations. Recall from p. 84 the notation $\{\boldsymbol{\Delta}_i \,|\, i \in I\}$ for the set of all nonzero jumps of the extensive variable $\boldsymbol{X}$. In order to appropriately modify the canonical form (4.45), assume that there are functions Φ_l, $l \in \mathbb{N}_0$, and a partition $I_1, \ldots, I_d$ of I such that for all $i \in I$ and $j = 1, \ldots, d$

$$W_N(t, \boldsymbol{X}, \boldsymbol{\Delta}_i) = f(N_j) \sum_{l=0}^{\infty} N_j^{-l} \Phi_l\left(t, \boldsymbol{M}^{-1}\boldsymbol{X}, \boldsymbol{\Delta}_i\right) \qquad \text{if } i \in I_j. \tag{4.65}$$

Plugging this in into the general form of the forward master equation

$$\frac{\partial P_N(t,\boldsymbol{X})}{\partial t} = \sum_{i\in I}\big(W_N(t,\boldsymbol{X}-\boldsymbol{\Delta}_i,\boldsymbol{\Delta}_i)P_N(t,\boldsymbol{X}-\boldsymbol{\Delta}_i) - W_N(t,\boldsymbol{X},\boldsymbol{\Delta})P_N(t,\boldsymbol{X})\big)$$

yields

$$\frac{\partial P_N(t,\boldsymbol{X})}{\partial t} = \sum_{v=1}^{d} f(N_v)\sum_{u\in I_v}\sum_{l=0}^{\infty} N_v^{-l}\Big[\Phi_l\big(t,\boldsymbol{M}^{-1}(\boldsymbol{X}-\boldsymbol{\Delta}_u),\boldsymbol{\Delta}_u\big)P_N(t,\boldsymbol{X}-\boldsymbol{\Delta}_u) - \Phi_l\big(t,\boldsymbol{M}^{-1}\boldsymbol{X},\boldsymbol{\Delta}_u\big)P_N(t,\boldsymbol{X})\Big]$$

as the equivalent of Eq. (4.46). Follow the transformations on pp. 78–80 to arrive at an expression corresponding to (4.50)–(4.52). That is

$$\frac{\partial \pi_N(s,\boldsymbol{z})}{\partial s} - \sum_{v=1}^{d} N_v^{\frac{1}{2}} \sum_{u\in J_v} \frac{\mathrm{d}\phi_u(s)}{\mathrm{d}s}\frac{\partial \pi_N(s,\boldsymbol{z})}{\partial z_u} \tag{4.66}$$

$$= -\sum_{v=1}^{d} N_v^{\frac{1}{2}} \sum_{j=1}^{n} \frac{\partial}{\partial z_j}\Big[\tilde a_{1,j,v}(s,\boldsymbol{\phi}(s)) + N_v^{-\frac{1}{2}}\sum_{i=1}^{n} z_i \tilde a_{1,j,v}^{(i)}(s,\boldsymbol{\phi}(s)) + \mathcal{O}(N_v^{-1})\Big]\pi_N(s,\boldsymbol{z}) \tag{4.67}$$

$$+\frac{1}{2}\sum_{v=1}^{d}\sum_{j=1}^{n}\sum_{k=1}^{n}\frac{\partial^2}{\partial z_j\partial z_k}\Big[\tilde a_{2,(j,k),v}(s,\boldsymbol{\phi}(s)) + \mathcal{O}(N_v^{-\frac{1}{2}})\Big]\pi_N(s,\boldsymbol{z}) + \sum_{v=1}^{d}\mathcal{O}(N_v^{-\frac{1}{2}}), \tag{4.68}$$

where

$$\tilde a_{1,j,v}(t,\boldsymbol{z}) = \sum_{u\in I_v}\Delta_{u,j}\Phi_0\big(t,\boldsymbol{z},\boldsymbol{\Delta}_u\big) \quad \text{and} \quad \tilde a_{2,(j,k),v}(t,\boldsymbol{z}) = \sum_{u\in I_v}\Delta_{u,j}\Delta_{u,k}\Phi_0\big(t,\boldsymbol{z},\boldsymbol{\Delta}_u\big)$$

for $j,k = 1,\dots,n$, $v = 1,\dots,d$ and $\boldsymbol{\Delta}_u = (\Delta_{u,1},\dots,\Delta_{u,n})'$, and $\tilde a_{1,j,v}^{(i)}$ denotes the first derivative of $\tilde a_{1,j,v}$ with respect to the ith component of the state variable. Furthermore, let

$$\tilde a_{1,j}(t,\boldsymbol{z}) = \sum_{u\in I}\Delta_{u,j}\Phi_0\big(t,\boldsymbol{z},\boldsymbol{\Delta}_u\big) = \sum_{v=1}^{d}\tilde a_{1,j,v}(t,\boldsymbol{z})$$

for $j = 1,\dots,n$ (compare with (4.49)). The terms of order $N_v^{1/2}$ in lines (4.66)–(4.68) cancel if

$$\sum_{u\in J_v}\frac{\mathrm{d}\phi_u(s)}{\mathrm{d}s}\frac{\partial \pi_N(s,\boldsymbol{z})}{\partial z_u} = \sum_{j=1}^{n}\tilde a_{1,j,v}(s,\boldsymbol{\phi}(s))\frac{\partial \pi_N(s,\boldsymbol{z})}{\partial z_j} \tag{4.69}$$

for all $v = 1,\dots,d$. Assume

$$\tilde a_{1,j,v}(t,\boldsymbol{z}) = 0 \quad \text{if } j \notin J_v. \tag{4.70}$$

This trivially implies

$$\tilde{a}_{1,j,v}(t, \boldsymbol{z}) = \tilde{a}_{1,j}(t, \boldsymbol{z}) \quad \text{if } j \in J_v,$$

which means that the jumps $\boldsymbol{\Delta}_u$ with $u \in I_v$ are sufficient to determine the first jump moment of the jth component of $\boldsymbol{X}$ if $j \in J_v$. Under this assumption, condition (4.69) turns into

$$\sum_{u \in J_v} \frac{\mathrm{d}\phi_u(s)}{\mathrm{d}s} \frac{\partial \pi_N(s, \boldsymbol{z})}{\partial z_u} = \sum_{u \in J_v} \tilde{a}_{1,u,v}(s, \boldsymbol{\phi}(s)) \frac{\partial \pi_N(s, \boldsymbol{z})}{\partial z_u},$$

i.e.

$$\frac{\mathrm{d}\phi_u(s)}{\mathrm{d}s} = \tilde{a}_{1,u,v}(s, \boldsymbol{\phi}(s)) \quad \text{for all } u \in J_v,$$

and that in turn is equivalent to

$$\frac{\mathrm{d}\phi_u(s)}{\mathrm{d}s} = \tilde{a}_{1,u}(s, \boldsymbol{\phi}(s)) \quad \text{for all } u = 1, \ldots, n.$$

This requirement again is fulfilled due to the definition of $\boldsymbol{\phi}(t)$; compare with Eq. (4.53). Hence, the expression (4.66)–(4.68) reduces to

$$\frac{\partial \pi_N(s, \boldsymbol{z})}{\partial s} = \sum_{v=1}^{d} \left[- \sum_{i,j=1}^{n} \tilde{a}^{(i)}_{1,j,v}(s, \boldsymbol{\phi}(s)) \frac{\partial z_i \, \pi_N(s, \boldsymbol{z})}{\partial z_j} \right.$$
$$\left. + \frac{1}{2} \sum_{j,k=1}^{n} \tilde{a}_{2,(j,k),v}(s, \boldsymbol{\phi}(s)) \frac{\partial^2 \pi_N(s, \boldsymbol{z})}{\partial z_j \partial z_k} \right]$$

as N_v tends to infinity for all $v = 1, \ldots, d$. Like Eq. (4.54), this is a linear forward diffusion equation for π_N. As shown on pp. 81–82, one can (heuristically) transform it to a forward diffusion equation for p, which is the density of the intended diffusion approximation process. The result is, according to (4.57),

$$\frac{\partial p(t, \boldsymbol{x})}{\partial t} = \sum_{v=1}^{d} \left[- \sum_{j=1}^{n} \frac{\partial}{\partial x_j} \tilde{a}_{1,j,v}(t, \boldsymbol{x}) p(t, \boldsymbol{x}) \right.$$
$$\left. + \frac{1}{2N_v} \sum_{j=1}^{n} \sum_{k=1}^{n} \frac{\partial^2}{\partial x_j \partial x_k} \tilde{a}_{2,(j,k),v}(t, \boldsymbol{x}) p(t, \boldsymbol{x}) \right].$$

Overall, provided that the diffusion matrix is positive definite, the diffusion approximation can be described by an SDE with drift vector

$$\boldsymbol{\mu}(\boldsymbol{x},t)=\left(\sum_{v=1}^{d}\tilde{a}_{1,j,v}(t,\boldsymbol{x})\right)_{j=1,\ldots,n}=\left(\tilde{a}_{1,j}(t,\boldsymbol{x})\right)_{j=1,\ldots,n}=\sum_{u\in I}\Phi_0(t,\boldsymbol{x},\boldsymbol{\Delta}_u)\boldsymbol{\Delta}_u \quad (4.71)$$

and diffusion matrix

$$\tilde{\boldsymbol{\Sigma}}(\boldsymbol{x},t)=\left(\sum_{v=1}^{d}N_v^{-1}\tilde{a}_{2,(j,k),v}(t,\boldsymbol{x})\right)_{j,k=1,\ldots,n}=\sum_{v=1}^{d}N_v^{-1}\sum_{u\in I_v}\Phi_0(t,\boldsymbol{x},\boldsymbol{\Delta}_u)\boldsymbol{\Delta}_u\boldsymbol{\Delta}_u'. \quad (4.72)$$

This result holds under the assumption of the existence of a canonical form (4.65), condition (4.70) and further rather weak regularity conditions as in Sect. 4.3.5.

4.5 Choice of Stochastic Integral

Sections 4.3 and 4.4 introduced several techniques for the approximation of Markov jump processes by solutions of stochastic differential equations. An immediate issue in this context is the question of appropriateness of different stochastic calculi with the Itô and Stratonovich calculus as their most prominent representatives, see Sect. 3.2.2. As a rule of thumb, one usually chooses the Itô interpretation as an appropriate model if the random force is assumed to be exactly Gaussian white noise. If the white noise process is only an idealisation, the Stratonovich representation should be employed (Arnold 1973, Chap. 10.3).

In applications in life sciences, the memory of a system is usually short but nonzero. In those cases the noise is called *coloured*, and the Stratonovich interpretation is the suitable choice of integral. On the other hand, in some models the underlying dynamics might be best described in discrete time with discrete but uncorrelated noise forces, for example in population dynamics if successive generations do not overlap in time. The white noise in the continuous model can then be considered as exact, and the Itô calculus applies (Horsthemke and Lefever 1984, Chap. 5.4; Kloeden and Platen 1999, Chap. 6.1).

An argument supporting the Stratonovich interpretation is the following *Wong-Zakai theorem* (Wong and Zakai 1965): Let $\boldsymbol{B}^{(n)}=(\boldsymbol{B}_t^{(n)})_{t\geq t_0}$, $n\in\mathbb{N}$, be a sequence of processes with continuous state space, bounded variation, piecewise continuous derivatives and Brownian motion as almost surely uniform limit as n tends to infinity. Then the solutions of the random differential equations

$$\mathrm{d}\boldsymbol{X}_t^{(n)}=\boldsymbol{\mu}\big(\boldsymbol{X}_t^{(n)},t\big)\mathrm{d}t+\boldsymbol{\sigma}\big(\boldsymbol{X}_t^{(n)},t\big)\mathrm{d}\boldsymbol{B}_t^{(n)}\quad,\ \boldsymbol{X}_{t_0}^{(n)}=\boldsymbol{x}_0,$$

converge sample-pathwise uniformly to the solution of the Stratonovich SDE with (sufficiently regular) drift $\boldsymbol{\mu}$, diffusion coefficient $\boldsymbol{\sigma}$ and initial value $\boldsymbol{x}_0$. In the context of approximating a given process by the solution of a stochastic differential equation, the Stratonovich interpretation hence seems more natural.

In the present chapter, the choice of calculus is superfluous as the interpretation is already fixed by construction: The approaches in Sects. 4.3.1, 4.3.2, 4.3.4 and 4.3.5 lead to forward or backward diffusion equations of the Itô type; compare with Sect. 3.2.8. In Sect. 4.3.3, the interpretation is determined by the assumption that the process satisfies an Itô SDE. The same holds for the extended methods in Sect. 4.4.

Howsoever, both the Itô and the Stratonovich interpretations are mathematically correct. Processes of these two types generally differ in their drifts but coincide in their random fluctuations; transformation from one to the other is straightforward (see also Sect. 3.2.2). Hence, the true question is not which calculus to follow but how to correctly determine the drift coefficient of the approximating process. Braumann (2007) illustrates this on the example of modelling random population growth. Results should be evaluated by comparison of analytical insight with experimental data. See van Kampen (1981a) for further discussion.

4.6 Discussion and Conclusion

This chapter motivates and explains the approximate representation of pure Markov jump processes by ordinary or stochastic differential equations: A jump process $\boldsymbol{X}$ occurs whenever numbers of countable objects are observed, which is frequently the case in life sciences applications such as genetics, systems biology, population dynamics or physics. Suppose the size of a system can satisfyingly be described by a single parameter N. If N is comparatively large, a state-continuous approximation for the evolution of the intensive process $\boldsymbol{x} = \boldsymbol{X}/N$ seems appropriate. Section 4.2 lists the benefits arising from such an approximation in detail. The model then changes its characteristics as follows: In the original discrete state space model, the probability for the process to stay in a given state during a short time interval of length Δt tends to one as Δt approaches zero. In the continuous state space approximation, on the other hand, this probability tends to zero.

Section 4.1 reviews the characteristics of jump processes, diffusions and deterministic processes with continuous sample paths, as the latter two are the solutions to the approximating differential equations. The three types of processes correspond to models on a microscopic, mesoscopic and macroscopic level, respectively. The macroscopic features of a process are determined by the average of all particles of the system. The mesoscopic description additionally takes into account internal fluctuations which are caused by the discrete nature of matter. These are small when the system is large. A microscopic model is exact but usually too expensive to work with except for small systems. Indisputably, the stochastic (mesoscopic) approximation is more realistic than the deterministic (macroscopic) one; the chapter hence concentrates on the derivation of approximating diffusion processes. A deterministic model is then again an approximation of the stochastic one.

The concrete derivation of such a diffusion model, however, is complicated especially for nonlinear fluctuations and has caused substantial confusion in the literature as authors obtained different, but all plausible, results for identical problems. The reason is that nonlinear processes cannot exactly be described by second order differential equations for their transition densities, i.e. the description by Kolmogorov equations is generally not free from error (van Kampen 1965). Under relatively mild regularity conditions, however, approximate descriptions of jump processes by diffusions are possible, and Sect. 4.3 introduces several approaches to obtain these. The framework is kept heuristic in order to achieve comprehensibility also for practitioners. The reader interested in more mathematical detail is referred to the according references in Sect. 4.2. In all approaches, one arrives for the intensive process $\boldsymbol{x}$ at a stochastic differential equation

$$\mathrm{d}\boldsymbol{x}_t = \boldsymbol{\mu}(\boldsymbol{x}_t, t)\mathrm{d}t + \frac{1}{\sqrt{N}}\boldsymbol{\sigma}(\boldsymbol{x}_t, t)\mathrm{d}\boldsymbol{B}_t\,, \quad \boldsymbol{x}_{t_0} = \boldsymbol{x}_0, \tag{4.73}$$

with some drift vector $\boldsymbol{\mu}$ and diffusion matrix $N^{-1}\boldsymbol{\Sigma} = N^{-1}\boldsymbol{\sigma}\boldsymbol{\sigma}'$. The Itô lemma implies

$$\mathrm{d}\boldsymbol{X}_t = N\mathrm{d}\boldsymbol{x}_t\,, \quad \boldsymbol{X}_{t_0} = N\boldsymbol{x}_{t_0},$$

in accordance with the results of the single algorithms when applied directly to the extensive process $\boldsymbol{X}$. The diffusion matrix of (4.73) scales with the inverse system size, losing ground in large systems. As mentioned before, Eq. (4.73) is a diffusion approximation rather than a diffusion limit. The latter corresponds to a deterministic model and can be obtained by ignoring the stochastic part of (4.73).

The results of the approaches in Sect. 4.3 are as follows: In the rearrangement of the master equation (Sect. 4.3.1) and of the infinitesimal generator (Sect. 4.3.2), the drift and diffusion matrix are assembled as sums of limits of difference quotients; explicit formulas for $\boldsymbol{\mu}$ and $\boldsymbol{\sigma}$ are not available except for special cases as in Example B.2 in the Appendix. However, Algorithm B.1 describes their derivation for the general case. The two approximation approaches assume that the transition rate w_N of $\boldsymbol{x}$ fulfils

$$w_N(t, \boldsymbol{x}, \boldsymbol{\Delta}) = Nw(t, \boldsymbol{x}, \boldsymbol{\Delta}) \tag{4.74}$$

for all t, $\boldsymbol{x}$ and $\boldsymbol{\Delta}$ and a function w which does not depend on N. This situation applies, at least approximately, in most examples in life sciences. In the Langevin approach (Sect. 4.3.3) and Kramers-Moyal expansion (Sect. 4.3.4), $\boldsymbol{\mu}$ and $\boldsymbol{\Sigma}$ are obtained as

$$\begin{aligned}\boldsymbol{\mu}(\boldsymbol{x}, t) &= N^{-1}\sum_{\boldsymbol{\Delta}} w_N(t, \boldsymbol{x}, \boldsymbol{\Delta})\boldsymbol{\Delta} \\ \boldsymbol{\Sigma}(\boldsymbol{x}, t) &= N^{-1}\sum_{\boldsymbol{\Delta}} w_N(t, \boldsymbol{x}, \boldsymbol{\Delta})\boldsymbol{\Delta}\boldsymbol{\Delta}',\end{aligned} \tag{4.75}$$

where the sum is over all nonzero jumps $\boldsymbol{\Delta}$ of the extensive jump process $\boldsymbol{X}$. With Eq. (4.74) fulfilled, these simplify to

$$\begin{aligned}\boldsymbol{\mu}(\boldsymbol{x},t) &= \sum_{\boldsymbol{\Delta}} w(t,\boldsymbol{x},\boldsymbol{\Delta})\boldsymbol{\Delta} \\ \boldsymbol{\Sigma}(\boldsymbol{x},t) &= \sum_{\boldsymbol{\Delta}} w(t,\boldsymbol{x},\boldsymbol{\Delta})\boldsymbol{\Delta}\boldsymbol{\Delta}'.\end{aligned}$$

Van Kampen's method (Sect. 4.3.5) replaces condition (4.74) by the less restrictive *canonical form*

$$w_N(t,\boldsymbol{x},\boldsymbol{\Delta}) = f(N)\sum_{l=0}^{\infty} N^{-l}\,\Phi_l\,(t,\boldsymbol{x},\boldsymbol{\Delta}) \tag{4.76}$$

for a positive function f and appropriate functions Φ_l. The expansion results in

$$\begin{aligned}\boldsymbol{\mu}(\boldsymbol{x},t) &= \sum_{\boldsymbol{\Delta}} \Phi_0(t,\boldsymbol{x},\boldsymbol{\Delta})\boldsymbol{\Delta} \\ \boldsymbol{\Sigma}(\boldsymbol{x},t) &= \sum_{\boldsymbol{\Delta}} \Phi_0(t,\boldsymbol{x},\boldsymbol{\Delta})\boldsymbol{\Delta}\boldsymbol{\Delta}'.\end{aligned} \tag{4.77}$$

Certainly, there are similarities in the results: The main difference between the representation of the master equation through difference operators and the Kramers-Moyal expansion lies in when to perform certain critical large N considerations which are possible only in a heuristic framework. For example, in the former method derivatives appear as late as possible, whereas in the Kramers-Moyal expansion they already form the first step. The same parallels apply for the approximation of the infinitesimal generator and a Kramers-Moyal backward expansion as contained in Risken (1984, Chap. 4.2). Example B.2 shows that under certain requirements on the possible jumps the techniques from Sects. 4.3.1–4.3.4 yield identical results.

Furthermore, if Eq. (4.74) is true, then $f(N) = N$, $\Phi_0 = w$ and $\Phi_l \equiv 0$ for $l \geq 1$ in (4.76), i.e. van Kampen's expansion yields the same result as the Langevin approach and the Kramers-Moyal expansion. However, there are cases where the outcomes (4.75) and (4.77) differ: Plugging in the canonical form (4.76) into Formula (4.75) from the Langevin and Kramers-Moyal approach produces

$$\begin{aligned}\boldsymbol{\mu}(\boldsymbol{x},t) &= f(N)N^{-1}\sum_{\boldsymbol{\Delta}}\Big[\Phi_0\,(t,\boldsymbol{x},\boldsymbol{\Delta}) + N^{-1}\Phi_1\,(t,\boldsymbol{x},\boldsymbol{\Delta}) + \ldots\Big]\boldsymbol{\Delta} \\ \boldsymbol{\Sigma}(\boldsymbol{x},t) &= f(N)N^{-1}\sum_{\boldsymbol{\Delta}}\Big[\Phi_0\,(t,\boldsymbol{x},\boldsymbol{\Delta}) + \ldots\Big]\boldsymbol{\Delta}\boldsymbol{\Delta}'.\end{aligned} \tag{4.78}$$

Horsthemke and Brenig (1977) cite an example where (4.77) and (4.78) yield different results: Consider the chemical reactions

$$\text{Ⓐ} + \text{Ⓧ} \xrightarrow{k_1} 2\,\text{Ⓧ}\,, \qquad 2\,\text{Ⓧ} \xrightarrow{k_2} \text{Ⓐ} + \text{Ⓧ}\,, \qquad \text{Ⓑ} + \text{Ⓧ} \xrightarrow{k_3} \text{Ⓒ}$$

with rates k_1, k_2 and k_3, where the numbers of particles of types A and B are kept constant and particles of type C are immediately removed. The de facto transitions are thus

$$\text{(A)} + \text{(X)} \xrightarrow{k_1} \text{(A)} + 2\,\text{(X)}\,, \qquad 2\,\text{(X)} \xrightarrow{k_2} \text{(X)}\,, \qquad \text{(B)} + \text{(X)} \xrightarrow{k_3} \text{(B)}\,.$$

Denote by a and b the fractions of type A and B particles in the system of size N. The resulting forward diffusion equations for the transition densities $p(t,x)$ for fractions x of type X particles at time t are

$$\frac{\partial p(t,x)}{\partial t} = -\frac{\partial}{\partial x}\big(k_1 a - k_2 x - k_3 b\big)x\,p(t,x) + \frac{1}{2N}\frac{\partial^2}{\partial x^2}\big(k_1 a + k_2 x + k_3 b\big)x\,p(t,x)$$

according to van Kampen, Formula (4.77), and

$$\begin{aligned}\frac{\partial p(t,x)}{\partial t} &= -\frac{\partial}{\partial x}\big(k_1 a - k_2 x - k_3 b + \frac{k_2}{N}\big)x\,p(t,x) \\ &\quad + \frac{1}{2N}\frac{\partial^2}{\partial x^2}\big(k_1 a + k_2 x + k_3 b\big)x\,p(t,x)\end{aligned} \tag{4.79}$$

due to Langevin and Kramers-Moyal, Formula (4.78) (and, by the way, also as a result of the approximation procedures from Sects. 4.3.1 and 4.3.2). The reason for this deviation is that the Langevin approach models the fluctuating part of x, whilst van Kampen considers the fluctuations around the deterministic solution $\phi(t)$. The above example illustrates that the Formulas (4.75) and (4.77) shall not be applied uncritically: Detailed analysis of the diffusion equation (4.79) shows that its only stationary solution is $p(t,x) = \delta(x)$ with the Dirac delta function δ; the state $x = 0$ is an absorbing boundary that is reached in finite time. This result agrees with the master equation description. Improvident transition to stochastic differential equations, on the other hand, erroneously suggests fluctuations around the nonzero stationary state of the macroscopic equation

$$\mathrm{d}x_t = \big(k_1 a - k_2 x_t - k_3 b\big)x_t \mathrm{d}t.$$

In fact, the regularity assumptions for van Kampen's expansion, which were mentioned on p. 80, are not fulfilled in the above example, i.e. the method is not applicable. All other approaches have to be applied with care as well. See Horsthemke and Brenig (1977) for a more detailed discussion and further examples. Section 4.3.6 covers some methods which are to be favoured if one wishes to determine quantities that sensitively depend on the equilibrium properties. Gitterman and Weiss (1991) however emphasise that no technique can reproduce *all* characteristics of the original model.

In the usual case, where the methods from Sects. 4.3.1–4.3.5 are applicable, all of them are asymptotically equivalent. Differences between the Kramers-Moyal expansion (4.36) and van Kampen's expansion occur only when higher moments are included in the model; see van Kampen (1997, Chap. 10.6)) for the univariate van Kampen expansion including higher moments. The representations of the master equation or infinitesimal generator through difference operators seem appealing if the forward or backward master equation are given and one does not want to reproduce the single transition rates W. Otherwise, the Langevin, Kramers-Moyal and van Kampen approach provide immediate formulas for the drift and diffusion matrix and are hence more convenient and widely used.

In some applications, the limitation to a single size parameter N does not suffice to completely characterise the dynamics of a system. Instead, multiple size parameters $N_1, \ldots, N_d$ are applied. As a new result, Sect. 4.4 performs the adjustment of the methods from Sects. 4.3.1–4.3.5 to the advanced setting.

Immediately involved with the application of stochastic differential equations is the choice of stochastic integral. Section 4.5 discusses this matter with the plain conclusion that the application of both the Itô and the Stratonovich interpretation is correct as long as an analysis of the SDE follows the same calculus as the approximation procedure.

In any case, the validity of a diffusion approximation should always be judged by comparison of numerical results from the master equation and the diffusion approximation model or, if available, by comparison of numerical and analytical characteristics. This has been done by Ewens (1963), Gillespie (1980), Hayot and Jayaprakash (2004), Ferm et al. (2008) and Sjöberg et al. (2009). A comparison between stochastic and deterministic models has been performed by Nåsell (2002). See Grasman and Ludwig (1983) for an investigation of the accuracy of diffusion approximations.

Diffusion approximations are not always possible; requirements to the original model are sketched at the beginning of Sect. 4.3. See Pollett (2001) for comments on cases in which diffusion models are inappropriate as certain assumptions are not fulfilled. In any circumstances, such approximations are only legitimate for large system sizes. For medium sized systems that are too small for diffusion approximations but too large for Monte Carlo evaluation of the master equation, different methods are proposed; see for example Ohkubo (2008) and the references therein.

To summarise, this chapter offers a survey of methods to model a pure Markov jump process by a diffusion approximation. It supplements the variety of known approaches by new formulations and fills the gap of general multi-dimensional formulas, which partly do not appear in the existing literature. Furthermore, all approximation techniques are extended to systems with multiple size parameters. Assumptions and derivations are provided for all approaches to allow for critical evaluation. Various references guide the reader to more detailed information. In all, this chapter allows scientists with a moderate mathematical background to easily apply diffusion approximation methods to a broad class of jump processes in order to gain full advantages from that modelling approach.

References

Allen L (2003) An introduction to stochastic processes with applications to biology. Pearson Prentice Hall, Upper Saddle River
Alonso D, McKane A, Pascual M (2007) Stochastic amplification in epidemics. J R Soc Interface 4:575–582
Anderson R, May R (1991) Infectious diseases of humans. Oxford University Press, Oxford
Andersson H, Britton T (2000) Stochastic epidemic models and their statistical analysis. Lecture notes in statistics, vol 151. Springer, New York
Arnold L (1973) Stochastische Differentialgleichungen. Oldenbourg, München
Bailey N (1975) The mathematical theory of infectious diseases, 2nd edn. Charles Griffin, London
Barbour A (1972) The principle of the diffusion of arbitrary constants. J Appl Probab 9:519–541
Barbour A (1974) On a functional central limit theorem for Markov population processes. Adv Appl Probab 6:21–39
Barbour A (1975a) The asymptotic behaviour of birth and death and some related processes. Adv Appl Probab 7:28–43
Barbour A (1975b) The duration of a closed stochastic epidemic. Biometrika 62:477–482
Barbour A (1975c) A note on the maximum size of a closed epidemic. J R Stat Soc Ser B 37:459–460
Billingsley P (1968) Convergence of probability measures. Wiley, New York
Bouchaud JP, Cont R (1998) A Langevin approach to stock market fluctuations and crashes. Eur Phys J B 6:543–550
Braumann C (2007) Itô versus Stratonovich calculus in random population growth. Math Biosci 206:81–107
Busenberg S, Martelli M (1990) Differential equations models in biology, epidemiology and ecology. Lecture notes in biomathematics, vol 92. Springer, Berlin
Capasso V, Morale D (2009) Stochastic modelling of tumour-induced angiogenesis. J Math Biol 58:219–233
Chaturvedi S, Gardiner C (1978) The Poisson representation. II. Two-time correlation functions. J Stat Phys 18:501–522
Chen WY, Bokka S (2005) Stochastic modeling of nonlinear epidemiology. J Theor Biol 234:455–470
Clancy D, French N (2001) A stochastic model for disease transmission in a managed herd, motivated by Neospora caninum amongst dairy cattle. Math Biosci 170:113–132
Clancy D, O'Neill P, Pollett P (2001) Approximations for the long-term behavior of an open-population epidemic model. Methodol Comput Appl Probab 3:75–95
Daley D, Gani J (1999) Epidemic modelling: an introduction. Cambridge studies in mathematical biology, vol 15. Cambridge University Press, Cambridge
Daley D, Kendall D (1965) Stochastic rumours. IMA J Appl Math 1:42–55
Daniels H (1974) The maximum size of a closed epidemic. Adv Appl Probab 6:607–621
de la Lama M, Szendro I, Iglesias J, Wio H (2006) Van Kampen's expansion approach in an opinion formation model. Eur Phys J B 51:435–442
Drummond P, Gardiner C, Walls D (1981) Quasiprobability methods for nonlinear chemical and optical systems. Phys Rev A 24:914–926
Eigen M (1971) Selforganization of matter and the evolution of biological macromolecules. Naturwissenschaften 58:465–523
Elf J, Ehrenberg M (2003) Fast evaluation of fluctuations in biochemical networks with the linear noise approximation. Genome Res 13:2475–2484
Ethier S, Kurtz T (1986) Markov processes. Characterization and convergence. Wiley, New York
Ewens W (1963) Numerical results and diffusion approximations in a genetic process. Biometrika 50:241–249

Feller W (1951) Diffusion processes in genetics. In: Proceedings of the second Berkeley symposium on mathematical statistics and probability. University of California Press, Berkeley, pp 227–246

Ferm L, Lötstedt P, Hellander A (2008) A hierarchy of approximations of the master equation scaled by a size parameter. J Sci Comput 34:127–151

Gardiner C (1983) Handbook of stochastic methods. Springer, Berlin/Heidelberg

Gardiner C, Chaturvedi S (1977) The Poisson representation. I. A new technique for chemical master equations. J Stat Phys 17:429–468

Gibson M, Mjolsness E (2001) Modeling of the activity of single genes. In: Bolouri H, Bower J (eds) Computational modeling of genetic and biochemical networks. Lecture notes in computer science, vol 4699. MIT, Cambridge, pp 1–48

Gillespie D (1976) A general method for numerically simulating the stochastic time evolution of coupled chemical reactions. J Comput Phys 22:403–434

Gillespie D (1977) Exact stochastic simulation of coupled chemical reactions. J Phys Chem 81:2340–2361

Gillespie D (1980) Approximating the master equation by Fokker-Planck-type equations for single-variable chemical systems. J Chem Phys 72:5363–5370

Gitterman M, Weiss G (1991) Some comments on approximations to the master equation. Phys A 170:503–510

Goel N, Richter-Dyn N (1974) Stochastic models in biology. Academic, New York

Golightly A, Wilkinson D (2005) Bayesian inference for stochastic kinetic models using a diffusion approximation. Biometrics 61:781–788

Golightly A, Wilkinson D (2006) Bayesian sequential inference for stochastic kinetic biochemical network models. J Comput Biol 13:838–851

Golightly A, Wilkinson D (2008) Bayesian inference for nonlinear multivariate diffusion models observed with error. Comput Stat Data Anal 52:1674–1693

Golightly A, Wilkinson D (2010) Markov chain Monte Carlo algorithms for SDE parameter estimation. In: Lawrence N, Girolami M, Rattray M, Sanguinetti G (eds) Introduction to learning and inference for computational systems biology. MIT, Cambridge, pp 253–275

Grabert H, Green M (1979) Fluctuations and nonlinear irreversible processes. Phys Rev A 19:1747–1756

Grabert H, Graham R, Green M (1980) Fluctuations and nonlinear irreversible processes II. Phys Rev A 21:2136–2146

Grabert H, Hänggi P, Oppenheim I (1983) Fluctuations in reversible chemical reactions. Phys A 117:300–316

Grasman J, Ludwig D (1983) The accuracy of the diffusion approximation to the expected time to extinction for some discrete stochastic processes. J Appl Probab 20:305–321

Green M (1952) Markoff random processes and the statistical mechanics of time-dependent phenomena. J Chem Phys 20:1281–1295

Guess H, Gillespie J (1977) Diffusion approximations to linear stochastic difference equations with stationary coefficients. J Appl Probab 14:58–74

Hänggi P (1982) Nonlinear fluctuations: the problem of deterministic limit and reconstruction of stochastic dynamics. Phys Rev A 25:1130–1136

Hänggi P, Jung P (1988) Bistability in active circuits: application of a novel Fokker-Planck approach. IBM J Res Dev 32:119–126

Hänggi P, Grabert H, Talkner P, Thomas H (1984) Bistable systems: master equation versus Fokker-Planck modeling. Phys Rev A 29:371–378

Haskey H (1954) A general expression for the mean in a simple stochastic epidemic. Biometrika 41:272–275

Hayot F, Jayaprakash C (2004) The linear noise approximation for molecular fluctuations within cells. Phys Biol 1:205–210

Horsthemke W, Brenig L (1977) Non-linear Fokker-Planck equation as an asymptotic representation of the master equation. Z Phys B 27:341–348

Horsthemke W, Lefever R (1984) Noise-induced transitions: theory and applications in physics, chemistry, and biology. Springer, Berlin

Hsu JP, Wang HH (1987) Kinetics of bacterial adhesion – a stochastic analysis. J Theor Biol 124:405–413

Hufnagel L, Brockmann D, Geisel T (2004) Forecast and control of epidemics in a globalized world. Proc Natl Acad Sci USA 101:15124–15129

Karth M, Peinke J (2003) Stochastic modeling of fat-tailed probabilities of foreign exchange rates. Complexity 8:34–42

Keeling M, Rohani P (2008) Modeling infectious disease in humans and animals. Princeton University Press, Princeton

Kepler T, Elston T (2001) Stochasticity in transcriptional regulation: origins, consequences, and mathematical representations. Biophys J 81:3116–3136

Kishida K, Kanemoto S, Sekiya T (1976) Reactor noise theory based on system size expansion. J Nucl Sci Technol 13:19–29

Kleinhans D, Friedrich R, Nawroth A, Peinke J (2005) An iterative procedure for the estimation of drift and diffusion coefficients of Langevin processes. Phys Lett A 346:42–46

Kloeden P, Platen E (1999) Numerical solution of stochastic differential equations, 3rd edn. Springer, Berlin/Heidelberg/New York

Kramers H (1940) Brownian motion in a field of force and the diffusion model of chemical reactions. Physica 7:284–304

Kubo R, Matsuo K, Kitahara K (1973) Fluctuation and relaxation of macrovariables. J Stat Phys 9:51–96

Kurtz T (1970) Solutions of ordinary differential equations as limits of pure jump Markov processes. J Appl Probab 7:49–58

Kurtz T (1971) Limit theorems for sequences of jump Markov processes approximating ordinary differential processes. J Appl Probab 8:344–356

Kurtz T (1981) Approximation of population processes. Society for Industrial and Applied Mathematics, Philadelphia

Lande R, Engen S, Sæther B (2003) Stochastic population dynamics in ecology and conservation. Oxford University Press, New York

Leung H (1985) Expansion of the master equation for a biomolecular selection model. Bull Math Biol 47:231–238

McKane AJ, Newman T (2004) Stochastic models in population biology and their deterministic analogs. Phys Rev E 70:041902

McNeil D (1973) Diffusion limits for congestion models. J Appl Probab 10:368–376

McQuarrie D (1967) Stochastic approach to chemical kinetics. J Appl Probab 4:413–478

Moyal J (1949) Stochastic processes and statistical physics. J R Stat Soc Ser B 11:150–210

Muñoz M, Garrido P (1994) Fokker-Planck equation for nonequilibrium competing dynamic models. Phys Rev E 50:2458–2466

Naert A, Friedrich R, Peinke J (1997) Fokker-Planck equation for the energy cascade in turbulence. Phys Rev E 56:6719–6722

Nåsell I (2002) Stochastic models of some endemic infections. Math Biosci 179:1–19

Norman M (1974) A central limit theorem for Markov processes that move by small steps. Ann Probab 2:1065–1074

Norman M (1975) Diffusion approximation of non-Markovian processes. Ann Probab 3:358–364

Ohkubo J (2008) Approximation scheme for master equations: variational approach to multivariate case. J Chem Phys 129:044108

Paulsson J (2004) Summing up the noise in gene networks. Nature 427:415–418

Pawula R (1967a) Approximation of the linear Boltzmann equation by the Fokker-Planck equation. Phys Rev 162:186–188

Pawula R (1967b) Generalizations and extensions of the Fokker-Planck-Kolmogorov equations. IEEE Trans Inf Theory 13:33–41

Pielou (1969) An introduction to mathematical ecology. Wiley, New York

Pierobon P, Parmeggiani A, von Oppen F, Frey E (2005) Dynamic correlation functions and Boltzmann Langevin approach for a driven one dimensional lattice gas. Phys Rev E 72:036123

Pollard D (1984) Convergence of stochastic processes. Springer, New York

Pollett P (1990) On a model for interference between searching insect parasites. J Aust Math Soc Ser B 32:133–150

Pollett P (2001) Diffusion approximations for ecological models. Proceedings of the international congress of modelling and simulation, Australian National University, Canberra

Ramshaw J (1985) Augmented Langevin approach to fluctuations in nonlinear irreversible processes. J Stat Phys 38:669–680

Rao C, Wolf D, Arkin A (2002) Control, exploitation and tolerance of intracellular noise. Nature 420:231–237

Renshaw E (1991) Modelling biological populations in space and time. Cambridge University Press, Cambridge

Risken H (1984) The Fokker-Planck equation. Springer, Berlin

Risken H, Vollmer H (1987) On solutions of truncated Kramers-Moyal expansions; continuum approximations to the Poisson process. Condens Matter 66:257–262

Robertson S, Pilling M, Green N (1996) Diffusion approximations of the two-dimensional master equation. Mol Phys 88:1541–1561

Sancho J, San Miguel M (1984) Unified theory of internal and external fluctuations. In: Casas-Vázquez J, Jou D, Lebon G (eds) Recent developments in nonequilibrium thermodynamics. Lecture notes in physics, vol 199. Springer, Berlin, pp 337–352

Seifert U (2008) Stochastic thermodynamics: principles and perspectives. Eur Phys J B 64:423–431

Shizgal B, Barrett J (1989) Time dependent nucleation. J Chem Phys 91:6505–6518

Sjöberg P, Lötstedt P, Elf J (2009) Fokker-Planck approximation of the master equation in molecular biology. Comput Vis Sci 12:37–50

Song X, Wang H, van Voorhis T (2008) A Langevin equation approach to electron transfer reactions in the diabatic basis. J Chem Phys 129:144502

Strumik M, Macek W (2008) Statistical analysis of transfer of fluctuations in solar wind turbulence. Nonlinear Proc Geophys 15:607–613

Tian T, Burrage K, Burrage P, Carletti M (2007) Stochastic delay differential equations for genetic regulatory networks. J Comput Appl Math 205:696–707

van Kampen N (1961) A power series expansion of the master equation. Can J Phys 39:551–567

van Kampen N (1965) Fluctuations in nonlinear systems. In: Burgess R (ed) Fluctuation phenomena in solids. Academic, New York, pp 139–177

van Kampen N (1981a) Itô versus Stratonovich. J Stat Phys 24:175–187

van Kampen N (1981b) The validity of nonlinear Langevin equations. J Stat Phys 25:431–442

van Kampen N (1997) Stochastic processes in physics and chemistry, 2nd edn. Elsevier, Amsterdam

Walsh J (1981) Well-timed diffusion approximations. Adv Appl Probab 13:352–368

Wong E, Zakai M (1965) On the convergence of ordinary integrals to stochastic integrals. Ann Math Stat 36:1560–1564

Chapter 5
Diffusion Models in Life Sciences

This chapter investigates representative models from life sciences which typically involve large populations. These models are in a first step formulated in terms of pure Markov jump processes. However, as motivated in Sect. 4.2, a more convenient representation is obtained by the transition to diffusion approximations. This facilitates simulation and statistical inference. Hence, in a second step, the jump processes are approximated by diffusions. The purpose of this chapter is on the one hand to illustrate the methods from Chap. 4. On the other hand, the presented models and their diffusion approximations are the basis for Chap. 8, where Bayesian inference is performed on them.

The considered models are from the field of epidemiology, which represents one important branch of life sciences. In particular, Sect. 5.1 covers the standard susceptible–infected–removed (SIR) model from Sect. 2.2.2 for describing the spread of infectious diseases. Section 5.2 proposes an extension of this standard model in order to allow for host heterogeneity. Further diffusion approximations are derived in Chap. 9, where the binding behaviour of proteins is investigated in living cells.

In both Sects. 5.1 and 5.2, the respective model is first introduced via a compartmental representation and then described in terms of a Markov jump process. Afterwards, the approximation approaches from Chap. 4 are applied. If one is only interested in the resulting diffusion approximations rather than in the approximation procedures, it is sufficient to only read the then following summaries. Each section concludes with some illustration of the respective model. Section 5.3 investigates the existence and uniqueness of solutions of the stochastic differential equations derived in Sects. 5.1 and 5.2. Section 5.4 concludes this chapter.

It has already been discovered in Sect. 4.6 that under certain conditions the results of different approximation procedures coincide. This is actually the case also for the models considered here; it is hence redundant to apply more than one approximation method. However, in order to provide examples for the theoretical investigations from Chap. 4, each approach is considered separately.

C. Fuchs, *Inference for Diffusion Processes: With Applications in Life Sciences*,
DOI 10.1007/978-3-642-25969-2_5,

The asymptotic behaviour of the standard SIR model as the population size tends to infinity has been treated by several authors, e.g. by Nagaev and Startsev (1970), Barbour (1974), Wang (1977), Kurtz (1981), Andersson and Britton (2000) and Allen (2003). The case of open populations is investigated by Clancy et al. (2001). Moreover, Alonso et al. (2007) take into account demographic changes.

Similar multitype SIR models have been considered in the literature as well, for instance by Ball (1986), Bailey (1975), Daley and Gani (1999) and Andersson and Britton (2000). In most cases, but not exclusively, these are formulated in terms of deterministic processes. Diffusion processes for non-standard SIR models have been treated by, for example, Hufnagel et al. (2004), Sani et al. (2007) and McCormack and Allen (2006). These models are however different from the one considered in this chapter.

5.1 Standard SIR Model

The following considerations introduce the standard SIR model in Sect. 5.1.1, characterise it as a jump process through its master equation in Sect. 5.1.2 and describe its approximation through a diffusion process in Sect. 5.1.3. The results are summarised in Sect. 5.1.4, and the diffusion process is illustrated in Sect. 5.1.5.

5.1.1 Model

Consider a population of size N in which individuals are either susceptible to a disease, infected, or removed. The population is assumed to be closed, i.e. the size parameter N remains fixed, ignoring demographical changes that are not related to the epidemic.

Recall from Sect. 2.2.2 the standard SIR model with the following transitions:

1. The contact between a susceptible and an infectious individual causes an infection:
$$\textcircled{\scriptsize S} + \textcircled{\scriptsize I} \xrightarrow{\alpha} 2\,\textcircled{\scriptsize I}, \tag{5.1}$$
where $\alpha \in \mathbb{R}_+$ is the contact number of an infectious individual sufficient to spread the disease.
2. An infective recovers:
$$\textcircled{\scriptsize I} \xrightarrow{\beta} \textcircled{\scriptsize R}, \tag{5.2}$$
where $\beta \in \mathbb{R}_+$ is the reciprocal average infectious period.

The variables above the arrows indicate which model parameters enter the probability for the respective event to occur. As explained in Sect. 2.2.2, an appropriate state space for a process following these transitions is

$$\mathcal{D}_N = \{(S,I)' \in [0,N]^2 \cap \mathbb{N}_0^2 \,|\, S+I \leq N\}.$$

The vector of model parameters is $\boldsymbol{\theta} = (\alpha, \beta)'$.

This model has been widely adopted in infectious disease modelling due to its simplicity and generality (see e.g. Keeling and Rohani 2008, for a monograph). However, a central assumption in this formulation is that the population mixes homogeneously. Surely, this situation is not given in many applications, for example when one considers the nationwide or even worldwide spread of a disease. The multitype SIR model in Sect. 5.2 corrects for this.

5.1.2 Jump Process

There are two possible nonzero jumps of the Markov process following the transitions (5.1) and (5.2) that can occur within an infinitesimal time interval. These are

$$\boldsymbol{\Delta}_1 = \begin{pmatrix} -1 \\ 1 \end{pmatrix} \quad \text{for an infection and} \quad \boldsymbol{\Delta}_2 = \begin{pmatrix} 0 \\ -1 \end{pmatrix} \quad \text{for a recovery.}$$

Throughout this section, let S and I denote the absolute numbers of susceptible and infective individuals. The process is considered time-homogeneous, and all individuals are assumed to be mutually independent. Given the current state $\boldsymbol{X} = (S,I)' \in \mathcal{D}_N$, the probability $P_N(\Delta t, \boldsymbol{X}, \boldsymbol{X} + \boldsymbol{\Delta}_1)$ for an infection to happen within time Δt is as follows: Each of the I infectives has α potentially infectious contacts per time unit. On average, $\alpha \cdot S/N$ of these contacts will be with a susceptible individual and actually cause an infection.[1] The probability of an infective contact in the considered time interval is hence $I \cdot \alpha S/N \cdot \Delta t + o(\Delta t)$, where $o(\Delta t)/\Delta t \to 0$ as $\Delta t \to 0$. Similarly, the probability $P_N(\Delta t, \boldsymbol{X}, \boldsymbol{X} + \boldsymbol{\Delta}_2)$ of a recovery is $\beta I \Delta t + o(\Delta t)$. The transition rates

$$W_{N,j}(\boldsymbol{X}) = W_N(\boldsymbol{X}, \boldsymbol{\Delta}_j) = \lim_{\Delta t \downarrow 0} \frac{1}{\Delta t} P_N(\Delta t, \boldsymbol{X}, \boldsymbol{X} + \boldsymbol{\Delta}_j)$$

are thus

$$W_{N,j}(\boldsymbol{X}) = W_{N,j}(S,I) = \begin{cases} \dfrac{\alpha}{N} SI & \text{if } j = 1, \\ \beta I & \text{if } j = 2. \end{cases}$$

[1] More precisely, the number is $\alpha \cdot S/(N-1)$ as self-infections are excluded. However, this difference is compensated by adequate choice of α and marginal for large N anyway. Division by N instead of $N-1$ is the standard notation.

Let $P_N(t, \boldsymbol{X}) = P_N(t; S, I)$ denote the probability that within time t the extensive process arrives at state $\boldsymbol{X} = (S, I)' \in \mathcal{D}_N$ (subject to some initial condition). Outside the state space, this probability is assumed zero. The master equation (4.8) of the jump process $\boldsymbol{X}$ is then given by

$$\begin{aligned}\frac{\partial P_N(t; S, I)}{\partial t} &= \frac{\alpha}{N}(S+1)(I-1)P_N(t; S+1, I-1) \\ &\quad + \beta(I+1)P_N(t; S, I+1) - \Big(\frac{\alpha}{N} SI + \beta I\Big) P_N(t; S, I).\end{aligned}$$

In terms of the intensive variable $\boldsymbol{x} = \boldsymbol{X}/N = (s, i)' \in \mathcal{C}_N = N^{-1}\mathcal{D}_N$, the transition rates read

$$w_N(\boldsymbol{x}, \boldsymbol{\Delta}_j) = w_{N,j}(\boldsymbol{x}) = w_{N,j}(s, i) = W_{N,j}(Ns, Ni) = \begin{cases} N\alpha si & \text{if } j = 1, \\ N\beta i & \text{if } j = 2, \end{cases} \tag{5.3}$$

i.e. one has $w_N = Nw$, where

$$w(\boldsymbol{x}, \boldsymbol{\Delta}_j) = w_j(\boldsymbol{x}) = w_j(s, i) = \begin{cases} \alpha si & \text{if } j = 1, \\ \beta i & \text{if } j = 2. \end{cases} \tag{5.4}$$

5.1.3 Diffusion Approximation

This section now applies the diffusion approximation methods from Chap. 4 to the standard SIR model. As the size of the system is completely characterised by the single parameter N, the appropriate techniques are those from Sect. 4.3. As indicated earlier, all approximation approaches yield the same diffusion process for the model considered here. It is hence sufficient to restrain this section to the application of one single approximation method. As an illustration for the theoretical derivations in Chap. 4, however, *all* techniques are applied here. The reader who is rather interested in the results than in the procedures can skip this section and continue with the summary in Sect. 5.1.4.

Convergence of the Master Equation

The first approximation approach to look at is the representation of the master equation through a collection of difference quotients (cf. Sect. 4.3.1). The standard SIR model is actually already covered by Example B.2 in Sect. B.1 in the Appendix. Nevertheless, the derivation is repeated here for illustration purposes.

Let $p_N(t; s, i) = p_N(t, \boldsymbol{x}) = P_N(t; N\boldsymbol{x})$ be the probability that the intensive process is in state $\boldsymbol{x} = (s, i)'$ after time t with respect to a fixed predefined initial state and initial time. For $\varepsilon = N^{-1}$, the forward master equation (4.18) from p. 67 for this process reads

$$\frac{\partial p_N(t,\boldsymbol{x})}{\partial t} = \frac{w_1(\boldsymbol{x}-\varepsilon\boldsymbol{\Delta}_1)p_N(t,\boldsymbol{x}-\varepsilon\boldsymbol{\Delta}_1) - w_1(\boldsymbol{x})p_N(t,\boldsymbol{x})}{\varepsilon} + \frac{w_2(\boldsymbol{x}-\varepsilon\boldsymbol{\Delta}_2)p_N(t,\boldsymbol{x}-\varepsilon\boldsymbol{\Delta}_2) - w_2(\boldsymbol{x})p_N(t,\boldsymbol{x})}{\varepsilon}.$$

Transferring this cohesion to the limit function p of p_N yields

$$\frac{\partial p(t;s,i)}{\partial t} = \frac{w_1(s+\varepsilon,i-\varepsilon)p(t;s+\varepsilon,i-\varepsilon) - w_1(s,i)p(t;s,i)}{\varepsilon} \tag{5.5}$$

$$+\frac{w_2(s,i+\varepsilon)p(t;s,i+\varepsilon) - w_2(s,i)p(t;s,i)}{\varepsilon}. \tag{5.6}$$

We seek to write the right hand side of this equation in terms of difference quotients in order to be able to approximate it in terms of derivatives of the respective functions. The master equation then attains the form of a Kolmogorov equation. The numerators of (5.5) and (5.6) are not yet of the required difference form as stated in Definition B.1 in Sect. B.1. However, Algorithm B.1 describes how they can neatly be expanded. For the numerator of the first term (5.5), this is as follows:

$$\begin{aligned}
&(w_1\cdot p)(t;s+\varepsilon,i-\varepsilon) - (w_1\cdot p)(t;s,i)\\
&\quad= (w_1\cdot p)(t;s+\varepsilon,i-\varepsilon) - (w_1\cdot p)(t;s+\varepsilon,i) - (w_1\cdot p)(t;s,i-\varepsilon)\\
&\quad+ (w_1\cdot p)(t;s,i) + (w_1\cdot p)(t;s+\varepsilon,i) - (w_1\cdot p)(t;s,i)\\
&\quad+ (w_1\cdot p)(t;s,i-\varepsilon) - (w_1\cdot p)(t;s,i),
\end{aligned} \tag{5.7}$$

where the notation $(w_1\cdot p)(t;s,i)$ is short for $w_1(s,i)p(t;s,i)$. With the difference operator notation from Definition B.1, (5.7) can be expressed as

$$\left(D^2_{(1,1)',(\varepsilon,-\varepsilon)'} + D^1_{(1,0)',(\varepsilon,\cdot)'} + D^1_{(0,1)',(\cdot,-\varepsilon)'}\right)(w_1\cdot p)(t;s,i). \tag{5.8}$$

Again, the dot in the subscript means that the respective component does not have to be specified. Following the remarks from Example B.1 and especially Eq. (B.7), rewrite Eq. (5.8) as

$$\begin{aligned}
&D^2_{(1,1)',(\varepsilon,-\varepsilon)'}(w_1\cdot p)(t;s,i)\\
&+\frac{1}{2}D^2_{(2,0)',(\varepsilon,\cdot)'}(w_1\cdot p)(t;s-\varepsilon,i) + \frac{1}{2}D^1_{(1,0)',(2\varepsilon,\cdot)'}(w_1\cdot p)(t;s-\varepsilon,i)\\
&+\frac{1}{2}D^2_{(0,2)',(\cdot,-\varepsilon)'}(w_1\cdot p)(t;s,i+\varepsilon) + \frac{1}{2}D^1_{(0,1)',(\cdot,-2\varepsilon)'}(w_1\cdot p)(t;s,i+\varepsilon)
\end{aligned}$$

$$
\begin{aligned}
= &- \varepsilon^2 \frac{D^2_{(1,1)',(\varepsilon,-\varepsilon)'}}{-\varepsilon^2} (w_1 \cdot p)(t;s,i) \\
&+ \frac{\varepsilon^2}{2} \frac{D^2_{(2,0)',(\varepsilon,\cdot)'}}{\varepsilon^2} (w_1 \cdot p)(t;s-\varepsilon,i) \; + \varepsilon \frac{D^1_{(1,0)',(2\varepsilon,\cdot)'}}{2\varepsilon} (w_1 \cdot p)(t;s-\varepsilon,i) \\
&+ \frac{\varepsilon^2}{2} \frac{D^2_{(0,2)',(\cdot,-\varepsilon)'}}{\varepsilon^2} (w_1 \cdot p)(t;s,i+\varepsilon) - \varepsilon \frac{D^1_{(0,1)',(\cdot,-2\varepsilon)'}}{-2\varepsilon} (w_1 \cdot p)(t;s,i+\varepsilon).
\end{aligned}
$$

All quotients in this expression have the difference quotient form (B.1). It can hence be approximated by

$$
\left(-\varepsilon^2 \frac{\partial^2}{\partial s \partial i} + \frac{\varepsilon^2}{2} \frac{\partial^2}{\partial s^2} + \varepsilon \frac{\partial}{\partial s} + \frac{\varepsilon^2}{2} \frac{\partial^2}{\partial i^2} - \varepsilon \frac{\partial}{\partial i}\right) (w_1 \cdot p)(t;s,i). \tag{5.9}
$$

The numerator of the second term (5.6) is simply

$$
\begin{aligned}
&(w_2 \cdot p)(t;s,i+\varepsilon) - (w_2 \cdot p)(t;s,i) \\
&= D^1_{(0,1)',(\cdot,\varepsilon)'} (w_2 \cdot p)(t;s,i) \\
&= \left(\frac{\varepsilon^2}{2} \frac{D^2_{(0,2)',(\cdot,\varepsilon)'}}{\varepsilon^2} + \varepsilon \frac{D^1_{(0,1)',(\cdot,2\varepsilon)'}}{2\varepsilon} \right) (w_2 \cdot p)(t;s,i-\varepsilon) \\
&\approx \left(\frac{\varepsilon^2}{2} \frac{\partial^2}{\partial i^2} + \varepsilon \frac{\partial}{\partial i} \right) (w_2 \cdot p)(t;s,i).
\end{aligned} \tag{5.10}
$$

Combining (5.9) and (5.10), an approximate representation of the forward master equation is the Kolmogorov forward equation

$$
\frac{\partial p(t,\boldsymbol{x})}{\partial t} = -\sum_{j=1}^{2} \frac{\partial \left[\mu_j(\boldsymbol{x}) p(t,\boldsymbol{x})\right]}{\partial x^{(j)}} + \frac{1}{2N} \sum_{j,k=1}^{2} \frac{\partial^2 \left[\Sigma_{jk}(\boldsymbol{x}) p(t,\boldsymbol{x})\right]}{\partial x^{(j)} \partial x^{(k)}}, \tag{5.11}
$$

where $\boldsymbol{x} = (x^{(1)}, x^{(2)})' = (s,i)'$, and μ_j and Σ_{jk} are the components of

$$
\boldsymbol{\mu}(\boldsymbol{x}) = \begin{pmatrix} -w_1(\boldsymbol{x}) \\ w_1(\boldsymbol{x}) - w_2(\boldsymbol{x}) \end{pmatrix} = \begin{pmatrix} -\alpha s i \\ \alpha s i - \beta i \end{pmatrix} \tag{5.12}
$$

and

$$
\boldsymbol{\Sigma}(\boldsymbol{x}) = \begin{pmatrix} w_1(\boldsymbol{x}) & -w_1(\boldsymbol{x}) \\ -w_1(\boldsymbol{x}) & w_1(\boldsymbol{x}) + w_2(\boldsymbol{x}) \end{pmatrix} = \begin{pmatrix} \alpha s i & -\alpha s i \\ -\alpha s i & \alpha s i + \beta i \end{pmatrix}. \tag{5.13}
$$

The diffusion matrix $N^{-1}\boldsymbol{\Sigma}$ is positive definite for all positive s and i. Hence, the original intensive Markov jump process can be approximated by a diffusion process which is the solution of

$$\mathrm{d}\boldsymbol{x}_t = \boldsymbol{\mu}(\boldsymbol{x}_t)\mathrm{d}t + \frac{1}{\sqrt{N}}\boldsymbol{\sigma}(\boldsymbol{x}_t)\mathrm{d}\boldsymbol{B}_t, \quad \boldsymbol{x}_{t_0} = \boldsymbol{x}_0,$$

where $\boldsymbol{x}_0$ is the initial value of the jump process at time t_0, and $\boldsymbol{\Sigma} = \boldsymbol{\sigma}\boldsymbol{\sigma}'$. The decomposition of $\boldsymbol{\Sigma}$ is not unique; one possible diffusion coefficient is given by

$$\frac{1}{\sqrt{N}}\boldsymbol{\sigma}(\boldsymbol{x}) = \frac{1}{\sqrt{N}}\begin{pmatrix} \sqrt{\alpha si} & 0 \\ -\sqrt{\alpha si} & \sqrt{\beta i} \end{pmatrix}. \tag{5.14}$$

A different diffusion coefficient is contained in Allen (2003, Chap. 8.11.3).

Convergence of the Infinitesimal Generator

In this paragraph, a diffusion approximation of the jump process is obtained by approximating the respective infinitesimal generator (cf. Sect. 4.3.2).

Let $f : \mathcal{C} \times T \to \mathbb{R}$ be a measurable twice continuously differentiable function, where $\mathcal{C} \supset \mathcal{C}_N$ denotes the continuous state space of the diffusion approximation, and $T \subseteq \mathbb{R}_0$ is the time set. According to Formula (4.24) on p. 70 and the considerations in Sect. 5.1.2, the infinitesimal generator $\mathcal{G}_N$ of the intensive jump process equals

$$\begin{aligned}\mathcal{G}_N f(\boldsymbol{x}, t) = \frac{\partial f(\boldsymbol{x}, t)}{\partial t} &+ w_1(\boldsymbol{x})\frac{f\big(\boldsymbol{x} + \varepsilon\boldsymbol{\Delta}_1, t\big) - f\big(\boldsymbol{x}, t\big)}{\varepsilon} \\ &+ w_2(\boldsymbol{x})\frac{f\big(\boldsymbol{x} + \varepsilon\boldsymbol{\Delta}_2, t\big) - f\big(\boldsymbol{x}, t\big)}{\varepsilon}\end{aligned}$$

for $\boldsymbol{x} = (s, i)' \in \mathcal{C}_N$, $t \in T$ and $\varepsilon = N^{-1}$. In terms of the variables s and i, this is

$$\begin{aligned}\mathcal{G}_N f(s, i; t) = \frac{\partial f(s, i; t)}{\partial t} &+ w_1(s, i)\,\frac{f\big(s - \varepsilon, i + \varepsilon; t\big) - f\big(s, i; t\big)}{\varepsilon} \\ &+ w_2(s, i)\,\frac{f\big(s, i - \varepsilon; t\big) - f\big(s, i; t\big)}{\varepsilon}.\end{aligned} \tag{5.15}$$

Analogously to the expansions in the previous approximation approach (or in Example B.2), one has

$$\begin{aligned}f(s{-}\varepsilon, i{+}\varepsilon; t){-}f(s, i; t) = &\Big(f(s{-}\varepsilon, i{+}\varepsilon; t){-}f(s{-}\varepsilon, i; t){-}f(s, i{+}\varepsilon; t){+}f(s, i; t)\Big) \\ &+\Big(f(s{-}\varepsilon, i; t){-}f(s, i; t)\Big){+}\Big(f(s, i{+}\varepsilon; t){-}f(s, i; t)\Big)\end{aligned}$$

$$= \left(D^2_{(1,1)',(-\varepsilon,\varepsilon)'} + D^1_{(1,0)',(-\varepsilon,\cdot)'} + D^1_{(0,1)',(\cdot,\varepsilon)'}\right) f(s,i;t)$$

$$\approx \left(-\varepsilon^2 \frac{\partial^2}{\partial s \partial i} + \frac{\varepsilon^2}{2}\frac{\partial^2}{\partial s^2} - \varepsilon\frac{\partial}{\partial s} + \frac{\varepsilon^2}{2}\frac{\partial^2}{\partial i^2} + \varepsilon\frac{\partial}{\partial i}\right) f(s,i;t)$$

and

$$f(s,i-\varepsilon;t) - f(s,i;t) = D^1_{(0,1)',(\cdot,-\varepsilon)'} f(s,i;t) \approx \left(\frac{\varepsilon^2}{2}\frac{\partial^2}{\partial i^2} - \varepsilon\frac{\partial}{\partial i}\right) f(s,i;t).$$

Altogether,

$$\begin{aligned}
\mathcal{G}_N &\approx \frac{\partial}{\partial t} + w_1(s,i)\left[-\frac{\partial}{\partial s} + \frac{\partial}{\partial i} + \frac{1}{2N}\frac{\partial^2}{\partial s^2} - \frac{1}{N}\frac{\partial^2}{\partial s \partial i} + \frac{1}{2N}\frac{\partial^2}{\partial i^2}\right] \\
&\quad + w_2(s,i)\left[-\frac{\partial}{\partial i} + \frac{1}{2N}\frac{\partial^2}{\partial i^2}\right] \\
&= \frac{\partial}{\partial t} + \sum_{j=1}^{2} \mu_j(\boldsymbol{x})\frac{\partial}{\partial x^{(j)}} + \frac{1}{2N}\sum_{j,k=1}^{2} \Sigma_{jk}(\boldsymbol{x})\frac{\partial^2}{\partial x^{(j)}\partial x^{(k)}} \qquad (5.16)
\end{aligned}$$

for $\boldsymbol{x} = (x^{(1)}, x^{(2)})' = (s,i)'$ and

$$\boldsymbol{\mu}(\boldsymbol{x}) = \begin{pmatrix} -\alpha si \\ \alpha si - \beta i \end{pmatrix} \quad \text{and} \quad \boldsymbol{\Sigma}(\boldsymbol{x}) = \begin{pmatrix} \alpha si & -\alpha si \\ -\alpha si & \alpha si + \beta i \end{pmatrix}. \qquad (5.17)$$

That means, the generator $\mathcal{G}_N$ of the considered jump process approximately coincides with $\partial/\partial t + \mathcal{L}^B_{\boldsymbol{\mu},\boldsymbol{\Sigma}/N}$, where the latter is the Kolmogorov backward operator. Regarding this as the generator of the approximating diffusion process, we again arrive at the SDE

$$\mathrm{d}\boldsymbol{x}_t = \boldsymbol{\mu}(\boldsymbol{x}_t)\mathrm{d}t + \frac{1}{\sqrt{N}}\boldsymbol{\sigma}(\boldsymbol{x}_t)\mathrm{d}\boldsymbol{B}_t, \quad \boldsymbol{x}_{t_0} = \boldsymbol{x}_0,$$

as a description of the diffusion approximation. Once more, $\boldsymbol{\sigma}$ is a square root of $\boldsymbol{\Sigma}$, compare with Eq. (5.14).

Langevin Approach, Kramers-Moyal Expansion and van Kampen Expansion

In the Langevin approach (Sect. 4.3.3), Kramers-Moyal expansion (Sect. 4.3.4) and van Kampen's expansion (Sect. 4.3.5), the diffusion approximation of the intensive jump process is given by the solution of

$$\mathrm{d}\boldsymbol{x}_t = \boldsymbol{\mu}(\boldsymbol{x}_t)\mathrm{d}t + \frac{1}{\sqrt{N}}\boldsymbol{\sigma}(\boldsymbol{x}_t)\mathrm{d}\boldsymbol{B}_t, \quad \boldsymbol{x}_{t_0} = \boldsymbol{x}_0, \tag{5.18}$$

where explicit formulas for the drift $\boldsymbol{\mu}$ and diffusion matrix $N^{-1}\boldsymbol{\sigma}\boldsymbol{\sigma}'$ are provided. In Sect. 5.1.2, Eqs. (5.3) and (5.4), it has been observed that the transition rate w_N of the jump process fulfils the condition $w_N = Nw$ for some function w that does not depend on the population size N. As already discovered in the discussion of Chap. 4 on p. 94, in this case the three approximation approaches yield the same result, which is

$$\boldsymbol{\mu}(\boldsymbol{x}) = \sum_{j=1,2} w_j(\boldsymbol{x})\boldsymbol{\Delta}_j \quad \text{and} \quad \boldsymbol{\Sigma}(\boldsymbol{x}) = \sum_{j=1,2} w_j(\boldsymbol{x})\boldsymbol{\Delta}_j\boldsymbol{\Delta}_j'$$

for $\boldsymbol{x} = (s,i)' \in \mathcal{C}_N$. Hence, in the standard SIR model, the diffusion approximation is given by Eq. (5.18) with

$$\boldsymbol{\mu}(\boldsymbol{x}) = \alpha si\begin{pmatrix}-1\\1\end{pmatrix} + \beta i\begin{pmatrix}0\\-1\end{pmatrix} = \begin{pmatrix}-\alpha si\\ \alpha si - \beta i\end{pmatrix}$$

and

$$\boldsymbol{\Sigma}(\boldsymbol{x}) = \alpha si\begin{pmatrix}1 & -1\\-1 & 1\end{pmatrix} + \beta i\begin{pmatrix}0 & 0\\0 & 1\end{pmatrix} = \begin{pmatrix}\alpha si & -\alpha si\\-\alpha si & \alpha si + \beta i\end{pmatrix}.$$

As announced earlier, that reproduces the result from the previous methods.

5.1.4 Summary

To summarise the results so far, Sect. 5.1.1 introduced the standard SIR model, Sect. 5.1.2 characterised it as a jump process through its master equation, and Sect. 5.1.3 applied the various approaches from Sect. 4.3 to derive a diffusion approximation for it. Let $\boldsymbol{x} = (s,i)'$, where s and i denote the fractions of susceptible and infectious individuals of the total population of size N. The master equation of the jump process with transitions (5.1) and (5.2) is

$$\frac{\partial p_N(t;s,i)}{\partial t} = N\alpha(s+\varepsilon)(i-\varepsilon)p_N(t;s+\varepsilon,i-\varepsilon) + N\beta(i+\varepsilon)p_N(t;s,i+\varepsilon)$$
$$-N(\alpha si + \beta i)p_N(t;s,i),$$

where p_N is the transition probability of $\boldsymbol{x}$, and $\varepsilon = N^{-1}$. All considered approximation approaches arrive at the same stochastic differential equation, the solution of which is the desired diffusion approximation process. This SDE reads

$$\begin{pmatrix}\mathrm{d}s\\ \mathrm{d}i\end{pmatrix} = \begin{pmatrix}-\alpha si\\ \alpha si - \beta i\end{pmatrix}\mathrm{d}t + \frac{1}{\sqrt{N}}\begin{pmatrix}\sqrt{\alpha si} & 0\\ -\sqrt{\alpha si} & \sqrt{\beta i}\end{pmatrix}\begin{pmatrix}\mathrm{d}B_1\\ \mathrm{d}B_2\end{pmatrix}, \tag{5.19}$$

where B_1 and B_2 are independent Brownian motions, and $\mathrm{d}B_1/\mathrm{d}t$ and $\mathrm{d}B_2/\mathrm{d}t$ can hence be interpreted as Gaussian white noise forces (see Sect. 3.1.3) accounting for fluctuations in transmission and recovery. The continuous state space of $\boldsymbol{x}$ is the simplex

$$\mathcal{C} = \{(s,i)' \in [0,1]^2 \cap \mathbb{R}_0^2 \,|\, s+i \leq 1\}. \tag{5.20}$$

The differential equation is subject to an appropriate initial condition $\boldsymbol{x}_{t_0} = (s_0, i_0)' \in \mathcal{C}$. Note that the process given by the solution of (5.19) is not a diffusion limit but a diffusion approximation as it still contains the size parameter N. In the limit $N \to \infty$, one obtains the ordinary differential equation

$$\begin{pmatrix} \mathrm{d}s \\ \mathrm{d}i \end{pmatrix} = \begin{pmatrix} -\alpha s i \\ \alpha s i - \beta i \end{pmatrix} \mathrm{d}t \tag{5.21}$$

as a deterministic description of the dynamics of the system. However, in the context of infectious disease epidemiology, one is dealing with processes that are highly sensitive to disturbances. Although the ODE (5.21) mirrors the macroscopic behaviour of the system, the stochastic and the deterministic process may differ substantially regarding single realisations. Stochasticity becomes particularly important when the initial fraction of infectives is small and the occurrence of an outbreak is not obvious. The SDE (5.19) is hence clearly to be preferred.

5.1.5 *Illustration*

Graphical illustrations of the standard SIR model were shown in Sect. 2.2.2, where Fig. 2.3 on p. 15 displayed sample paths for the three considered types of Markov processes, and Figs. 2.4 and 2.5 demonstrated the role of the basic reproductive ratio $\mathcal{R}_0 = \alpha/\beta$ and the impact of stochasticity. Figure 5.1 in this section contrasts the trajectories of the diffusion process defined through (5.19) and the deterministic process given by (5.21). Figure 5.1a shows how the stochastic sample paths (thin lines) fluctuate around the deterministic course (thick lines). Figures 5.1b, c display empirical pointwise 95 %-confidence bands for the trajectories of the diffusion process, where the population size equals $N = 1{,}000$ and $N = 10{,}000$, repectively. As obvious from (5.19), the width of the confidence band decreases for larger N. Figure 5.1b particularly elucidates that the paths of a diffusion process do generally not fluctuate around their deterministic counterpart in a symmetric manner. Moreover, variation is obviously non-constant. This reveals the weaknesses of an estimation approach where one assumes independent and identically distributed deviations of the observations from the deterministic prediction.

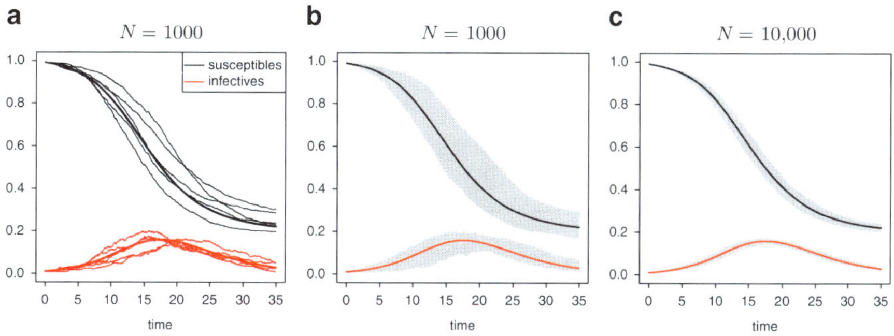

Fig. 5.1 Simulation of the standard SIR diffusion process with $\alpha = 0.5$ and $\beta = 0.25$. The trajectories have been obtained by application of the Euler scheme from Sect. 3.3.2 with time step 0.025 and initial value $(s_0, i_0)' = (0.99, 0.01)'$ at time zero. (**a**) Five realisations of the diffusion process (*thin lines*) in comparison to the deterministic course (*thick lines*) for $N = 1{,}000$. The sample paths for the fractions of susceptibles are plotted in *black*, the paths for the fractions of infectives in *red*. (**b**) Empirical pointwise 95 %-confidence bands for the trajectories, represented by the *grey areas*. These have been obtained from another 100 realisations of the diffusion process with $N = 1{,}000$. The *black* and *red lines* show the paths of the deterministic model. (**c**) Confidence bands for $N = 10{,}000$

5.2 Multitype SIR Model

After having extensively considered the standard SIR model, we now turn to a multitype extension of it. The standard SIR model does not allow for host heterogeneity. It is therefore appropriately modified in what follows. The extended model is introduced in Sect. 5.2.1, formulated as a jump process through its master equation in Sect. 5.2.2 and approximated by a diffusion process in Sect. 5.2.3. Again, a summary is provided in Sect. 5.2.4, and the model is further analysed in Sect. 5.2.5.

5.2.1 *Model*

In the multitype modelling approach, the population under consideration is partitioned into pairwise disjoint clusters $j = 1, \ldots, n$ of sizes N_j, i.e. $N = \sum_{j=1}^n N_j$. Such clusters might for example correspond to different geographic regions or age classes. Individuals of each type are divided into n groups according to their cluster such that S_j, I_j and R_j denote the respective numbers of susceptible, infective and removed individuals in cluster j. Define $S = \sum_{j=1}^n S_j$, $I = \sum_{j=1}^n I_j$ and $R = \sum_{j=1}^n R_j$.

Within each cluster the population is assumed to mix homogeneously—i.e. the infection dynamics within the cluster follows again the standard SIR model (5.1)–(5.2)—but with a certain rate γ_{jk}^N individuals from cluster j are involved in the infection dynamics of cluster k rather than of their own cluster j. These rates are summarised in a network matrix $\gamma^N = (\gamma_{jk}^N)_{j,k=1,\ldots,n}$ with row sums equal to one. The entries on the main diagonal are the rates with which individuals are part of the infection processes of their own clusters.

In case the clusters represent geographical regions like rural and urban districts, individuals might be divided into n groups according to their (unique) home region. The network matrix might then describe the traffic of commuters having their social environment in their region of residence but being away from home while working in a different region. If the clusters refer to different age classes, they might represent homogeneous groups that gather for example in school, at work or in homes for the aged. The network matrix then stands for social contacts between these age groups.

Since the travelling or contact behaviour of individuals may depend on their medical state, we further introduce the network matrices $\gamma^S = (\gamma_{jk}^S)_{j,k=1,\ldots,n}$ and $\gamma^I = (\gamma_{jk}^I)_{j,k=1,\ldots,n}$ for susceptibles and infectives, respectively.

The transitions in this model are for all clusters $j, k, m = 1, \ldots, n$:

1. A susceptible from cluster k gets infected in cluster j by an infective from cluster m (where k and m might be equal to j):

$$\text{Ⓢ}_k + \text{Ⓘ}_m \xrightarrow{\alpha_j, \boldsymbol{\gamma}_{\bullet j}} \text{Ⓘ}_k + \text{Ⓘ}_m. \tag{5.22}$$

   The parameter $\alpha_j \in \mathbb{R}_+$ is the contact number in cluster j, and $\boldsymbol{\gamma}_{\bullet j}$ is short for the jth columns $\gamma_{\bullet j}^S$, $\gamma_{\bullet j}^I$ and $\gamma_{\bullet j}^N$ of γ^S, γ^I and γ^N, respectively.
2. An infective individual from cluster j recovers:

$$\text{Ⓘ}_j \xrightarrow{\beta_j} \text{Ⓡ}_j, \tag{5.23}$$

   where $\beta_j \in \mathbb{R}_+$ is the reciprocal average infectious period in cluster j.

Note that the critical contact number depends on the cluster where the infection takes place. For example, if the clusters represent geographical regions, the risk for a contagious contact strongly depends on parameters such as the population density or the use of public transport. In case the clusters refer to age groups, infectious contacts depend for example on contact behaviour of the individuals at a certain age. In comparison, the average infectious period is determined by the cluster of the recovering individual. To illustrate this on the above examples, in case of geographical clusters recovery is subject to the medical standards at the place of residence. For age groups, infectious periods depend on physical shapes.

In order to keep N_j constant for all $j = 1, \ldots, n$, changes in cluster affiliation—such as changes of places of residence or ageing—are ignored during the presumably relatively short period of an epidemic outbreak. The numbers of removed individuals in each cluster j can then be obtained as $R_j = N_j - S_j - I_j$ at any time point such that

$$\mathcal{D}_N^{(n)} := \left\{(S_1, \ldots, S_n, I_1, \ldots, I_n)' \in [0, N]^{2n} \cap \mathbb{N}_0^{2n} \,\middle|\, S_j + I_j \leq N_j \text{ for all } j = 1, \ldots, n\right\}$$

is an eligible state space for a process following the multitype SIR model. The model parameter is $\boldsymbol{\theta} = (\boldsymbol{\alpha}, \boldsymbol{\beta}, \boldsymbol{\gamma}^S, \boldsymbol{\gamma}^I, \boldsymbol{\gamma}^N)$ with $\boldsymbol{\alpha} = (\alpha_1, \dots, \alpha_n)'$ and $\boldsymbol{\beta} = (\beta_1, \dots, \beta_n)'$.

5.2.2 Jump Process

The transitions of the multitype SIR model are given by Eqs. (5.22) and (5.23). For $j = 1, \dots, n$, let S_j and I_j be the numbers of susceptible and infective individuals in cluster j, which has a total population of size N_j. Then, the state space of the jump process is $\mathcal{D}_N^{(n)}$ as defined above. The size of the system is described by the set of population sizes $\{N_1, \dots, N_n\}$. Denote by $\boldsymbol{e}_j = (0, \dots, 1, \dots, 0)'$ the jth unit vector and by $\mathbf{0}$ the null vector of dimension n. Assuming that at most one event can happen within a short time interval of length Δt, possible steps of the jump process are

$$\boldsymbol{\Delta}_{1,j} = \begin{pmatrix} -\boldsymbol{e}_j \\ \boldsymbol{e}_j \end{pmatrix} \quad \text{for an infection and} \quad \boldsymbol{\Delta}_{2,j} = \begin{pmatrix} \mathbf{0} \\ -\boldsymbol{e}_j \end{pmatrix} \quad \text{for a recovery}$$

of an individual from cluster $j \in \{1, \dots, n\}$. All transition probabilities are considered homogeneous in time. Given the current state $\boldsymbol{X} = (S_1, \dots, S_n, I_1, \dots, I_n)' \in \mathcal{D}_N^{(n)}$, the probability of an infection of a susceptible from cluster j in cluster k (where $j, k = 1, \dots, n$) within time Δt is $\Pi^{jk}(\boldsymbol{X})\Delta t + o(\Delta t)$ with

$$\Pi^{jk}(\boldsymbol{X}) = \alpha_k \frac{(\text{number of infectives in cluster } k) \cdot (\text{number of susceptibles from } j \text{ in } k)}{\text{total number of individuals in cluster } k},$$

that is

$$\Pi^{jk}(\boldsymbol{X}) = \alpha_k \frac{\sum_{m=1}^{n} \gamma_{mk}^I I_m}{\sum_{m=1}^{n} \gamma_{mk}^N N_m} \gamma_{jk}^S S_j.$$

The probability of the recovery of an infective from cluster j is $\Upsilon^j(\boldsymbol{X})\Delta t + o(\Delta t)$ with

$$\Upsilon^j(\boldsymbol{X}) = \beta_j I_j.$$

Therefore, the transition rates of the process $\boldsymbol{X}$ are for all j

$$W_{N,r,j}(\boldsymbol{X}) = W_N(\boldsymbol{X}, \boldsymbol{\Delta}_{r,j}) = \begin{cases} \sum_{k=1}^{n} \Pi^{jk}(\boldsymbol{X}) & \text{if } r = 1, \\ \Upsilon^j(\boldsymbol{X}) & \text{if } r = 2. \end{cases}$$

Let $P_N(t, \boldsymbol{X})$ denote the probability that within time t the extensive process arrives at state $\boldsymbol{X} \in \mathcal{D}_N^{(n)}$ conditioned on a prespecified initial state. The master equation of the extensive process is

$$\frac{\partial P_N(t,\boldsymbol{X})}{\partial t} = \sum_{j=1}^{n}\Bigg[\sum_{k=1}^{n} \Pi^{jk}(S_j+1, I_j-1)P_N(t; S_j+1, I_j-1)$$

$$+ \Upsilon^j(S_j, I_j+1)P_N(t; S_j, I_j+1)$$

$$- \left(\sum_{k=1}^{n} \Pi^{jk}(S_j, I_j) + \Upsilon^j(S_j, I_j)\right) P_N(t; S_j, I_j)\Bigg].$$

For better readability, only the relevant components of $\boldsymbol{X}$ are displayed here as arguments of Π^{jk}, Υ^j and P_N. Now consider the intensive variable

$$\boldsymbol{x} = (s_1, \ldots, s_n, i_1, \ldots, i_n)' = (N_1^{-1}S_1, \ldots, N_n^{-1}S_n, N_1^{-1}I_1, \ldots, N_n^{-1}I_n)' = \boldsymbol{M}^{-1}\boldsymbol{X}$$

with $\boldsymbol{M} = \operatorname{diag}(N_1, \ldots, N_n, N_1, \ldots, N_n)'$. The state space of the according intensive jump process is $\mathcal{C}_N^{(n)} = \boldsymbol{M}^{-1}\mathcal{D}_N^{(n)} = \{\boldsymbol{M}^{-1}\boldsymbol{X} \mid \boldsymbol{X} \in \mathcal{D}_N^{(n)}\}$, i.e.

$$\mathcal{C}_N^{(n)} = \left\{(s_1, \ldots, s_n, i_1, \ldots, i_n)' \in [0,1]^{2n} \cap \boldsymbol{M}^{-1}\mathbb{N}_0^{2n} \mid s_j + i_j \le 1 \text{ for all } j = 1, \ldots, n\right\}.$$

Possible nonzero jumps of $\boldsymbol{x}$ are

$$\tilde{\boldsymbol{\Delta}}_{1,j} = \frac{1}{N_j}\begin{pmatrix} -\boldsymbol{e}_j \\ \boldsymbol{e}_j \end{pmatrix} \quad \text{and} \quad \tilde{\boldsymbol{\Delta}}_{2,j} = \frac{1}{N_j}\begin{pmatrix} \boldsymbol{0} \\ -\boldsymbol{e}_j \end{pmatrix}$$

for $j = 1, \ldots, n$. Define

$$\pi^{jk}(\boldsymbol{x}) = N_j^{-1}\Pi^{jk}(\boldsymbol{M}\boldsymbol{x}) = \alpha_k \frac{\sum_{m=1}^{n} \gamma_{mk}^{I} \frac{N_m}{N_k} i_m}{\sum_{m=1}^{n} \gamma_{mk}^{N} \frac{N_m}{N_k}} \gamma_{jk}^{S} s_j$$

and

$$\upsilon^j(\boldsymbol{x}) = N_j^{-1}\Upsilon^j(\boldsymbol{M}\boldsymbol{x}) = \beta_j i_j.$$

Then, for $j = 1, \ldots, n$,

$$w_N(\boldsymbol{x}, \boldsymbol{\Delta}_{r,j}) = w_{N,r,j}(\boldsymbol{x}) = W_{N,r,j}(\boldsymbol{M}\boldsymbol{x}) = \begin{cases} N_j \sum_{k=1}^{n} \pi^{jk}(\boldsymbol{x}) & \text{if } r = 1, \\ N_j \upsilon^j(\boldsymbol{x}) & \text{if } r = 2. \end{cases}$$

Thus $w_{N,r,j} = N_j w_{r,j}$ with

$$w(\boldsymbol{x}, \boldsymbol{\Delta}_{r,j}) = w_{r,j}(\boldsymbol{x}) = \begin{cases} \sum\limits_{k=1}^{n} \pi^{jk}(\boldsymbol{x}) & \text{if } r = 1, \\ \upsilon^j(\boldsymbol{x}) & \text{if } r = 2. \end{cases}$$

This function depends on the population sizes $N_1, \ldots, N_n$ only through their mutual ratios N_m/N_k, where $k, m = 1, \ldots, n$.

5.2.3 *Diffusion Approximation*

We now want to approximate the multitype SIR model by a diffusion. Clearly, this model does not fit in the rather simple framework of the approximation methods from Sect. 4.3; in order to completely describe its dynamics, one needs to employ a whole set of size parameters $N_1, \ldots, N_n$. Hence, for the derivation of a diffusion approximation for the multitype SIR model, the extended techniques from Sect. 4.4 are applied.

Convergence of the Master Equation

The master equation of the multitype SIR process can be approximated by a Kolmogorov forward equation as follows (cf. Sect. 4.4.1).

As before, denote by p the transition density of the approximating diffusion process $\boldsymbol{x}$. Following Eq. (4.60), it roughly fulfils

$$\frac{\partial p(t,\boldsymbol{x})}{\partial t} = \sum_{j=1}^{n} \Big[N_j \Big(w_{1,j}(\boldsymbol{x} - \tilde{\boldsymbol{\Delta}}_{1,j}) p(t, \boldsymbol{x} - \tilde{\boldsymbol{\Delta}}_{1,j}) - w_{1,j}(\boldsymbol{x}) p(t,\boldsymbol{x}) \Big)$$
$$+ N_j \Big(w_{2,j}(\boldsymbol{x} - \tilde{\boldsymbol{\Delta}}_{2,j}) p(t, \boldsymbol{x} - \tilde{\boldsymbol{\Delta}}_{2,j}) - w_{2,j}(\boldsymbol{x}) p(t,\boldsymbol{x}) \Big) \Big].$$

For the sake of better readability, suppress non-involved components of $\boldsymbol{x}$. With $\varepsilon_j = N_j^{-1}$, one then has

$$\frac{\partial p(t,\boldsymbol{x})}{\partial t}$$
$$= \sum_{j=1}^{n} \Bigg[\frac{w_{1,j}(s_j+\varepsilon_j, i_j-\varepsilon_j) p(t; s_j+\varepsilon_j, i_j-\varepsilon_j) - w_{1,j}(s_j, i_j) p(t; s_j, i_j)}{\varepsilon_j} \tag{5.24}$$
$$+ \frac{w_{2,j}(s_j, i_j + \varepsilon_j) p(t; s_j, i_j + \varepsilon_j) - w_{2,j}(s_j, i_j) p(t; s_j, i_j)}{\varepsilon_j} \Bigg]. \tag{5.25}$$

Close similarity to Eqs. (5.5) and (5.6) on p. 105 is unmistakable. Hence, in complete analogy to Eqs. (5.11)–(5.13), a diffusion approximation of the multitype SIR model is described by the Kolmogorov forward equation

$$\frac{\partial p(t,\boldsymbol{x})}{\partial t} = -\sum_{j=1}^{n}\left[\frac{\partial \mu_j^S(\boldsymbol{x})p(t,\boldsymbol{x})}{\partial s_j} + \frac{\partial \mu_j^I(\boldsymbol{x})p(t,\boldsymbol{x})}{\partial i_j}\right]$$
$$+\frac{1}{2}\sum_{j=1}^{n}\left[\frac{\partial^2 \tilde{\Sigma}_{jj}^{SS}(\boldsymbol{x})p(t,\boldsymbol{x})}{\partial s_j^2} + \frac{\partial^2 \tilde{\Sigma}_{jj}^{II}(\boldsymbol{x})p(t,\boldsymbol{x})}{\partial i_j^2} + 2\frac{\partial^2 \tilde{\Sigma}_{jj}^{SI}(\boldsymbol{x})p(t,\boldsymbol{x})}{\partial s_j \partial i_j}\right]$$

with the following coefficients: The drift vector and diffusion matrix of the diffusion approximation are given by

$$\boldsymbol{\mu}(\boldsymbol{x}) = \begin{pmatrix} \boldsymbol{\mu}^S(\boldsymbol{x}) \\ \boldsymbol{\mu}^I(\boldsymbol{x}) \end{pmatrix} \quad \text{and} \quad \tilde{\boldsymbol{\Sigma}}(\boldsymbol{x}) = \begin{pmatrix} \tilde{\boldsymbol{\Sigma}}^{SS}(\boldsymbol{x}) & \tilde{\boldsymbol{\Sigma}}^{SI}(\boldsymbol{x}) \\ \tilde{\boldsymbol{\Sigma}}^{IS}(\boldsymbol{x}) & \tilde{\boldsymbol{\Sigma}}^{II}(\boldsymbol{x}) \end{pmatrix}. \tag{5.26}$$

The components of $\boldsymbol{\mu}^S(\boldsymbol{x}) = (\mu_j^S(\boldsymbol{x}))_{j=1,\ldots,n}$ and $\boldsymbol{\mu}^I(\boldsymbol{x}) = (\mu_j^I(\boldsymbol{x}))_{j=1,\ldots,n}$ are in turn

$$\mu_j^S(\boldsymbol{x}) = -w_{1,j}(\boldsymbol{x}) = -\sum_{k=1}^{n}\pi^{jk}(\boldsymbol{x}), \tag{5.27}$$

$$\mu_j^I(\boldsymbol{x}) = w_{1,j}(\boldsymbol{x}) - w_{2,j}(\boldsymbol{x}) = \sum_{k=1}^{n}\pi^{jk}(\boldsymbol{x}) - \upsilon^j(\boldsymbol{x}), \tag{5.28}$$

and $\tilde{\boldsymbol{\Sigma}}$ consists of the diagonal matrices $\tilde{\boldsymbol{\Sigma}}^{SS}$, $\tilde{\boldsymbol{\Sigma}}^{II}$ and $\tilde{\boldsymbol{\Sigma}}^{SI} = \tilde{\boldsymbol{\Sigma}}^{IS}$ with main diagonal elements

$$\tilde{\Sigma}_{jj}^{SS}(\boldsymbol{x}) = N_j^{-1}\, w_{1,j}(\boldsymbol{x}) = N_j^{-1}\sum_{k=1}^{n}\pi^{jk}(\boldsymbol{x}), \tag{5.29}$$

$$\tilde{\Sigma}_{jj}^{II}(\boldsymbol{x}) = N_j^{-1}\big(w_{1,j}(\boldsymbol{x}) + w_{2,j}(\boldsymbol{x})\big) = N_j^{-1}\left(\sum_{k=1}^{n}\pi^{jk}(\boldsymbol{x}) + \upsilon^j(\boldsymbol{x})\right), \tag{5.30}$$

$$\tilde{\Sigma}_{jj}^{SI}(\boldsymbol{x}) = -N_j^{-1}\, w_{1,j}(\boldsymbol{x}) = -N_j^{-1}\sum_{k=1}^{n}\pi^{jk}(\boldsymbol{x}) \tag{5.31}$$

for $j = 1,\ldots,n$. The matrix

$$\tilde{\boldsymbol{\sigma}}(\boldsymbol{x}) = \begin{pmatrix} \tilde{\boldsymbol{\sigma}}^{SS}(\boldsymbol{x}) & \boldsymbol{0} \\ \tilde{\boldsymbol{\sigma}}^{SI}(\boldsymbol{x}) & \tilde{\boldsymbol{\sigma}}^{II}(\boldsymbol{x}) \end{pmatrix}$$

with diagonal matrices $\tilde{\boldsymbol{\sigma}}^{SS}, \tilde{\boldsymbol{\sigma}}^{II}, \tilde{\boldsymbol{\sigma}}^{SI}$, zero matrix $\mathbf{0}$ and

$$\tilde{\sigma}^{SS}_{jj}(\boldsymbol{x}) = \sqrt{\sum_{k=1}^{n} \frac{\pi^{jk}(\boldsymbol{x})}{N_j}}, \quad \tilde{\sigma}^{II}_{jj}(\boldsymbol{x}) = \sqrt{\frac{\upsilon^j(\boldsymbol{x})}{N_j}}, \quad \tilde{\sigma}^{SI}_{jj}(\boldsymbol{x}) = -\sqrt{\sum_{k=1}^{n} \frac{\pi^{jk}(\boldsymbol{x})}{N_j}}$$

for all j is a square root of $\tilde{\boldsymbol{\Sigma}}$. Denote by $\mathcal{C}^{(n)}$ the continuous analogue of $\mathcal{C}_N^{(n)}$. The diffusion matrix $\tilde{\boldsymbol{\Sigma}}(\boldsymbol{x})$ is positive semi-definite for all $\boldsymbol{x} \in \mathcal{C}^{(n)}$ as

$$\boldsymbol{y}'\tilde{\boldsymbol{\Sigma}}(\boldsymbol{x})\boldsymbol{y} = \big(\tilde{\boldsymbol{\sigma}}'(\boldsymbol{x})\boldsymbol{y}\big)'\big(\tilde{\boldsymbol{\sigma}}'(\boldsymbol{x})\boldsymbol{y}\big) \geq 0 \quad \text{for all } \boldsymbol{y} \in \mathbb{R}^{2n}.$$

$\tilde{\boldsymbol{\Sigma}}(\boldsymbol{x})$ is positive definite if furthermore all s_j and i_j are nonzero since

$$\big(\tilde{\boldsymbol{\sigma}}'(\boldsymbol{x})\boldsymbol{y}\big)'\big(\tilde{\boldsymbol{\sigma}}'(\boldsymbol{x})\boldsymbol{y}\big) = 0 \quad \Leftrightarrow \quad \tilde{\boldsymbol{\sigma}}'(\boldsymbol{x})\boldsymbol{y} = \mathbf{0} \quad \Leftrightarrow \quad \boldsymbol{y} = \mathbf{0}.$$

The last equivalence is true because $\tilde{\boldsymbol{\sigma}}'(\boldsymbol{x})$ has nonzero determinant (in case all components of $\boldsymbol{x}$ are positive) and is hence of full rank. Therefore, the intensive jump process can be approximated by a diffusion process that is the solution of the SDE

$$\mathrm{d}\boldsymbol{x}_t = \boldsymbol{\mu}(\boldsymbol{x}_t)\mathrm{d}t + \tilde{\boldsymbol{\sigma}}(\boldsymbol{x}_t)\mathrm{d}\boldsymbol{B}_t, \quad \boldsymbol{x}_{t_0} = \boldsymbol{x}_0,$$

where $\boldsymbol{x}_0$ is the state of the jump process at time t_0. Note that, in contrast to the SDE (5.19) for the approximation of the standard SIR model, there is no universal scaling factor $N^{-1/2}$ for the diffusion coefficient here. Instead, individual factors $N_j^{-1/2}$ are included directly in the components of $\tilde{\boldsymbol{\sigma}}$.

Example 5.1. In case of one group, i.e. $n = 1$, the network matrices γ^S, γ^I and γ^N consist of the single entry $\gamma_{11} = 1$. The diffusion approximation for the multitype SIR has drift

$$\boldsymbol{\mu}(s_1, i_1) = \begin{pmatrix} -\pi^{11}(s_1, i_1) \\ \pi^{11}(s_1, i_1) - \upsilon^1(s_1, i_1) \end{pmatrix} = \begin{pmatrix} -\alpha_1 s_1 i_1 \\ \alpha_1 s_1 i_1 - \beta_1 i_1 \end{pmatrix}$$

and diffusion coefficient

$$\tilde{\boldsymbol{\sigma}}(s_1, i_1) = \begin{pmatrix} \sqrt{\dfrac{\pi^{11}(s_1, i_1)}{N_1}} & 0 \\ -\sqrt{\dfrac{\pi^{11}(s_1, i_1)}{N_1}} & \sqrt{\dfrac{\upsilon^1(s_1, i_1)}{N_1}} \end{pmatrix} = \frac{1}{\sqrt{N_1}} \begin{pmatrix} \sqrt{\alpha_1 s_1 i_1} & 0 \\ -\sqrt{\alpha_1 s_1 i_1} & \sqrt{\beta_1 i_1} \end{pmatrix}.$$

This complies with the standard SIR model in Sect. 5.1. For $n = 2$, the drift and diffusion coefficient are

$$\boldsymbol{\mu}(\boldsymbol{x}) = \begin{pmatrix} -\pi^{11}(\boldsymbol{x}) - \pi^{12}(\boldsymbol{x}) \\ -\pi^{21}(\boldsymbol{x}) - \pi^{22}(\boldsymbol{x}) \\ \pi^{11}(\boldsymbol{x}) + \pi^{12}(\boldsymbol{x}) - \beta_1 i_1 \\ \pi^{21}(\boldsymbol{x}) + \pi^{22}(\boldsymbol{x}) - \beta_2 i_2 \end{pmatrix}$$

and

$$\tilde{\boldsymbol{\sigma}}(\boldsymbol{x}) = \begin{pmatrix} \sqrt{\dfrac{\pi^{11}(\boldsymbol{x}) + \pi^{12}(\boldsymbol{x})}{N_1}} & 0 & 0 & 0 \\ 0 & \sqrt{\dfrac{\pi^{21}(\boldsymbol{x}) + \pi^{22}(\boldsymbol{x})}{N_2}} & 0 & 0 \\ -\sqrt{\dfrac{\pi^{11}(\boldsymbol{x}) + \pi^{12}(\boldsymbol{x})}{N_1}} & 0 & \sqrt{\dfrac{\beta_1 i_1}{N_1}} & 0 \\ 0 & -\sqrt{\dfrac{\pi^{21}(\boldsymbol{x}) + \pi^{22}(\boldsymbol{x})}{N_2}} & 0 & \sqrt{\dfrac{\beta_2 i_2}{N_2}} \end{pmatrix}$$

with

$$\pi^{11}(\boldsymbol{x}) = \alpha_1 \frac{\gamma_{11}^I i_1 + \gamma_{21}^I \frac{N_2}{N_1} i_2}{\gamma_{11}^N + \gamma_{21}^N \dfrac{N_2}{N_1}} \gamma_{11}^S s_1, \qquad \pi^{12}(\boldsymbol{x}) = \alpha_2 \frac{\gamma_{12}^I \frac{N_1}{N_2} i_1 + \gamma_{22}^I i_2}{\gamma_{12}^N \frac{N_1}{N_2} + \gamma_{22}^N} \gamma_{12}^S s_1,$$

$$\pi^{21}(\boldsymbol{x}) = \alpha_1 \frac{\gamma_{11}^I i_1 + \gamma_{21}^I \frac{N_2}{N_1} i_2}{\gamma_{11}^N + \gamma_{21}^N \dfrac{N_2}{N_1}} \gamma_{21}^S s_2, \qquad \pi^{22}(\boldsymbol{x}) = \alpha_2 \frac{\gamma_{12}^I \frac{N_1}{N_2} i_1 + \gamma_{22}^I i_2}{\gamma_{12}^N \dfrac{N_1}{N_2} + \gamma_{22}^N} \gamma_{22}^S s_2.$$

An illustration of the multitype SIR model follows in Sect. 5.2.5.

Convergence of the Infinitesimal Generator

This section deals with the approximation of the infinitesimal generator of the jump process $\boldsymbol{x}$ as described in Sect. 4.4.2.

Consider a measurable twice continuously differentiable function $f : \mathcal{C}^{(n)} \times T \to \mathbb{R}$, where $\mathcal{C}^{(n)}$ is the state space of the diffusion approximation, and T is the time set. Equation (4.63) from p. 86 reads for the multitype SIR model

$$\mathcal{G}_N f(\boldsymbol{x}, t) = \frac{\partial f(\boldsymbol{x}, t)}{\partial t} + \sum_{j=1}^{n} \sum_{r=1}^{2} w_{r,j}(\boldsymbol{x}) \frac{f(\boldsymbol{x} + \tilde{\boldsymbol{\Delta}}_{r,j}, t) - f(\boldsymbol{x}, t)}{\varepsilon_j}$$

for $\boldsymbol{x} \in \mathcal{C}_N^{(n)}$, $t \in T$ and $\varepsilon_j = N_j^{-1}$. In order to simplify notation, non-involved components of $\boldsymbol{x}$ are dropped. Then

$$\mathcal{G}_N f(\boldsymbol{x},t) = \frac{\partial f(\boldsymbol{x},t)}{\partial t} + \sum_{j=1}^{n} \Bigg[w_{1,j}(\boldsymbol{x})\, \frac{f(s_j - \varepsilon_j, i_j + \varepsilon_j; t) - f(s_j, i_j; t)}{\varepsilon_j} + w_{2,j}(\boldsymbol{x})\, \frac{f(s_j, i_j - \varepsilon_j; t) - f(s_j, i_j; t)}{\varepsilon_j} \Bigg].$$

Once more, this resembles the derivation of a diffusion approximation for the standard SIR model, specifically Eq. (5.15) on p. 107. Hence, the results (5.16) and (5.17) can be adopted for the present model. That yields

$$\mathcal{G}_N \approx \frac{\partial}{\partial t} + \sum_{j=1}^{n} \left[\mu_j^S(\boldsymbol{x})\, \frac{\partial}{\partial s_j} + \mu_j^I(\boldsymbol{x})\, \frac{\partial}{\partial i_j} \right] + \frac{1}{2} \sum_{j=1}^{n} \left[\tilde{\Sigma}_{jj}^{SS}(\boldsymbol{x})\, \frac{\partial^2}{\partial s_j^2} + \tilde{\Sigma}_{jj}^{II}(\boldsymbol{x})\, \frac{\partial^2}{\partial i_j^2} + 2\tilde{\Sigma}_{jj}^{SI}(\boldsymbol{x})\, \frac{\partial^2}{\partial s_j \partial i_j} \right],$$

where

$$\boldsymbol{\mu}(\boldsymbol{x}) = \begin{pmatrix} \boldsymbol{\mu}^S(\boldsymbol{x}) \\ \boldsymbol{\mu}^I(\boldsymbol{x}) \end{pmatrix} \quad \text{and} \quad \tilde{\boldsymbol{\Sigma}}(\boldsymbol{x}) = \begin{pmatrix} \tilde{\boldsymbol{\Sigma}}^{SS}(\boldsymbol{x}) & \tilde{\boldsymbol{\Sigma}}^{SI}(\boldsymbol{x}) \\ \tilde{\boldsymbol{\Sigma}}^{SI}(\boldsymbol{x}) & \tilde{\boldsymbol{\Sigma}}^{II}(\boldsymbol{x}) \end{pmatrix}.$$

The vector $\boldsymbol{\mu}$ has components

$$\mu_j^S(\boldsymbol{x}) = -w_{1,j}(\boldsymbol{x}), \quad \mu_j^I(\boldsymbol{x}) = w_{1,j}(\boldsymbol{x}) - w_{2,j}(\boldsymbol{x}),$$

and $\tilde{\boldsymbol{\Sigma}}$ consists of the diagonal matrices $\tilde{\boldsymbol{\Sigma}}^{SS}$, $\tilde{\boldsymbol{\Sigma}}^{II}$ and $\tilde{\boldsymbol{\Sigma}}^{SI}$ with main diagonal elements

$$\tilde{\Sigma}_{jj}^{SS}(\boldsymbol{x}) = N_j^{-1} w_{1,j}(\boldsymbol{x}), \quad \tilde{\Sigma}_{jj}^{II}(\boldsymbol{x}) = N_j^{-1}\big(w_{1,j}(\boldsymbol{x}) + w_{2,j}(\boldsymbol{x})\big), \quad \tilde{\Sigma}_{jj}^{SI}(\boldsymbol{x}) = -N_j^{-1} w_{1,j}(\boldsymbol{x}).$$

Apply the approximation of $\mathcal{G}_N$ as the generator of the diffusion approximation. One thus obtains a diffusion process with drift $\boldsymbol{\mu}$ and positive definite diffusion matrix $\tilde{\boldsymbol{\Sigma}}$. This is the same result as obtained in the previous approximation approach.

Langevin Approach and Kramers-Moyal Expansion

Also in the framework of multiple size variables $N_1, \ldots, N_n$, the Langevin approach (Sect. 4.4.3) and Kramers-Moyal expansion (Sect. 4.4.4) give explicit formulas for the drift $\boldsymbol{\mu}$ and diffusion matrix $\tilde{\boldsymbol{\Sigma}}$ of a diffusion approximation of a given jump process. These are for the multitype SIR model

$$\boldsymbol{\mu}(\boldsymbol{x}) = \sum_{j=1}^{n} N_j \left(w_{1,j}(\boldsymbol{x})\tilde{\boldsymbol{\Delta}}_{1,j} + w_{2,j}(\boldsymbol{x})\tilde{\boldsymbol{\Delta}}_{2,j} \right)$$
$$= \sum_{j=1}^{n} \left(w_{1,j}(\boldsymbol{x}) \begin{pmatrix} -\boldsymbol{e}_j \\ \boldsymbol{e}_j \end{pmatrix} + w_{2,j}(\boldsymbol{x}) \begin{pmatrix} \boldsymbol{0} \\ -\boldsymbol{e}_j \end{pmatrix} \right)$$

and

$$\tilde{\boldsymbol{\Sigma}}(\boldsymbol{x}) = \sum_{j=1}^{n} N_j \left(w_{1,j}(\boldsymbol{x})\tilde{\boldsymbol{\Delta}}_{1,j}\tilde{\boldsymbol{\Delta}}_{1,j}' + w_{2,j}(\boldsymbol{x})\tilde{\boldsymbol{\Delta}}_{2,j}\tilde{\boldsymbol{\Delta}}_{2,j}' \right)$$
$$= \sum_{j=1}^{n} N_j^{-1} \left(w_{1,j}(\boldsymbol{x}) \begin{pmatrix} \mathrm{diag}\,(\boldsymbol{e}_j) & -\mathrm{diag}\,(\boldsymbol{e}_j) \\ -\mathrm{diag}\,(\boldsymbol{e}_j) & \mathrm{diag}\,(\boldsymbol{e}_j) \end{pmatrix} + w_{2,j}(\boldsymbol{x}) \begin{pmatrix} \boldsymbol{0} & \boldsymbol{0} \\ \boldsymbol{0} & \mathrm{diag}\,(\boldsymbol{e}_j) \end{pmatrix} \right)$$

for $\boldsymbol{x} \in \mathcal{C}$, where $\mathcal{C}$ is the appropriate state space of thc diffusion approximation. These findings agree with those from the two preceding procedures, i.e. with Eqs. (5.26)–(5.31).

Van Kampen Expansion

Finally, consider the approximation of the multitype SIR model using the extended version of van Kampen's expansion as developed in Sect. 4.4.5.

This technique is applicable if the transition rate W_N can be written in the canonical form (4.65) and if condition (4.70) holds. The former requirement is fulfilled as

$$W_N(\boldsymbol{X}, \boldsymbol{\Delta}_{r,j}) = N_j\, w(\boldsymbol{M}^{-1}\boldsymbol{X}, \boldsymbol{\Delta}_{r,j})$$

for all $r = 1, 2$ and $j = 1, \ldots, n$, i.e. the terms in Formula (4.65) are to be chosen as $\Phi_0(\boldsymbol{x}, \boldsymbol{\Delta}) = w(\boldsymbol{x}, \boldsymbol{\Delta})$, $\Phi_l = 0$ for $l \geq 1$,

$$I_j = \{(r, j) \,|\, r = 1, 2\} \quad \text{with} \quad I = \biguplus_{j=1}^{n} I_j = \{(r, j) \,|\, r = 1, 2 \text{ and } j = 1, \ldots, n\},$$

and f is the identity function. Because of $\boldsymbol{M} = \mathrm{diag}\,(N_1, \ldots, N_n, N_1, \ldots, N_n)'$, one has $J_v = \{v, v+n\}$ (compare with definition (4.64)). Hence, the second condition is also true since

$$\tilde{a}_{1,j,v}(\boldsymbol{x}) = \sum_{u \in I_v} (\boldsymbol{\Delta}_u)_j\, \Phi_0(\boldsymbol{x}, \boldsymbol{\Delta}_u) = (\boldsymbol{\Delta}_{1,v})_j\, w_{1,v}(\boldsymbol{x}) + (\boldsymbol{\Delta}_{2,v})_j\, w_{2,v}(\boldsymbol{x})$$
$$= \begin{pmatrix} -\boldsymbol{e}_v \\ \boldsymbol{e}_v \end{pmatrix}_j w_{1,v}(\boldsymbol{x}) + \begin{pmatrix} \boldsymbol{0} \\ -\boldsymbol{e}_v \end{pmatrix}_j w_{2,v}(\boldsymbol{x}),$$

which equals zero if $j \notin \{v, v+n\}$. The drift vector and diffusion matrix of the diffusion approximation can thus be obtained by using Formulas (4.71) and (4.72), that is

$$\begin{aligned}\boldsymbol{\mu}(\boldsymbol{x}) &= \sum_{u\in I} \Phi_0(\boldsymbol{x}, \boldsymbol{\Delta}_u)\boldsymbol{\Delta}_u \\ &= \sum_{j=1}^{n} \left(w_{1,j}(\boldsymbol{x})\boldsymbol{\Delta}_{1,j} + w_{2,j}(\boldsymbol{x})\boldsymbol{\Delta}_{2,j}\right) \\ &= \sum_{j=1}^{n} \left(w_{1,j}(\boldsymbol{x}) \begin{pmatrix} -\boldsymbol{e}_j \\ \boldsymbol{e}_j \end{pmatrix} + w_{2,j}(\boldsymbol{x}) \begin{pmatrix} \boldsymbol{0} \\ -\boldsymbol{e}_j \end{pmatrix}\right)\end{aligned}$$

and

$$\begin{aligned}\tilde{\boldsymbol{\Sigma}}(\boldsymbol{x}) &= \sum_{v=1}^{n} N_v^{-1} \sum_{u\in I_v} \Phi_0(\boldsymbol{x}, \boldsymbol{\Delta}_u)\boldsymbol{\Delta}_u\boldsymbol{\Delta}_u' \\ &= \sum_{j=1}^{n} N_j^{-1} \left(w_{1,j}(\boldsymbol{x})\boldsymbol{\Delta}_{1,j}\boldsymbol{\Delta}_{1,j}' + w_{2,j}(\boldsymbol{x})\boldsymbol{\Delta}_{2,j}\boldsymbol{\Delta}_{2,j}'\right) \\ &= \sum_{j=1}^{n} N_j^{-1} \left(w_{1,j}(\boldsymbol{x}) \begin{pmatrix} \operatorname{diag}(\boldsymbol{e}_j) & -\operatorname{diag}(\boldsymbol{e}_j) \\ -\operatorname{diag}(\boldsymbol{e}_j) & \operatorname{diag}(\boldsymbol{e}_j) \end{pmatrix} + w_{2,j}(\boldsymbol{x}) \begin{pmatrix} \boldsymbol{0} & \boldsymbol{0} \\ \boldsymbol{0} & \operatorname{diag}(\boldsymbol{e}_j) \end{pmatrix}\right)\end{aligned}$$

in line with the results from all other approximation methods considered in this section.

The fact that van Kampen's expansion resembles the result of the Langevin approach and the Kramers-Moyal expansion is not only—as in the single size parameter case—in consequence of the special canonical form

$$w_N(\boldsymbol{x}, \boldsymbol{\Delta}_{r,j}) = N_j\, \Phi_0(\boldsymbol{x}, \boldsymbol{\Delta}_{r,j}),$$

but also due to the structure of the possible jumps: Because of $\boldsymbol{\Delta}_u = N_v\tilde{\boldsymbol{\Delta}}_u$ for $u \in I_v$, Formula (4.72) turns into

$$\tilde{\boldsymbol{\Sigma}}(\boldsymbol{x}) = \sum_{v=1}^{d} \sum_{u\in I_v} N_v^{-2} w_N(\boldsymbol{x}, \boldsymbol{\Delta}_u)\boldsymbol{\Delta}_u\boldsymbol{\Delta}_u' = \sum_{u\in I} w_N(\boldsymbol{x}, \boldsymbol{\Delta}_u)\tilde{\boldsymbol{\Delta}}_u\tilde{\boldsymbol{\Delta}}_u'.$$

5.2.4 Summary

The previous paragraphs dealt with the formulation as a jump process and the derivation of a diffusion approximation for the multitype SIR model. As the size of this system is best characterised through a collection of size parameters $N_1, \ldots, N_n$, the appropriate approximation techniques are the modified ones from Sect. 4.4.

Let $\boldsymbol{x} = (s_1, \ldots, s_n, i_1, \ldots, i_n)'$ denote the vector of fractions of susceptible and infectious individuals in the n distinct clusters. The master equation of the jump process with transitions (5.22) and (5.23) equals

$$\frac{\partial p_N(t,\boldsymbol{x})}{\partial t} = \sum_{j=1}^{n} N_j \left[\sum_{k=1}^{n} \pi^{jk}(s_j+\varepsilon_j, i_j-\varepsilon_j) p_N(t; s_j+\varepsilon_j, i_j-\varepsilon_j) \right.$$

$$+ \upsilon^j(s_j, i_j+\varepsilon_j) p_N(t; s_j, i_j+\varepsilon_j)$$

$$\left. - \left(\sum_{k=1}^{n} \pi^{jk}(s_j, i_j) + \upsilon^j(s_j, i_j) \right) p_N(t; s_j, i_j) \right],$$

where $\varepsilon_j = N_j^{-1}$,

$$\pi^{jk}(\boldsymbol{x}) = \alpha_k \frac{\sum_{m=1}^{n} \gamma_{mk}^{I} \frac{N_m}{N_k} i_m}{\sum_{m=1}^{n} \gamma_{mk}^{N} \frac{N_m}{N_k}} \gamma_{jk}^{S} s_j \qquad \text{and} \qquad \upsilon^j(\boldsymbol{x}) = \beta_j i_j$$

for all j and k. Note that for clarity only the relevant arguments of π^{jk}, υ^j and of the transition probability p_N of $\boldsymbol{x}$ are displayed.

As for the standard SIR model, all approximation methods yield identical diffusion processes. Together with an appropriate initial condition, this is the solution of the SDE

$$\mathrm{d}\boldsymbol{x}_t = \boldsymbol{\mu}(\boldsymbol{x}_t)\mathrm{d}t + \tilde{\boldsymbol{\sigma}}(\boldsymbol{x}_t)\mathrm{d}\boldsymbol{B}_t, \tag{5.32}$$

where

$$\boldsymbol{\mu}(\boldsymbol{x}) = \begin{pmatrix} \boldsymbol{\mu}^S(\boldsymbol{x}) \\ \boldsymbol{\mu}^I(\boldsymbol{x}) \end{pmatrix} \quad \text{and} \quad \tilde{\boldsymbol{\Sigma}}(\boldsymbol{x}) = \tilde{\boldsymbol{\sigma}}(\boldsymbol{x})\tilde{\boldsymbol{\sigma}}'(\boldsymbol{x}) = \begin{pmatrix} \tilde{\boldsymbol{\Sigma}}^{SS}(\boldsymbol{x}) & \tilde{\boldsymbol{\Sigma}}^{SI}(\boldsymbol{x}) \\ \tilde{\boldsymbol{\Sigma}}^{SI}(\boldsymbol{x}) & \tilde{\boldsymbol{\Sigma}}^{II}(\boldsymbol{x}) \end{pmatrix}$$

for vectors $\boldsymbol{\mu}^S$ and $\boldsymbol{\mu}^I$ and diagonal matrices $\tilde{\boldsymbol{\Sigma}}^{SS}$, $\tilde{\boldsymbol{\Sigma}}^{II}$ and $\tilde{\boldsymbol{\Sigma}}^{SI}$. The single components of these are

$$\begin{aligned} \mu_j^S(\boldsymbol{x}) &= -\sum_{k=1}^{n} \pi^{jk}(\boldsymbol{x}), \\ \mu_j^I(\boldsymbol{x}) &= \sum_{k=1}^{n} \pi^{jk}(\boldsymbol{x}) - \upsilon^j(\boldsymbol{x}) \end{aligned}$$

and

$$\tilde{\Sigma}_{jj}^{SS}(\boldsymbol{x}) = N_j^{-1} \sum_{k=1}^{n} \pi^{jk}(\boldsymbol{x}),$$

$$\tilde{\Sigma}_{jj}^{II}(\boldsymbol{x}) = N_j^{-1}\left(\sum_{k=1}^{n}\pi^{jk}(\boldsymbol{x}) + \upsilon^j(\boldsymbol{x})\right),$$

$$\tilde{\Sigma}_{jj}^{SI}(\boldsymbol{x}) = -N_j^{-1}\sum_{k=1}^{n}\pi^{jk}(\boldsymbol{x})$$

for $j = 1, \ldots, n$. The matrix

$$\tilde{\boldsymbol{\sigma}}(\boldsymbol{x}) = \begin{pmatrix} \tilde{\boldsymbol{\sigma}}^{SS}(\boldsymbol{x}) & \boldsymbol{0} \\ \tilde{\boldsymbol{\sigma}}^{SI}(\boldsymbol{x}) & \tilde{\boldsymbol{\sigma}}^{II}(\boldsymbol{x}) \end{pmatrix}$$

with diagonal matrices $\tilde{\boldsymbol{\sigma}}^{SS}, \tilde{\boldsymbol{\sigma}}^{II}, \tilde{\boldsymbol{\sigma}}^{SI}$ and

$$\tilde{\sigma}_{jj}^{SS}(\boldsymbol{x}) = \sqrt{\sum_{k=1}^{n}\frac{\pi^{jk}(\boldsymbol{x})}{N_j}}, \quad \tilde{\sigma}_{jj}^{II}(\boldsymbol{x}) = \sqrt{\frac{\upsilon^j(\boldsymbol{x})}{N_j}}, \quad \tilde{\sigma}_{jj}^{SI}(\boldsymbol{x}) = -\sqrt{\sum_{k=1}^{n}\frac{\pi^{jk}(\boldsymbol{x})}{N_j}}$$

for all j is a square root of $\tilde{\boldsymbol{\Sigma}}$. The state space of the diffusion approximation is

$$\mathcal{C}^{(n)} = \left\{(s_1, \ldots, s_n, i_1, \ldots, i_n)' \in [0,1]^{2n} \cap \mathbb{R}_0^{2n} \,\middle|\, s_j + i_j \le 1 \forall j = 1, \ldots, n\right\}. \quad (5.33)$$

The $2n$-dimensional Brownian motion $\boldsymbol{B}$ in Eq. (5.32) represents disturbances in transmission, recovery, and migration. A corresponding deterministic description of the model dynamics is given by

$$\mathrm{d}\boldsymbol{x}_t = \boldsymbol{\mu}(\boldsymbol{x}_t)\mathrm{d}t.$$

An illustration of the multitype SIR model follows in the next section.

5.2.5 Illustration and Further Remarks

In order to briefly demonstrate the dynamics of the multitype SIR model, the course of an epidemic is simulated for network matrices

$$\boldsymbol{\gamma}^N = \boldsymbol{\gamma}^S = \boldsymbol{\gamma}^I = \begin{pmatrix} 1-(n-1)a & a & \cdots & a \\ a & 1-(n-1)a & \cdots & a \\ \vdots & \vdots & \ddots & \vdots \\ a & a & \cdots & 1-(n-1)a \end{pmatrix} \in \mathbb{R}^{n \times n} \quad (5.34)$$

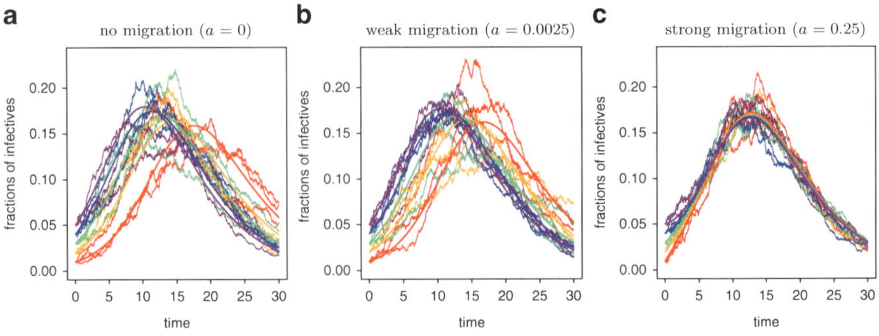

Fig. 5.2 Evolution of the fractions of infectives in $n = 5$ clusters. These agree in all parameters but the initial fractions of infectives. In particular, $\alpha_j = 0.5$, $\beta_j = 0.25$ and $N_j = 1{,}000$ for $j \in \{1, \dots, 5\}$. The initial numbers of infectives vary from 1 to 5 % of the population. Contacts between clusters occur according to the network matrix (5.34). There is no connection ($a = 0$) between clusters in (**a**), weak contact ($a = 0.0025$) in (**b**), and strong influence ($a = 0.25$) in (**c**). The *thick lines* show the deterministic evolution, the *thin lines* are three independent realisations of the diffusion process. All paths have been obtained by application of the Euler scheme with time step 0.025, see Sect. 3.3.2

with $0 \le a \le (n-1)^{-1}$ describing the strength of contacts between clusters. Figure 5.2 shows the evolution of the fractions of infectives during an epidemic with $n = 5$ clusters which agree in all parameters but the initial numbers of infectives. In the graphic on the very left there is no contact between clusters ($a = 0$), while there is strong exchange on the right ($a = 0.25$). Apparently, with increasing contacts of individuals between clusters, the courses of the epidemics synchronise. This fact is again illustrated in Fig. 5.3, where the dotted vertical lines mark the instants at which the fractions of susceptibles in the deterministic course fall below $\mathcal{R}_0^{-1}$, while the dashed lines indicate the actual turning points of the deterministic course of the epidemics, defined as the instants where the maximum amounts of infectives are reached. For clusters with initially high fractions of infectives, the actual turning point lies before the one that is valid for the model without exchange; for clusters with relatively few cases, the opposite situation applies.

The definition of a multitype counterpart to the basic reproductive ratio $\mathcal{R}_0$ in the standard SIR model with one homogeneous population is for example discussed by Andersson and Britton (2000) and Isham (2004). Moreover, Roberts and Heesterbeek (2003) and Heesterbeek and Roberts (2007) define and analyse a *type-reproduction number* as an alternative threshold quantity. This number coincides with $\mathcal{R}_0$ for homogeneous populations.

A possible modification of the multitype SIR model in this section is to consider movement of individuals between clusters instead of cross-infection. That means, individuals can change the cluster which they are associated with, and infection occurs only within clusters. This case is for example investigated by Dargatz et al. (2006). A disadvantage of that approach, however, is that

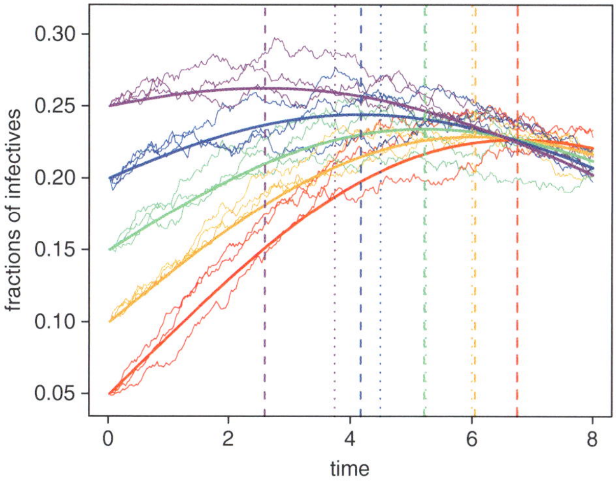

Fig. 5.3 Evolution of the fractions of infectives in $n = 5$ clusters between which people have contacts according to the network matrix (5.34) with $a = 0.075$. The clusters agree in all parameters but the initial numbers of infectives, which vary from 5 to 25 % of the population. As in Fig. 5.2, one has $\alpha_j = 0.5$, $\beta_j = 0.25$ and N_j =1,000 for $j \in \{1, \ldots, 5\}$. The *thick curves* show the deterministic evolution, the *thin lines* are three independent stochastic simulations of the diffusion process. The *dotted vertical lines* indicate the instants at which the fractions of susceptibles in the deterministic course fall below $\mathcal{R}_0^{-1}$. The *dashed vertical lines* mark the actual turning points of the deterministic course of the epidemic, that are the time instants where the fractions of infectives reach their maximums. Without contacts between clusters, these lines would agree within each community. This is actually the case for the *red lines* here (up to numerical inaccuracies). The sample paths have been obtained by application of the Euler scheme with time step 0.025, introduced in Sect. 3.3.2

population sizes of the distinct clusters do not remain constant, and the model is not immediately applicable to, for example, the case where clusters represent age groups.

In Chap. 8, the multitype SIR model is applied for modelling the spatial spread of influenza in Germany. Other models involving local and global infection dynamics are developed in Hufnagel et al. (2004), Germann et al. (2006), Débarre et al. (2007), Dybiec et al. (2009) and Ball et al. (2010). Watts et al. (2005) consider mixing on even more than two scales.

5.3 Existence and Uniqueness of Solutions

When considering an SDE as a model for some natural phenomenon, one implicitly assumes the existence of a solution of this SDE. Section 3.2.3 specified the Lipschitz condition (3.10) under which a strong solution of an SDE exists pathwise uniquely. This solution is non-explosive when it satisfies the growth condition (3.11).

For the standard and the multitype SIR models, the Lipschitz condition is actually not fulfilled as demonstrated in Sect. B.2 in the Appendix. Importantly, conditions (3.10) and (3.11) are sufficient but not necessary for the unique existence and non-explosiveness. Some authors describe weaker conditions, see e.g. Kloeden and Platen (1999, pp. 134–135). Further references include Kushner (1972), who studies the existence of a solution of an SDE when the drift function is not Lipschitz continuous, Abundo (1991), who considers the existence of solutions for a predator-prey model, and Kusuoka (2010), who investigates the existence of densities of solutions in case the Lipschitz condition is not fulfilled. Related to this general problem, Kaneko and Nakao (1988), Marion et al. (2002) and Berkaoui et al. (2005) deal with conditions under which numerical schemes converge to the true but unknown solution in case the Lipschitz condition is violated. Alternatively, one could settle for weak instead of strong solutions as distinguished in Sect. 3.2.3; this requires weaker assumptions.

In this book, the question of the existence of a strong solution for the considered SIR models on the entire state space is not completely answered as it is not the focus of this work. For our purposes, it suffices to consider the standard and multitype SIR models on a slightly restricted state space such that all fractions of susceptible and infectious individuals are bounded from below by an arbitrarily small but fixed positive constant ε. This does not limit the practical applicability of the diffusion models. The original state spaces $\mathcal{C}$ and $\mathcal{C}^{(n)}$ from Eqs. (5.20) and (5.33) on pp. 110 and 123 then become

$$\mathcal{C}_\varepsilon = \big\{(s,i)' \in [\varepsilon,1]^2 \cap \mathbb{R}_0^2 \,|\, s+i \leq 1\big\}$$

and

$$\mathcal{C}_\varepsilon^{(n)} = \big\{(s_1,\ldots,s_n,i_1,\ldots,i_n)' \in [\varepsilon,1]^{2n} \cap \mathbb{R}_0^{2n} \,\big|\, s_j + i_j \leq 1 \text{ for all } j=1,\ldots,n\big\},$$

respectively. This modification has the effect that the drift vectors and diffusion coefficients fulfil the Lipschitz and growth bound conditions as shown in Sect. B.2, i.e. there uniquely exist non-explosive strong solutions of the SDEs on the modified state spaces.

Independently of the investigation of the existence of a solution, diffusion approximations for the SIR model are considered problematic anyway when there are only few infectious individuals (e.g. Andersson and Britton 2000). The above proposed restriction of the state spaces of the diffusion approximations does hence not impose a serious constraint. An alternative approximation of the general stochastic epidemic during the initial and final phase of an epidemic is for example provided by Barbour (1976) and Andersson and Britton (2000, Chap. 3.3).

5.4 Conclusion

The description of the spread of infectious diseases in terms of diffusion processes enables convenient simulation of the random course of an epidemic even for large populations. In this chapter, diffusion approximations for the standard SIR model and a multitype extension were derived. On the one hand, these served as illustrations for the theoretical investigations in Chap. 4. On the other hand, the present chapter provides the basis for Chap. 8, where an influenza outbreak in a boarding school and the geographical spread of influenza in Germany are statistically analysed. Another application of diffusion approximations in life sciences is presented in Chap. 9. There, the in vivo binding behaviour of proteins is investigated as an example from molecular biology.

When applying the multitype SIR model in practice, several difficulties arise: First of all, one will typically want to prespecify the network matrices γ^N, γ^S and γ^I, or at least supply some information on their structure. That requires knowledge about, for example, transportation or social networks, depending on the definition of the clusters. In Chap. 8, commuter data from Germany is taken in order to estimate the geographical dispersal of the population. References for further examples for the utilisation of transportation networks are given in that chapter. Social contact networks may, for example, be approximated by the evaluation of contact diaries of similar surveys (Edmunds et al. 1997, 2006; Beutels et al. 2006; Wallinga et al. 2006; Mossong et al. 2008).

Another issue concerns the data about disease counts which is most often incomplete as many cases are not reported. In general, one also does not know the exact times at which infections occurred, and data is aggregated over periods of time. This is, of course, also problematic in case of one homogeneous population, but worsens in case of multiple communities. For example, Uphoff et al. (2004) summarise several difficulties arising from data aggregation over large geographical areas, ranging from dissimilar consultation behaviour to differences in physicians' opening hours, which limit the comparability of disease counts in distinct regions. These examples represent only some out of many challenges which epidemiologists are facing. Dealing with them is the subject of active research.

References

Abundo M (1991) A stochastic model for predator-prey systems: basic properties, stability and computer simulation. J Math Biol 29:495–511

Allen L (2003) An introduction to stochastic processes with applications to biology. Pearson Prentice Hall, Upper Saddle River

Alonso D, McKane A, Pascual M (2007) Stochastic amplification in epidemics. J R Soc Interface 4:575–582

Andersson H, Britton T (2000) Stochastic epidemic models and their statistical analysis. Lecture notes in statistics, vol 151. Springer, New York

Bailey N (1975) The mathematical theory of infectious diseases, 2nd edn. Charles Griffin, London

Ball F (1986) A unified approach to the distribution of total size and total area under the trajectory of infectives in epidemic models. Adv Appl Probab 18:289–310

Ball F, Sirl D, Trapman P (2010) Analysis of a stochastic SIR epidemic on a random network incorporating household structure. Math Biosci 224:53–73

Barbour A (1974) On a functional central limit theorem for Markov population processes. Adv Appl Probab 6:21–39

Barbour A (1976) Quasi-stationary distributions in Markov population processes. Adv Appl Probab 8:296–314

Berkaoui A, Bossy M, Diop A (2005) Euler scheme for SDEs with non-Lipschitz diffusion coefficient: strong convergence. Working paper 5637, INRIA Sophia Antipolis

Beutels P, Shkedy Z, Aerts M, van Damme P (2006) Social mixing patterns for transmission models of close contact infections: exploring self-evaluation and diary-based data collection through a web-based interface. Epidemiol Infect 134:1158–1166

Clancy D, O'Neill P, Pollett P (2001) Approximations for the long-term behavior of an open-population epidemic model. Methodol Comput Appl Probab 3:75–95

Daley D, Gani J (1999) Epidemic modelling: an introduction. Cambridge studies in mathematical biology, vol 15. Cambridge University Press, Cambridge

Dargatz C, Georgescu V, Held L (2006) Stochastic modelling of the spatial spread of influenza in Germany. Austrian J Stat 35:5–20

Débarre F, Bonhoeffer S, Regoes R (2007) The effect of population structure on the emergence of drug resistance during influenza pandemics. J R Soc Interface 4:893–906

Dybiec B, Kleczkowski A, Gilligan C (2009) Modelling control of epidemics spreading by long-range interactions. J R Soc Interface 6:941–950

Edmunds W, O'Callaghan C, Nokes D (1997) Who mixes with whom? A method to determine the contact patterns of adults that may lead to the spread of airborne infections. Proc R Soc Lond Ser B 264:949–957

Edmunds W, Kafatos G, Wallinga J, Mossong J (2006) Mixing patterns and the spread of close-contact infectious diseases. Emerg Themes Epidemiol 3:14847–14852

Germann T, Kadau K, Longini I, Macken C (2006) Mitigation strategies for pandemic influenza in the United States. Proc Natl Acad Sci USA 103:5935–5940

Heesterbeek J, Roberts M (2007) The type-reproduction number T in models for infectious disease control. Math Biosci 206:3–10

Hufnagel L, Brockmann D, Geisel T (2004) Forecast and control of epidemics in a globalized world. Proc Natl Acad Sci USA 101:15124–15129

Isham V (2004) Stochastic models for epidemics. Research Report No 263, Department of Statistical Science, University College London

Kaneko H, Nakao S (1988) A note on approximation for stochastic differential equations. In: Séminaire de Probabilités XXII. Lecture notes in mathematics. Springer, Berlin/Heidelberg

Keeling M, Rohani P (2008) Modeling infectious disease in humans and animals. Princeton University Press, Princeton

Kloeden P, Platen E (1999) Numerical solution of stochastic differential equations, 3rd edn. Springer, Berlin/Heidelberg/New York

Kurtz T (1981) Approximation of population processes. Society for Industrial and Applied Mathematics, Philadelphia

Kushner H (1972) Stability and existence of diffusions with discontinuous or rapidly growing drift terms. J Differ Equ 11:156–168

Kusuoka S (2010) Existence of densities of solutions of stochastic differential equations by Malliavin calculus. J Funct Anal 258:758–784

Marion G, Mao X, Renshaw E (2002) Convergence of the Euler scheme for a class of stochastic differential equation. Int Math J 1:9–22

McCormack R, Allen L (2006) Stochastic SIS and SIR multihost epidemic models. In: Agarwal R, Perera K (eds) Proceedings of the conference on differential and difference equations and applications. Hindawi, Cairo, pp 775–785

Mossong J, Hens N, Jit M, Beutels P, Auranen K, Mikolajczyk R, Massari M, Salmaso S, Scalia Tomba G, Wallinga J, Heijne J, Sadkowska-Todys M, Rosinska M, Edmunds W (2008) Social contacts and mixing patterns relevant to the spread of infectious diseases. PLoS Med 5:381–391

Nagaev A, Startsev A (1970) The asymptotic analysis of a stochastic model of an epidemic. Theory Probab Appl 15:98–107

Roberts M, Heesterbeek J (2003) A new method for estimating the effort required to control an infectious disease. Proc R Soc Lond Ser B 270:1359–1364

Sani A, Kroese D, Pollett P (2007) Stochastic models for the spread of HIV in a mobile heterosexual population. Math Biosci 208:98–124

Uphoff H, Stalleicken I, Bartelds A, Phiesel B, Kistemann B (2004) Are influenza surveillance data useful for mapping presentations? Virus Res 103:35–46

Wallinga J, Teunis P, Kretzschmar M (2006) Using data on social contacts to estimate age-specific transmission parameters for respiratory-spread infectious agents. Am J Epidemiol 164:936–944

Wang F (1977) Gaussian approximation of some closed stochastic epidemic models. J Appl Probab 14:221–231

Watts D, Muhamad R, Medina D, Dodds P (2005) Multiscale, resurgent epidemics in a hierarchical metapopulation model. Proc Natl Acad Sci USA 102:11157–11162

Part II
Statistical Inference

Chapter 6
Parametric Inference for Discretely-Observed Diffusions

As we have seen in Chap. 3, diffusion processes provide a widely-used and powerful modelling tool, and their mathematics is well understood. Chapter 4 described how to construct a diffusion approximation to a given stochastic phenomenon. This diffusion model is then known in parametric form. In practice, one usually wishes to furthermore estimate the parameters of this model. Statistical inference for diffusion processes, however, is a challenging problem. Difficulties arise from the fact that observations are typically discrete while the underlying diffusion model is continuous in time. In case of time-discrete observations, the likelihood function for the model parameters is generally unknown, and hence maximum likelihood estimation is not immediately possible.

This chapter provides a review on more sophisticated approaches to parametric inference for discretely-observed diffusion processes. The literature already provides a variety of different estimation techniques, but this subject is also still a highly developing research area. The present chapter concentrates on frequentist methodology and serves as an overview and introduction to statistical inference for diffusions. The emphasis of this book, however, lies on Bayesian techniques, which show even more attractive characteristics. These are presented and further developed in Chap. 7.

Throughout this chapter, we consider the time-homogeneous Itô diffusion $\boldsymbol{X} = (\boldsymbol{X}_t)_{t \geq 0}$ satisfying the stochastic differential equation

$$\mathrm{d}\boldsymbol{X}_t = \boldsymbol{\mu}(\boldsymbol{X}_t, \boldsymbol{\theta})\mathrm{d}t + \boldsymbol{\sigma}(\boldsymbol{X}_t, \boldsymbol{\theta})\mathrm{d}\boldsymbol{B}_t, \quad \boldsymbol{X}_{t_0} = \boldsymbol{x}_0, \tag{6.1}$$

with state space $\mathcal{X} \subseteq \mathbb{R}^d$, starting value $\boldsymbol{x}_0 \in \mathcal{X}$ at time $t_0 = 0$ and m-dimensional standard Brownian motion $\boldsymbol{B} = (\boldsymbol{B}_t)_{t \geq 0}$. The drift function $\boldsymbol{\mu} : \mathcal{X} \times \Theta \to \mathbb{R}^d$ and diffusion coefficient $\boldsymbol{\sigma} : \mathcal{X} \times \Theta \to \mathbb{R}^{d \times m}$ are assumed to be known in a parametric form. The statistical estimation of the possibly vector-valued parameter $\boldsymbol{\theta}$ from an open set $\Theta \subseteq \mathbb{R}^p$ is the objective of the methods introduced in this chapter.

We assume that $\boldsymbol{\mu}$, $\boldsymbol{\sigma}$ and the diffusion matrix $\boldsymbol{\Sigma} = \boldsymbol{\sigma}\boldsymbol{\sigma}'$ fulfil the regularity conditions stated in Sect. 3.4 for all $\boldsymbol{\theta} \in \Theta$; in particular, it is provided that an almost surely pathwise unique solution of the differential equation (6.1) exists for all

C. Fuchs, *Inference for Diffusion Processes: With Applications in Life Sciences*,
DOI 10.1007/978-3-642-25969-2_6,

parameters on a respective filtered probability space $(\Omega, \mathcal{F}^*, \mathcal{F}, \mathbb{P}_{\boldsymbol{\theta}})$, cf. Sect. 3.2.3. The state space $\mathcal{X}$ is the same for all values of $\boldsymbol{\theta}$. The true parameter value is denoted by $\boldsymbol{\theta}_0 \in \Theta$, and $\mathbb{E}_{\boldsymbol{\theta}}$ and $\mathrm{Var}_{\boldsymbol{\theta}}$ stand for the expectation and variance with respect to $\mathbb{P}_{\boldsymbol{\theta}}$, respectively. For some estimation approaches it is furthermore required that the diffusion process is ergodic. Such assumptions are indicated in the respective sections. Observations of the diffusion path are always considered to be measured without error.

This chapter is organised as follows: In order to provide the theoretical background, Sect. 6.1 starts with the formulation of the estimation problem for continuous-time observations and then goes over to discrete time under the assumption that the likelihood function of the parameter is known. Both scenarios are not directly applicable in practice. Section 6.2 hence presents a first attempt to obtain a feasible approximate maximum likelihood estimator. This approach, however, leads to asymptotically biased estimators. The remaining techniques covered in this chapter are more elaborate. They are grouped into three categories, in particular into approximations of the likelihood function in Sect. 6.3, alternatives to maximum likelihood estimation in Sect. 6.4 and a recent approach called the Exact Algorithm in Sect. 6.5. A comparison of the presented estimation techniques by means of a simulation study is beyond the scope of this book. However, a discussion follows in Sect. 6.6 including a summary and references to evaluation studies from the literature.

Other surveys on inference for discretely-observed diffusion processes are given by Prakasa Rao (1999), Nielsen et al. (2000), Sørensen (2004), Jimenez et al. (2006), Hurn et al. (2007) and Iacus (2008). Whenever an estimation technique is formulated for multi-dimensional diffusion processes in the original work, or the extension to multi-dimensional diffusions is obvious, this chapter presents the more general multi-dimensional case. Observation times are assumed non-equidistant even though the simpler equidistant setting is common in the original literature. Overall, the emphasis of this chapter is on the presentation of ideas and not on technical detail. For the latter, the reader is referred to the references given along the way.

The present review omits nonparametric inference. References for this topic include Florens-Zmirou (1993), Aït-Sahalia (1996), Jiang and Knight (1997), Soulier (1998), Jacod (2000), Hurn et al. (2003), Nicolau (2003) and Comte et al. (2007). An introduction to the subject is given in Iacus (2008, Chap. 4.2), a detailed overview by Prakasa Rao (1999).

6.1 Preliminaries

Crucially different situations occur depending on whether a diffusion process is observed continuously or discretely in time. Time-continuous observation is obviously impossible in practical applications. Still, the corresponding well-established theory is discussed in Sect. 6.1.1 for the sake of completeness and further understanding of subsequent asymptotic considerations. It forms the basis for the investigations in Sect. 7.3 in the next chapter, for example. In real data situations, one naturally has to deal with time-discrete observations. Section 6.1.2 briefly presents the challenges of parameter estimation for this setting. This is the starting point for the remainder of this chapter. Finally, Sect. 6.1.3 specifies the data situation which is considered in subsequent sections.

6.1.1 Time-Continuous Observation

Facing the hypothetical situation of continuous observation of a trajectory of $\boldsymbol{X}$ on a finite time interval $[s,t]$, parameter estimation can be carried out in two steps. This procedure has been described by Le Breton (1974) for linear SDEs and is explained for general SDEs in what follows: Split $\boldsymbol{\theta}$ into one part $\boldsymbol{\theta}_1$ already uniquely determined by the value of $\boldsymbol{\Sigma}(\cdot,\boldsymbol{\theta})$ and the remaining part $\boldsymbol{\theta}_2$. That means, if $\boldsymbol{\Sigma}(\boldsymbol{X}_t,\boldsymbol{\theta}) = \boldsymbol{S}$ for some matrix $\boldsymbol{S}$, then there exists a unique deterministic function g such that $\boldsymbol{\theta}_1 = g(\boldsymbol{S},\boldsymbol{X}_t)$. This does not necessarily imply that $\boldsymbol{\theta}_1$ enters only the diffusion matrix and $\boldsymbol{\theta}_2$ enters only the drift function.

Without loss of generality, let $\boldsymbol{\theta} = (\boldsymbol{\theta}_1', \boldsymbol{\theta}_2')'$. Since $\boldsymbol{X}$ has been observed continuously, it is straightforward to calculate its quadratic variation

$$\langle \boldsymbol{X},\boldsymbol{X}\rangle_{[s,t]} = \lim_{n\to\infty} \sum_{k=1}^{2^n} \Big(\boldsymbol{X}_{t_k^{(n)}} - \boldsymbol{X}_{t_{k-1}^{(n)}}\Big)\Big(\boldsymbol{X}_{t_k^{(n)}} - \boldsymbol{X}_{t_{k-1}^{(n)}}\Big)' = \int_s^t \boldsymbol{\Sigma}(\boldsymbol{X}_\tau,\boldsymbol{\theta})\mathrm{d}\tau,$$

where $t_k^{(n)} = s + k\,2^{-n}(t-s)$ for $k = 0,\ldots,2^n$, and the second equality holds in probability and almost surely (see Sect. 3.2.6). As a first step of the estimation procedure, the parameter $\boldsymbol{\theta}_1$ can then be determined through the limits in probability

$$\begin{aligned}\boldsymbol{\Sigma}(\boldsymbol{X}_t,\boldsymbol{\theta}) = \frac{\mathrm{d}\langle \boldsymbol{X},\boldsymbol{X}\rangle_{[s,t]}}{\mathrm{d}t} &= \lim_{n\to\infty} \frac{\langle \boldsymbol{X},\boldsymbol{X}\rangle_{[s,t]} - \langle \boldsymbol{X},\boldsymbol{X}\rangle_{[s,t-2^{-n}]}}{2^{-n}} \\ &= \lim_{n\to\infty} 2^n \left(\boldsymbol{X}_t - \boldsymbol{X}_{t-2^{-n}}\right)\left(\boldsymbol{X}_t - \boldsymbol{X}_{t-2^{-n}}\right)'\end{aligned}$$

by definition of $\boldsymbol{\theta}_1$ (see also Polson and Roberts 1994). Figure 7.25 on p. 232 in the next chapter illustrates how the diffusion coefficient of an Ornstein-Uhlenbeck process can precisely be determined from a sample path with small inter-observation time intervals.

In a second step, the remaining parameter $\boldsymbol{\theta}_2$ is now usually estimated using likelihood inference. In Sect. 3.2.4, the likelihood function with respect to Lebesgue measure was already considered for discrete observations. This function is generally unknown. If, however, the parameter $\boldsymbol{\theta}_1$ has already been determined as described above, and hence the diffusion coefficient is known as a function of $\boldsymbol{X}_t$, one can change the dominating measure such that the likelihood becomes available. In that case, the log-likelihood function of $\boldsymbol{\theta}_2$ reads

$$\begin{aligned}\ell^{\text{cont}}(\boldsymbol{\theta}_2) = &\int_s^t \boldsymbol{\mu}'(\boldsymbol{X}_\tau, \boldsymbol{\theta})\boldsymbol{\Sigma}^{-1}(\boldsymbol{X}_\tau, \boldsymbol{\theta})\mathrm{d}\boldsymbol{X}_\tau \\ &- \frac{1}{2}\int_s^t \boldsymbol{\mu}'(\boldsymbol{X}_\tau, \boldsymbol{\theta})\boldsymbol{\Sigma}^{-1}(\boldsymbol{X}_\tau, \boldsymbol{\theta})\boldsymbol{\mu}(\boldsymbol{X}_\tau, \boldsymbol{\theta})\mathrm{d}\tau,\end{aligned} \tag{6.2}$$

where $\boldsymbol{\theta} = (\boldsymbol{\theta}_1', \boldsymbol{\theta}_2')'$ is composed of the fixed $\boldsymbol{\theta}_1$ and the argument $\boldsymbol{\theta}_2$ of the log-likelihood function. Equation (6.2) is the Radon-Nikodym derivative $\mathrm{d}\mathbb{P}_{\boldsymbol{\sigma}}/\mathrm{d}\mathbb{W}_{\boldsymbol{\sigma}}$ from Girsanov's formula in Sect. 3.2.12, where $\mathbb{P}_{\boldsymbol{\sigma}}$ is the law of $\boldsymbol{X}$ defined by (6.1), and $\mathbb{W}_{\boldsymbol{\sigma}}$ is the law of the respective driftless process. The crucial point why it is possible to employ expression (6.2) as the log-likelihood is that the dominating measure $\mathbb{W}_{\boldsymbol{\sigma}}$ does not depend on $\boldsymbol{\theta}_2$ by definition of $\boldsymbol{\theta}_2$. Then maximisation of either $\mathrm{d}\mathbb{P}_{\boldsymbol{\sigma}}/\mathrm{d}\mathbb{W}_{\boldsymbol{\sigma}}$ or $\mathrm{d}\mathbb{P}_{\boldsymbol{\sigma}}/\mathrm{d}\mathbb{L}$, where $\mathbb{L}$ denotes Lebesgue measure, yields the same estimate for $\boldsymbol{\theta}_2$ irrespectively of the dominating measure. See also Kutoyants (2004, Chap. 1.1) or Liptser and Shiryayev (1977, Chap. 7, 1978, Chap. 17) on this topic.

In practice, Eq. (6.2) would be replaced by its discretisation

$$\begin{aligned}&\sum_{k=0}^{n-1} \boldsymbol{\mu}'(\boldsymbol{X}_{t_k}, \boldsymbol{\theta})\boldsymbol{\Sigma}^{-1}(\boldsymbol{X}_{t_k}, \boldsymbol{\theta})\big(\boldsymbol{X}_{t_{k+1}} - \boldsymbol{X}_{t_k}\big) \\ &- \frac{1}{2}\sum_{k=0}^{n-1} \boldsymbol{\mu}'(\boldsymbol{X}_{t_k}, \boldsymbol{\theta})\boldsymbol{\Sigma}^{-1}(\boldsymbol{X}_{t_k}, \boldsymbol{\theta})\boldsymbol{\mu}(\boldsymbol{X}_{t_k}, \boldsymbol{\theta})\big(t_{k+1} - t_k\big)\end{aligned} \tag{6.3}$$

according to the Itô interpretation of stochastic integrals, where $s = t_0 < t_1 < \ldots < t_n = t$ are observation times.

6.1.2 Time-Discrete Observation

In practice, however, the paths of a diffusion process cannot be observed continuously in time; due to the extremely wiggly trajectories (cf. Sect. 3.2.5), observations can never be complete but always have a smoothing character. Estimation of $\boldsymbol{\theta}$ will hence be based on observed states $\boldsymbol{x}_1, \ldots, \boldsymbol{x}_n$ of $\boldsymbol{X}$ at discrete times $t_1 < \ldots < t_n$ as well as on the starting value $\boldsymbol{x}_0$ at time $t_0 = 0$. The Kullback-Leibler distance between the continuous-time and the discrete-time model has been investigated by Dacunha-Castelle and Florens-Zmirou (1986) as a function of the time step between observations.

The focus of interest for discrete-time observations now lies on the transition density $p_{\boldsymbol{\theta}}(s, \boldsymbol{x}, t, \boldsymbol{y})$ with respect to Lebesgue measure, introduced in Sect. 3.2.4, which is defined by

$$\mathbb{P}_{\boldsymbol{\theta}}(\boldsymbol{X}_t \in A | \boldsymbol{X}_s = \boldsymbol{x}) = \int\limits_A p_{\boldsymbol{\theta}}(s, \boldsymbol{x}, t, \boldsymbol{y}) \mathrm{d}\boldsymbol{y}$$

for all measurable sets A, $t > s \geq 0$ and $\boldsymbol{x}, \boldsymbol{y} \in \mathcal{X}$. As diffusion processes are Markovian, the log-likelihood function of $\boldsymbol{\theta}$ with respect to Lebesgue measure is

$$\ell_n(\boldsymbol{\theta}) = \sum_{k=0}^{n-1} \log p_{\boldsymbol{\theta}}\big(\Delta t_k; \boldsymbol{x}_k, \boldsymbol{x}_{k+1}\big) \tag{6.4}$$

with $\Delta t_k = t_{k+1} - t_k$ for $k = 0, \ldots, n-1$. Under regularity conditions, Dacunha-Castelle and Florens-Zmirou (1986) prove consistency, asymptotic normality and asymptotic efficiency of the corresponding maximum likelihood estimator as n tends to infinity for one-dimensional ergodic diffusion processes and arbitrary equidistant time step.

However, the transition probability and hence the log-likelihood function are intractable unless the diffusion process is analytically explicitly solvable, which is rarely the case. Hence, in most situations, alternative methods need to be employed; this chapter gives an overview of the most established ones.

6.1.3 Time Scheme

In the remainder of this chapter, we assume that the diffusion process under consideration is observed at non-random discrete instants $t_1 < \ldots < t_n$ yielding a dataset $\{\boldsymbol{x}_1, \ldots, \boldsymbol{x}_n\}$. Furthermore, the initial state $\boldsymbol{x}_0$ at time $t_0 = 0$ is required to be known.

Let $\Delta := \max_k \Delta t_k$ be the maximum time step and $T = \sum_{k=0}^{n-1} \Delta t_k = t_n$ the time horizon. Three different experimental designs have been regarded in the literature for increasing number of observations, i.e. $n \to \infty$; the following names are adopted from Iacus (2008):

1. *Large-sample scheme:* The inter-observation times Δt_k remain fixed and T tends to infinity.
2. *High-frequency scheme:* Observations become denser, i.e. Δ goes to zero, and T remains constant.
3. *Rapidly increasing design:* The maximum time step Δ tends to zero while T grows to infinity at the same time.

From a theoretical point of view, the high-frequency scheme and the rapidly increasing design appear most convenient because they correspond to continuous observation in the limit. The setup of consistent estimators for the model parameter is often facilitated in these situations. For example, in some cases one can abandon regularity assumptions such as ergodicity of the diffusion process. However, the more complicated large-sample scheme seems to be most realistic in practice since observations typically arrive at fixed intervals. Like most authors cited in this review, we will hence base the following sections on that design. Some considerations of the other two schemes can for example be found in Prakasa Rao (1999) and Iacus (2008).

6.2 Naive Maximum Likelihood Approach

As discussed in Sect. 6.1.2, the exact log-likelihood function (6.4) of the parameter $\boldsymbol{\theta}$ for a discretely-observed diffusion process is usually unknown. Approximate maximum likelihood estimation would, however, be possible if an appropriate approximation of the transition density was available. A first attempt to implement this idea is described in the following.

Section 3.3.2 introduced the Euler scheme

$$\boldsymbol{Y}_{k+1} = \boldsymbol{Y}_k + \boldsymbol{\mu}(\boldsymbol{Y}_k, \boldsymbol{\theta})\Delta t_k + \boldsymbol{\sigma}(\boldsymbol{Y}_k, \boldsymbol{\theta})\mathcal{N}(\boldsymbol{0}, \Delta t_k \boldsymbol{I}), \tag{6.5}$$

where $k = 0, \ldots, n-1$, for approximately sampling the process $(\boldsymbol{X}_{t_k})_{k\in\mathbb{N}_0} = (\boldsymbol{Y}_k)_{k\in\mathbb{N}_0}$ at discrete time points $t_1 < \ldots < t_n$ for given parameter $\boldsymbol{\theta}$ and initial value $\boldsymbol{X}_{t_0} = \boldsymbol{Y}_0 = \boldsymbol{x}_0$. This scheme becomes more accurate as the maximum distance between two consecutive time instants tends to zero. Hence, for small Δt_k, we can assume $\boldsymbol{Y}_{k+1}$ conditional on $\boldsymbol{Y}_k$ to be approximately normally distributed. The conditional mean and variance can be obtained from (6.5) as

$$\mathbb{E}_{\boldsymbol{\theta}}\left(\boldsymbol{Y}_{k+1}\middle|\,\boldsymbol{Y}_k = \boldsymbol{x}_k\right) = \boldsymbol{x}_k + \boldsymbol{\mu}(\boldsymbol{x}_k, \boldsymbol{\theta})\Delta t_k \tag{6.6}$$

and

$$\mathrm{Var}_{\boldsymbol{\theta}}\left(\boldsymbol{Y}_{k+1}\middle|\,\boldsymbol{Y}_k = \boldsymbol{x}_k\right) = \boldsymbol{\Sigma}(\boldsymbol{x}_k, \boldsymbol{\theta})\Delta t_k. \tag{6.7}$$

The probability density $p_{\boldsymbol{\theta}}(\Delta t_k; \boldsymbol{x}_k, \boldsymbol{x}_{k+1})$ can thus be approximated by a Gaussian density with mean and variance according to (6.6) and (6.7). In case $\boldsymbol{\Sigma}$ does not depend on $\boldsymbol{\theta}$, the so-resulting log-likelihood function

$$\begin{aligned} \ell_n^{\mathrm{Euler}}(\boldsymbol{\theta}) = {} & \sum_{k=0}^{n-1} \boldsymbol{\mu}'(\boldsymbol{x}_k, \boldsymbol{\theta})\boldsymbol{\Sigma}^{-1}(\boldsymbol{x}_k)(\boldsymbol{x}_{k+1} - \boldsymbol{x}_k) \\ & - \frac{1}{2}\sum_{k=0}^{n-1} \boldsymbol{\mu}'(\boldsymbol{x}_k, \boldsymbol{\theta})\boldsymbol{\Sigma}^{-1}(\boldsymbol{x}_k)\boldsymbol{\mu}(\boldsymbol{x}_k, \boldsymbol{\theta})\Delta t_k \end{aligned} \tag{6.8}$$

corresponds to the Riemann-Itō approximation (6.3) of the log-likelihood (6.2) based on continuous observation. As a general convention, additive constants not depending on $\boldsymbol{\theta}$ are suppressed in the log-likelihood function.

Maximisation of the approximated log-likelihood function leads to an *approximate* or *naive maximum likelihood estimator*, sometimes also referred to as *quasi maximum likelihood estimator* (e.g. Honoré 1997). This estimator has good asymptotic properties in case of decreasing time step, in particular in the rapidly increasing design as defined in Sect. 6.1.3, see for example Florens-Zmirou (1989) or Yoshida (1992). The more realistic case, however, is that the time step is fixed. Lo (1988) provides a simple example where the naive maximum likelihood estimator is inconsistent for fixed observation intervals. More generally, Florens-Zmirou (1989) shows for ergodic diffusion processes with constant diffusion coefficient that the naive maximum likelihood estimator for the drift parameter has an asymptotic bias of the order of the equidistant fixed time step. This deficiency is not due to the Gaussian nature of the approximated transition density but because of the generally misspecified mean and variance of this normal density.

Unfortunately, in many applications in life sciences the time steps Δt_k are rather large. The fairly simple maximum likelihood approach considered in this section is hence not expected to yield satisfactory results in those cases. More advanced estimation procedures are required in order to address this problem. The following sections present such techniques.

6.3 Approximation of the Likelihood Function

The previous section concluded that in practical applications, where time steps between observations are large, the transition density of a diffusion process cannot satisfyingly be approximated by plain application of one of the standard numerical schemes from Sect. 3.3.2. This section hence introduces several more advanced approaches to approximate the transition density. These can be utilised to derive approximations of the log-likelihood (6.4). Maximisation of the so-obtained approximate log-likelihood then leads to an approximate maximum likelihood estimator.

6.3.1 Analytical Approximation of the Likelihood Function

The first more advanced approach considered in this review was originated by Aït-Sahalia (2002) and involves the expansion of the transition density in a Gram-Charlier series, which will be specified below. The result is a closed-form expression which is shown to converge to the true likelihood as more and more correction terms are included.

The method works for one-dimensional diffusion processes under fairly weak regularity conditions; see the original paper for details. Suppose the target process X satisfies the SDE

$$\mathrm{d}X_t = \mu_X(X_t, \boldsymbol{\theta})\mathrm{d}t + \sigma_X(X_t, \boldsymbol{\theta})\mathrm{d}B_t\ , \quad X_0 = x_0,$$

for $t \geq 0$. In general, the transition density of this process is not suitable for the expansion that is intended in this section as particularised below. The original process X is hence transformed to an appropriate process Z. The approximation of the transition density of Z can then be transferred to the transition density of X. The transformation from X to Z takes place in two invertible steps as follows.

The first operation transforms the diffusion X to a diffusion Y with unit diffusion coefficient. This is done with Lamperti's transformation described in Sect. 3.2.11. Then Y fulfils the SDE

$$\mathrm{d}Y_t = \left(\frac{\mu_X(X_t, \boldsymbol{\theta})}{\sigma_X(X_t, \boldsymbol{\theta})} - \frac{1}{2} \frac{\partial \sigma_X}{\partial x}(X_t, \boldsymbol{\theta}) \right) \mathrm{d}t + \mathrm{d}B_t, \; Y_0 = y_0, \tag{6.9}$$

for $t \geq 0$, where $\partial/\partial x$ denotes differentiation with respect to the state variable. Let $p_{X,\boldsymbol{\theta}}$ and $p_{Y,\boldsymbol{\theta}}$ denote the transition densities of X and Y, respectively. Aït-Sahalia (2002) demonstrates that the tails of $p_{Y,\boldsymbol{\theta}}$ are thin enough for the considered expansion. Overall, however, the density $p_{Y,\boldsymbol{\theta}}(\Delta t; y_0, y)$ is still not suitable as the function is peaked around $y = y_0$ for small Δt. Hence, one performs a second transformation

$$Z_t = \frac{Y_t - y_0}{\sqrt{t}}$$

for all $t \geq 0$. Naturally, the initial value of this new process equals $z_0 = 0$. Aït-Sahalia shows that for fixed Δt the transition density $p_{Z,\boldsymbol{\theta}}(\Delta t; z_0, z)$ of Z fulfils the necessary criteria; specifically, it can appropriately be expanded in a convergent series around a standard normal density.

Hence, one writes the function $p_{Z,\boldsymbol{\theta}}$ as a Gram-Charlier series (e.g. Kendall et al. 1987, Chap. 6), that is

$$p_{Z,\boldsymbol{\theta}}\big(\Delta t; z_0, z\big) = \phi(z) \sum_{j=0}^{\infty} \eta_j\big(\Delta t, \boldsymbol{\theta}, y_0\big) H_j(z). \tag{6.10}$$

In this expression, ϕ is the standard normal density, H_j are Hermite polynomials

$$H_j(z) = \exp\left(\frac{z^2}{2}\right) \frac{\partial^j}{\partial z^j} \exp\left(-\frac{z^2}{2}\right) \quad \text{for } j \in \mathbb{N}_0,$$

and

$$\begin{aligned} \eta_j(\Delta t, \boldsymbol{\theta}, y_0) &= \frac{1}{j!} \int_{-\infty}^{\infty} p_{Z,\boldsymbol{\theta}}\big(\Delta t; z_0, z\big) H_j(z) \mathrm{d}z \\ &= \frac{1}{j!} \mathbb{E}_{\boldsymbol{\theta}}\big(H_j(Z_{\Delta t}) \,\big|\, Z_0 = z_0\big). \end{aligned} \tag{6.11}$$

Kendall et al. actually define the H_j with alternating sign; that, however, does not change (6.10). The notation here follows Aït-Sahalia (2002).

The expected value in (6.11) can be rewritten via Taylor expansion (e.g. Gard 1988, Chap. 7) such that

$$\begin{aligned}\eta_j(\Delta t, \boldsymbol{\theta}, y_0) &= \frac{1}{j!}\,\mathbb{E}_{\boldsymbol{\theta}}\left(H_j\left(\frac{Y_{\Delta t} - y_0}{\sqrt{\Delta t}}\right)\middle|\, Y_0 = y_0\right)\\ &= \frac{1}{j!}\sum_{k=0}^{\infty}\frac{(\Delta t)^k}{k!}\left[\mathcal{G}_{\boldsymbol{\theta}}^k H_j\left(\frac{Y_{\Delta t} - y_0}{\sqrt{\Delta t}}\right)\right]_{Y_{\Delta t}=y_0},\end{aligned} \tag{6.12}$$

where $\mathcal{G}_{\boldsymbol{\theta}}$ is the infinitesimal generator (cf. Sect. 3.2.9) of the diffusion process Y with parameter $\boldsymbol{\theta}$, i.e.

$$\mathcal{G}_{\boldsymbol{\theta}} f = \mu_Y(\cdot, \boldsymbol{\theta}) f' + \frac{1}{2} f''$$

for any sufficiently regular function f. The function μ_Y denotes the drift of Y as apparent from (6.9).

Because of $\eta_0 \equiv 1$ and $H_0 \equiv 1$, the expansion (6.10) has leading term $\phi(z)$, i.e. the transition density of Z is expanded around a standard normal density. The change of variables theorem yields

$$p_{Y,\boldsymbol{\theta}}(\Delta t; y_0, y) = (\Delta t)^{-\frac{1}{2}} p_{Z,\boldsymbol{\theta}}(\Delta t; z_0, z) \tag{6.13}$$

and

$$p_{X,\boldsymbol{\theta}}(\Delta t; x_0, x) = (\sigma_X(x, \boldsymbol{\theta}))^{-1} p_{Y,\boldsymbol{\theta}}(\Delta t; y_0, y). \tag{6.14}$$

Findings for $p_{Z,\boldsymbol{\theta}}$ can thus be transferred to $p_{X,\boldsymbol{\theta}}$.

Equation (6.10) provides an explicit closed-form expression for the transition density $p_{Z,\boldsymbol{\theta}}$. The infinite sums in (6.10) and (6.12), however, can certainly not be computed in practice. Thus truncate these sums to obtain

$$p_{Z,\boldsymbol{\theta}}^{(J,K)}(\Delta t; z_0, z) = \phi(z)\sum_{j=0}^{J}\eta_j^{(K)}(\Delta t, \boldsymbol{\theta}, y_0) H_j(z) \tag{6.15}$$

with

$$\eta_j^{(K)}(\Delta t, \boldsymbol{\theta}, y_0) = \frac{1}{j!}\sum_{k=0}^{K}\frac{(\Delta t)^k}{k!}\left[\mathcal{G}_{\boldsymbol{\theta}}^k H_j\left(\frac{Y_{\Delta t} - y_0}{\sqrt{\Delta t}}\right)\right]_{Y_{\Delta t}=y_0}$$

as approximations to the true density $p_{Z,\boldsymbol{\theta}}(\Delta; z_0, z)$. Define $p_{Y,\boldsymbol{\theta}}^{(J,K)}$ and $p_{X,\boldsymbol{\theta}}^{(J,K)}$ as transformations of $p_{Z,\boldsymbol{\theta}}^{(J,K)}$ analogously to (6.13) and (6.14). Aït-Sahalia proves that there exists $\tilde{\Delta} > 0$ such that for all $\Delta t \in (0, \tilde{\Delta})$, $\boldsymbol{\theta} \in \Theta$ and $x_0, x \in \mathcal{X}$ one has

$$p_{X,\boldsymbol{\theta}}^{(J,\infty)}(\Delta t; x_0, x) \longrightarrow p_{X,\boldsymbol{\theta}}(\Delta t; x_0, x) \quad \text{as } J \to \infty.$$

Equation (6.15) provides a closed-form approximation to the transition density of Z but involves the fairly complex coefficients $\eta_j^{(K)}$. For example, one has

$$\begin{aligned}\eta_1^{(3)}(\Delta t, \boldsymbol{\theta}, y_0) = & -(\Delta t)^{\frac{1}{2}} \mu_Y - \frac{1}{2}(\Delta t)^{\frac{3}{2}} \left(\mu_Y \mu_Y' + \frac{1}{2}\mu_Y''\right) \\ & - \frac{1}{6}(\Delta t)^{\frac{5}{2}} \left(\mu_Y (\mu_Y')^2 + \mu_Y^2 \mu_Y'' + \mu_Y \mu_Y''' + \frac{3}{2}\mu_Y' \mu_Y'' + \frac{1}{4}\mu_Y''''\right),\end{aligned}$$

where the μ_Y are all evaluated at $(y_0, \boldsymbol{\theta})$. Aït-Sahalia however demonstrates that the approximation is sufficiently accurate already for a small number of terms.

An extension of the above approximation procedure for multi-dimensional diffusion processes is described by Aït-Sahalia (2008). It is applicable whenever the process can be transformed to one with unit diffusion; cf. the remarks at the end of Sect. 3.2.11. Singer (2004) chooses an approach for one-dimensional processes which is related to the one described here but expresses the coefficients of the expansion in terms of conditional moments of the diffusion process.

6.3.2 Numerical Solutions of the Kolmogorov Forward Equation

Section 3.2.8 introduced the Kolmogorov forward equation which uniquely determines the transition density of a diffusion process with respect to a given initial condition. Poulsen (1999) makes use of this description and approximates the transition density by numerically solving this deterministic partial differential equation. This approach has already been pursued by Lo (1988) who applies this idea to particular (jump-)diffusion processes but does not develop a general procedure.

The following considerations assume a one-dimensional diffusion process whose transition density fulfils the Kolmogorov forward equation

$$\frac{\partial p_{\boldsymbol{\theta}}(t; x_0, x)}{\partial t} = -\frac{\partial \left[\mu(x, \boldsymbol{\theta}) p_{\boldsymbol{\theta}}(t; x_0, x)\right]}{\partial x} + \frac{1}{2}\frac{\partial^2 \left[\sigma^2(x, \boldsymbol{\theta}) p_{\boldsymbol{\theta}}(t; x_0, x)\right]}{\partial x^2} \tag{6.16}$$

for $t \geq 0$ and $x_0, x \in \mathcal{X}$. The diffusion is assumed stationary and ergodic (cf. Sect. 3.2.7). Indications for handling multi-dimensional diffusion processes are given in the paper by Poulsen (1999).

By the product rule, Eq. (6.16) is identical with

$$\begin{aligned}&\frac{\partial p_{\boldsymbol{\theta}}(t; x_0, x)}{\partial t} \\ &\quad = a(x, \boldsymbol{\theta}) p_{\boldsymbol{\theta}}(t; x_0, x) + b(x, \boldsymbol{\theta}) \frac{\partial p_{\boldsymbol{\theta}}(t; x_0, x)}{\partial x} + c(x, \boldsymbol{\theta}) \frac{\partial^2 p_{\boldsymbol{\theta}}(t; x_0, x)}{\partial x^2},\end{aligned} \tag{6.17}$$

where

$$a(x, \boldsymbol{\theta}) = -\frac{\partial \mu(x, \boldsymbol{\theta})}{\partial x} + \frac{1}{2}\frac{\partial^2 \sigma^2(x, \boldsymbol{\theta})}{\partial x^2},$$

$$b(x,\boldsymbol{\theta}) = -\mu(x,\boldsymbol{\theta}) + 2\sigma(x,\boldsymbol{\theta})\,\frac{\partial\sigma(x,\boldsymbol{\theta})}{\partial x},$$
$$c(x,\boldsymbol{\theta}) = \frac{1}{2}\,\sigma^2(x,\boldsymbol{\theta}).$$

These coefficients are known as functions of x and $\boldsymbol{\theta}$ since μ and σ are known in parametric form. Poulsen (1999) approximates Eq. (6.17) by application of the *Crank-Nicolson method* (Crank and Nicolson 1947). In the following, some more detail on this is given than in Poulsen (1999). The reader who is rather interested in the conceptual idea of the estimation procedure, however, may directly proceed to the last paragraph of this section.

In the Crank-Nicolson technique, Eq. (6.17) is approximated by

$$\frac{p^x_{t+\Delta t} - p^x_t}{\Delta t}$$
$$= \frac{1}{2}\left(ap^x_t + b\,\frac{p^{x+\Delta x}_t - p^{x-\Delta x}_t}{2\Delta x} + c\,\frac{p^{x+\Delta x}_t - 2p^x_t + p^{x-\Delta x}_t}{(\Delta x)^2}\right) \tag{6.18}$$
$$+\frac{1}{2}\left(ap^x_{t+\Delta t} + b\frac{p^{x+\Delta x}_{t+\Delta t} - p^{x-\Delta x}_{t+\Delta t}}{2\Delta x} + c\frac{p^{x+\Delta x}_{t+\Delta t} - 2p^x_{t+\Delta t} + p^{x-\Delta x}_{t+\Delta t}}{(\Delta x)^2}\right). \tag{6.19}$$

In this equation, $p^x_t = p_{\boldsymbol{\theta}}(t;x_0,x)$ for all $t \geq 0$ and $x \in \mathcal{X}$, i.e. the lower index of p^x_t denotes the time variable, the upper index denotes the state at this time, and $\boldsymbol{\theta}$ and x_0 are kept fixed. Furthermore, $a = a(x,\boldsymbol{\theta})$, $b = b(x,\boldsymbol{\theta})$ and $c = c(x,\boldsymbol{\theta})$. The bracketed term in line (6.18) corresponds to an Euler forward approximation of the right hand side of (6.17), and line (6.19) stems from an Euler backward approximation. The Crank-Nicolson method is hence the average of these two schemes. With the notation from Definition B.1 on p. 379 in the Appendix, as a side note, the last equation reads

$$\frac{D^1_{(1,0)',(\Delta t,\cdot)'}p^x_t}{\Delta t} = \frac{1}{2}\left(ap^x_t + b\,\frac{D^1_{(0,1)',(\cdot,2\Delta x)'}p^{x-\Delta x}_t}{2\Delta x} + c\,\frac{D^2_{(0,2)',(\cdot,\Delta x)'}p^{x-\Delta x}_t}{(\Delta x)^2}\right)$$
$$+\frac{1}{2}\left(ap^x_{t+\Delta t} + b\frac{D^1_{(0,1)',(\cdot,2\Delta x)'}p^{x-\Delta x}_{t+\Delta t}}{2\Delta x} + c\frac{D^2_{(0,2)',(\cdot,\Delta x)'}p^{x-\Delta x}_{t+\Delta t}}{(\Delta x)^2}\right),$$

where t is considered the first and x the second argument of p^x_t. Now assume that states $x_0, x_1, \ldots, x_n$ of the diffusion process have been observed at times $t_0, t_1, \ldots, t_n$. Adapted to this setting, the expressions (6.18)–(6.19) read

$$\frac{p^{x_i}_{t_j} - p^{x_i}_{t_{j-1}}}{t_j - t_{j-1}} = \frac{1}{2}\left(a(x_i,\boldsymbol{\theta})p^{x_i}_{t_{j-1}} + b(x_i,\boldsymbol{\theta})\,\frac{p^{x_{i+1}}_{t_{j-1}} - p^{x_{i-1}}_{t_{j-1}}}{x_{i+1} - x_{i-1}} + c(x_i,\boldsymbol{\theta})\,\frac{p^{x_{i+1}}_{t_{j-1}} - 2p^{x_i}_{t_{j-1}} + p^{x_{i-1}}_{t_{j-1}}}{(x_{i+1}-x_i)(x_i - x_{i-1})}\right)$$
$$+\frac{1}{2}\left(a(x_i,\boldsymbol{\theta})p^{x_i}_{t_j} + b(x_i,\boldsymbol{\theta})\,\frac{p^{x_{i+1}}_{t_j} - p^{x_{i-1}}_{t_j}}{x_{i+1} - x_{i-1}} + c(x_i,\boldsymbol{\theta})\,\frac{p^{x_{i+1}}_{t_j} - 2p^{x_i}_{t_j} + p^{x_{i-1}}_{t_j}}{(x_{i+1}-x_i)(x_i - x_{i-1})}\right)$$

for $i = 1, \ldots, n-1$ and $j = 1, \ldots, n$. Rearrangement yields

$$A_{ij} p_{t_j}^{x_{i-1}} + B_{ij} p_{t_j}^{x_i} + C_{ij} p_{t_j}^{x_{i+1}} = q_{i,j-1}, \tag{6.20}$$

where

$$\begin{aligned}
A_{ij} &= \frac{b(x_i, \boldsymbol{\theta})}{2(x_{i+1} - x_{i-1})} - \frac{c(x_i, \boldsymbol{\theta})}{2(x_{i+1} - x_i)(x_i - x_{i-1})}, \\
B_{ij} &= \frac{1}{t_j - t_{j-1}} - \frac{a(x_i, \boldsymbol{\theta})}{2} + \frac{c(x_i, \boldsymbol{\theta})}{(x_{i+1} - x_i)(x_i - x_{i-1})}, \\
C_{ij} &= -\frac{b(x_i, \boldsymbol{\theta})}{2(x_{i+1} - x_{i-1})} - \frac{c(x_i, \boldsymbol{\theta})}{2(x_{i+1} - x_i)(x_i - x_{i-1})}, \\
q_{i,j-1} &= \left(-\frac{b(x_i, \boldsymbol{\theta})}{2(x_{i+1} - x_{i-1})} + \frac{c(x_i, \boldsymbol{\theta})}{2(x_{i+1} - x_i)(x_i - x_{i-1})} \right) p_{t_{j-1}}^{x_{i-1}} \\
&\quad + \left(\frac{1}{t_j - t_{j-1}} + \frac{a(x_i, \boldsymbol{\theta})}{2} - \frac{c(x_i, \boldsymbol{\theta})}{(x_{i+1} - x_i)(x_i - x_{i-1})} \right) p_{t_{j-1}}^{x_i} \\
&\quad + \left(\frac{b(x_i, \boldsymbol{\theta})}{2(x_{i+1} - x_{i-1})} + \frac{c(x_i, \boldsymbol{\theta})}{2(x_{i+1} - x_i)(x_i - x_{i-1})} \right) p_{t_{j-1}}^{x_{i+1}}.
\end{aligned}$$

In order to approximate the log-likelihood function $\ell_n(\boldsymbol{\theta})$ as shown in (6.4) for given $\boldsymbol{\theta}$, one has to approximately determine all elements of $\{p_{\boldsymbol{\theta}}(\Delta t_k; x_k, x_{k+1}) | \, k = 0, \ldots, n-1\}$. For $k = 0$, this can be done as follows: Summarise Eq. (6.20) as the tridiagonal system

$$\begin{pmatrix}
D_j^{(1)} & D_j^{(2)} & 0 & 0 & \cdots & 0 & 0 & 0 \\
A_{1j} & B_{1j} & C_{1j} & 0 & \cdots & 0 & 0 & 0 \\
0 & A_{2j} & B_{2j} & C_{2j} & \cdots & 0 & 0 & 0 \\
\vdots & \vdots & \vdots & \vdots & \ddots & \vdots & \vdots & \vdots \\
0 & 0 & 0 & 0 & \cdots & A_{n-1,j} & B_{n-1,j} & C_{n-1,j} \\
0 & 0 & 0 & 0 & \cdots & 0 & D_j^{(3)} & D_j^{(4)}
\end{pmatrix}
\begin{pmatrix}
p_{t_j}^{x_0} \\ p_{t_j}^{x_1} \\ p_{t_j}^{x_2} \\ p_{t_j}^{x_3} \\ \vdots \\ p_{t_j}^{x_{n-1}} \\ p_{t_j}^{x_n}
\end{pmatrix}
=
\begin{pmatrix}
q_{0,j-1} \\ q_{1,j-1} \\ q_{2,j-1} \\ q_{3,j-1} \\ \vdots \\ q_{n-1,j-1} \\ q_{n,j-1}
\end{pmatrix}, \tag{6.21}$$

where $D_j^{(1)}$, $D_j^{(2)}$, $D_j^{(3)}$, $D_j^{(4)}$, $q_{0,j-1}$ and $q_{n,j-1}$ have to be determined separately from the boundary conditions. Derive $p_{t_0}^{x_i}$ and q_{i0} for all i according to the initial conditions. Finally, solve (6.21) for $j = 1$. For different values of k, i.e. different initial states and times, adapt (6.21) accordingly and successively solve the resulting system for $j = 1, \ldots, k+1$. See Poulsen (1999) for technical details considering the initial and boundary conditions. Note that this numerical procedure determines several more values of the transition density than actually needed for the approximation of $\ell_n(\boldsymbol{\theta})$.

Poulsen (1999) shows that the so-obtained approximation $\hat{\ell}_n(\boldsymbol{\theta})$ of the log-likelihood function $\ell_n(\boldsymbol{\theta})$ satisfies

$$\hat{\ell}_n(\boldsymbol{\theta}) = \ell_n(\boldsymbol{\theta}) + h^2 g_n^{(1)}(\boldsymbol{\theta}, x_0, \ldots, x_n) + o(h^2) g_n^{(2)}(\boldsymbol{\theta}, x_0, \ldots, x_n),$$

where $g_n^{(1)}$ and $g_n^{(2)}$ are appropriate functions and $h > 0$ is chosen such that the computing time for the approximated log-likelihood is at most of order n/h^2.

6.3.3 Simulated Maximum Likelihood Estimation

This section describes an approach by Pedersen (1995b) and Santa-Clara (1995) which is known as *simulated maximum likelihood estimation (SMLE)*. It is based on the observation that by the Chapman-Kolmogorov equation the transition density can be expressed as

$$\begin{aligned} p_{\boldsymbol{\theta}}(s, \boldsymbol{x}, t, \boldsymbol{y}) &= \int_{\mathcal{X}} p_{\boldsymbol{\theta}}(s, \boldsymbol{x}, t-\delta, \boldsymbol{z}) p_{\boldsymbol{\theta}}(t-\delta, \boldsymbol{z}, t, \boldsymbol{y}) \mathrm{d}\boldsymbol{z} \\ &= \mathbb{E}_{\boldsymbol{\theta}} \left(p_{\boldsymbol{\theta}}(t-\delta, \boldsymbol{X}_{t-\delta}, t, \boldsymbol{y}) \middle| \boldsymbol{X}_s = \boldsymbol{x} \right) \end{aligned}$$

for all $\boldsymbol{x}, \boldsymbol{y} \in \mathcal{X}$, $t > s \geq 0$ and $0 < \delta < t - s$. For small δ, usually $\delta \ll t - s$, the function $p_{\boldsymbol{\theta}}(t-\delta, \cdot, t, \cdot)$ can be replaced by a Gaussian density, and hence an approximation of the above expectation can be obtained by Monte Carlo integration through repeated (approximate) simulation of $\boldsymbol{X}_{t-\delta} | \{\boldsymbol{X}_s = \boldsymbol{x}\}$.

In the following, we concentrate on the work by Pedersen (1995b) who defines the first-order approximation

$$p_{\boldsymbol{\theta}}^{(1)}(s, \boldsymbol{x}, t, \boldsymbol{y}) = \phi\Big(\boldsymbol{y} \,\big|\, \boldsymbol{x} + (t-s)\boldsymbol{\mu}(\boldsymbol{x}, \boldsymbol{\theta}), (t-s)\boldsymbol{\Sigma}(\boldsymbol{x}, \boldsymbol{\theta})\Big) \tag{6.22}$$

of $p_{\boldsymbol{\theta}}$ like in the naive maximum likelihood approach in Sect. 6.2. Once again, the notation $\phi(\boldsymbol{z}|\boldsymbol{\nu}, \boldsymbol{\Lambda})$ refers to a multivariate Gaussian density with mean $\boldsymbol{\nu}$ and variance $\boldsymbol{\Lambda}$ evaluated at $\boldsymbol{z}$. As further refinements of $p_{\boldsymbol{\theta}}^{(1)}$, Pedersen introduces for numbers $N \geq 2$

$$p_{\boldsymbol{\theta}}^{(N)}(s, \boldsymbol{x}, t, \boldsymbol{y}) = \int_{\mathcal{X}} \cdots \int_{\mathcal{X}} \prod_{k=0}^{N-1} p_{\boldsymbol{\theta}}^{(1)}(\tau_k, \boldsymbol{\xi}_k, \tau_{k+1}, \boldsymbol{\xi}_{k+1}) \, \mathrm{d}\boldsymbol{\xi}_1 \ldots \mathrm{d}\boldsymbol{\xi}_{N-1}, \tag{6.23}$$

where $\tau_k = s + k(t-s)/N$ for $k = 0, \ldots, N$, $\boldsymbol{\xi}_0 = \boldsymbol{x}$ and $\boldsymbol{\xi}_N = \boldsymbol{y}$. Pedersen proves that, under weak regularity conditions,

$$\lim_{N \to \infty} p_{\boldsymbol{\theta}}^{(N)}(s, \boldsymbol{x}, t, \boldsymbol{y}) = p_{\boldsymbol{\theta}}(s, \boldsymbol{x}, t, \boldsymbol{y}) \quad \text{in } L^1.$$

Then, for observed states $\boldsymbol{x}_1, \ldots, \boldsymbol{x}_n$ at times $t_1 < \ldots < t_n$, the so-approximated log-likelihood function converges in probability to the true log-likelihood function (6.4):

$$\lim_{N\to\infty} \ell_n^{(N)}(\boldsymbol{\theta}) := \lim_{N\to\infty} \sum_{k=0}^{n-1} \log p_{\boldsymbol{\theta}}^{(N)}\big(t_k, \boldsymbol{x}_k, t_{k+1}, \boldsymbol{x}_{k+1}\big)$$
$$= \ell_n(\boldsymbol{\theta}) \quad \text{in probability under } \mathbb{P}_{\boldsymbol{\theta}_0}.$$

Pedersen (1995a) proves consistency and asymptotic normality of the estimator $\hat{\boldsymbol{\theta}}_n^{(N)}$ which is obtained through maximisation of $\ell_n^{(N)}$. Note that $N \to \infty$ refers to decreasing time steps due to *imputed* intermediate states $\boldsymbol{\xi}_1, \ldots, \boldsymbol{\xi}_{N-1}$, i.e. it does *not* correspond to the high-frequency time scheme defined in Sect. 6.1.3.

It would be computationally too costly to integrate out all unobserved variables $\boldsymbol{\xi}_1, \ldots, \boldsymbol{\xi}_{N-1}$ in (6.23), but, as indicated before, we can alternatively write the integral as

$$\begin{aligned} p_{\boldsymbol{\theta}}^{(N)}(s, \boldsymbol{x}, t, \boldsymbol{y}) &= \mathbb{E}_{\boldsymbol{\theta}}\left(p_{\boldsymbol{\theta}}^{(1)}(\tau_{N-1}, \boldsymbol{X}_{\tau_{N-1}}, t, \boldsymbol{y}) \middle| \boldsymbol{X}_s = \boldsymbol{x}\right) \\ &= \int_{\mathcal{X}} p_{\boldsymbol{\theta}}^{(1)}(\tau_{N-1}, \boldsymbol{z}_{N-1}, t, \boldsymbol{y}) \mathrm{d}\mathbb{P}_{N-1}^{(N)}(\boldsymbol{z}_{N-1}), \end{aligned} \tag{6.24}$$

where $\mathbb{P}_{N-1}^{(N)}$ is the law of a random variable that is generated by $N-1$ Euler steps with equidistant time step $(t-s)/N$ and starting point $\boldsymbol{x}$ at time s. Pedersen (1995b) hence proposes to draw M independent random variables $\boldsymbol{z}_{N-1}^m$, $m = 1, \ldots, M$, from $\mathbb{P}_{N-1}^{(N)}$ and to estimate $p_{\boldsymbol{\theta}}^{(N)}(s, \boldsymbol{x}, t, \boldsymbol{y})$ by

$$\hat{p}_{\boldsymbol{\theta}}^{(N,M)} = \frac{1}{M} \sum_{m=1}^{M} p_{\boldsymbol{\theta}}^{(1)}(\tau_{N-1}, \boldsymbol{z}_{N-1}^m, t, \boldsymbol{y}). \tag{6.25}$$

Since each of the M realisations $\boldsymbol{z}_{N-1}^m$ requires $N-1$ Euler steps, the computational demand of this estimation is of order $O(MN)$. It is hence desirable that (6.25) converges quickly. Unfortunately, this is not the case for the just proposed sampling scheme.

The reason for the poor convergence has been pointed out by Durham and Gallant (2002) as follows: Expression (6.24) can be rewritten as

$$p_{\boldsymbol{\theta}}^{(N)}(s, \boldsymbol{x}, t, \boldsymbol{y}) = \int_{\mathcal{X}} p_{\boldsymbol{\theta}}^{(1)}(\tau_{N-1}, \boldsymbol{z}_{N-1}, t, \boldsymbol{y}) \rho(\boldsymbol{z}_{N-1})\, \mathrm{d}\mathbb{Q}^{(N)}(\boldsymbol{z}_{N-1}), \tag{6.26}$$

where $\rho = \mathrm{d}\mathbb{P}_{N-1}^{(N)}/\mathrm{d}\mathbb{Q}^{(N)}$ is the Radon-Nikodym derivative of $\mathbb{P}_{N-1}^{(N)}$ with respect to a probability measure $\mathbb{Q}^{(N)}$, where $\mathbb{P}_{N-1}^{(N)}$ is absolutely continuous with respect to $\mathbb{Q}^{(N)}$. (6.26) can then be estimated by importance sampling as

$$\frac{1}{M} \sum_{m=1}^{M} p_{\boldsymbol{\theta}}^{(1)}(\tau_{N-1}, \tilde{\boldsymbol{z}}_{N-1}^m, t, \boldsymbol{y}) \rho(\tilde{\boldsymbol{z}}_{N-1}^m),$$

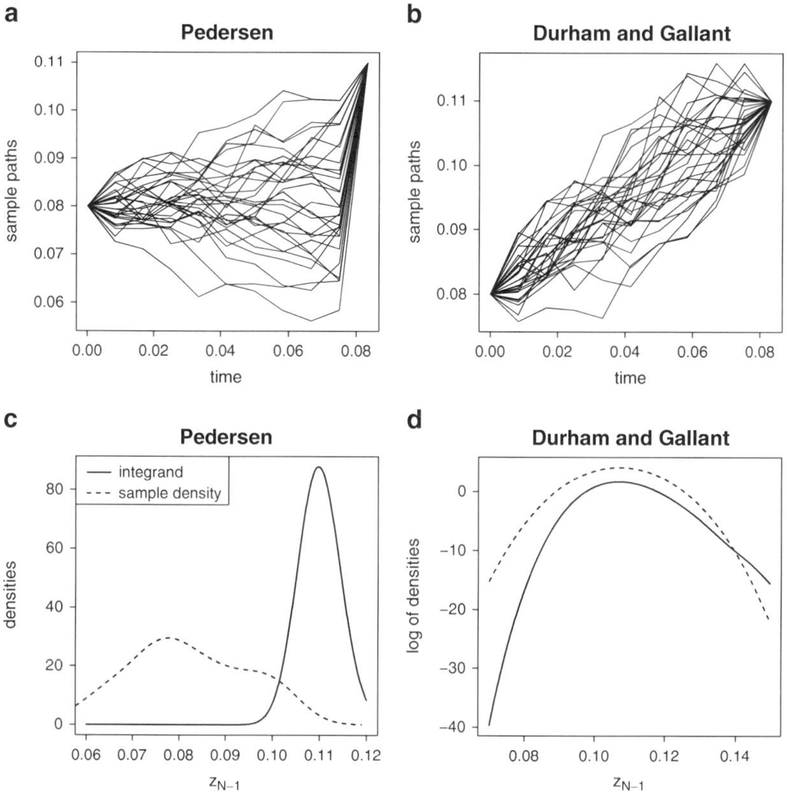

Fig. 6.1 Illustration of importance sampling for intermediate states of Cox-Ingersoll-Ross (CIR) process $\mathrm{d}X_t = 0.06(0.5 - X_t)\mathrm{d}t + 0.15\sqrt{X_t}\mathrm{d}B_t$ starting in $x = 0.08$ at time $s = 0$ and ending in $y = 0.11$ at time $t = 1/12 = 0.08$. The time interval $[s, t]$ is further divided into $N = 10$ equidistant subintervals. This setting corresponds to the example in Sect. 5 of Durham and Gallant (2002). The CIR process is introduced in Sect. A.3 in the Appendix of this book. (**a**) Simulation of $M = 30$ independent discretised sample paths $\{z_1^m, \ldots, z_{N-1}^m\}$, $m = 1, \ldots, M$, by application of the Euler scheme, i.e. as in Pedersen (1995b). Apparently, there occur relatively large jumps between the states z_{N-1}^m and y. (**b**) Simulation of $M = 30$ independent discretised sample paths as in (6.27), i.e. following Durham and Gallant (2002). These appear more likely. (**c**) Comparison of the integrand $p_{\boldsymbol{\theta}}^{(1)}(9/120, z_{N-1}, 1/12, y)$ in (6.24) (*solid line*) and the empirical sampling density corresponding to the M realisations z_{N-1}^m in Fig. 6.1a (*dashed line*). (**d**) Comparison of the integrand $p_{\boldsymbol{\theta}}^{(1)}(9/120, z_{N-1}, 1/12, y)\rho(z_{N-1})$ in (6.26) and the empirical sampling density corresponding to the M realisations z_{N-1}^m in Fig. 6.1b. The density ρ has been obtained by Monte Carlo estimation with sample sizes 10^5. This graphic shows the logarithms of the densities

where $\tilde{z}_{N-1}^1, \ldots, \tilde{z}_{N-1}^M$ are independent draws from $\mathbb{Q}^{(N)}$. Good results are obtained in finite time if $\mathbb{Q}^{(N)}$ has large probability mass where the integrand is large. Figure 6.1 displays the results of a small simulation study which indicate that this property is not met by Pedersen's choice $\mathbb{Q}^{(N)} = \mathbb{P}_{N-1}^{(N)}$: Since $\mathbb{P}_{N-1}^{(N)}$ is not conditioned on the end point $\boldsymbol{X}_t = \boldsymbol{y}$, it produces trajectories which are usually quite unlikely due to relatively large jumps between the states at times τ_{N-1} and t. This becomes apparent in Fig. 6.1a, c.

One proposal for $\mathbb{Q}^{(N)}$ by Durham and Gallant (2002) is to replace the normal densities

$$\pi(\boldsymbol{z}_{k+1}|\boldsymbol{z}_k) = \phi\big(\boldsymbol{z}_{k+1}\big|\boldsymbol{z}_k + \boldsymbol{\mu}(\boldsymbol{z}_k,\boldsymbol{\theta})\delta, \boldsymbol{\Sigma}(\boldsymbol{z}_k,\boldsymbol{\theta})\delta\big)$$

in the Euler scheme by densities which are further conditioned on the end point, i.e. by

$$\begin{aligned}
\pi(\boldsymbol{z}_{k+1}|\,\boldsymbol{z}_k,\boldsymbol{z}_N) &\propto \pi\,(\boldsymbol{z}_{k+1}|\,\boldsymbol{z}_k)\,\pi\,(\boldsymbol{z}_N|\boldsymbol{z}_{k+1})\\
&= \phi\big(\boldsymbol{z}_{k+1}|\,\boldsymbol{z}_k + \boldsymbol{\mu}(\boldsymbol{z}_k,\boldsymbol{\theta})\delta, \boldsymbol{\Sigma}(\boldsymbol{z}_k,\boldsymbol{\theta})\delta\big)\\
&\quad\cdot \phi\big(\boldsymbol{z}_N|\,\boldsymbol{z}_{k+1} + (N-k-1)\boldsymbol{\mu}(\boldsymbol{z}_{k+1},\boldsymbol{\theta})\delta, (N-k-1)\boldsymbol{\Sigma}(\boldsymbol{z}_{k+1},\boldsymbol{\theta})\delta\big)\\
&\approx \phi\big(\boldsymbol{z}_{k+1}|\,\boldsymbol{z}_k + \boldsymbol{\mu}(\boldsymbol{z}_k,\boldsymbol{\theta})\delta, \boldsymbol{\Sigma}(\boldsymbol{z}_k,\boldsymbol{\theta})\delta\big)\\
&\quad\cdot \phi\big(\boldsymbol{z}_N|\,\boldsymbol{z}_{k+1} + (N-k-1)\boldsymbol{\mu}(\boldsymbol{z}_k,\boldsymbol{\theta})\delta, (N-k-1)\boldsymbol{\Sigma}(\boldsymbol{z}_k,\boldsymbol{\theta})\delta\big)\\
&\propto \phi\left(\boldsymbol{z}_{k+1}\middle|\,\boldsymbol{z}_k + \frac{\boldsymbol{z}_N - \boldsymbol{z}_k}{\tau_N - \tau_k}\,\delta, \frac{\tau_N - \tau_{k+1}}{\tau_N - \tau_k}\,\boldsymbol{\Sigma}(\boldsymbol{z}_k,\boldsymbol{\theta})\delta\right),
\end{aligned}$$

where $\delta = (t-s)/N$. Then, if trajectories are sampled by setting $\boldsymbol{z}_0 = \boldsymbol{x}$ and successively drawing

$$\boldsymbol{z}_{k+1} \sim \mathcal{N}\left(\boldsymbol{z}_k + \frac{\boldsymbol{y} - \boldsymbol{z}_k}{N-k}, \frac{N-k-1}{N-k}\,\boldsymbol{\Sigma}(\boldsymbol{z}_k,\boldsymbol{\theta})\,\delta\right) \tag{6.27}$$

for $k = 0,\ldots,N-2$, the resulting $\boldsymbol{z}_{N-1}$ is a realisation from $\mathbb{Q}^{(N)}$. Durham and Gallant call this sampling pattern the *modified bridge*. The modified bridge will also play a central role in the Bayesian estimation approaches in Chap. 7.

Figure 6.1b displays trajectories from this improved scheme. They seem more likely than the sample paths in Fig. 6.1a which are simulated as proposed by Pedersen (1995b). Figure 6.1c, d confirms this impression: These show the empirical sampling densities corresponding to Fig. 6.1a, b, respectively. Whereas the empirical sampling density in Fig. 6.1c clearly differs from the integrand $p_{\boldsymbol{\theta}}^{(1)}$, the sampling density in Fig. 6.1d obviously draws from regions where the integrand $p_{\boldsymbol{\theta}}^{(1)}\rho$ is large.

There is a number of suggestions how to further improve the SMLE approach. One idea is to change the first-order approximation (6.22) in order to reduce the estimation bias without having to increase the number of subintervals N. Elerian (1998), for example, replaces the Euler scheme by the Milstein scheme as introduced in Sect. 3.3.2. Durham and Gallant (2002) consider the application of a higher order Itï-Taylor expansion as in Kessler (1997) and the local linearisation method described in Sect. 6.3.4 below. They furthermore suggest various variance-reduction techniques like the use of antithetic variates. The latter is also applied by Brandt and Santa-Clara (2001). Stramer and Yan (2007) investigate the trade-off between increasing the number of auxiliary time points and increasing the number of simulated diffusion paths.

6.3.4 Local Linearisation

The idea of the *local linearisation method* (Shoji and Ozaki 1998a) is to approximate the considered diffusion process by a linear one. Like all linear diffusions, the resulting approximation is explicitly solvable (e.g. Kloeden and Platen 1999, Chap. 4.2), i.e. its transition density is available and can serve as an approximation to the true transition density of the original process. The local linearisation constitutes an improvement of the Euler scheme: While the Euler approximation sets the drift and diffusion coefficient piecewise *constant*, the local linearisation method considers them piecewise *linear*.

For one-dimensional diffusions, the local linearisation is performed as follows (Shoji 1998; Shoji and Ozaki 1998a): Assume that the process X of interest fulfils the SDE

$$\mathrm{d}X_t = \mu(X_t, \boldsymbol{\theta}) + \sigma(\boldsymbol{\theta})\mathrm{d}B_t, \quad X_0 = x_0. \tag{6.28}$$

In case the diffusion coefficient σ depends on the state variable, the process can be converted to this form using Lamperti's transform from Sect. 3.2.11. The local linearisation method usually allows the drift to also depend on time; the focus of this chapter, however, is on time-homogeneous diffusion processes.

For X satisfying the SDE (6.28), Itô's formula from Sect. 3.2.10 yields

$$\mathrm{d}\mu(X_t, \boldsymbol{\theta}) = \left(\frac{\partial \mu}{\partial x}\right)(X_t, \boldsymbol{\theta})\,\mathrm{d}X_t + \frac{1}{2}\left(\frac{\partial^2 \mu}{\partial x^2}\right)(X_t, \boldsymbol{\theta})\sigma^2(\boldsymbol{\theta})\,\mathrm{d}t,$$

where $\partial/\partial x$ denotes differentiation with respect to the state variable. Assume that $\partial\mu/\partial x$ and $\partial^2\mu/\partial x^2$ are constant in X_t for $t \in [s, s + \Delta s)$, where $s \geq 0$ and $\Delta s > 0$. Then

$$\mu(X_t, \boldsymbol{\theta}) - \mu(X_s, \boldsymbol{\theta}) = \left(\frac{\partial \mu}{\partial x}\right)(X_s, \boldsymbol{\theta})\,(X_t - X_s) + \frac{1}{2}\left(\frac{\partial^2 \mu}{\partial x^2}\right)(X_s, \boldsymbol{\theta})\sigma^2(\boldsymbol{\theta})\,(t - s)$$

for all $t \in [s, s + \Delta s)$. With this, the drift function of X can be written as

$$\mu(X_t, \boldsymbol{\theta}) = C_s^{(1)} X_t + C_s^{(2)} t + C_s^{(3)}$$

for appropriate constants $C_s^{(1)}$, $C_s^{(2)}$ and $C_s^{(3)}$ that depend only on s and X_s. The resulting SDE

$$\mathrm{d}X_t = \left(C_s^{(1)} X_t + C_s^{(2)} t + C_s^{(3)}\right)\mathrm{d}t + \sigma(\boldsymbol{\theta})\mathrm{d}B_t \tag{6.29}$$

for $t \in [s, s + \Delta s)$ has a linear drift function and a constant diffusion coefficient. As indicated above, an explicit solution to such an SDE is generally available. See Shoji (1998), Shoji and Ozaki (1998a) or Kloeden and Platen (1999, Chap. 4.2) for the specific solution of (6.29).

Shoji and Ozaki (1998b) describe this linearisation procedure also for multi-dimensional processes. The method is, however, only applicable where a transformation to constant diffusion coefficient is possible; cf. the remarks in Sect. 3.2.11.

6.4 Alternatives to Maximum Likelihood Estimation

Unlike all estimating techniques investigated so far in this chapter, the approaches in this section do not try to set up or approximate the likelihood function of the parameter. Instead, they match certain statistics of the model with that of the data. These statistics may be the moments of the process (as in Sect. 6.4.1) or their sample analogues (Sects. 6.4.2 and 6.4.3) or some functions derived from auxiliary models (Sects. 6.4.4 and 6.4.5). The model parameter is then estimated by the candidate which produces the best conformity.

6.4.1 Estimating Functions

This section briefly describes how the general concept of estimating functions (Godambe 1991; Heyde 1997) can be applied to diffusion processes. For a detailed review on this topic see Bibby et al. (2009).

Let $\boldsymbol{X}$ be an ergodic diffusion which is the solution of the SDE (6.1), and assume one has observations $\boldsymbol{x}_0, \boldsymbol{x}_1, \ldots, \boldsymbol{x}_n$ of $\boldsymbol{X}$ at times $0 = t_0 < t_1 < \ldots < t_n$. An *estimating function* for the parameter $\boldsymbol{\theta}$ is a function $G_n(\boldsymbol{\theta}; \boldsymbol{x}_0, \ldots, \boldsymbol{x}_n)$ which depends on the parameter and the data. When dependence on the observations is clear, we simply write $G_n(\boldsymbol{\theta})$. An estimate $\hat{\boldsymbol{\theta}}_n$ for $\boldsymbol{\theta}$ is obtained as a solution of

$$G_n(\boldsymbol{\theta}) = \mathbf{0}.$$

Once more, let $\boldsymbol{\theta}_0$ denote the true parameter value. One usually requires that the estimating function is (at least asymptotically) *unbiased*, that means

$$\mathbb{E}_{\boldsymbol{\theta}}(G_n(\boldsymbol{\theta})) = \mathbf{0},$$

and that the parameter is *uniquely identifiable*, i.e.

$$\mathbb{E}_{\boldsymbol{\theta}_0}(G_n(\boldsymbol{\theta})) = \mathbf{0} \quad \Leftrightarrow \quad \boldsymbol{\theta} = \boldsymbol{\theta}_0.$$

The most prominent representative for an unbiased estimating function is the score function

$$s_n(\boldsymbol{\theta}) = \frac{\partial \ell_n(\boldsymbol{\theta})}{\partial \boldsymbol{\theta}} = \sum_{k=0}^{n-1} \frac{\partial}{\partial \boldsymbol{\theta}} \log p_{\boldsymbol{\theta}}\big(\Delta t_k; \boldsymbol{x}_k, \boldsymbol{x}_{k+1}\big),$$

where the equation $s_n(\boldsymbol{\theta}) = \mathbf{0}$ is solved by the maximum likelihood estimator. However, as discussed before, the score function is usually not available. In that case, one tries to imitate it by adapting the general form

$$G_n(\boldsymbol{\theta}; \boldsymbol{x}_0, \ldots, \boldsymbol{x}_n) = \sum_{k=0}^{n-1} g(\boldsymbol{\theta}, \Delta t_k, \boldsymbol{x}_k, \boldsymbol{x}_{k+1}). \tag{6.30}$$

In all estimating functions introduced below, the function g, in turn, is of type

$$g(\boldsymbol{\theta}, \Delta t, \boldsymbol{x}, \boldsymbol{y}) = \sum_{j=1}^{J} a_j(\boldsymbol{\theta}, \Delta t, \boldsymbol{x}) h_j(\boldsymbol{\theta}, \Delta t, \boldsymbol{x}, \boldsymbol{y}) \tag{6.31}$$

for some $J \in \mathbb{N}$, where the a_j are called the *weights* of the functions h_j. The most common estimating functions that appear in the literature can be categorised in the following non-disjoint classes.

Martingale Estimating Functions

Martingale estimating functions satisfy the martingale property

$$\mathbb{E}_{\boldsymbol{\theta}}\big(G_n(\boldsymbol{\theta}; \boldsymbol{X}_{t_0}, \ldots, \boldsymbol{X}_{t_n}) \,\big|\, \mathcal{F}_{n-1}\big) = G_{n-1}(\boldsymbol{\theta}; \boldsymbol{X}_{t_0}, \ldots, \boldsymbol{X}_{t_{n-1}}) \tag{6.32}$$

for all $n \in \mathbb{N}$, where $\mathcal{F}_n$ is the σ-algebra generated by $\{\boldsymbol{X}_{t_0}, \ldots, \boldsymbol{X}_{t_n}\}$. If the function g in (6.30) attains the form (6.31), the condition (6.32) is fulfilled if and only if

$$\mathbb{E}_{\boldsymbol{\theta}}\big(h_j(\boldsymbol{\theta}, \Delta t, \boldsymbol{x}, \boldsymbol{X}_{\Delta t}) \,\big|\, \boldsymbol{X}_0 = \boldsymbol{x}\big) = \mathbf{0} \tag{6.33}$$

for all $j \in \{1, \ldots, J\}$, $\boldsymbol{\theta} \in \Theta$, $\Delta t \in \mathbb{R}_+$ and $\boldsymbol{x} \in \mathcal{X}$. An obvious choice is

$$h_j(\boldsymbol{\theta}, \Delta t, \boldsymbol{x}, \boldsymbol{y}) = f_j(\boldsymbol{y}, \boldsymbol{\theta}) - \mathbb{E}_{\boldsymbol{\theta}}\big(f_j(\boldsymbol{X}_{\Delta t}, \boldsymbol{\theta}) \,\big|\, \boldsymbol{X}_0 = \boldsymbol{x}\big) \tag{6.34}$$

for *base functions* f_j which are regular enough such that the expected values exist. Then (6.33) is trivially fulfilled. Examples for functions of type (6.34) are given in the polynomial estimating functions below.

Martingale estimating functions appear as a natural choice of an estimating function since the score function possesses the martingale property as well. Furthermore, the well-known martingale theory allows for immediate asymptotic results as $n \to \infty$. Unbiasedness is directly implied by (6.33).

For given h_j, $j = 1, \ldots, J$, the weights α_j can be chosen in an optimal way such that the resulting estimator of $\boldsymbol{\theta}$ has smallest asymptotic variance within the class of estimating functions satisfying (6.31) with the specified h_j (e.g. Sørensen 2007; Bibby et al. 2009). The choice of the h_j is, however, more subtle. For asymptotic properties of martingale estimating functions, and in particular for asymptotic variances, which are typically of sandwich type and depend on the true parameter, see Sørensen (2008) or Bibby et al. (2009).

Polynomial Estimating Functions

Polynomial estimating functions employ the form (6.34) with f_j being a polynomial. They hence form a subgroup of the above martingale estimating functions. Unlike the true score function, polynomial estimating functions do not require knowledge of the whole transition density but only of the first few conditional moments. They are hence more robust to misspecification. When the conditional moments are analytically not available, they may be obtained e.g. by Monte Carlo simulation. Kessler and Paredes (2002) describe the impact of such simulation on the resulting estimator of $\boldsymbol{\theta}$.

A *linear estimating function* utilises (6.31) with $J = 1$, h_1 as in (6.34) and $f_1(\boldsymbol{y}, \boldsymbol{\theta}) = \boldsymbol{y}$, i.e.

$$h_1\big(\boldsymbol{\theta}, \Delta t, \boldsymbol{x}, \boldsymbol{y}\big) = \boldsymbol{y} - \mathbb{E}_{\boldsymbol{\theta}}\big(\boldsymbol{X}_{\Delta t} \,\big|\, \boldsymbol{X}_0 = \boldsymbol{x}\big).$$

This estimating function is appropriate when the diffusion coefficient does not depend on $\boldsymbol{\theta}$; otherwise, a higher order polynomial should be employed. A *quadratic estimating function* uses $J = 2$, h_1 as above and $f_2(\boldsymbol{y}, \boldsymbol{\theta}) = h_1(\boldsymbol{\theta}, \Delta t, \boldsymbol{x}, \boldsymbol{y}) h_1'(\boldsymbol{\theta}, \Delta t, \boldsymbol{x}, \boldsymbol{y})$, i.e.

$$h_2\big(\boldsymbol{\theta}, \Delta t, \boldsymbol{x}, \boldsymbol{y}\big) = h_1\big(\boldsymbol{\theta}, \Delta t, \boldsymbol{x}, \boldsymbol{y}\big) h_1'\big(\boldsymbol{\theta}, \Delta t, \boldsymbol{x}, \boldsymbol{y}\big) - \mathrm{Var}_{\boldsymbol{\theta}}\big(\boldsymbol{X}_{\Delta t} \,\big|\, \boldsymbol{X}_0 = \boldsymbol{x}\big).$$

When the diffusion process is one-dimensional, higher order polynomial estimating functions usually employ

$$f_j(y, \boldsymbol{\theta}) = y^{k_j}$$

for suitable $k_j \in \mathbb{N}_0$. Examples for particular linear and quadratic estimating functions are given in Examples 6.1 and 6.2 below.

Estimating Functions Based on Eigenfunctions

The class of *estimating functions based on eigenfunctions* has been investigated e.g. by Kessler and Sørensen (1999). Let $\mathcal{G}_{\boldsymbol{\theta}}$ denote the infinitesimal generator, introduced in Sect. 3.2.9, of a one-dimensional diffusion process X which solves the SDE (6.1). A twice differentiable function $\eta(x, \boldsymbol{\theta})$ is an *eigenfunction* of $\mathcal{G}_{\boldsymbol{\theta}}$ with *eigenvalue* $\lambda(\boldsymbol{\theta}) \in \mathbb{R}_0$ if

$$\mathcal{G}_{\boldsymbol{\theta}} \eta(x, \boldsymbol{\theta}) = -\lambda(\boldsymbol{\theta}) \eta(x, \boldsymbol{\theta})$$

for all $x \in \mathcal{X}$. Kessler and Sørensen (1999) show that under mild regularity conditions one has

$$\mathbb{E}_{\boldsymbol{\theta}}\big(\eta(X_{\Delta t}, \boldsymbol{\theta}) \,\big|\, X_0 = x\big) = \exp\big(-\lambda(\boldsymbol{\theta})\Delta t\big)\eta(x, \boldsymbol{\theta}).$$

Hence, if $\eta_1, \ldots, \eta_K$ are eigenfunctions of $\mathcal{G}_{\boldsymbol{\theta}}$ with eigenvalues $\lambda_1, \ldots, \lambda_K$, the choice (6.31) with $J = K$ and

$$h_j(\boldsymbol{\theta}, \Delta t, x, y) = \eta_j(y, \boldsymbol{\theta}) - \exp\big(-\lambda_j(\boldsymbol{\theta})\Delta t\big)\eta_j(x, \boldsymbol{\theta}),$$

i.e. h_j being as in (6.34) with

$$f_j(y, \boldsymbol{\theta}) = \eta_j(y, \boldsymbol{\theta}),$$

yields a martingale estimating function. This estimating function is explicit in the sense that the h_j are known in explicit form. Kessler and Sørensen (1999) determine the optimal weights for this function and prove consistency and asymptotic normality for the resulting estimators. For eigenfunctions of the generator of a multi-dimensional diffusion process see Bibby and Sørensen (1995).

Simple Estimating Functions

In the class of *simple estimating functions*, the function g in (6.30) depends on one state variable only, in particular

$$g(\boldsymbol{\theta}, \Delta t, \boldsymbol{x}, \boldsymbol{y}) = \tilde{g}(\boldsymbol{\theta}, \boldsymbol{x}) \qquad \text{or} \qquad g(\boldsymbol{\theta}, \Delta t, \boldsymbol{x}, \boldsymbol{y}) = \bar{g}(\boldsymbol{\theta}, \boldsymbol{y})$$

for some functions $\tilde{g}$ and $\bar{g}$. Simple estimating functions have the advantage that they are often explicitly available. On the other hand, they do not take into account the dependence structure between successive observations.

For example, let $\pi_{\boldsymbol{\theta}}$ denote the invariant density of the ergodic diffusion process as introduced in Sect. 3.2.7. Then (6.30) with

$$g(\boldsymbol{\theta}, \Delta t, \boldsymbol{x}, \boldsymbol{y}) = \bar{g}(\boldsymbol{\theta}, \boldsymbol{y}) = \frac{\partial \log \pi_{\boldsymbol{\theta}}(\boldsymbol{y})}{\partial \boldsymbol{\theta}} \tag{6.35}$$

forms a simple estimating function which is based on the assumption that $\boldsymbol{x}_1, \ldots, \boldsymbol{x}_n$ are independent and identically distributed draws from $\pi_{\boldsymbol{\theta}}$ (Kessler 2000). In that case, G_n would equal the score function of $\boldsymbol{\theta}$. Utilisation of this estimating function is only reasonable when the process has reached stationarity. Furthermore, it is only applicable for the estimation of those parameters that enter the invariant measure.

As another example, Kessler (2000) constructs simple estimating functions by application of the infinitesimal generator $\mathcal{G}_{\boldsymbol{\theta}}$ for one-dimensional diffusion processes. In particular, he sets

$$\tilde{g}(\boldsymbol{\theta}, x) = \mathcal{G}_{\boldsymbol{\theta}}\rho(x, \boldsymbol{\theta}) = \mu(x, \boldsymbol{\theta})\frac{\partial \rho(x, \boldsymbol{\theta})}{\partial x} + \frac{1}{2}\Sigma(x, \boldsymbol{\theta})\frac{\partial^2 \rho(x, \boldsymbol{\theta})}{\partial x^2} \tag{6.36}$$

for a sufficiently regular function ρ and shows that under certain assumptions the resulting estimating function leads to a consistent and asymptotically normal estimator of $\boldsymbol{\theta}$. Estimating functions with g of type (6.36) are also discussed by Sørensen (2001).

That completes the collection of those classes of estimating functions which are considered in this section. Combinations are possible, that means different estimating functions can be used for different parameter components as for example for drift parameters and diffusion parameters. See Bibby and Sørensen (2001) for several examples.

Jacobsen (2001) investigates the problem of finding optimal or asymptotically optimal estimating functions in the sense that the resulting parameter estimate has smallest variance within a certain class of functions.

This section is concluded with the following two examples as an illustration of the just introduced classes of estimating functions.

Example 6.1. In what follows, a linear martingale estimating function is constructed as in Bibby and Sørensen (1995): For $\boldsymbol{\Sigma}$ not depending on $\boldsymbol{\theta}$, the Euler approximation of the log-likelihood function is shown in (6.8). Derivation of this function with respect to $\boldsymbol{\theta}$ yields the according score function. For n observations $\boldsymbol{x}_1, \ldots, \boldsymbol{x}_n$ in addition to the initial value $\boldsymbol{x}_0$, it reads

$$s_n^{\text{Euler}}(\boldsymbol{\theta}) = \sum_{k=0}^{n-1} \left(\frac{\partial \boldsymbol{\mu}(\boldsymbol{x}_k, \boldsymbol{\theta})}{\partial \boldsymbol{\theta}}\right)' \boldsymbol{\Sigma}^{-1}(\boldsymbol{x}_k)\big(\boldsymbol{x}_{k+1} - \boldsymbol{x}_k - \boldsymbol{\mu}(\boldsymbol{x}_k, \boldsymbol{\theta})\Delta t_k\big),$$

where $\partial\boldsymbol{\mu}/\partial\boldsymbol{\theta}$ is a $d \times p$-dimensional matrix. This score function is biased and hence not appropriate as an estimating function. An unbiased function can however be obtained as

$$G_n(\boldsymbol{\theta}) = s_n^{\text{Euler}}(\boldsymbol{\theta}) - C_n(\boldsymbol{\theta}),$$

where C_n is the compensator of s_n^{Euler}. This can be constructed as $C_0(\boldsymbol{\theta}) = \mathbf{0}$ and

$$\begin{aligned} C_n(\boldsymbol{\theta}) &= \sum_{k=0}^{n-1} \mathbb{E}_{\boldsymbol{\theta}}\big(s_{k+1}^{\text{Euler}}(\boldsymbol{\theta}) - s_k^{\text{Euler}}(\boldsymbol{\theta}) \,\big|\, \mathcal{F}_k\big) \\ &= \sum_{k=0}^{n-1} \left(\frac{\partial \boldsymbol{\mu}(\boldsymbol{x}_k, \boldsymbol{\theta})}{\partial \boldsymbol{\theta}}\right)' \boldsymbol{\Sigma}^{-1}(\boldsymbol{x}_k)\Big(\mathbb{E}_{\boldsymbol{\theta}}\big(\boldsymbol{X}_{t_{k+1}} \,\big|\, \mathcal{F}_k\big) - \boldsymbol{x}_k - \boldsymbol{\mu}(\boldsymbol{x}_k, \boldsymbol{\theta})\Delta t_k\Big) \end{aligned}$$

for $n \in \mathbb{N}$, where $s_0^{\text{Euler}}(\boldsymbol{\theta}) = \mathbf{0}$. Overall, one has

$$G_n(\boldsymbol{\theta}) = \sum_{k=0}^{n-1} \left(\frac{\partial \boldsymbol{\mu}(\boldsymbol{x}_k, \boldsymbol{\theta})}{\partial \boldsymbol{\theta}}\right)' \boldsymbol{\Sigma}^{-1}(\boldsymbol{x}_k)\Big(\boldsymbol{X}_{t_{k+1}} - \mathbb{E}_{\boldsymbol{\theta}}\big(\boldsymbol{X}_{t_{k+1}} \,\big|\, \mathcal{F}_k\big)\Big),$$

which is an unbiased linear martingale estimating function with weight

$$a_1(\boldsymbol{\theta}, \Delta t, \boldsymbol{x}) = \left(\frac{\partial \boldsymbol{\mu}(\boldsymbol{x}, \boldsymbol{\theta})}{\partial \boldsymbol{\theta}}\right)' \boldsymbol{\Sigma}^{-1}(\boldsymbol{x}). \tag{6.37}$$

Unbiasedness does not rely on the fact that the diffusion matrix does not depend on the parameter. Hence, the same estimating function can be employed when it does.

Bibby and Sørensen (1995) show that the optimal estimating function within this class, i.e. the one leading to an estimator with smallest asymptotic variance, is the one with weight

$$\tilde{a}_1(\boldsymbol{\theta}, \Delta t, \boldsymbol{x}) = \left(\frac{\partial}{\partial \boldsymbol{\theta}} \mathbb{E}_{\boldsymbol{\theta}}\big(\boldsymbol{X}_{\Delta t} \,\big|\, \boldsymbol{X}_0 = \boldsymbol{x}\big)\right)' \left(\mathrm{Var}_{\boldsymbol{\theta}}\big(\boldsymbol{X}_{\Delta t} \,\big|\, \boldsymbol{X}_0 = \boldsymbol{x}\big)\right)^{-1}.$$

The linear estimating function with the previous weight (6.37) is hence not optimal, but it is approximately optimal for small time steps. Both estimating functions lead to consistent and asymptotically normal estimators (Bibby and Sørensen 1995).

Example 6.2. A well-known quadratic martingale estimating function is derived as follows (see e.g. Sørensen 2008): Assume that the increments of a one-dimensional diffusion process with parameter $\boldsymbol{\theta}$ are approximately Gaussian, in particular

$$X_{t_{k+1}} \,\big|\, \{X_{t_k} = x_k\} \sim \mathcal{N}\big(E_k(\boldsymbol{\theta}), V_k(\boldsymbol{\theta})\big),$$

where

$$E_k(\boldsymbol{\theta}) = \mathbb{E}_{\boldsymbol{\theta}}\big(X_{t_{k+1}} \,\big|\, X_{t_k} = x_k\big) \qquad \text{and} \qquad V_k(\boldsymbol{\theta}) = \mathrm{Var}_{\boldsymbol{\theta}}\big(X_{t_{k+1}} \,\big|\, X_{t_k} = x_k\big).$$

The according score function for data $x_0, \ldots, x_n$ equals

$$s_n(\boldsymbol{\theta}) = \sum_{k=0}^{n-1} \left[\frac{\left(\frac{\partial}{\partial \boldsymbol{\theta}} E_k(\boldsymbol{\theta})\right)}{V_k(\boldsymbol{\theta})} \big(x_{k+1} - E_k(\boldsymbol{\theta})\big) + \frac{1}{2} \frac{\left(\frac{\partial}{\partial \boldsymbol{\theta}} V_k(\boldsymbol{\theta})\right)}{V_k^2(\boldsymbol{\theta})} \left((x_{k+1} - E_k(\boldsymbol{\theta}))^2 - V_k(\boldsymbol{\theta})\right) \right].$$

This is a quadratic martingale estimating function of the form (6.30) to (6.31) with $J = 2$, h_1 and h_2 as in (6.34), weights

$$a_1(\boldsymbol{\theta}, \Delta t, x) = \frac{\frac{\partial}{\partial \boldsymbol{\theta}} \mathbb{E}_{\boldsymbol{\theta}}(X_{\Delta t} | X_0 = x)}{\mathrm{Var}_{\boldsymbol{\theta}}(X_{\Delta t} | X_0 = x)} \text{ and } a_2(\boldsymbol{\theta}, \Delta t, x) = \frac{\frac{\partial}{\partial \boldsymbol{\theta}} \mathrm{Var}_{\boldsymbol{\theta}}(X_{\Delta t} | X_0 = x)}{2\mathrm{Var}_{\boldsymbol{\theta}}(X_{\Delta t} | X_0 = x)^2}$$

and polynomials

$$f_1(y, \boldsymbol{\theta}) = y \qquad \text{and} \qquad f_2(y, \boldsymbol{\theta}) = \big(y - \mathbb{E}_{\boldsymbol{\theta}}(X_{\Delta t} | X_0 = x)\big)^2.$$

This estimating function is approximately optimal (Sørensen 2008).

6.4.2 Generalised Method of Moments

Related to the theory of estimating functions in the previous section is the *generalised method of moments (GMM)* as developed by Hansen (1982). In this approach, one considers functions $\psi_1, \ldots, \psi_J$ depending on the parameter $\boldsymbol{\theta}$ and the process $\boldsymbol{X}$, where

$$\mathbb{E}_{\boldsymbol{\theta}}\big(\psi_j(\boldsymbol{\theta}; \boldsymbol{X})\big) = \mathbf{0} \tag{6.38}$$

for $j = 1, \ldots, J$. Equation (6.38) for all j are called the *moment conditions*. These are employed in order to estimate the parameter $\boldsymbol{\theta}$. For p-dimensional parameter $\boldsymbol{\theta}$, one requires $J \geq p$ moment conditions. The functions ψ_j usually depend on only one or two state variables in addition to the parameter. This case is also assumed in the following.

Consider a one-dimensional ergodic diffusion process with true parameter $\boldsymbol{\theta}_0$ starting in x_0 at time t_0, and assume that there are observations $x_1, \ldots, x_n$ at times $t_1 < \ldots < t_n$ of this process which are i.i.d. draws from its invariant measure $\pi_{\boldsymbol{\theta}_0}$. Denote by X_s and X_t two independent random variables with density $\pi_{\boldsymbol{\theta}_0}$. The expected value of $\psi_j(\boldsymbol{\theta}; X_s, X_t)$ for $\boldsymbol{\theta} \in \Theta$ is usually not available but can be approximated through the *method of moments estimator*

$$\frac{1}{n}\sum_{k=0}^{n-1} \psi_j\big(\boldsymbol{\theta}; x_k, x_{k+1}\big). \tag{6.39}$$

The *GMM estimator* $\hat{\boldsymbol{\theta}}_n$ is obtained as the minimiser of a norm of this expression. In particular,

$$\hat{\boldsymbol{\theta}}_n = \operatorname*{argmin}_{\boldsymbol{\theta}\in\Theta} \left(\frac{1}{n}\sum_{k=0}^{n-1} \boldsymbol{\psi}\big(\boldsymbol{\theta}; x_k, x_{k+1}\big)\right)' \boldsymbol{C}_n \left(\frac{1}{n}\sum_{k=0}^{n-1} \boldsymbol{\psi}\big(\boldsymbol{\theta}; x_k, x_{k+1}\big)\right), \tag{6.40}$$

where $\boldsymbol{\psi} = (\psi_1, \ldots, \psi_J)'$ and $\boldsymbol{C}_n \in \mathbb{R}^{J\times J}$ is a positive semi-definite weight matrix. Under certain conditions (Hansen 1982), $\hat{\boldsymbol{\theta}}_n$ is consistent and asymptotically normal. Hansen (1982) determines the optimal weight matrix yielding an asymptotically efficient estimator, but as this matrix involves the unknown parameter $\boldsymbol{\theta}_0$, it cannot be used in practice.

In what follows, we look at two examples. Once again, denote by $\mathcal{G}_{\boldsymbol{\theta}}$ the infinitesimal generator of the considered diffusion process with parameter $\boldsymbol{\theta}$ as introduced in Sect. 3.2.9. Let g and h be two functions for which $\mathcal{G}_{\boldsymbol{\theta}}\, g$ and $\mathcal{G}_{\boldsymbol{\theta}}\, h$ are well-defined. Hansen and Scheinkman (1995) utilise the two moment conditions

$$\mathbb{E}_{\boldsymbol{\theta}}\big(\mathcal{G}_{\boldsymbol{\theta}}\, g(X_t)\big) = 0 \tag{6.41}$$

and

$$\mathbb{E}_{\boldsymbol{\theta}}\big(\mathcal{G}_{\boldsymbol{\theta}}\, g(X_t)h(X_s) - g(X_t)\mathcal{G}_{\boldsymbol{\theta}}\, h(X_s)\big) = 0 \tag{6.42}$$

for all $t > s \geq 0$. The latter equation holds for time-reversible processes (Kent 1978), which includes all one-dimensional ergodic diffusions. The above moment conditions (6.41) and (6.42) lead to method of moments estimators (6.39) with functions

$$\psi(\boldsymbol{\theta}; x, y) = \mathcal{G}_{\boldsymbol{\theta}}\, g(y)$$

and

$$\psi(\boldsymbol{\theta}; x, y) = \mathcal{G}_{\boldsymbol{\theta}}\, g(\boldsymbol{\theta}, y)h(\boldsymbol{\theta}, x) - g(\boldsymbol{\theta}, y)\mathcal{G}_{\boldsymbol{\theta}}\, h(\boldsymbol{\theta}, x),$$

respectively. These are investigated for example by Jacobsen (2001).

The above elucidations refer to one-dimensional diffusions only. For the generalisation to multi-dimensional processes see Hansen and Scheinkman (1995) or Jacobsen (2001).

6.4.3 Simulated Moments Estimation

Simulated moments estimation (SME) as carried out by Duffie and Singleton (1993) bases on the same idea as the just considered generalised method of moments: In the previous section, a norm of (6.39) was minimised. Due to the moment conditions (6.38), this is exactly the same as minimising a norm of

$$\frac{1}{n}\sum_{k=0}^{n-1} \psi_j\big(\boldsymbol{\theta}; x_k, x_{k+1}\big) - \mathbb{E}_{\boldsymbol{\theta}}\big(\psi_j(\boldsymbol{\theta}; X_s, X_t)\big) \tag{6.43}$$

for some $t > s \geq 0$. The generalised method of moments could hence theoretically be extended to functions $\psi_j(\boldsymbol{\theta}; X_s, X_t)$ with nonzero expectation as long as this expectation is known. When it is unknown, the SME is still appropriate as it replaces the analytical expected value in (6.43) by the sample mean based on $m+1$ additionally simulated values $z_0, z_1, \ldots, z_m$ from the invariant density $\pi_{\boldsymbol{\theta}}$. The *simulated moments estimator* is obtained as the minimiser of the norm of

$$\frac{1}{n}\sum_{k=0}^{n-1} \psi_j\big(\boldsymbol{\theta}; x_k, x_{k+1}\big) - \frac{1}{m}\sum_{i=0}^{m-1} \psi_j\big(\boldsymbol{\theta}; z_i, z_{i+1}\big)$$

in an analogous manner as in (6.40) above. Duffie and Singleton (1993) supply conditions for the consistency and asymptotic normality of the simulated moments estimator.

6.4.4 Indirect Inference

In *indirect inference*, the parameter of a model of interest is estimated indirectly via the parameter of an auxiliary model (Gourieroux et al. 1993). This is convenient whenever statistical inference for the original model is complicated and the auxiliary model is chosen such that its parameter can easily be estimated.

Let $\mathcal{M}_{\boldsymbol{\theta}}$ denote the original model with parameter $\boldsymbol{\theta} \in \Theta \subseteq \mathbb{R}^p$, and $\mathcal{A}_{\boldsymbol{\rho}}$ be the auxiliary model with parameter $\boldsymbol{\rho} \in \mathcal{R} \subseteq \mathbb{R}^h$, where $h \geq p$. Both $\mathcal{M}_{\boldsymbol{\theta}}$ and $\mathcal{A}_{\boldsymbol{\rho}}$ are assumed to be known in parametric form. Suppose one has observations $\boldsymbol{x}_0, \ldots, \boldsymbol{x}_n$ from $\mathcal{M}_{\boldsymbol{\theta}_0}$ for some unknown $\boldsymbol{\theta}_0 \in \Theta$. The objective is the estimation of $\boldsymbol{\theta}_0$. If simulation from $\mathcal{M}_{\boldsymbol{\theta}}$ is (at least approximately) possible, this can be performed by indirect inference as follows:

In a first step, obtain an estimator $\hat{\boldsymbol{\rho}}_n^{\text{obs}}$ of $\boldsymbol{\rho}$ by treating $\boldsymbol{x}_0, \ldots, \boldsymbol{x}_n$ as observations from the auxiliary model $\boldsymbol{A}_{\boldsymbol{\rho}}$. Under regularity conditions, this estimator $\hat{\boldsymbol{\rho}}_n^{\text{obs}}$ tends to a parameter $\boldsymbol{\rho}_0 = g(\boldsymbol{\theta}_0)$ for some invertible unknown function g as $n \to \infty$.

In a second step, determine $\boldsymbol{\theta}$ such that simulated data from $\mathcal{M}_{\boldsymbol{\theta}}$ associated with $\mathcal{A}_{\boldsymbol{\rho}}$ leads to an estimate of $\boldsymbol{\rho}$ that is close to $\hat{\boldsymbol{\rho}}_n^{\text{obs}}$: For fixed $\boldsymbol{\theta}$, (approximately) simulate K datasets $\boldsymbol{z}^{(\boldsymbol{\theta},k)} = \{\boldsymbol{z}_0^{(\boldsymbol{\theta},k)}, \ldots, \boldsymbol{z}_n^{(\boldsymbol{\theta},k)}\}$, $k = 1, \ldots, K$, from $\mathcal{M}_{\boldsymbol{\theta}}$. Denote by $\hat{\boldsymbol{\rho}}_n^{(\boldsymbol{\theta},k)}$ the estimator of $\boldsymbol{\rho}$ that is obtained when $\boldsymbol{z}^{(\boldsymbol{\theta},k)}$ is treated as observed from $\mathcal{A}_{\boldsymbol{\rho}}$. Then, if certain assumptions are fulfilled, $\hat{\boldsymbol{\rho}}_n^{(\boldsymbol{\theta},k)}$ tends to $g(\boldsymbol{\theta})$ for all k and $n \to \infty$. The *indirect estimator* $\hat{\boldsymbol{\theta}}_n$ of $\boldsymbol{\theta}$ is now chosen such that $\hat{\boldsymbol{\rho}}_n^{\text{sim}}(\boldsymbol{\theta}) = K^{-1} \sum_{k=1}^{K} \hat{\boldsymbol{\rho}}_n^{(\boldsymbol{\theta},k)}$ is close to $\hat{\boldsymbol{\rho}}_n^{\text{obs}}$. In particular,

$$\hat{\boldsymbol{\theta}}_n = \underset{\boldsymbol{\theta} \in \Theta}{\operatorname{argmin}} \left(\hat{\boldsymbol{\rho}}_n^{\text{obs}} - \hat{\boldsymbol{\rho}}_n^{\text{sim}}(\boldsymbol{\theta})\right)' \boldsymbol{D}_n \left(\hat{\boldsymbol{\rho}}_n^{\text{obs}} - \hat{\boldsymbol{\rho}}_n^{\text{sim}}(\boldsymbol{\theta})\right), \tag{6.44}$$

where $\boldsymbol{D}_n$ is a positive definite matrix converging to a deterministic positive definite matrix $\boldsymbol{D}$ as $n \to \infty$. Under assumptions stated in Gourieroux et al. (1993), $\hat{\boldsymbol{\theta}}_n$ consistently estimates $\boldsymbol{\theta}_0$.

In the context of estimating the parameter $\boldsymbol{\theta}$ of a stationary and ergodic diffusion process, the auxiliary model is most conveniently chosen as a time-discretisation of the original SDE as considered in Sect. 3.3.2. The parameters $\boldsymbol{\theta}$ and $\boldsymbol{\rho}$ then have the same dimension and interpretation. The auxiliary parameter $\boldsymbol{\rho}$ is estimated by maximum likelihood methodology. In that case, the indirect estimator is independent of $\boldsymbol{D}_n$ in (6.44), and $\hat{\boldsymbol{\rho}}_n^{\text{obs}} = \hat{\boldsymbol{\rho}}_n^{\text{sim}}(\hat{\boldsymbol{\theta}}_n)$ (Gourieroux et al. 1993).

The following describes the indirect inference procedure for the parameter $\boldsymbol{\theta}$ of an SDE

$$\mathrm{d}\boldsymbol{X}_t = \boldsymbol{\mu}(\boldsymbol{X}_t, \boldsymbol{\theta})\mathrm{d}t + \boldsymbol{\sigma}(\boldsymbol{X}_t, \boldsymbol{\theta})\mathrm{d}\boldsymbol{B}_t, \quad \boldsymbol{X}_{t_0} = \boldsymbol{x}_0, \tag{6.45}$$

for observations $\boldsymbol{x}_0, \ldots, \boldsymbol{x}_n$ at times $0 = t_0 < t_1 < \ldots < t_n = T$. The auxiliary model is chosen to be the Euler discretisation

$$\boldsymbol{Y}_{k+1} = \boldsymbol{Y}_k + \boldsymbol{\mu}(\boldsymbol{Y}_k, \boldsymbol{\rho})\Delta t_k + \boldsymbol{\sigma}(\boldsymbol{Y}_k, \boldsymbol{\rho})\mathcal{N}(\boldsymbol{0}, \Delta t_k \boldsymbol{I}), \quad \boldsymbol{Y}_0 = \boldsymbol{x}_0,$$

with $\Delta t_k = t_{k+1} - t_k$. For observations $\boldsymbol{x}_0 = \boldsymbol{y}_0, \boldsymbol{y}_1, \ldots, \boldsymbol{y}_n$, the log-likelihood function of $\boldsymbol{\rho}$ for the auxiliary model is

$$q(\boldsymbol{\rho}; \boldsymbol{y}_0, \ldots, \boldsymbol{y}_n) = \sum_{k=0}^{n-1} \log \phi\left(\boldsymbol{y}_{k+1} \,\middle|\, \boldsymbol{y}_k + \boldsymbol{\mu}(\boldsymbol{y}_k, \boldsymbol{\rho})\Delta t_k, \boldsymbol{\Sigma}(\boldsymbol{y}_k, \boldsymbol{\rho})\Delta t_k\right),$$

where $\phi(\boldsymbol{z}|\boldsymbol{\nu}, \boldsymbol{\Lambda})$ is the Gaussian density with mean $\boldsymbol{\nu}$ and variance $\boldsymbol{\Lambda}$ evaluated at $\boldsymbol{z}$, and $\boldsymbol{\Sigma} = \boldsymbol{\sigma}\boldsymbol{\sigma}'$. Now proceed as follows.

Step 1: Calculate the maximum likelihood estimator $\hat{\boldsymbol{\rho}}_n^{\text{obs}}$ of $\boldsymbol{\rho}$ given the observations from the original model,

$$\hat{\boldsymbol{\rho}}_n^{\text{obs}} = \underset{\boldsymbol{\rho} \in \mathcal{R}}{\operatorname{argmax}}\ q\big(\boldsymbol{\rho}; \boldsymbol{x}_0, \ldots, \boldsymbol{x}_n\big).$$

Step 2: Determine the indirect estimator $\hat{\boldsymbol{\theta}}_n$ of $\boldsymbol{\theta}_0$ by (numerically) solving

$$\hat{\boldsymbol{\rho}}_n^{\text{obs}} = \hat{\boldsymbol{\rho}}_n^{\text{sim}}\big(\hat{\boldsymbol{\theta}}_n\big),$$

where $\hat{\boldsymbol{\rho}}_n^{\text{sim}}(\boldsymbol{\theta})$ is determined for all $\boldsymbol{\theta} \in \Theta$ as follows:

(i) For $k = 1, \ldots, K$, simulate the original diffusion process $\boldsymbol{X}$ with parameter $\boldsymbol{\theta}$ at times $t_0, \ldots, t_n$. Denote the simulated values by $\boldsymbol{z}_0^{(\boldsymbol{\theta},k)}, \ldots, \boldsymbol{z}_n^{(\boldsymbol{\theta},k)}$. If exact simulation from (6.45) is inconvenient or impossible, apply one of the numerical approximation schemes from Sect. 3.3.2 on a time grid that is much finer than the grid of observation times. Gourieroux et al. (1993) emphasise that in this simulation, the same random seeds should be employed for all values of $\boldsymbol{\theta}$.

(ii) For $k = 1, \ldots, K$, obtain the maximum likelihood estimators

$$\hat{\boldsymbol{\rho}}_n^{(\boldsymbol{\theta},k)} = \underset{\boldsymbol{\rho} \in \mathcal{R}}{\operatorname{argmax}}\, q\left(\boldsymbol{\rho}; \boldsymbol{z}_0^{(\boldsymbol{\theta},k)}, \ldots, \boldsymbol{z}_n^{(\boldsymbol{\theta},k)}\right). \tag{6.46}$$

(iii) Calculate the average of these estimators,

$$\hat{\boldsymbol{\rho}}_n^{\text{sim}}(\boldsymbol{\theta}) = \frac{1}{K}\sum_{k=1}^{K} \hat{\boldsymbol{\rho}}_n^{(\boldsymbol{\theta},k)}.$$

Under fairly general assumptions, the indirect estimator $\hat{\boldsymbol{\theta}}_n$ is a consistent and asymptotically normal estimator of $\boldsymbol{\theta}$. See Gourieroux et al. (1993) for further asymptotic properties and fields of application.

Broze et al. (1998) discuss the fact that approximate instead of exact simulation in item (6.4.4) introduces a simulation bias. In case of approximate simulation, they hence refer to the above method as *quasi-indirect inference*. Monte Carlo experiments, however, show good performance of this approach for moderate sample sizes K. Furthermore, the authors remark that estimation results do not improve if the Milstein instead of the Euler scheme is employed for the approximate simulation. Overall, the resulting estimator is asymptotically unbiased also for quasi-indirect inference (Gourieroux et al. 1993; Broze et al. 1998).

6.4.5 Efficient Method of Moments

A conceptually similar approach to the indirect inference in the previous section is the *efficient method of moments (EMM)* (Gallant and Tauchen 1996). In this technique, the criterion (6.44) is replaced by

$$\hat{\boldsymbol{\theta}}_n = \underset{\boldsymbol{\theta}\in\Theta}{\operatorname{argmin}}\, \boldsymbol{Q}(\boldsymbol{\theta})'\boldsymbol{E}_n\,\boldsymbol{Q}(\boldsymbol{\theta}) \tag{6.47}$$

for an appropriate positive definite matrix $\boldsymbol{E}_n$, where

$$\boldsymbol{Q}(\boldsymbol{\theta}) = \frac{1}{K}\sum_{k=1}^{K} \frac{\partial q}{\partial \boldsymbol{\rho}}\left(\hat{\boldsymbol{\rho}}_n^{\text{obs}}; \boldsymbol{z}_0^{(\boldsymbol{\theta},k)}, \ldots, \boldsymbol{z}_n^{(\boldsymbol{\theta},k)}\right)$$

and q is the log-likelihood function of $\boldsymbol{\rho}$ in the auxiliary model. For certain choices of $\boldsymbol{D}_n$ and $\boldsymbol{E}_n$, the estimators $\hat{\boldsymbol{\theta}}_n$ obtained through (6.44) and (6.47) are asymptotically equivalent (Gourieroux et al. 1993). The EMM is generally more efficient than indirect inference as the computationally possibly demanding calculation of the maximum likelihood estimator $\hat{\boldsymbol{\rho}}_n^{(\boldsymbol{\theta},k)}$ in Eq. (6.46) is not required. For details on the EMM, see Gallant and Tauchen (1996).

6.5 Exact Algorithm

A recent development in the simulation and estimation of diffusion processes is the introduction of the *Exact Algorithm* which enables exact simulation of diffusion paths without any time discretisation error. By now, the algorithm is available in the different variants EA1 (Beskos and Roberts 2005), EA2 (Beskos et al. 2006a) and EA3 (Beskos et al. 2008). Its implementation is easy and computationally efficient. A drawback, however, is its limited applicability which is formulated in detail below. This section explains the EA1, which is the earliest and simplest version of the Exact Algorithm. The following first discusses the simulation of diffusion paths and then the estimation of parameters.

Consider a one-dimensional diffusion process $X = (X_t)_{t\geq 0}$ with unit diffusion coefficient, i.e. satisfying an SDE

$$\mathrm{d}X_t = \mu(X_t, \boldsymbol{\theta})\mathrm{d}t + \mathrm{d}B_t, \quad X_0 = x_0. \tag{6.48}$$

The EA1 applies only to this class of diffusions. However, transformation of a general one-dimensional diffusion to (6.48) can be obtained with Lamperti's transform, see Sect. 3.2.11.

Simulation

The EA1 aims to draw exact time-discrete skeletons of the diffusion process on the time interval $[0,T]$, where $T \in \mathbb{R}_+$ is fixed. The algorithm is based on the rejection sampling scheme (see e.g. Grimmett and Stirzaker 2001) which works as follows:

Algorithm 6.1 (Rejection Sampling on $\mathbb{R}$). *Consider two equivalent probability measures ρ and ν on $\mathbb{R}$ with bounded Radon-Nikodym derivative, i.e. there exists $\kappa \in \mathbb{R}_+$ such that for all $z \in \mathbb{R}$ one has $\kappa \cdot (d\rho/d\nu)(z) \leq 1$. Suppose one is able to sample from ν. Perform the following steps:*

1. Draw $Z \sim \nu$.
2. Accept Z with probability $\kappa \cdot (d\rho/d\nu)(Z)$. Otherwise, return to step 1.

Then $Z \sim \mu$.

In our case, we do not wish to sample a real random variable from ρ but a diffusion process from a probability measure $\mathbb{P}_{\boldsymbol{\theta}}$ induced by (6.48). In order to apply the rejection sampling algorithm to this situation, we are looking for a probability measure $\mathbb{Z}_{\boldsymbol{\theta}}$ which fulfils the following requirements:

(i) It is possible to sample from $\mathbb{Z}_{\boldsymbol{\theta}}$.
(ii) $\mathbb{P}_{\boldsymbol{\theta}}$ and $\mathbb{Z}_{\boldsymbol{\theta}}$ are equivalent.
(iii) The Radon-Nikodym derivative $\mathrm{d}\mathbb{P}_{\boldsymbol{\theta}}/\mathrm{d}\mathbb{Z}_{\boldsymbol{\theta}}$ is bounded from above.
(iv) It is possible to exactly apply the acceptance probability in step 2 of Algorithm 6.1.

Beskos and Roberts (2005) found out that an appropriate candidate for $\mathbb{Z}_{\boldsymbol{\theta}}$ is the law of Brownian motion starting in x_0 and conditioned on an end point which is drawn from a probability distribution with density

$$h(u) \propto \exp\left(A(u) - \frac{u^2}{2T}\right),$$

where $A(u) = \int_0^u \mu(y)\mathrm{d}y$ for all $u \in \mathbb{R}$. In addition to the general assumptions at the beginning of this chapter, they require the following conditions to hold:

- The drift coefficient μ is everywhere differentiable.
- The integral $\int_{\mathbb{R}} \exp(A(u) - u^2/2T)\mathrm{d}u$ is finite.
- There exist constants $k_1, k_2 \in \mathbb{R}$ such that $k_1 \leq 0.5(\mu^2(u) + \mu'(u)) \leq k_2$ for all $u \in \mathbb{R}$.
- The time horizon T is small enough such that $0 \leq \varphi(u) \leq T^{-1}$ for all $u \in \mathbb{R}$, where $\varphi(u) = 0.5(\mu^2(u) + \mu'(u)) - k_1$.

Beskos and Roberts show that the above choice of $\mathbb{Z}_{\boldsymbol{\theta}}$ fulfils the requirements (6.5) and (6.5). In particular,

$$\frac{\mathrm{d}\mathbb{P}_{\boldsymbol{\theta}}}{\mathrm{d}\mathbb{Z}_{\boldsymbol{\theta}}}(X_{[0,T]}) = \kappa \exp\bigl(-H(X_{[0,T]})\bigr), \tag{6.49}$$

where $H(X_{[0,T]}) = \int_0^T \varphi(X_t)\mathrm{d}t$. In a rejection sampling algorithm, the right hand side of Eq. (6.49) can be taken as the acceptance probability with $\kappa = 1$ because of $\exp(-H(X_{[0,T]})) \leq 1$.

Naturally, it is not possible to sample infinite-dimensional objects from $\mathbb{Z}_{\boldsymbol{\theta}}$. However, assumption (6.5) is fulfilled in the sense that one can obtain exact finite skeletons from $\mathbb{Z}_{\boldsymbol{\theta}}$ by first drawing the end point X_T from the density h and then constructing a Brownian bridge skeleton at discrete time points as described in Sect. 3.3.3.

This skeleton from $\mathbb{Z}_{\boldsymbol{\theta}}$ then has to be accepted or rejected as a draw from $\mathbb{P}_{\boldsymbol{\theta}}$ with probability $\exp(-H(X_{[0,T]}))$. Assumption (6.5) requires that this is possible. The value of $H(X_{[0,T]})$ cannot be calculated as this requires knowledge of the full path $X_{[0,T]}$. One can however circumvent this calculation; to that end, note that a decision with acceptance probability $H(X_{[0,T]}) = \int_0^T \varphi(X_t)\mathrm{d}t$ can be made as follows: First, draw a uniformly distributed point $(t, y) \sim U([0,T] \times [0,M])$, where M is an upper bound of φ (e.g. $M = T^{-1}$). Next, simulate the value X_t of the diffusion path at time t. Accept if $y \leq \varphi(X_t)$, reject otherwise. In order to simulate an event with probability $\exp(-H(X_{[0,T]}))$, Beskos and Roberts expand the probability in a Taylor series and express the event as the countable union of a sequence of increasing events and as the countable intersection of another sequence of decreasing events. With this construction, they are able to come to an accept or reject decision on the basis of a finite skeleton of the diffusion path. For details, see Beskos and Roberts (2005).

Beskos et al. (2006a) replace this last mechanism by a simpler and more efficient procedure that is based on the following observation: Let Ψ be a homogeneous marked Poisson process of unit intensity on $[0,T] \times [0,M]$. That means $\Psi = \{(t_1, y_1), \ldots, (t_k, y_k)\}$, where $t_1 < \ldots < t_k$ are the jump times of a homogeneous Poisson process with intensity one on the time interval $[0,T]$, and $y_1, \ldots, y_k \sim U([0,M])$ are i.i.d. marks at these instants. Given a diffusion path $X_{[0,T]}$, let N be the number of marks below the graph $(t, \varphi(X_t))$, where $t \in [0,T]$. The total number of marks in the rectangle $[0,T] \times [0,M]$ is Poisson distributed with intensity one. Thus N is Poisson distributed with intensity $H(X_{[0,T]}) = \int_0^T \varphi(X_t)\mathrm{d}t$, and

$$\mathbb{P}\bigl(N = 0 \,\big|\, X_{[0,T]}\bigr) = \exp\bigl(-H(X_{[0,T]})\bigr).$$

The number N can be determined given the discrete skeleton of $X_{[0,T]}$. That means, there is a simple possibility to make an accept/reject decision with the required acceptance probability (6.49). The resulting algorithm is the following.

Algorithm 6.2 (EA1).

1. *Simulate a homogeneous marked Poisson process* $\Psi = \{(t_1, y_1), \ldots, (t_k, y_k)\}$ *of unit intensity on* $[0, T] \times [0, M]$, *i.e.* $t_1, \ldots, t_k \in [0, T]$ *are the jump times of the Poisson process and* $y_1, \ldots, y_k \sim U([0, M])$ *are the i.i.d. marks.*
2. *Draw a skeleton from* $\mathbb{Z}_{\boldsymbol{\theta}}$ *at times* $t_1, \ldots, t_k$, *i.e.*
   a. *Simulate the ending point* $X_T \sim h$ *of the diffusion path.*
   b. *Simulate the values* $Y_1, \ldots, Y_k$ *at times* $t_1, \ldots, t_k$ *of a Brownian* $(0, x_0, T, X_T)$*-bridge as described in Sect. 3.3.3.*
3. *If* $\varphi(Y_i) \leq y_i$ *for all* $i \in \{1, \ldots, k\}$, *accept the skeleton. Otherwise, reject and return to step 1.*

Then $S = \{(0, x_0), (t_1, Y_1), \ldots, (t_k, Y_k), (T, X_T)\}$ *can be regarded as a time-discrete sample from* $\mathbb{P}_{\boldsymbol{\theta}}$.

Once a skeleton is accepted as a sample from $\mathbb{P}_{\boldsymbol{\theta}}$, it can be amended by further draws from $\mathbb{Z}_{\boldsymbol{\theta}}$ at additional time instants (Beskos et al. 2006a).

The EA1 requires the function φ to be bounded. The EA2 (Beskos et al. 2006a,b) extends the above methodology such that it is applicable also to diffusions where either $\limsup_{u \to -\infty} \varphi(u) < \infty$ or $\limsup_{u \to \infty} \varphi(u) < \infty$. The algorithm starts by simulating the infimum of the diffusion path and the (maximum) time when this infimum is achieved. Then, the diffusion path is composed of two Bessel processes. This construction makes sure that these path segments do not fall below the infimum. The EA3 (Beskos et al. 2008) even removes the just stated requirements on the bounds of φ by including in the analysis not only the infimum but the whole range of the diffusion path. The decision about acceptance or rejection in both EA2 and EA3 is again under consideration of a marked Poisson process. Generalisations to time-inhomogeneous and multivariate diffusions are discussed in Beskos et al. (2008).

Estimation

The Exact Algorithm enables simple Monte Carlo maximum likelihood estimation for those diffusions where the algorithm is applicable. This is described in the following for the EA1 (Algorithm 6.2). Hence assume that the assumptions required by EA1 are fulfilled.

The objective is the estimation of the transition density $p_{\boldsymbol{\theta}}(t; x_0, x)$ of the diffusion process solving (6.48) for any $x \in \mathbb{R}$. Inference based on EA1 utilises the fact that

$$p_{\boldsymbol{\theta}}(t; x_0, x) = \mathbb{E}_{\boldsymbol{\theta}}\big(p_{\boldsymbol{\theta}}(t; x_0, x|S)\big),$$

where $p_{\boldsymbol{\theta}}(t; x_0, x|S)$ is the usual transition density $p_{\boldsymbol{\theta}}$ further conditioned on a skeleton $S = ((0, x_0), (t_1, Y_1), \ldots, (t_k, Y_k), (T, X_T))$ which has been constructed with EA1 for $T > t$. Because of the Markov property of diffusion processes, conditioning on the whole skeleton reduces to conditioning on the left and right neighbours (t_l, Y_l) and (t_r, Y_r) of (t, x), i.e.

$$p_{\boldsymbol{\theta}}(t; x_0, x|S) = p_{\boldsymbol{\theta}}(t; x_0, x|Y_l, Y_r),$$

where Y_l and Y_r are the values in the skeleton at times

$$t_l = \max\{t_i \,|\, t_i < t, i = 0, \dots, k+1\} \tag{6.50}$$

and

$$t_r = \min\{t_i \,|\, t_i > t, i = 0, \dots, k+1\} \tag{6.51}$$

with $t_0 = 0$ and $t_{k+1} = T$. It has already been noted above that, conditioned on the skeleton, the diffusion process has the same law as a Brownian bridge. Hence,

$$p_{\boldsymbol{\theta}}(t; x_0, x|Y_l, Y_r) = \phi\left(x \,\Big|\, Y_l + \frac{t - t_l}{t_r - t_l}(Y_r - Y_l)\,, \frac{(t - t_l)(t_r - t)}{t_r - t_l}\right) \tag{6.52}$$

(Beskos et al. 2006a); this formula can be derived from the construction of Brownian bridges in Sect. 3.3.3. That means, the transition density $p_{\boldsymbol{\theta}}(t; x_0, x)$ can be estimated by Monte Carlo evaluation of the expected value of (6.52). This is described by the following algorithm.

Algorithm 6.3 (Monte Carlo Likelihood Estimation using EA1).

1. *For $j = 1, \dots, J$, perform the following steps:*
   a. *Using EA1, draw a skeleton S from $\mathbb{P}_{\boldsymbol{\theta}}$ on $[0, T]$, where $T > t$.*
   b. *Identify t_l and t_r as defined in* (6.50) *and* (6.51) *and the corresponding values Y_l and Y_r.*
   c. *Compute* (6.52) *and store the result in p^j.*
2. *Calculate the mean of all p^j to obtain a Monte Carlo estimate of $p_{\boldsymbol{\theta}}(t; x_0, x)$.*

Algorithm 6.3 yields an unbiased estimate of the transition density. It can be utilised to obtain $p_{\boldsymbol{\theta}}$ for different values of the possibly multi-dimensional parameter $\boldsymbol{\theta}$. The maximum likelihood estimator can then for example be found by grid search methods.

More advanced techniques for statistical inference on the basis of the Exact Algorithm are extensively discussed in Beskos et al. (2006b) and Beskos et al. (2009).

6.6 Discussion and Conclusion

This chapter reviews a variety of frequentist methods for the parameter estimation of discretely-observed diffusion processes. Sections 6.1 and 6.2 introduce techniques which are applicable only in an ideal situation where the diffusion process is observed continuously in time, or the exact transition density is known, or observations are available at very dense time points. Sections 6.3–6.5, in contrast, cover more sophisticated estimation approaches which are capable to cope with larger observation intervals even when the transition density of the diffusion process in unknown.

An ultimate grading of the various approaches is not clear cut as each technique has its own strengths and weaknesses. In practice, the choice of an appropriate method is typically problem-specific. First of all, it may depend on the form of the drift or diffusion coefficient, or on the knowledge of eigenfunctions, or on the fact whether the observed diffusion process is ergodic. Furthermore, the number of observations, the data frequency, the dimension of the parameter and of the process, or available computing power possibly influence the decision for or against a certain estimating technique.

However, some general advantages and disadvantages of the presented techniques can be identified: Approximations of the likelihood function as considered in Sect. 6.3 yield the approximated function as a convenient by-product. The Hermite expansion from Sect. 6.3.1 is generally appraised to be fairly efficient; unfortunately it is also quite complex and barely transparent. The latter property also applies to the Crank-Nicolson method from Sect. 6.3.2. Conveniently, the Crank-Nicolson method and the simulated maximum likelihood estimation from Sect. 6.3.3 are generic approaches, i.e. the drift function and diffusion coefficient just have to be plugged in. In practice that means that these approaches need to be implemented only once in order to apply them to different models. However, simulation-based techniques such as the simulated maximum likelihood approach, the simulated moments estimation from Sect. 6.4.3, the indirect inference from Sect. 6.4.4 and the efficient method of moments from Sect. 6.4.5 are computationally demanding. The estimating functions from Sect. 6.4.1 are less hard to compute. Moreover, they are robust to misspecification when only the moments of the diffusion process are matched. On the other hand, important information may be wasted when only moments are considered. The quality of estimates obtained by the indirect inference from Sect. 6.4.4 and the efficient method of moments from Sect. 6.4.5 severely depends on the choice of the auxiliary model. Finally, the Exact Algorithm from Sect. 6.5 yields unbiased estimators and is the most efficient technique of this chapter unless the exact likelihood is available. Unfortunately, its applicability is yet quite restricted; the EA1, for example, requires the diffusion process to be univariate with unit diffusion coefficient and a drift function fulfilling the assumptions listed on p. 161.

A critical comparison of the presented estimation approaches by means of a simulation study is beyond the scope of this book. Some evaluations can, however, be found in the literature, shortly summarised in the next three paragraphs. A documentation of the R-package sde, which implements several of the techniques considered in this chapter, is contained in Iacus (2008).

Jensen and Poulsen (2002) numerically evaluate several estimation approaches on the example of specific one-dimensional diffusion processes. The considered techniques are the Hermite expansion from Sect. 6.3.1, the Crank-Nicolson method from Sect. 6.3.2, a binomial approximation technique as for example considered by Nelson and Ramaswamy (1990), and the naive maximum likelihood approach from Sect. 6.2. Concerning the trade-off between speed and accuracy of the approximations, these approaches turn out to be clearly ranked in the above order with the Hermite expansion showing the best performance.

Honoré (1997) applies the naive maximum likelihood approach from Sect. 6.2, the generalised method of moments from Sect. 6.4.2 and the simulated maximum likelihood estimation from Sect. 6.3.3 to a specific model from financial economics. Based on the outcomes of a simulation study, he labels the first two methods as inappropriate due to large estimation bias, whereas the third approach is found to be practical.

In another simulation study, Hurn et al. (2007) evaluate most of the approaches considered in this chapter with respect to the ease of the implementation, time exposure of the estimation method and accuracy of the resulting estimates. Briefly summarised, the estimating functions based on eigenfunctions from Sect. 6.4.1 and the Hermite expansion from Sect. 6.3.1 are most satisfying considering time exposure and accuracy at the same time. The authors however emphasise that the diffusion models in the simulation study suit these two estimation techniques.

The application of most methods in this chapter becomes problematic as soon as a diffusion process is only partially observed, i.e. some components of the state vector are latent, or observations are measured with error. For example, in case of latent variables, the Markov property of consecutive observations may no longer hold, and the approximations of the likelihood function in Sect. 6.3 may not be applicable anymore. Partial and/or noisy observations are for example considered by Gloter and Jacod (2001a,b) and Jimenez et al. (2006).

A powerful approach to overcome this problem is to estimate the model parameters in a Bayesian framework. Such a procedure is able to handle multi-dimensional diffusion processes which are partially observed and measured with error. This way, it outperforms the majority (if not all) of the methods presented in this chapter and enables the statistical analysis of the complex applications in Chaps. 8 and 9. Large samples or stationarity of the underlying process are not required. As an appreciated by-product, the technique also estimates the sample path at intermediate observation times. The presentation and further development of such a method is the subject of the next chapter.

References

Aït-Sahalia Y (1996) Nonparametric pricing of interest rate derivative securities. Econometrica 64:527–560

Aït-Sahalia Y (2002) Maximum likelihood estimation of discretely sampled diffusions: a closed-form approximation approach. Econometrica 70:223–262

Aït-Sahalia Y (2008) Closed-form likelihood expansions for multivariate diffusions. Ann Stat 36:906–937

Beskos A, Roberts G (2005) Exact simulation of diffusions. Ann Appl Probab 15:2422–2444

Beskos A, Papaspiliopoulos O, Roberts G (2006a) Retrospective exact simulation of diffusion sample paths with applications. Bernoulli 12:1077–1098

Beskos A, Papaspiliopoulos O, Roberts G, Fearnhead P (2006b) Exact and computationally efficient likelihood-based estimation for discretely observed diffusion processes (with comments). J R Stat Soc Ser B 68:333–382

Beskos A, Papaspiliopoulos O, Roberts G (2008) A factorisation of diffusion measure and finite sample path constructions. Methodol Comput Appl Probab 10:85–104

Beskos A, Papaspiliopoulos O, Roberts G (2009) Monte-Carlo maximum likelihood estimation for discretely observed diffusion processes. Ann Stat 37:223–245

Bibby B, Sørensen M (1995) Martingale estimation functions for discretely observed diffusion processes. Bernoulli 1:17–39

Bibby B, Sørensen M (2001) Simplified estimating functions for diffusion models with a high-dimensional parameter. Scand J Stat 28:99–112

Bibby B, Jacobsen M, Sørensen M (2009) Estimating functions for discretely sampled diffusion-type models. In: Aït-Sahalia Y, Hansen L (eds) Handbook of financial econometrics. North-Holland, Amsterdam, pp 203–268

Brandt M, Santa-Clara P (2001) Simulated likelihood estimation of diffusions with an application to exchange rate dynamics in incomplete markets. Working paper 274, National Bureau of Economic Research

Broze L, Scaillet O, Zakoïan JM (1998) Quasi-indirect inference for diffusion processes. Econometric Theory 14:161–186

Comte F, Genon-Catalot V, Rozenholc Y (2007) Penalized nonparametric mean square estimation of the coefficients of diffusion processes. Bernoulli 13:514–543

Crank J, Nicolson E (1947) A practical method for numerical evaluation of solutions of partial differential equations of the heat-conduction type. Math Proc Camb Philos Soc 43:50–67

Dacunha-Castelle D, Florens-Zmirou D (1986) Estimation of the coefficients of a diffusion from discrete observations. Stochastics 19:263–284

Duffie D, Singleton K (1993) Simulated moments estimation of Markov models of asset prices. Econometrica 61:929–952

Durham G, Gallant A (2002) Numerical techniques for maximum likelihood estimation of continuous-time diffusion processes (with comments). J Bus Econom Stat 20:297–316

Elerian O (1998) A note on the existence of a closed form conditional transition density for the Milstein scheme. Working paper, Nuffield College, University of Oxford

Florens-Zmirou D (1989) Approximate dicrete-time schemes for statistics of diffusion processes. Statistics 20:547–557

Florens-Zmirou D (1993) On estimating the diffusion coefficient from discrete observations. J Appl Probab 30:790–804

Gallant A, Tauchen G (1996) Which moments to match? Econom Theory 12:657–681

Gard T (1988) Introduction to stochastic differential equations. Monographs and textbooks in pure and applied mathematics, vol 114. M. Dekker, New York

Gloter A, Jacod J (2001a) Diffusions with measurement errors. I. Local asymptotic normality. ESAIM Probab Stat 5:225–242

Gloter A, Jacod J (2001b) Diffusions with measurement errors. II. Optimal estimators. ESAIM Probab Stat 5:243–260
Godambe V (1991) Estimating functions. Oxford University Press, Oxford
Gourieroux C, Monfort A, Renault E (1993) Indirect inference. J Appl Econom 8:85–118
Grimmett G, Stirzaker D (2001) Probability and random processes, 3rd edn. Oxford University Press, Oxford
Hansen L (1982) Large sample properties of generalized method of moments estimators. Econometrica 50:1029–1054
Hansen L, Scheinkman J (1995) Back to the future: Generating moment implications for continuous-time Markov processes. Econometrica 63:767–804
Heyde C (1997) Quasi-likelihood and its application: a general approach to optimal parameter estimation. Springer, New York
Honoré P (1997) Maximum likelihood estimation of non-linear continuous-time term-structure models. Aarhus School of Business, Aarhus
Hurn A, Lindsay K, Martin V (2003) On the efficacy of simulated maximum likelihood for estimating the parameters of stochastic differential equations. J Time Ser Anal 24:45–63
Hurn A, Jeisman J, Lindsay K (2007) Seeing the wood for the trees: a critical evaluation of methods to estimate the parameters of stochastic differential equations. J Financ Econom 5:390–455
Iacus S (2008) Simulation and inference for stochastic differential equations. Springer series in statistics. Springer, New York
Jacobsen M (2001) Discretely observed diffusions: classes of estimating functions and small δ-optimality. Scand J Stat 28:123–149
Jacod J (2000) Non-parametric kernel estimation of the coefficient of a diffusion. Scand J Stat 27:83–96
Jensen B, Poulsen R (2002) Transition densities of diffusion processes: numerical comparison of approximation techniques. J Derivatives 9:18–32
Jiang G, Knight J (1997) A nonparametric approach to the estimation of diffusion processes, with an application to a short-term interest rate model. Econometric Theory 13:615–645
Jimenez J, Biscay R, Ozaki T (2006) Inference methods for discretely observed continuous-time stochastic volatility models: a commented overview. Asia Pac Financ Mark 12:109–141
Kendall M, Stuart A, Ord J (1987) Kendall's advanced theory of statistics. Volume I: distribution theory, 5th edn. Charles Griffin, London
Kent J (1978) Time-reversible diffusions. Adv Appl Probab 10:819–835
Kessler M (1997) Estimation of an ergodic diffusion from discrete observations. Scand J Stat 24:211–229
Kessler M (2000) Simple and explicit estimating functions for a discretely observed diffusion process. Scand J Stat 27:65–82
Kessler M, Paredes S (2002) Computational aspects related to martingale estimating functions for a discretely observed diffusion. Scand J Stat 29:425–440
Kessler M, Sørensen M (1999) Estimating equations based on eigenfunctions for a discretely observed diffusion process. Bernoulli 5:299–314
Kloeden P, Platen E (1999) Numerical solution of stochastic differential equations, 3rd edn. Springer, Berlin/Heidelberg/New York
Kutoyants Y (2004) Statistical inference for ergodic diffusion processes. Springer series in statistics. Springer, London
Le Breton A (1974) Parameter estimation in a linear stochastic differential equation. In: Transactions of the seventh Prague conference and of the European meeting of statisticians, Prague, pp 353–366
Liptser R, Shiryayev A (1977) Statistics of random processes, vol 1. Springer, New York
Liptser R, Shiryayev A (1978) Statistics of random processes, vol 2. Springer, New York
Lo A (1988) Maximum likelihood estimation of generalized Itô processes with discretely sampled data. Econom Theory 4:231–247

Nelson D, Ramaswamy K (1990) Simple binomial processes as diffusion approximations in financial models. Rev Financ Stud 3:393–430
Nicolau J (2003) Bias reduction in nonparametric diffusion coefficient estimation. Econom Theory 19:754–777
Nielsen J, Madsen H, Young P (2000) Parameter estimation in stochastic differential equations: an overview. Annu Rev Control 24:83–94
Pedersen A (1995a) Consistency and asymptotic normality of an approximate maximum likelihood estimator for discretely observed diffusion processes. Bernoulli 1:257–279
Pedersen A (1995b) A new approach to maximum likelihood estimation for stochastic differential equations based on discrete observations. Scand J Stat 22:55–71
Polson N, Roberts G (1994) Bayes factors for discrete observations from diffusion processes. Biometrika 81:11–26
Poulsen R (1999) Approximate maximum likelihood estimation of discretely observed diffuson processes. Working paper 29, Center for Analytical Finance, Aarhus
Prakasa Rao B (1999) Statistical inference for diffusion type processes. Arnold, London
Santa-Clara P (1995) Simulated likelihood estimation of diffusions with an application to the short term interest rate. Working paper, Anderson Graduate School of Management, UCLA
Shoji I (1998) Approximation of continuous time stochastic processes by a local linearization method. Math Comp 67:287–298
Shoji I, Ozaki T (1998a) Estimation for nonlinear stochastic differential equations by a local linearization method. Stoch Anal Appl 16:733–752
Shoji I, Ozaki T (1998b) A statistical method of estimation and simulation for systems of stochastic differential equations. Biometrika 85:240–243
Singer H (2004) Moment equations and Hermite expansion for nonlinear stochastic differential equations with application to stock price models. In: The 6th international conference on social science methodology, Amsterdam
Sørensen H (2001) Dicretely observed diffusions: approximation of the continuous-time score function. Scand J Stat 28:113–121
Sørensen H (2004) Parametric inference for diffusion processes observed at discrete points in time: a survey. Int Stat Rev 72:337–354
Sørensen M (2007) Efficient estimation for ergodic diffusions sampled at high frequency. Department of Economics and Business, Aarhus University. CREATES Research Paper No. 2007–46
Sørensen M (2008) Parametric inference for discretely sampled stochastic differential equations. Department of Economics and Business, Aarhus University. CREATES Research Paper No 2008–18
Soulier P (1998) Non parametric estimation of the diffusion coefficient of a diffusion process. Stoch Anal Appl 16:185–200
Stramer O, Yan J (2007) Asymptotics of an efficient Monte Carlo estimation for the transition density of diffusion processes. Methodol Comput Appl Probab 9:483–496
Yoshida N (1992) Estimation for diffusion processes from discrete observation. J Multivar Anal 41:220–242

Chapter 7
Bayesian Inference for Diffusions with Low-Frequency Observations

The previous chapter considered a variety of frequentist procedures to infer on the parameters of a diffusion process. The difficulty that underlies all approaches is the general intractability of the transition density for discrete-time observations. Most techniques struggle when inter-observation times are large. Datasets in life sciences, however, may well be of low-frequency type. Examples are plant surveys with yearly assessment, epidemics where public health reporting considers new infections per week, or cost-intensive and hence infrequent measurements in genetics. Even in finance, where high-frequency measurements are often available, it may be advantageous to work with a thinned dataset; for instance, asset price data can be corrected for certain microstructure effects this way (Jones 1998).

The present chapter introduces Bayesian inference methods which all base on introducing missing data such that the union of missing values and observations forms a high-frequency dataset. This facilitates approximation of the transition density and hence enables parametric inference even for large inter-observation times. Moreover, the techniques are suitable for irregularly spaced observation intervals, multivariate diffusions with possibly latent components and for observations that are subject to measurement error. They even apply when different components of the state space are observed nonsynchronously. Stationarity and ergodicity of the diffusion are generally not required. As a Bayesian method, the estimation procedure is not indispensably dependent on large samples.

The introduction of intermediate data between every two observations implies the estimation of the missing values in addition to the model parameters, where both the missing data and the parameters are treated as random variables. This task is performed by application of Markov chain Monte Carlo (MCMC) techniques which alternately update the imputed data and the model parameter and are usually feasible within moderate computing time. The following considerations require familiarity with basic MCMC ideas. Introductory texts on this topic can be found in Gilks et al. (1996), Robert and Casella (2004) and Gamerman and Lopes (2006).

C. Fuchs, *Inference for Diffusion Processes: With Applications in Life Sciences*, DOI 10.1007/978-3-642-25969-2_7,

As in Chap. 6, the focus of interest lies on the time-homogeneous Itô diffusion process $\boldsymbol{X} = (\boldsymbol{X}_t)_{t \geq 0}$ satisfying

$$\mathrm{d}\boldsymbol{X}_t = \boldsymbol{\mu}(\boldsymbol{X}_t, \boldsymbol{\theta})\mathrm{d}t + \boldsymbol{\sigma}(\boldsymbol{X}_t, \boldsymbol{\theta})\mathrm{d}\boldsymbol{B}_t\,, \quad \boldsymbol{X}_{t_0} = \boldsymbol{x}_0, \tag{7.1}$$

with state space $\mathcal{X} \subseteq \mathbb{R}^d$, starting value $\boldsymbol{x}_0 \in \mathcal{X}$ at time $t_0 = 0$ and m-dimensional standard Brownian motion $\boldsymbol{B} = (\boldsymbol{B}_t)_{t \geq 0}$. The drift function $\boldsymbol{\mu} : \mathcal{X} \times \Theta \to \mathbb{R}^d$ and diffusion coefficient $\boldsymbol{\sigma} : \mathcal{X} \times \Theta \to \mathbb{R}^{d \times m}$ are assumed to be known in a parametric form, and $\boldsymbol{\theta} \in \Theta$ for an open set $\Theta \subseteq \mathbb{R}^p$ is the model parameter. $\mathcal{X}$ is the same for all $\boldsymbol{\theta} \in \Theta$. Once again, we assume that $\boldsymbol{\mu}$, $\boldsymbol{\sigma}$ and the diffusion matrix $\boldsymbol{\Sigma} = \boldsymbol{\sigma}\boldsymbol{\sigma}'$ fulfil the regularity conditions stated in Sect. 3.4 for all $\boldsymbol{\theta} \in \Theta$.

This chapter is organised as follows: Sect. 7.1 comprehensively explains the basic concept of Bayesian data augmentation for diffusions. Proposal distributions for both the diffusion path and the parameter are introduced and illustrated in a simulation study. This is done under the assumption of complete observations at discrete time points without measurement error. As the latter assumption is not necessarily fulfilled in applications in life sciences, Sect. 7.2 extends the methodology to a latent data framework which also allows for observation error. This section is especially helpful for practitioners, but it is not a premise for the comprehension of the remaining chapter and may hence be skipped. Whilst Sects. 7.1 and 7.2 treat the imputed path segments as countable sets of discrete data points, Sects. 7.3 and 7.4 are dedicated to the consideration of continuous data. This reveals a well-known convergence problem caused by a close link between the model parameters and the quadratic variation of the diffusion path, pointed out in Sect. 7.3. In practice, this dependency causes arbitrarily slow mixing of the Markov chains when large amounts of auxiliary data are imputed. In Sect. 7.4, different approaches are hence presented which aim at overcoming this difficulty. Special focus is on the innovation scheme, newly developed in a continuous observation framework, in Sect. 7.4.4.

This chapter brings together approaches from different authors on estimation via Bayesian data augmentation in a multivariate framework and evaluates them both analytically and computationally. A new sampling scheme is suggested where existing methods do not lead to success, and its universal applicability is proven. The contents of this chapter address both practicioners who wish to implement the estimation schemes and theoreticians who are interested in convergence proofs. The methods are deployed in Chaps. 8 and 9 where they enable statistical inference in complex models in life sciences.

7.1 Concepts of Bayesian Data Augmentation for Diffusions

The idea of parameter estimation based on Bayesian data augmentation is similar to the concept of the simulated maximum likelihood estimation (SMLE) approach which was introduced in Sect. 6.3.3: In order to perform inference on the model parameter $\boldsymbol{\theta}$, one tries to approximate the true transition density $p_{\boldsymbol{\theta}}$ of the diffusion process by the Euler scheme or one of the higher-order numerical schemes from Sect. 3.3.2. This is eligible only if inter-observation times of the observed data $\boldsymbol{X}^{\text{obs}}$ are small. Since such a requirement is usually not fulfilled in applications in life sciences, additional data $\boldsymbol{X}^{\text{imp}}$ at intermediate time points is introduced. To this end, a Markov chain Monte Carlo (MCMC) approach is employed to construct a Markov chain $\{\boldsymbol{\theta}^{(i)}, \boldsymbol{X}^{\text{imp}(i)}\}_{i=1,\dots,L}$ of length L whose elements are samples from the joint posterior density $\pi(\boldsymbol{\theta}, \boldsymbol{X}^{\text{imp}}|\boldsymbol{X}^{\text{obs}})$ of the parameter $\boldsymbol{\theta}$ and the imputed data $\boldsymbol{X}^{\text{imp}}$ conditioned on the sample path observations $\boldsymbol{X}^{\text{obs}}$. The Markov chain $\{\boldsymbol{\theta}^{(i)}\}_{i=1,\dots,L}$ is then regarded as a draw from the marginal density $\pi(\boldsymbol{\theta}|\boldsymbol{X}^{\text{obs}})$. The imputed data $\{\boldsymbol{X}^{\text{obs}(i)}\}_{i=1,\dots,L}$ is a convenient by-product of the estimation procedure.

For the construction of the Markov chain $\{\boldsymbol{\theta}^{(i)}, \boldsymbol{X}^{\text{imp}(i)}\}_{i=1,\dots,L}$, the following two steps are alternately executed:

$$\begin{aligned} &\textit{Path Update:} && \text{Draw}\,\boldsymbol{X}^{\text{imp}(i)} \sim \pi\big(\boldsymbol{X}^{\text{imp}(i)} \,\big|\, \boldsymbol{X}^{\text{obs}}, \boldsymbol{\theta}^{(i-1)}\big). \\ &\textit{Parameter Update:} && \text{Draw }\ \boldsymbol{\theta}^{(i)} \sim \pi\big(\boldsymbol{\theta}^{(i)} \,\big|\, \boldsymbol{X}^{\text{obs}}, \boldsymbol{X}^{\text{imp}(i)}\big). \end{aligned} \tag{7.2}$$

This procedure has been proposed and shown to converge by Tanner and Wong (1987), though not in the context of diffusions. The underlying idea is similar to the one for the expectation-maximisation (EM) algorithm by Dempster et al. (1977). In general, however, direct sampling is possible neither from $\pi(\boldsymbol{X}^{\text{imp}}|\boldsymbol{X}^{\text{obs}}, \boldsymbol{\theta})$ nor from $\pi(\boldsymbol{\theta}|\boldsymbol{X}^{\text{obs}}, \boldsymbol{X}^{\text{imp}})$. Hence, in both steps the Metropolis-Hastings algorithm is used. This is further specified in Sect. 7.1.2 for the path update and in Sect. 7.1.3 for the parameter update.

The concept of Bayesian data imputation as a tool in inference for diffusions has been utilised by a number of authors including Jones (1998), Elerian et al. (2001), Eraker (2001), Roberts and Stramer (2001), Chib et al. (2004) and Golightly and Wilkinson (2005, 2006a,b, 2008).

7.1.1 Preliminaries and Notation

As it is common practice in Bayesian analysis, let π generically denote all posterior densities. In particular, the exact meaning of π is implied by the occurrence, order and number of its arguments. If these differ for two densities, the two functions are generally not the same. However, for notational brevity, according subscripts are suppressed. The interpretation of π depends on the context but is always apparent from its arguments. Analogously, let p generically denote all prior densities and q all proposal densities. In sampling instructions such as (7.2), the variables on which one conditions are usually not shown on the left of the tilde if their appearance is clear.

We basically aim to approximate the posterior density

$$\pi(\boldsymbol{\theta}|\boldsymbol{X}_{\tau_1}, \dots, \boldsymbol{X}_{\tau_M}) \propto \pi(\boldsymbol{X}_{\tau_1}, \dots, \boldsymbol{X}_{\tau_M}|\boldsymbol{\theta})p(\boldsymbol{\theta})$$

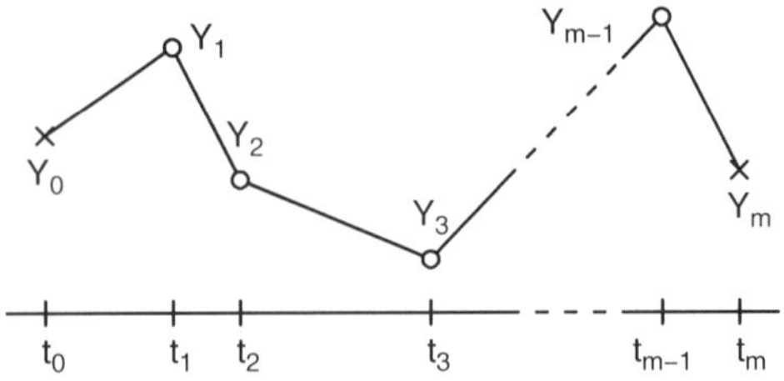

Fig. 7.1 Illustration of a one-dimensional path segment consisting of discrete data points $Y_0, \ldots, Y_m$ at times $t_0, \ldots, t_m$. The observed data Y_0, Y_m is labelled with *crosses*, the imputed data $Y_1, \ldots, Y_{m-1}$ with *circles*

of the parameter $\boldsymbol{\theta}$ based on discrete observations $\boldsymbol{X}_{\tau_1}, \ldots, \boldsymbol{X}_{\tau_M}$ of a diffusion process. In the present and the following subsections we assume that all observations are complete, i.e. there are no latent or unobserved components for all observations $\boldsymbol{X}_{\tau_i} = (X_{\tau_i,1}, \ldots, X_{\tau_i,d})'$. Since diffusion processes possess the Markov property, such complete observations divide a sample path into segments that are mutually independent conditioned on $\boldsymbol{\theta}$. The likelihood of $\boldsymbol{\theta}$ factorises as

$$\pi\big(\boldsymbol{X}_{\tau_1}, \ldots, \boldsymbol{X}_{\tau_M} \,\big|\, \boldsymbol{\theta}\big) = \pi\big(\boldsymbol{X}_{\tau_1} \,\big|\, \boldsymbol{\theta}\big) \prod_{i=2}^{M} \pi\big(\boldsymbol{X}_{\tau_i} \,\big|\, \boldsymbol{X}_{\tau_{i-1}}, \boldsymbol{\theta}\big).$$

It is hence sufficient to consider the theory of Bayesian data imputation for a single path segment between two consecutive complete observations; the generalisation to more observed data points is then straightforward and clarified in Sect. 7.1.4. In the following, we will hence restrict our attention to diffusions on a time interval $[0, T]$, where the starting value $\boldsymbol{X}_0 = \boldsymbol{x}_0$ and the final value $\boldsymbol{X}_T = \boldsymbol{x}$ are completely observed and all intermediate data is unknown. As we consider time-homogeneous diffusions here, the starting time zero is not a constraint.

As motivated above, the time interval is divided into m subintervals which are not necessarily equidistant. The end points of these intervals are $0 = t_0 < t_1 < \ldots < t_{m-1} < t_m = T$, implying the time steps $\Delta t_k = t_{k+1} - t_k$ for $k = 0, \ldots, m-1$. The diffusion process $\boldsymbol{X}$ is in state $\boldsymbol{X}_{t_k}$ at time t_k, but these values are unknown for $k = 1, \ldots, m-1$ and are hence treated as missing data. An example for a path segment consisting of discretely observed and imputed data is shown in Fig. 7.1. For shorter notation, introduce $\boldsymbol{Y}_k = \boldsymbol{X}_{t_k}$ for $k = 0, \ldots, m$. In particular, $\boldsymbol{Y}_0 = \boldsymbol{x}_0$ and $\boldsymbol{Y}_m = \boldsymbol{x}$. Collect the observed data as $\boldsymbol{Y}^{\text{obs}} = \{\boldsymbol{Y}_0, \boldsymbol{Y}_m\}$ and the imputed data as $\boldsymbol{Y}^{\text{imp}} = \{\boldsymbol{Y}_1, \ldots, \boldsymbol{Y}_{m-1}\}$. Furthermore, refer to subsets of the imputed data by $\boldsymbol{Y}^{\text{imp}}_{(a,b)} = \{\boldsymbol{Y}_{a+1}, \ldots, \boldsymbol{Y}_{b-1}\}$ for $a, b \in \{0, \ldots, m\}$ and $a \leq b$. For $b - a < 2$, $\boldsymbol{Y}^{\text{imp}}_{(a,b)}$ is the empty set. Define the complement of $\boldsymbol{Y}^{\text{imp}}_{(a,b)}$ as $\boldsymbol{Y}^{\text{imp}}_{-(a,b)} = \boldsymbol{Y}^{\text{imp}} \setminus \boldsymbol{Y}^{\text{imp}}_{(a,b)}$. Note that later in this chapter $\boldsymbol{X}^{\text{imp}}_{(t_a,t_b)}$ will refer to continuous observation $(\boldsymbol{X}_t)_{t\in(t_a,t_b)}$ and will hence substantially differ from the countable set $\boldsymbol{Y}^{\text{imp}}_{(a,b)}$.

7.1.2 Path Update

We now investigate how to appropriately perform the path update step in (7.2). As indicated above, direct sampling from the posterior distribution of the imputed data given the observed data and the parameter is usually not possible; this option comes into question only if the underlying SDE is analytically solvable or when the conditions for the Exact Algorithm in Sect. 6.5 apply. We hence utilise the Metropolis-Hastings algorithm for the general implementation of this step.

Satisfying convergence results are often achieved by application of update strategies where at each iteration only a subset of the imputed data instead of the whole path segment is renewed. For this, as a first step of the path update, one chooses a time interval (t_a, t_b) with $a, b \in \{0, \ldots, m\}$ and $b - a \geq 2$ in whose interior the path is to be updated. An update of the entire imputed data corresponds to $(t_a, t_b) = (0, T)$, i.e. $a = 0$ and $b = m$. The choice of (t_a, t_b) may be deterministic or random with fixed or varying interval length $t_b - t_a$. Possible options are considered in Sect. 7.1.6.

Having decided about the block update strategy, select an appropriate proposal distribution for $\boldsymbol{Y}^{\mathrm{imp}}_{(a,b)}$ with density q which possibly depends on the observed data $\boldsymbol{Y}^{\mathrm{obs}}$, the current imputed data $\boldsymbol{Y}^{\mathrm{imp}}$ and the parameter $\boldsymbol{\theta}$. In particular, q may be conditioned on (subsets of) both the previously imputed data $\boldsymbol{Y}^{\mathrm{imp}}_{(a,b)}$ and the unaltered imputed data $\boldsymbol{Y}^{\mathrm{imp}}_{-(a,b)}$. From this distribution, draw a proposal $\boldsymbol{Y}^{\mathrm{imp}*}_{(a,b)} = \{\boldsymbol{Y}^*_{a+1}, \ldots, \boldsymbol{Y}^*_{b-1}\}$ for the subset of the imputed data which is to be updated. Accept $\boldsymbol{Y}^{\mathrm{imp}*}_{(a,b)}$ with probability

$$\zeta\big(\boldsymbol{Y}^{\mathrm{imp}*}_{(a,b)}, \boldsymbol{Y}^{\mathrm{imp}}_{(a,b)}\big) = 1 \wedge \frac{\pi\big(\boldsymbol{Y}^{\mathrm{imp}*}_{(a,b)}, \boldsymbol{Y}^{\mathrm{imp}}_{-(a,b)} \,\big|\, \boldsymbol{Y}^{\mathrm{obs}}, \boldsymbol{\theta}\big) q\big(\boldsymbol{Y}^{\mathrm{imp}}_{(a,b)} \,\big|\, \boldsymbol{Y}^{\mathrm{imp}*}_{(a,b)}, \boldsymbol{Y}^{\mathrm{imp}}_{-(a,b)}, \boldsymbol{Y}^{\mathrm{obs}}, \boldsymbol{\theta}\big)}{\pi\big(\boldsymbol{Y}^{\mathrm{imp}}_{(a,b)}, \boldsymbol{Y}^{\mathrm{imp}}_{-(a,b)} \,\big|\, \boldsymbol{Y}^{\mathrm{obs}}, \boldsymbol{\theta}\big) q\big(\boldsymbol{Y}^{\mathrm{imp}*}_{(a,b)} \,\big|\, \boldsymbol{Y}^{\mathrm{imp}}_{(a,b)}, \boldsymbol{Y}^{\mathrm{imp}}_{-(a,b)}, \boldsymbol{Y}^{\mathrm{obs}}, \boldsymbol{\theta}\big)}.$$

Otherwise, discard the proposal and keep the previous data $\boldsymbol{Y}^{\mathrm{imp}}_{(a,b)}$. Due to the Markov property of diffusions, one has

$$\frac{\pi\big(\boldsymbol{Y}^{\mathrm{imp}*}_{(a,b)}, \boldsymbol{Y}^{\mathrm{imp}}_{-(a,b)} \,\big|\, \boldsymbol{Y}^{\mathrm{obs}}, \boldsymbol{\theta}\big)}{\pi\big(\boldsymbol{Y}^{\mathrm{imp}}_{(a,b)}, \boldsymbol{Y}^{\mathrm{imp}}_{-(a,b)} \,\big|\, \boldsymbol{Y}^{\mathrm{obs}}, \boldsymbol{\theta}\big)} = \prod_{k=a}^{b-1} \frac{\pi\big(\boldsymbol{Y}^*_{k+1} \,\big|\, \boldsymbol{Y}^*_k, \boldsymbol{\theta}\big)}{\pi\big(\boldsymbol{Y}_{k+1} \,\big|\, \boldsymbol{Y}_k, \boldsymbol{\theta}\big)} = \prod_{k=a}^{b-1} \frac{p_{\boldsymbol{\theta}}\big(\Delta t_k, \boldsymbol{Y}^*_k, \boldsymbol{Y}^*_{k+1}\big)}{p_{\boldsymbol{\theta}}\big(\Delta t_k, \boldsymbol{Y}_k, \boldsymbol{Y}_{k+1}\big)},$$

where $\boldsymbol{Y}^*_a = \boldsymbol{Y}_a$ and $\boldsymbol{Y}^*_b = \boldsymbol{Y}_b$. The time steps Δt_k in $p_{\boldsymbol{\theta}}$ are now supposed to be small enough such that an approximation with the Euler scheme is allowed, i.e. $p_{\boldsymbol{\theta}}$ may be replaced by

$$\pi^{\mathrm{Euler}}\big(\boldsymbol{Y}_{k+1} \,\big|\, \boldsymbol{Y}_k, \boldsymbol{\theta}\big) = \phi\big(\boldsymbol{Y}_{k+1} \,\big|\, \boldsymbol{Y}_k + \boldsymbol{\mu}(\boldsymbol{Y}_k, \boldsymbol{\theta})\Delta t_k\,,\ \boldsymbol{\Sigma}(\boldsymbol{Y}_k, \boldsymbol{\theta})\Delta t_k\big). \quad (7.3)$$

Here, as before, $\phi(\boldsymbol{z}|\boldsymbol{\nu}, \boldsymbol{\Lambda})$ denotes the possibly multivariate normal density with mean $\boldsymbol{\nu}$ and covariance $\boldsymbol{\Lambda}$ at $\boldsymbol{z}$. In the following, we hence apply

$$\zeta\big(\boldsymbol{Y}^{\text{imp}*}_{(a,b)}, \boldsymbol{Y}^{\text{imp}}_{(a,b)}\big)$$

$$= 1 \wedge \left(\prod_{k=a}^{b-1} \frac{\pi^{\text{Euler}}\big(\boldsymbol{Y}^*_{k+1} \big| \boldsymbol{Y}^*_k, \boldsymbol{\theta}\big)}{\pi^{\text{Euler}}\big(\boldsymbol{Y}_{k+1} \big| \boldsymbol{Y}_k, \boldsymbol{\theta}\big)} \right) \frac{q\big(\boldsymbol{Y}^{\text{imp}}_{(a,b)} \,\big|\, \boldsymbol{Y}^{\text{imp}*}_{(a,b)}, \boldsymbol{Y}^{\text{imp}}_{-(a,b)}, \boldsymbol{Y}^{\text{obs}}, \boldsymbol{\theta}\big)}{q\big(\boldsymbol{Y}^{\text{imp}*}_{(a,b)} \,\big|\, \boldsymbol{Y}^{\text{imp}}_{(a,b)}, \boldsymbol{Y}^{\text{imp}}_{-(a,b)}, \boldsymbol{Y}^{\text{obs}}, \boldsymbol{\theta}\big)} \tag{7.4}$$

with $\boldsymbol{Y}^*_a = \boldsymbol{Y}_a$ and $\boldsymbol{Y}^*_b = \boldsymbol{Y}_b$ as the acceptance probability for the proposal $\boldsymbol{Y}^{\text{imp}*}_{(a,b)}$. The choice of the proposal density q is discussed in what follows, where a number of possible schemes is presented.

Euler Proposal

The most naive proposal for $\boldsymbol{Y}^{\text{imp}}_{(a,b)}$ is to simply apply the Euler sampling scheme from Sect. 3.3.2, i.e. to successively draw

$$\boldsymbol{Y}^*_{k+1} \sim \mathcal{N}\big(\boldsymbol{Y}^*_k + \boldsymbol{\mu}(\boldsymbol{Y}^*_k, \boldsymbol{\theta})\Delta t_k\,,\ \boldsymbol{\Sigma}(\boldsymbol{Y}^*_k, \boldsymbol{\theta})\Delta t_k\big) \tag{7.5}$$

for $k = a, \dots, b-2$, where $\boldsymbol{Y}^*_a = \boldsymbol{Y}_a$. In this case the proposal density equals

$$q_{\text{E}}\big(\boldsymbol{Y}^{\text{imp}*}_{(a,b)} \,\big|\, \boldsymbol{Y}_a, \boldsymbol{\theta}\big) = \prod_{k=a}^{b-2} q_{\text{E}}\big(\boldsymbol{Y}^*_{k+1} \,\big|\, \boldsymbol{Y}^*_k, \boldsymbol{\theta}\big) = \prod_{k=a}^{b-2} \pi^{\text{Euler}}\big(\boldsymbol{Y}^*_{k+1} \,\big|\, \boldsymbol{Y}^*_k, \boldsymbol{\theta}\big),$$

and the acceptance probability (7.4) for the proposal $\boldsymbol{Y}^{\text{imp}*}_{(a,b)}$ reduces to

$$\zeta\big(\boldsymbol{Y}^{\text{imp}*}_{(a,b)}, \boldsymbol{Y}^{\text{imp}}_{(a,b)}\big) = 1 \wedge \left(\prod_{k=a}^{b-1} \frac{\pi^{\text{Euler}}(\boldsymbol{Y}^*_{k+1} | \boldsymbol{Y}^*_k, \boldsymbol{\theta})}{\pi^{\text{Euler}}(\boldsymbol{Y}_{k+1} | \boldsymbol{Y}_k, \boldsymbol{\theta})} \right) \left(\prod_{k=a}^{b-2} \frac{\pi^{\text{Euler}}(\boldsymbol{Y}_{k+1} | \boldsymbol{Y}_k, \boldsymbol{\theta})}{\pi^{\text{Euler}}(\boldsymbol{Y}^*_{k+1} | \boldsymbol{Y}^*_k, \boldsymbol{\theta})} \right)$$

$$= 1 \wedge \frac{\pi^{\text{Euler}}(\boldsymbol{Y}_b | \boldsymbol{Y}^*_{b-1}, \boldsymbol{\theta})}{\pi^{\text{Euler}}(\boldsymbol{Y}_b | \boldsymbol{Y}_{b-1}, \boldsymbol{\theta})},$$

where $\boldsymbol{Y}^*_b = \boldsymbol{Y}_b$. The *Euler proposal* (7.5) conditions on the starting point $\boldsymbol{Y}_a$ of the path segment but is independent of its end point $\boldsymbol{Y}_b$. Hence, a problematic situation arises which is similar to the difficulties in the SMLE approach by Pedersen (1995) described in Sect. 6.3.3: Transitions from $\boldsymbol{Y}^*_{b-1}$ to $\boldsymbol{Y}_b$ are improbable if the according jumps are large. This is most likely the case if one does not condition on the end point; see also the typical path proposals in Figs. 7.2 and 7.3 on pp. 178 and 179. The acceptance probability for $\boldsymbol{Y}^{\text{imp}*}_{(a,b)}$ is then typically small, leading to low acceptance probabilities, i.e. inefficient MCMC samplers due to large numbers of rejections. The following proposal densities condition on both $\boldsymbol{Y}_a$ and $\boldsymbol{Y}_b$.

Double-Sided Euler Proposal

One way to obtain more likely path proposals is to update $\boldsymbol{Y}_{(a,b)}^{\text{imp}}$ from the left to the right, where for all k the proposal distribution of $\boldsymbol{Y}_{k+1}^*$ is conditioned on the already updated preceding value $\boldsymbol{Y}_k^*$ and the subsequent value $\boldsymbol{Y}_{k+2}$. This approach is referred to as *double-sided Euler proposal* in the following. It has been employed by Golightly and Wilkinson (2005) for equidistant time steps and, with some further modification, by Eraker (2001).

In order to derive an appropriate proposal density, consider

$$\begin{aligned}
&\pi(\boldsymbol{Y}_{k+1}|\boldsymbol{Y}_k, \boldsymbol{Y}_{k+2}, \boldsymbol{\theta}) \\
&\quad \propto \pi(\boldsymbol{Y}_{k+2}\,|\,\boldsymbol{Y}_{k+1}, \boldsymbol{\theta})\pi(\boldsymbol{Y}_{k+1}|\boldsymbol{Y}_k, \boldsymbol{\theta}) \\
&\quad \approx \phi(\boldsymbol{Y}_{k+2}\,|\,\boldsymbol{Y}_{k+1}+\boldsymbol{\mu}(\boldsymbol{Y}_{k+1}, \boldsymbol{\theta})\Delta t_{k+1}, \boldsymbol{\Sigma}(\boldsymbol{Y}_{k+1}, \boldsymbol{\theta})\Delta t_{k+1}) \\
&\quad \cdot \phi(\boldsymbol{Y}_{k+1}\,|\,\boldsymbol{Y}_k + \boldsymbol{\mu}(\boldsymbol{Y}_k, \boldsymbol{\theta})\Delta t_k, \boldsymbol{\Sigma}(\boldsymbol{Y}_k, \boldsymbol{\theta})\Delta t_k),
\end{aligned}$$

which follows by exploitation of the Markov property of diffusion processes and application of the Euler approximation. Replace $\boldsymbol{\mu}(\boldsymbol{Y}_{k+1}, \boldsymbol{\theta})$ and $\boldsymbol{\Sigma}(\boldsymbol{Y}_{k+1}, \boldsymbol{\theta})$ by $\boldsymbol{\mu}(\boldsymbol{Y}_k, \boldsymbol{\theta})$ and $\boldsymbol{\Sigma}(\boldsymbol{Y}_k, \boldsymbol{\theta})$, respectively, which is especially justified for small Δt_k. Then, after some calculation, one obtains that $\pi(\boldsymbol{Y}_{k+1}|\boldsymbol{Y}_k, \boldsymbol{Y}_{k+2}, \boldsymbol{\theta})$ is approximately proportional to

$$\begin{aligned}
\exp\Bigg(-\frac{\Delta t_{k+1}^{-1} + \Delta t_k^{-1}}{2}\Bigg[&\boldsymbol{Y}_{k+1}'\boldsymbol{\Sigma}(\boldsymbol{Y}_k, \boldsymbol{\theta})^{-1}\boldsymbol{Y}_{k+1} \\
&- 2\boldsymbol{Y}_{k+1}'\boldsymbol{\Sigma}(\boldsymbol{Y}_k, \boldsymbol{\theta})^{-1}\left(\boldsymbol{Y}_k + \frac{\boldsymbol{Y}_{k+2} - \boldsymbol{Y}_k}{\Delta t_{k+1} + \Delta t_k}\,\Delta t_k\right)\Bigg]\Bigg).
\end{aligned}$$

The obtained expression is an unnormalised Gaussian density. The according proposal for $\boldsymbol{Y}_{(a,b)}^{\text{imp}}$ is to successively draw

$$\boldsymbol{Y}_{k+1}^* \sim \mathcal{N}\left(\boldsymbol{Y}_k^* + \frac{\boldsymbol{Y}_{k+2} - \boldsymbol{Y}_k^*}{t_{k+2} - t_k}\,\Delta t_k\,,\; \frac{t_{k+2} - t_{k+1}}{t_{k+2} - t_k}\,\boldsymbol{\Sigma}(\boldsymbol{Y}_k^*, \boldsymbol{\theta})\Delta t_k\right) \tag{7.6}$$

for $k = a, \ldots, b-2$ and $\boldsymbol{Y}_a^* = \boldsymbol{Y}_a$. The acceptance probability for a so-proposed path is

$$\zeta\big(\boldsymbol{Y}_{(a,b)}^{\text{imp}*}, \boldsymbol{Y}_{(a,b)}^{\text{imp}}\big) = 1 \wedge \left(\prod_{k=a}^{b-1} \frac{\pi^{\text{Euler}}\big(\boldsymbol{Y}_{k+1}^*|\boldsymbol{Y}_k^*, \boldsymbol{\theta}\big)}{\pi^{\text{Euler}}\big(\boldsymbol{Y}_{k+1}|\boldsymbol{Y}_k, \boldsymbol{\theta}\big)}\right) \frac{q_{\text{E2}}\big(\boldsymbol{Y}_{(a,b)}^{\text{imp}}\,|\,\boldsymbol{Y}_a, \boldsymbol{Y}_b, \boldsymbol{\theta}\big)}{q_{\text{E2}}\big(\boldsymbol{Y}_{(a,b)}^{\text{imp}*}\,|\,\boldsymbol{Y}_a, \boldsymbol{Y}_b, \boldsymbol{\theta}\big)}$$

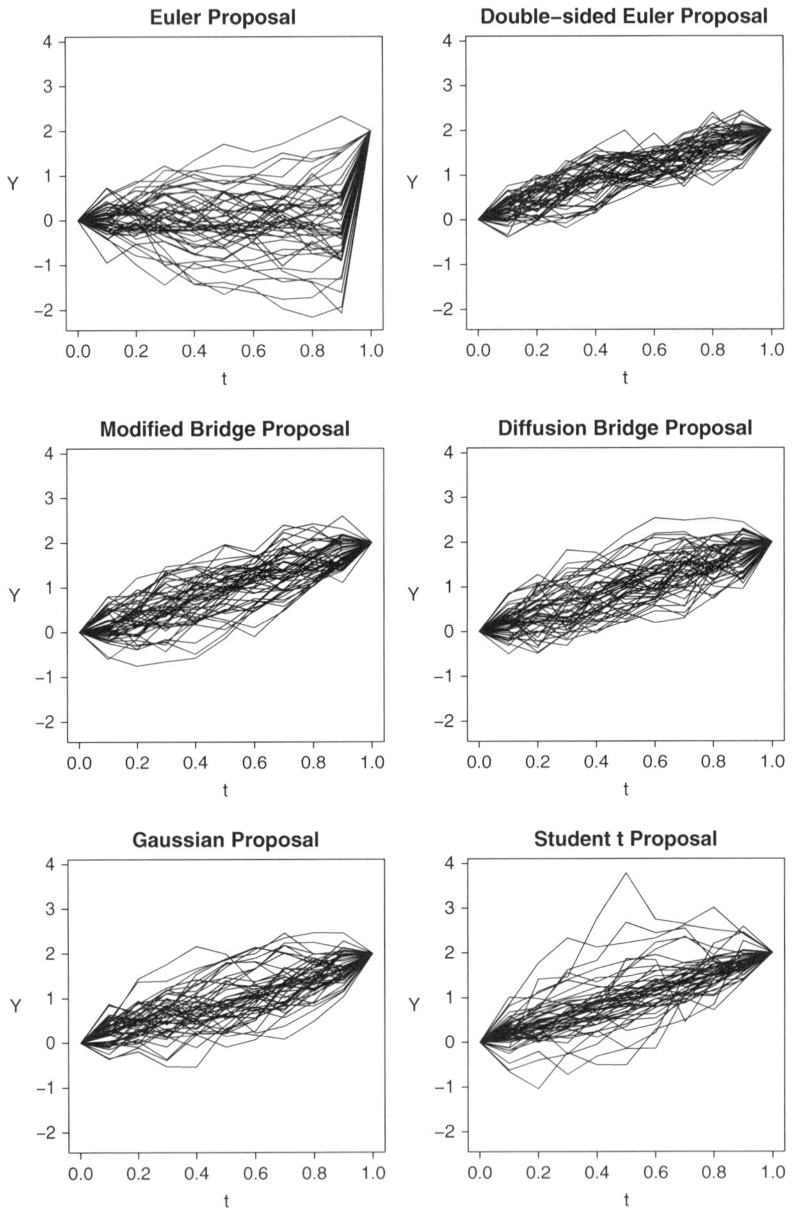

Fig. 7.2 Fifty path proposals for the Ornstein-Uhlenbeck process (A.2) on the time interval $[0, 1]$ fulfilling the SDE $\mathrm{d}X_t = -0.5\mathrm{d}t + \mathrm{d}B_t$. All proposals are conditioned on $X_0 = 0$ and $X_1 = 2$. The number of subintervals of $[0, 1]$ is $m = 10$. In row-wise order, the paths are proposed according to the Euler proposal (7.5), the double-sided Euler proposal (7.6), the modified bridge proposal (7.7), the diffusion bridge proposal (7.9), the Gaussian proposal (7.11) and the Student t proposal (7.13). The double-sided Euler proposal starts with a linear interpolation between X_0 and X_1 and then iteratively conditions on the previously proposed path. The t proposal uses $\nu = 3$ degrees of freedom

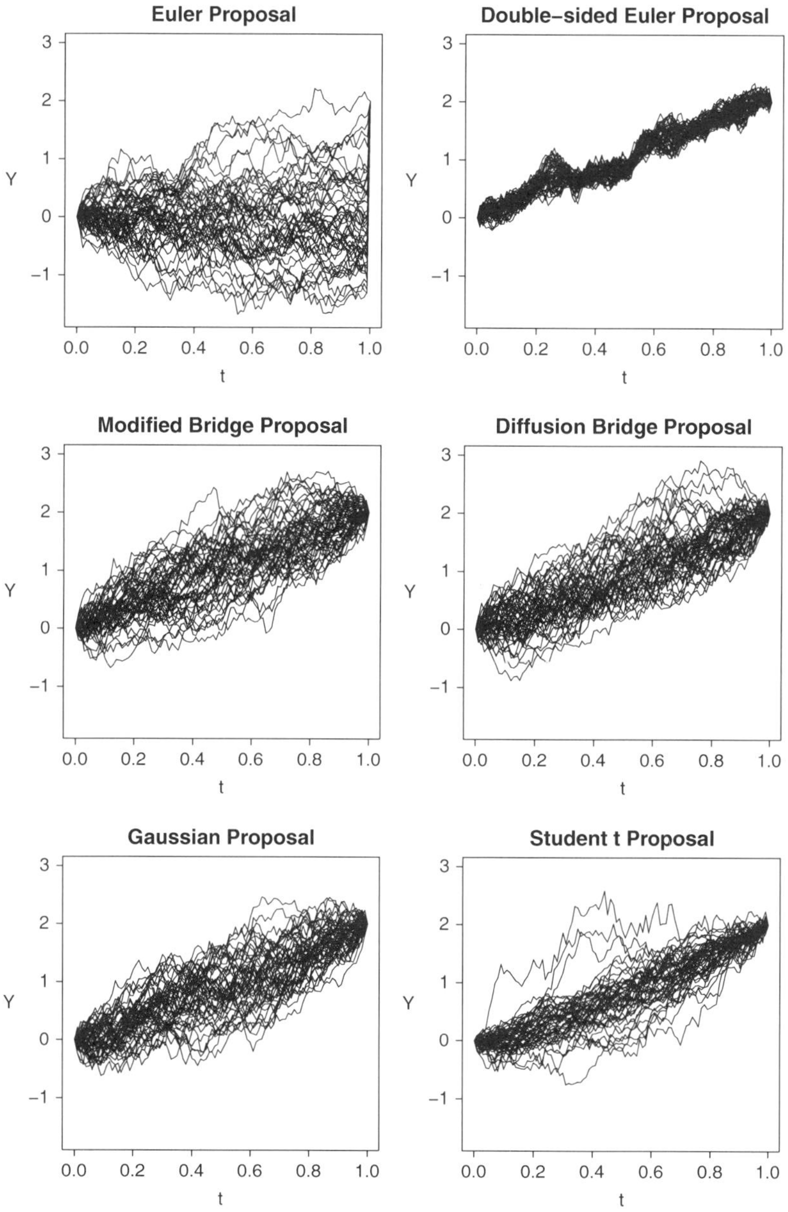

Fig. 7.3 Fifty path proposals as in Fig. 7.2, this time with $m = 100$ subintervals

with $\boldsymbol{Y}_b^* = \boldsymbol{Y}_b$ and proposal density

$$q_{\mathrm{E2}}\left(\boldsymbol{Y}_{(a,b)}^{\mathrm{imp}*}\middle|\boldsymbol{Y}_a, \boldsymbol{Y}_b, \boldsymbol{Y}_{(a,b)}^{\mathrm{imp}}, \boldsymbol{\theta}\right) = \prod_{k=a}^{b-2} q_{\mathrm{E2}}\left(\boldsymbol{Y}_{k+1}^*\middle|\boldsymbol{Y}_k^*, \boldsymbol{Y}_{k+2}, \boldsymbol{\theta}\right)$$

$$= \prod_{k=a}^{b-2} \phi\left(\boldsymbol{Y}_{k+1}^*\middle|\boldsymbol{Y}_k^* + \frac{\boldsymbol{Y}_{k+2}-\boldsymbol{Y}_k^*}{t_{k+2}-t_k}\Delta t_k, \frac{t_{k+2}-t_{k+1}}{t_{k+2}-t_k}\boldsymbol{\Sigma}(\boldsymbol{Y}_k^*, \boldsymbol{\theta})\Delta t_k\right).$$

For evenly spaced time intervals, i.e. $t_k = t_0 + k\Delta t$ for some Δt and all $k = 1, \dots, m$, the density q_{E2} simplifies to

$$q_{\mathrm{E2}}\left(\boldsymbol{Y}_{k+1}^* \,\middle|\, \boldsymbol{Y}_k^*, \boldsymbol{Y}_{k+2}, \boldsymbol{\theta}\right) = \phi\left(\boldsymbol{Y}_{k+1}^* \,\middle|\, \frac{1}{2}\left(\boldsymbol{Y}_k^* + \boldsymbol{Y}_{k+2}\right), \frac{1}{2}\boldsymbol{\Sigma}(\boldsymbol{Y}_k^*, \boldsymbol{\theta})\Delta t\right).$$

A possible variant of this Metropolis-Hastings update of $\boldsymbol{Y}_{(a,b)}^{\mathrm{imp}}$ is the following Metropolis-within-Gibbs procedure: Starting with $k = a$, propose $\boldsymbol{Y}_{k+1}^*$ as in (7.6) above. Immediately after this proposal, accept or reject $\boldsymbol{Y}_{k+1}^*$ with acceptance probability

$$\zeta\left(\boldsymbol{Y}_{k+1}^*, \boldsymbol{Y}_{k+1}\right)$$

$$= 1 \wedge \frac{\pi^{\mathrm{Euler}}\left(\boldsymbol{Y}_{k+2}|\boldsymbol{Y}_{k+1}^*, \boldsymbol{\theta}\right)\pi^{\mathrm{Euler}}\left(\boldsymbol{Y}_{k+1}^*|\boldsymbol{Y}_k^*, \boldsymbol{\theta}\right)}{\pi^{\mathrm{Euler}}\left(\boldsymbol{Y}_{k+2}|\boldsymbol{Y}_{k+1}, \boldsymbol{\theta}\right)\pi^{\mathrm{Euler}}\left(\boldsymbol{Y}_{k+1}|\boldsymbol{Y}_k^*, \boldsymbol{\theta}\right)} \frac{q_{\mathrm{E2}}\left(\boldsymbol{Y}_{k+1} \,\middle|\, \boldsymbol{Y}_k^*, \boldsymbol{Y}_{k+2}, \boldsymbol{\theta}\right)}{q_{\mathrm{E2}}\left(\boldsymbol{Y}_{k+1}^* \,\middle|\, \boldsymbol{Y}_k^*, \boldsymbol{Y}_{k+2}, \boldsymbol{\theta}\right)},$$

where $\boldsymbol{Y}_a^* = \boldsymbol{Y}_a$. Proceed similarly for $k = a+1, \dots, b-2$. At the end, accept the entire so-constructed path $\boldsymbol{Y}_{(a,b)}^{\mathrm{imp}*}$.

With both the pure Metropolis-Hastings and the Metropolis-within-Gibbs version of this double-sided Euler proposal one obtains high acceptance rates as each proposed data point only means a minor change. On the other hand, the proposed paths are quite stiff as there is not much tolerance for major changes. This lack of flexibility is evident from Figs. 7.2 and 7.3 on pp. 178 and 179, which display representative trajectories that have successively been sampled from (7.6) without any intermediate acceptance or rejection. The result is again slow convergence and high serial correlation of the elements of the Markov chain.

The following proposals dispose of the difficulty of high dependency between $\boldsymbol{Y}_{(a,b)}^{\mathrm{imp}}$ and $\boldsymbol{Y}_{(a,b)}^{\mathrm{imp}*}$ as they neglect the previously imputed data $\boldsymbol{Y}_{(a,b)}^{\mathrm{imp}}$. Instead, the attempt is to appropriately bridge the gap between $\boldsymbol{Y}_a$ and $\boldsymbol{Y}_b$. As it is generally not possible to exactly sample diffusion bridges, that are diffusion processes conditioned on a starting and an end point, the proposals are approximations to such processes.

Modified Bridge Proposal

A flexible way to propose a diffusion bridge is to condition the proposal distribution of $\boldsymbol{Y}_{k+1}^*$ on the preceding value $\boldsymbol{Y}_k^*$ and on the right end point $\boldsymbol{Y}_b$ of the path segment to be updated. The resulting proposal (7.7) has been applied by Durham and Gallant (2002), though not in a Bayesian framework, who call it the *modified bridge*. Chib and Shephard (2002) discuss its utilisation in Bayesian analysis.

In analogy to the derivations for the double-sided Euler proposal above, regard

$$\begin{aligned}\pi(\boldsymbol{Y}_{k+1}|\boldsymbol{Y}_k,\boldsymbol{Y}_b,\boldsymbol{\theta}) &\propto \pi(\boldsymbol{Y}_b\,|\,\boldsymbol{Y}_{k+1},\boldsymbol{\theta})\pi(\boldsymbol{Y}_{k+1}|\boldsymbol{Y}_k,\boldsymbol{\theta})\\ &\approx \phi(\boldsymbol{Y}_b\,|\,\boldsymbol{Y}_{k+1}+\boldsymbol{\mu}(\boldsymbol{Y}_{k+1},\boldsymbol{\theta})\Delta_+,\boldsymbol{\Sigma}(\boldsymbol{Y}_{k+1},\boldsymbol{\theta})\Delta_+)\\ &\quad\cdot\phi(\boldsymbol{Y}_{k+1}\,|\,\boldsymbol{Y}_k+\boldsymbol{\mu}(\boldsymbol{Y}_k,\boldsymbol{\theta})\Delta t_k,\boldsymbol{\Sigma}(\boldsymbol{Y}_k,\boldsymbol{\theta})\Delta t_k),\end{aligned}$$

where $\Delta_+ = t_b - t_{k+1}$ is the distance between the right end point of the update interval and the time point of the currently considered imputed value. The approximation of $\pi(\boldsymbol{Y}_b\,|\,\boldsymbol{Y}_{k+1},\boldsymbol{\theta})$ by the Euler density is rough unless Δ_+ is small, and hence the length of the interval $[t_a,t_b]$ should not be chosen too large. As before, approximate $\boldsymbol{\mu}(\boldsymbol{Y}_{k+1},\boldsymbol{\theta})$ and $\boldsymbol{\Sigma}(\boldsymbol{Y}_{k+1},\boldsymbol{\theta})$ by $\boldsymbol{\mu}(\boldsymbol{Y}_k,\boldsymbol{\theta})$ and $\boldsymbol{\Sigma}(\boldsymbol{Y}_k,\boldsymbol{\theta})$. Then $\pi(\boldsymbol{Y}_{k+1}|\boldsymbol{Y}_k,\boldsymbol{Y}_b,\boldsymbol{\theta})$ is approximately proportional to

$$\exp\left(-\frac{\Delta_+^{-1}+\Delta t_k^{-1}}{2}\left[\boldsymbol{Y}_{k+1}'\boldsymbol{\Sigma}(\boldsymbol{Y}_k,\boldsymbol{\theta})^{-1}\left(\boldsymbol{Y}_{k+1}-2\left(\boldsymbol{Y}_k+\frac{\boldsymbol{Y}_b-\boldsymbol{Y}_k}{\Delta_++\Delta t_k}\Delta t_k\right)\right)\right]\right),$$

i.e. we again obtain a Gaussian density. The corresponding proposal for $\boldsymbol{Y}_{(a,b)}^{\text{imp}}$ is to iteratively draw

$$\boldsymbol{Y}_{k+1}^* \sim \mathcal{N}\left(\boldsymbol{Y}_k^* + \frac{\boldsymbol{Y}_b-\boldsymbol{Y}_k^*}{t_b-t_k}\Delta t_k\,,\ \frac{t_b-t_{k+1}}{t_b-t_k}\boldsymbol{\Sigma}(\boldsymbol{Y}_k^*,\boldsymbol{\theta})\Delta t_k\right) \tag{7.7}$$

for $k = a,\dots,b-2$ and $\boldsymbol{Y}_a^* = \boldsymbol{Y}_a$. The proposed path will be accepted with probability

$$\zeta\big(\boldsymbol{Y}_{(a,b)}^{\text{imp}*},\boldsymbol{Y}_{(a,b)}^{\text{imp}}\big) = 1\wedge\left(\prod_{k=a}^{b-1}\frac{\pi^{\text{Euler}}\big(\boldsymbol{Y}_{k+1}^*|\boldsymbol{Y}_k^*,\boldsymbol{\theta}\big)}{\pi^{\text{Euler}}\big(\boldsymbol{Y}_{k+1}|\boldsymbol{Y}_k,\boldsymbol{\theta}\big)}\right)\frac{q_{\text{MB}}\big(\boldsymbol{Y}_{(a,b)}^{\text{imp}}\,|\,\boldsymbol{Y}_a,\boldsymbol{Y}_b,\boldsymbol{\theta}\big)}{q_{\text{MB}}\big(\boldsymbol{Y}_{(a,b)}^{\text{imp}*}\,|\,\boldsymbol{Y}_a,\boldsymbol{Y}_b,\boldsymbol{\theta}\big)},$$

where $\boldsymbol{Y}_b^* = \boldsymbol{Y}_b$ and

$$\begin{aligned}q_{\text{MB}}\Big(\boldsymbol{Y}_{(a,b)}^{\text{imp}*}\,\Big|\,\boldsymbol{Y}_a,\boldsymbol{Y}_b,\boldsymbol{\theta}\Big) &= \prod_{k=a}^{b-2} q_{\text{MB}}\Big(\boldsymbol{Y}_{k+1}^*\,\Big|\,\boldsymbol{Y}_k^*,\boldsymbol{Y}_b,\boldsymbol{\theta}\Big)\\ &= \prod_{k=a}^{b-2}\phi\Big(\boldsymbol{Y}_{k+1}^*\,\Big|\,\boldsymbol{Y}_k^*+\frac{\boldsymbol{Y}_b-\boldsymbol{Y}_k^*}{t_b-t_k}\Delta t_k\,,\ \frac{t_b-t_{k+1}}{t_b-t_k}\boldsymbol{\Sigma}(\boldsymbol{Y}_k^*,\boldsymbol{\theta})\Delta t_k\Big).\end{aligned}$$

For equidistant time intervals with $t_k = t_0 + k\Delta t$, the proposal density reduces to

$$q_{\text{MB}}\left(\boldsymbol{Y}^*_{k+1} \,\middle|\, \boldsymbol{Y}^*_k, \boldsymbol{Y}_b, \boldsymbol{\theta}\right) = \phi\left(\boldsymbol{Y}^*_{k+1} \,\middle|\, \boldsymbol{Y}^*_k + \frac{\boldsymbol{Y}_b - \boldsymbol{Y}^*_k}{b-k},\ \frac{b-k-1}{b-k}\boldsymbol{\Sigma}(\boldsymbol{Y}^*_k, \boldsymbol{\theta})\Delta t\right).$$

Diffusion Bridge Proposal

Apart from the prefactor of the diffusion matrix, the modified bridge proposal (7.7) corresponds to the Euler sampling scheme for the SDE

$$\mathrm{d}\boldsymbol{X}_t = \frac{\boldsymbol{X}_{t_b} - \boldsymbol{X}_t}{t_b - t}\,\mathrm{d}t + \boldsymbol{\sigma}(\boldsymbol{X}_t, \boldsymbol{\theta})\,\mathrm{d}\boldsymbol{B}_t\,, \quad \boldsymbol{X}_0 = \boldsymbol{x}_0, \tag{7.8}$$

where $\boldsymbol{\sigma}\boldsymbol{\sigma}' = \boldsymbol{\Sigma}$. This scheme has been applied by Chib et al. (2004) as a proposal for $\boldsymbol{Y}^{\text{imp}}_{(a,b)}$, i.e. they successively sample

$$\boldsymbol{Y}^*_{k+1} \sim \mathcal{N}\left(\boldsymbol{Y}^*_k + \frac{\boldsymbol{Y}_b - \boldsymbol{Y}^*_k}{t_b - t_k}\Delta t_k\,,\ \boldsymbol{\Sigma}(\boldsymbol{Y}^*_k, \boldsymbol{\theta})\Delta t_k\right) \tag{7.9}$$

for $k = a, \ldots, b-2$ and $\boldsymbol{Y}^*_a = \boldsymbol{Y}_a$. Method (7.9) is termed *diffusion bridge* hereafter. The so-obtained candidate $\boldsymbol{Y}^{\text{imp}*}_{(a,b)}$ is accepted with probability

$$\zeta\left(\boldsymbol{Y}^{\text{imp}*}_{(a,b)}, \boldsymbol{Y}^{\text{imp}}_{(a,b)}\right) = 1 \wedge \left(\prod_{k=a}^{b-1} \frac{\pi^{\text{Euler}}\left(\boldsymbol{Y}^*_{k+1} | \boldsymbol{Y}^*_k, \boldsymbol{\theta}\right)}{\pi^{\text{Euler}}\left(\boldsymbol{Y}_{k+1} | \boldsymbol{Y}_k, \boldsymbol{\theta}\right)}\right) \frac{q_{\text{DB}}\left(\boldsymbol{Y}^{\text{imp}}_{(a,b)} \,\middle|\, \boldsymbol{Y}_a, \boldsymbol{Y}_b, \boldsymbol{\theta}\right)}{q_{\text{DB}}\left(\boldsymbol{Y}^{\text{imp}*}_{(a,b)} \,\middle|\, \boldsymbol{Y}_a, \boldsymbol{Y}_b, \boldsymbol{\theta}\right)},$$

where $\boldsymbol{Y}^*_b = \boldsymbol{Y}_b$ and

$$\begin{aligned} q_{\text{DB}}\left(\boldsymbol{Y}^{\text{imp}*}_{(a,b)} \,\middle|\, \boldsymbol{Y}_a, \boldsymbol{Y}_b, \boldsymbol{\theta}\right) &= \prod_{k=a}^{b-2} q_{\text{DB}}\left(\boldsymbol{Y}^*_{k+1} \,\middle|\, \boldsymbol{Y}^*_k, \boldsymbol{Y}_b, \boldsymbol{\theta}\right) \\ &= \prod_{k=a}^{b-2} \phi\left(\boldsymbol{Y}^*_{k+1} \,\middle|\, \boldsymbol{Y}^*_k + \frac{\boldsymbol{Y}_b - \boldsymbol{Y}^*_k}{t_b - t_k}\Delta t_k\,,\ \boldsymbol{\Sigma}(\boldsymbol{Y}^*_k, \boldsymbol{\theta})\Delta t_k\right). \end{aligned}$$

For equidistant time intervals with $t_k = t_0 + k\Delta t$, the proposal density equals

$$q_{\text{DB}}\left(\boldsymbol{Y}^*_{k+1} \,\middle|\, \boldsymbol{Y}^*_k, \boldsymbol{Y}_b, \boldsymbol{\theta}\right) = \phi\left(\boldsymbol{Y}^*_{k+1} \,\middle|\, \boldsymbol{Y}^*_k + \frac{\boldsymbol{Y}_b - \boldsymbol{Y}^*_k}{b-k},\ \boldsymbol{\Sigma}(\boldsymbol{Y}^*_k, \boldsymbol{\theta})\Delta t\right).$$

Remark 7.1. In this special case where the true process—satisfying the SDE (7.1)—and the proposal process—satisfying (7.8)—coincide in their diffusion matrices, the acceptance probability is available in explicit form by application of a generalisation of Girsanov's formula from Sect. 3.2.12. Let $\tilde{\mathbb{P}}_{\boldsymbol{\theta}}$ be the law induced by (7.1) conditioned on $\boldsymbol{X}_{t_a}$, $\boldsymbol{X}_{t_b}$ and $\boldsymbol{\theta}$, and let $\mathbb{Q}_{\boldsymbol{\theta}}$ be the law induced by (7.8). Then

$$\zeta\big(\boldsymbol{X}^{\mathrm{imp}*}_{(t_a,t_b)}, \boldsymbol{X}^{\mathrm{imp}}_{(t_a,t_b)}\big) = 1 \wedge \left(\frac{\mathrm{d}\tilde{\mathbb{P}}_{\boldsymbol{\theta}}}{\mathrm{d}\mathbb{Q}_{\boldsymbol{\theta}}}\big(\boldsymbol{X}^{\mathrm{imp}*}_{(t_a,t_b)}\big)\right) \Big/ \left(\frac{\mathrm{d}\tilde{\mathbb{P}}_{\boldsymbol{\theta}}}{\mathrm{d}\mathbb{Q}_{\boldsymbol{\theta}}}\big(\boldsymbol{X}^{\mathrm{imp}}_{(t_a,t_b)}\big)\right), \tag{7.10}$$

where $\boldsymbol{X}^{\mathrm{imp}}_{(t_a,t_b)}$ and $\boldsymbol{X}^{\mathrm{imp}*}_{(t_a,t_b)}$ refer to continuous path segments on (t_a, t_b). Delyon and Hu (2006) show that $\tilde{\mathbb{P}}_{\boldsymbol{\theta}}$ is absolutely continuous with respect to $\mathbb{Q}_{\boldsymbol{\theta}}$ (notation: $\tilde{\mathbb{P}}_{\boldsymbol{\theta}} \ll \mathbb{Q}_{\boldsymbol{\theta}}$) for all $\boldsymbol{\theta} \in \Theta$, i.e. the above Radon-Nikodym derivatives exist and are finite. In practice, a time-discretisation of (7.10) is used. Similar considerations follow in Sects. 7.3 and 7.4.

Gaussian and Student t Proposal

For one-dimensional diffusion processes, Elerian et al. (2001) suggest to find the mode $\boldsymbol{y}$ of the Euler approximated log-density of $Y^{\mathrm{imp}}_{(a,b)}$ given Y_a and Y_b, that is

$$\boldsymbol{y} = \underset{Y^{\mathrm{imp}}_{(a,b)}}{\operatorname{argmax}} \left(\sum_{k=a}^{b-1} \log \pi^{\mathrm{Euler}}\big(Y_{k+1} \,\big|\, Y_k, \boldsymbol{\theta}\big)\right),$$

and to work with the *Gaussian proposal*

$$Y^{\mathrm{imp}*}_{(a,b)} \sim \mathcal{N}\big(\boldsymbol{y}, \boldsymbol{V}(\boldsymbol{y})\big), \tag{7.11}$$

where $\boldsymbol{V}(\boldsymbol{y})$ is the negative inverse Hessian of the above density evaluated at $\boldsymbol{y}$,

$$\boldsymbol{V}(\boldsymbol{y}) = -\left(\left.\frac{\partial^2 \sum_{k=a}^{b-1} \log \pi^{\mathrm{Euler}}\big(Y_{k+1} \,\big|\, Y_k, \boldsymbol{\theta}\big)}{\partial Y^{\mathrm{imp}}_{(a,b)} \partial Y^{\mathrm{imp}\,\prime}_{(a,b)}}\right|_{Y^{\mathrm{imp}}_{(a,b)} = \boldsymbol{y}}\right)^{-1}. \tag{7.12}$$

The mode $\boldsymbol{y}$ can for example be computed by numerical schemes such as the Newton-Raphson method. Naturally, the according proposal density is

$$q_{\mathrm{G}}\big(Y^{\mathrm{imp}*}_{(a,b)} \,\big|\, Y_a, Y_b, \boldsymbol{\theta}\big) = \phi\big(Y^{\mathrm{imp}*}_{(a,b)} \,\big|\, \boldsymbol{y}, \boldsymbol{V}(\boldsymbol{y})\big).$$

A major advantage of this proposal distribution is that it allows simultaneous sampling of all components of $Y^{\mathrm{imp}*}_{(a,b)}$. In case of thin tails of q_{G}, Elerian et al. (2001) and Chib et al. (2004) propose to replace the Gaussian by Student's t distribution, resulting in the *Student t proposal*

$$Y^{\mathrm{imp}*}_{(a,b)} \sim t_\nu\left(\boldsymbol{y}, \frac{\nu-2}{\nu} \boldsymbol{V}(\boldsymbol{y})\right) \tag{7.13}$$

with proposal density

$$q_{\mathrm{t}}\big(Y_{(a,b)}^{\mathrm{imp}*}\big|Y_a,Y_b,\boldsymbol{\theta}\big)$$

$$=\frac{\Gamma\big(\frac{\nu+b-a-1}{2}\big)|\boldsymbol{V}(\boldsymbol{y})|^{-\frac{1}{2}}}{\Gamma\big(\frac{\nu}{2}\big)(\pi(\nu-2))^{\frac{b-a-1}{2}}}\left(1+\frac{\big(Y_{(a,b)}^{\mathrm{imp}*}-\boldsymbol{y}\big)'\boldsymbol{V}(\boldsymbol{y})^{-1}\big(Y_{(a,b)}^{\mathrm{imp}*}-\boldsymbol{y}\big)}{\nu-2}\right)^{-\frac{\nu+b-a-1}{2}},$$

where $\nu > 2$ denotes the degrees of freedom and $|\boldsymbol{A}|$ is the determinant of a square matrix $\boldsymbol{A}$.

Hurn et al. (2007, Sect. 2.5) point out that in practice the Hessian matrix in (7.12) might computationally not be positive-definite and propose an appropriate numerical correction.

Other Proposals

Recent further approaches include the following: Delyon and Hu (2006) suggest to draw path proposals from the Euler discretisation of the SDE

$$\mathrm{d}\boldsymbol{X}_t=\left(\boldsymbol{\mu}(\boldsymbol{X}_t,\boldsymbol{\theta})+\frac{\boldsymbol{X}_{t_b}-\boldsymbol{X}_t}{t_b-t}\right)\mathrm{d}t+\boldsymbol{\sigma}(\boldsymbol{X}_t,\boldsymbol{\theta})\,\mathrm{d}\boldsymbol{B}_t\,,\quad \boldsymbol{X}_0=\boldsymbol{x}_0. \tag{7.14}$$

The motivation of this choice is as follows: The proposed process should imitate the behaviour of the original process satisfying the SDE (7.1) with an appropriate end point condition as closely as possible. For $\boldsymbol{\mu}\equiv\boldsymbol{0}$ and $\boldsymbol{\sigma}\boldsymbol{\sigma}'\equiv\boldsymbol{I}$, the SDE (7.14) describes a Brownian bridge starting in $\boldsymbol{X}_0$ and ending in $\boldsymbol{X}_{t_b}$, and (7.1) refers to Brownian motion. Hence in that case the two SDEs induce the same law if the target SDE (7.1) is further conditioned on the end point $\boldsymbol{X}_{t_b}$.

For geometrically ergodic (Gilks et al. 1996, Chap. 3.3; Roberts and Rosenthal 1997) diffusions, Fearnhead (2008) introduces a mixture of the Euler proposal (7.5) and the modified bridge proposal (7.7),

$$q_{\mathrm{F}}\big(\boldsymbol{Y}_{k+1}^*\,\big|\,\boldsymbol{Y}_k^*,\boldsymbol{Y}_b,\boldsymbol{\theta}\big)$$

$$=c_1\left(1-\mathrm{e}^{-c_2\Delta_\circ}\right)q_{\mathrm{E}}\big(\boldsymbol{Y}_{k+1}^*\,\big|\,\boldsymbol{Y}_k^*,\boldsymbol{\theta}\big)+\mathrm{e}^{-c_2\Delta_\circ}q_{\mathrm{MB}}\big(\boldsymbol{Y}_{k+1}^*\,\big|\,\boldsymbol{Y}_k^*,\boldsymbol{Y}_b,\boldsymbol{\theta}\big),$$

where c_1 and c_2 are constants and $\Delta_\circ = t_b - t_{k+1}$. This construction puts large weight on the Euler proposal for k close to a, and for k close to b it puts more weight on the modified bridge proposal. For $c_1 = 1$, the according proposal scheme is

$$\boldsymbol{Y}_{k+1}^*\sim\mathcal{N}\left(\boldsymbol{\eta}_k,\boldsymbol{\Lambda}_k\right)$$

with

$$\boldsymbol{\eta}_k=\boldsymbol{Y}_k^*+\left(\left(1-\mathrm{e}^{-c_2\Delta_\circ}\right)\boldsymbol{\mu}(\boldsymbol{Y}_k^*,\boldsymbol{\theta})+\mathrm{e}^{-c_2\Delta_\circ}\frac{\boldsymbol{Y}_b-\boldsymbol{Y}_k^*}{t_b-t_k}\right)\Delta t_k$$

and

$$\boldsymbol{\Lambda}_k = \left(1 - \mathrm{e}^{-c_2 \Delta_\circ} \frac{\Delta t_k}{t_b - t_k}\right) \boldsymbol{\Sigma}(\boldsymbol{Y}_k^*, \boldsymbol{\theta}) \Delta t_k.$$

Similarly, a respective mixture of the Euler proposal and the diffusion bridge proposal (7.7) leads to

$$\boldsymbol{Y}_{k+1}^* \sim \mathcal{N}\left(\boldsymbol{\eta}_k, \boldsymbol{\Sigma}(\boldsymbol{Y}_k^*, \boldsymbol{\theta}) \Delta t_k\right).$$

That is the Euler discretisation of the SDE

$$\mathrm{d}\boldsymbol{X}_t = \left(\left(1 - \mathrm{e}^{-c_2 \Delta_\circ}\right) \boldsymbol{\mu}(\boldsymbol{X}_t, \boldsymbol{\theta}) + \mathrm{e}^{-c_2 \Delta_\circ} \frac{\boldsymbol{X}_{t_b} - \boldsymbol{X}_t}{t_b - t} \right) \mathrm{d}t + \boldsymbol{\sigma}(\boldsymbol{X}_t, \boldsymbol{\theta})\, \mathrm{d}\boldsymbol{B}_t, \quad \boldsymbol{X}_0 = \boldsymbol{x}_0$$

(Suda 2009). The proposal variants in this paragraph are not further considered in this chapter as the previous ones already form a representative selection.

As an illustration, Figs. 7.2 and 7.3 on pp. 178 and 179 show path proposals for the Ornstein-Uhlenbeck process, introduced in Sect. A.2, that are generated according to the above methods for $m = 10$ and $m = 100$ intermediate time intervals. The different proposals are applied in Sect. 7.1.7 in a simulation study to estimate the parameters of an Ornstein-Uhlenbeck process. A discussion follows in Sect. 7.1.8.

7.1.3 Parameter Update

We now turn to the second of the two alternating steps in the scheme (7.2): the parameter update. In most cases, direct sampling from the posterior distribution of the parameter $\boldsymbol{\theta}$ is impossible, thus once more the Metropolis-Hastings algorithm is utilised. To that end, choose a suitable proposal distribution with density q for the parameter $\boldsymbol{\theta}$ which may be conditioned on the observed and imputed data $\boldsymbol{Y}^{\text{obs}}$ and $\boldsymbol{Y}^{\text{imp}}$ and on the current value of the parameter. From this distribution, draw a parameter proposal $\boldsymbol{\theta}^*$ and accept it with probability

$$\zeta(\boldsymbol{\theta}^*, \boldsymbol{\theta}) = 1 \wedge \frac{\pi\left(\boldsymbol{\theta}^* \,\middle|\, \boldsymbol{Y}^{\text{obs}}, \boldsymbol{Y}^{\text{imp}}\right) q\left(\boldsymbol{\theta} \,\middle|\, \boldsymbol{\theta}^*, \boldsymbol{Y}^{\text{obs}}, \boldsymbol{Y}^{\text{imp}}\right)}{\pi\left(\boldsymbol{\theta} \,\middle|\, \boldsymbol{Y}^{\text{obs}}, \boldsymbol{Y}^{\text{imp}}\right) q\left(\boldsymbol{\theta}^* \,\middle|\, \boldsymbol{\theta}, \boldsymbol{Y}^{\text{obs}}, \boldsymbol{Y}^{\text{imp}}\right)}. \tag{7.15}$$

Otherwise, reject the proposal $\boldsymbol{\theta}^*$ and keep the previous value $\boldsymbol{\theta}$. As in the path update, one may decide to only update parts of the components of $\boldsymbol{\theta}$ at a time. In that case, the argument of the proposal density q might be adjusted respectively. However, the so-obtained proposal density is proportional to the proposal density for the whole parameter. Thus, we in the following denote by $\boldsymbol{\theta}^*$ the proposal for the entire parameter, even if some components agree with the according parts of the previous value $\boldsymbol{\theta}$. With Bayes' theorem, the probability (7.15) becomes

$$\zeta(\boldsymbol{\theta}^*, \boldsymbol{\theta}) = 1 \wedge \frac{\pi\big(\boldsymbol{Y}^{\text{obs}}, \boldsymbol{Y}^{\text{imp}} \,\big|\, \boldsymbol{\theta}^*\big) p(\boldsymbol{\theta}^*) q\big(\boldsymbol{\theta} \,\big|\, \boldsymbol{\theta}^*, \boldsymbol{Y}^{\text{obs}}, \boldsymbol{Y}^{\text{imp}}\big)}{\pi\big(\boldsymbol{Y}^{\text{obs}}, \boldsymbol{Y}^{\text{imp}} \,\big|\, \boldsymbol{\theta}\big)\, p(\boldsymbol{\theta})\, q\big(\boldsymbol{\theta}^* \,\big|\, \boldsymbol{\theta}, \boldsymbol{Y}^{\text{obs}}, \boldsymbol{Y}^{\text{imp}}\big)}$$

$$= 1 \wedge \left(\prod_{k=0}^{m-1} \frac{p_{\boldsymbol{\theta}^*}\big(\Delta t_k, \boldsymbol{Y}_k, \boldsymbol{Y}_{k+1}\big)}{p_{\boldsymbol{\theta}}\big(\Delta t_k, \boldsymbol{Y}_k, \boldsymbol{Y}_{k+1}\big)} \right) \cdot \frac{p(\boldsymbol{\theta}^*)}{p(\boldsymbol{\theta})} \cdot \frac{q\big(\boldsymbol{\theta} \,\big|\, \boldsymbol{\theta}^*, \boldsymbol{Y}^{\text{obs}}, \boldsymbol{Y}^{\text{imp}}\big)}{q\big(\boldsymbol{\theta}^* \,\big|\, \boldsymbol{\theta}, \boldsymbol{Y}^{\text{obs}}, \boldsymbol{Y}^{\text{imp}}\big)},$$

where p denotes the prior density of the parameter and $p_{\boldsymbol{\theta}}$ is the exact transition density of the diffusion process given the model parameter. As in the path update, one can approximate $p_{\boldsymbol{\theta}}$ with the Euler scheme since the time steps Δt_k are assumed to be small for all k. Hence, in the following we employ the acceptance probability

$$\zeta(\boldsymbol{\theta}^*, \boldsymbol{\theta}) = 1 \wedge \left(\prod_{k=0}^{m-1} \frac{\pi^{\text{Euler}}\big(\boldsymbol{Y}_{k+1} \big| \boldsymbol{Y}_k, \boldsymbol{\theta}^*\big)}{\pi^{\text{Euler}}\big(\boldsymbol{Y}_{k+1} \big| \boldsymbol{Y}_k, \boldsymbol{\theta}\big)} \right) \cdot \frac{p(\boldsymbol{\theta}^*)}{p(\boldsymbol{\theta})} \cdot \frac{q\big(\boldsymbol{\theta} \big| \boldsymbol{\theta}^*, \boldsymbol{Y}^{\text{obs}}, \boldsymbol{Y}^{\text{imp}}\big)}{q\big(\boldsymbol{\theta}^* \big| \boldsymbol{\theta}, \boldsymbol{Y}^{\text{obs}}, \boldsymbol{Y}^{\text{imp}}\big)} \tag{7.16}$$

for the parameter proposal $\boldsymbol{\theta}^*$. The prior density p may be proper or improper and usually depends on the considered diffusion model. For improper priors one however has to ensure that the joint posterior distribution of all parameters is well-defined. The choice of the proposal density q is model-specific and discussed in the following.

Full Conditional Proposal

An often favoured choice of proposal density is the *exact full conditional proposal*

$$q_{\text{eFC}}\big(\boldsymbol{\theta} \big| \boldsymbol{Y}^{\text{obs}}, \boldsymbol{Y}^{\text{imp}}\big) = \pi\big(\boldsymbol{\theta} \big| \boldsymbol{Y}^{\text{obs}}, \boldsymbol{Y}^{\text{imp}}\big) \propto p(\boldsymbol{\theta}) \prod_{k=0}^{m-1} p_{\boldsymbol{\theta}}\big(\Delta t_k, \boldsymbol{Y}_k, \boldsymbol{Y}_{k+1}\big). \tag{7.17}$$

If the normalising constant of this expression can be determined, one can perhaps sample a proposal $\boldsymbol{\theta}^*$ from q_{eFC}. However, as $p_{\boldsymbol{\theta}}$ is usually unknown, one may rather utilise the *approximate full conditional proposal*

$$q_{\text{aFC}}\big(\boldsymbol{\theta} \big| \boldsymbol{Y}^{\text{obs}}, \boldsymbol{Y}^{\text{imp}}\big) \propto p(\boldsymbol{\theta}) \prod_{k=0}^{m-1} \pi^{\text{Euler}}\big(\boldsymbol{Y}_{k+1} \big| \boldsymbol{Y}_k, \boldsymbol{\theta}\big), \tag{7.18}$$

which possibly results in a known distribution. If the exact transition density $p_{\boldsymbol{\theta}}$ is available and sampling from the exact full conditional proposal is performed, one can replace π^{Euler} by $p_{\boldsymbol{\theta}}$ in (7.16). Otherwise, Eq. (7.16) remains unchanged. Hence, for both exact and approximate full conditional proposals the acceptance probability is equal to one, i.e. Gibbs sampling is performed.

Cano et al. (2006) show the weak convergence of the approximate posterior density to the true posterior under fairly general assumptions. They however also give an example where the requirements are not fulfilled; that is the Ornstein-Uhlenbeck process satisfying the SDE (7.20) displayed on p. 192 with $\beta = 0$, $\sigma^2 = 1$ and a non-informative prior for α.

An example where both the exact and approximate full conditional densities can be obtained and sampling from them is uncomplicated is shown in Sect. 7.1.7. If sampling from neither q_{eFC} nor q_{aFC} is possible, different proposal schemes like the following one have to be considered.

Random Walk Proposal

A frequently used idea is a *random walk proposal* which is independent of the imputed and observed data and works as follows: Without loss of generality, assume that for some $r \in \{0, \ldots, p\}$ the components $\theta_1, \ldots, \theta_r$ take values on the real line, and $\theta_{r+1}, \ldots, \theta_p$ are strictly positive. For $j = 1, \ldots, r$, simply propose

$$\theta_j^* \sim \mathcal{N}(\theta_j, \gamma_j^2)$$

for some predefined $\gamma_j \in \mathbb{R}_+$. Then

$$q_{\text{RW}}\big(\theta_j^* \,|\, \theta_j\big) = \phi\big(\theta_j^* \,|\, \theta_j, \gamma_j^2\big) = \phi\big(\theta_j^* - \theta_j \,|\, 0, \gamma_j^2\big).$$

For $j = r + 1, \ldots, p$, draw

$$\log \theta_j^* \sim \mathcal{N}\big(\log \theta_j, \gamma_j^2\big).$$

This corresponds to the log-normal distribution, i.e.

$$\theta_j^* \sim \mathcal{LN}\big(\log \theta_j, \gamma_j^2\big)$$

and

$$q_{\text{RW}}\big(\theta_j^* \,|\, \theta_j\big) = \frac{1}{\theta_j^*}\,\phi\big(\log \theta_j^* \,|\, \log \theta_j, \gamma_j^2\big) = \frac{1}{\theta_j^*}\,\phi\big(\log(\theta_j^*/\theta_j) \,|\, 0, \gamma_j^2\big)$$

for $j = r + 1, \ldots, p$. Altogether, one has

$$\frac{q_{\text{RW}}(\boldsymbol{\theta} \,|\, \boldsymbol{\theta}^*)}{q_{\text{RW}}(\boldsymbol{\theta}^* \,|\, \boldsymbol{\theta})} = \left(\prod_{j=1}^{r} \frac{\phi(\theta_j - \theta_j^* \,|\, 0, \gamma_j^2)}{\phi(\theta_j^* - \theta_j \,|\, 0, \gamma_j^2)}\right)\left(\prod_{j=r+1}^{p} \frac{\theta_j^*\ \phi(\log(\theta_j/\theta_j^*) \,|\, 0, \gamma_j^2)}{\theta_j\ \phi(\log(\theta_j^*/\theta_j) \,|\, 0, \gamma_j^2)}\right).$$

Because of the symmetry of $\phi(z \,|\, 0, \gamma^2)$ around $z = 0$, the functions ϕ cancel in this expression. The acceptance probability (7.16) reduces to

$$\zeta(\boldsymbol{\theta}^*, \boldsymbol{\theta}) = 1 \wedge \left(\prod_{k=0}^{m-1} \frac{\pi^{\text{Euler}}(\boldsymbol{Y}_{k+1} | \boldsymbol{Y}_k, \boldsymbol{\theta}^*)}{\pi^{\text{Euler}}(\boldsymbol{Y}_{k+1} | \boldsymbol{Y}_k, \boldsymbol{\theta})} \right) \cdot \frac{p(\boldsymbol{\theta}^*)}{p(\boldsymbol{\theta})} \cdot \left(\prod_{j=r+1}^{p} \frac{\theta_j^*}{\theta_j} \right).$$

The parameters γ_j^2 of the proposal distributions should be chosen deliberately: A small variance usually causes higher acceptance rates; the resulting Markov chain may exhibit high autocorrelation though. A large variance may induce many rejections, but the Markov chain generally shows better mixing.

The above assumption about the components of $\boldsymbol{\theta}$ being either real or positive applies in most applications. However, generalisations are possible and often straightforward. For example, if for some j the component θ_j is negative, consider $-\theta_j$ and proceed as above. In case of $\theta_j \in [u, v]$, one might apply the generalised logit function and its inverse, that is

$$\text{logit} : \begin{cases} [u, v] \rightarrow \mathbb{R} \\ x \quad \mapsto \log\left(\dfrac{x-u}{v-x}\right) \end{cases} \tag{7.19}$$

and

$$\text{logit}^{-1} : \begin{cases} \mathbb{R} \rightarrow [u, v] \\ y \mapsto u + (v-u)\,\dfrac{\exp(y)}{1+\exp(y)} \end{cases}.$$

With this,

$$\theta_j^* \sim \text{logit}^{-1}\left(\mathcal{N}\left(\text{logit}(\theta_j), \gamma_j^2\right)\right)$$

would be an appropriate proposal. Furthermore, all proposals can certainly be extended by introducing dependencies between the single components of $\boldsymbol{\theta}$. The update of the parameters can also be performed blockwise, i.e. the components of $\boldsymbol{\theta}$ are divided into subsets which are proposed and accepted or rejected separately. This may lead to better mixing, but the repeated evaluation of the acceptance probability also implies an additional computational effort. Such strategies are not treated here.

A simulation study and evaluation of the three parameter proposals introduced above follows in Sects. 7.1.7 and 7.1.8.

7.1.4 Generalisation to Several Observation Times

As argued in Sect. 7.1.1, the imputation concepts considered so far are easily extendable to the general case where more observations are available than just the starting and the end point of a sample path of a diffusion process. Suppose there are—in addition to the initial value $\boldsymbol{x}_{\tau_0}$—M complete observations $\boldsymbol{x}_{\tau_1}, \ldots, \boldsymbol{x}_{\tau_M}$ at times $0 = \tau_0 < \tau_1 < \ldots < \tau_M = T$. For $i = 0, \ldots, M-1$, divide each inter-observation

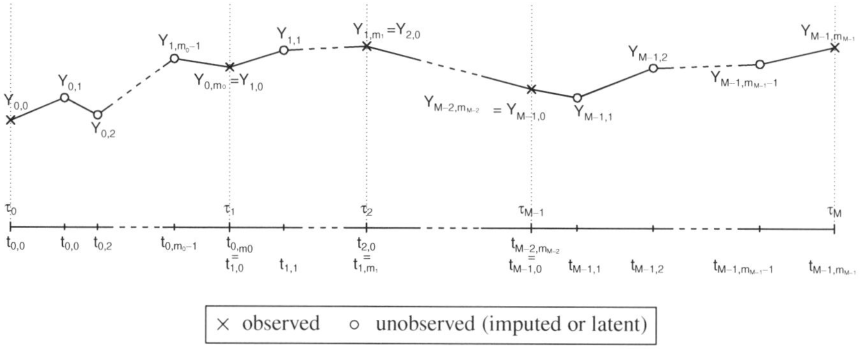

Fig. 7.4 Illustration of a one-dimensional path consisting of the initial value and M discrete observations at times $\tau_0, \ldots, \tau_M$ (*labelled with crosses*) and imputed data (*labelled with circles*)

interval $[\tau_i, \tau_{i+1}]$ into m_i sufficiently small subintervals with boundaries $\tau_i = t_{i,0} < t_{i,1} < \ldots < t_{i,m_i-1} < t_{i,m_i} = \tau_{i+1}$. Impute auxiliary data at the newly introduced time points. In the following, observations on $[\tau_i, \tau_{i+1}]$ are labelled $\boldsymbol{Y}_{i,0} = \boldsymbol{x}_{\tau_i}$ and $\boldsymbol{Y}_{i,m_i} = \boldsymbol{x}_{\tau_{i+1}}$, and the imputed data is referred to as $\boldsymbol{Y}_{i,1}, \ldots, \boldsymbol{Y}_{i,m_i-1}$. Thus, one has observed data $\boldsymbol{Y}^{\text{obs}} = \{\boldsymbol{Y}_{0,0}, \boldsymbol{Y}_{1,0}, \ldots, \boldsymbol{Y}_{M-1,0}, \boldsymbol{Y}_{M-1,m_{M-1}}\}$ and overall imputed data $\boldsymbol{Y}^{\text{imp}} = \{\boldsymbol{Y}_{i,1}, \ldots, \boldsymbol{Y}_{i,m_i-1} \,|\, i = 0, \ldots, M-1\}$. This notation is illustrated in Fig. 7.4.

The update scheme (7.2) is adapted to the generalised setting as follows: The likelihood of the entire discretely observed diffusion path changes to

$$\prod_{i=0}^{M-1} \prod_{k=0}^{m_i-1} p_{\boldsymbol{\theta}}\big(t_{i,k+1} - t_{i,k}, \boldsymbol{Y}_{i,k}, \boldsymbol{Y}_{i,k+1}\big) \approx \prod_{i=0}^{M-1} \prod_{k=0}^{m_i-1} \pi^{\text{Euler}}\big(\boldsymbol{Y}_{i,k+1} \,\big|\, \boldsymbol{Y}_{i,k}, \boldsymbol{\theta}\big)$$

and is to be deployed accordingly in all occurring acceptance probabilities. In case an interval (t_a, t_b) contains one or more observation times $\tau_i, \ldots, \tau_j$, a path proposal on (t_a, t_b) decomposes into independent path proposals on $(t_a, \tau_i), (\tau_i, \tau_{i+1}), \ldots, (\tau_j, t_b)$ with the data at times $t_a, \tau_i, \tau_{i+1}, \ldots, \tau_j, t_b$ remaining fixed. These proposals are either collectively accepted or rejected.

7.1.5 Generalisation to Several Observed Diffusion Paths

Assume one has $K \in \mathbb{N}$ independent observation sets $\boldsymbol{X}^{\text{obs},1}, \ldots, \boldsymbol{X}^{\text{obs},K}$ of a diffusion process fulfilling the SDE

$$\mathrm{d}\boldsymbol{X}_t = \boldsymbol{\mu}(\boldsymbol{X}_t, \boldsymbol{\theta})\mathrm{d}t + \boldsymbol{\sigma}(\boldsymbol{X}_t, \boldsymbol{\theta})\mathrm{d}\boldsymbol{B}_t\,, \quad \boldsymbol{X}_{t_0} = \boldsymbol{x}_0,$$

with identical parameter $\boldsymbol{\theta} \in \Theta$. This is for example the case in the application in Chap. 9 where a biological experiment is carried out several times under the same conditions and hence there are multiple series of observations of the same dynamics available. The observation sets may differ with respect to the numbers of observations, observation times and lengths of inter-observation intervals. In this case, each observation set should be augmented with auxiliary data at appropriate auxiliary time points, and inference on $\boldsymbol{\theta}$ can be performed by repeated execution of

Update of Path 1: Draw $\boldsymbol{X}^{\text{imp},1} \sim \pi\big(\boldsymbol{X}^{\text{imp},1} \,\big|\, \boldsymbol{X}^{\text{obs},1}, \boldsymbol{\theta}\big)$.

$\vdots$

Update of Path K: Draw $\boldsymbol{X}^{\text{imp},K} \sim \pi\big(\boldsymbol{X}^{\text{imp},K} \,\big|\, \boldsymbol{X}^{\text{obs},K}, \boldsymbol{\theta}\big)$.

Update of Parameter: Draw $\boldsymbol{\theta} \sim \pi\big(\boldsymbol{\theta} \,\big|\, \boldsymbol{X}^{\text{obs},1}, \boldsymbol{X}^{\text{imp},1}, \ldots, \boldsymbol{X}^{\text{obs},K}, \boldsymbol{X}^{\text{imp},K}\big)$.

Due to the assumption of independent paths, one has

$$\pi\big(\boldsymbol{\theta}\big|\boldsymbol{X}^{\text{obs},1}, \boldsymbol{X}^{\text{imp},1}, \ldots, \boldsymbol{X}^{\text{obs},K}, \boldsymbol{X}^{\text{imp},K}\big) \propto p(\boldsymbol{\theta}) \prod_{h=1}^{K} \pi\big(\boldsymbol{X}^{\text{obs},h}, \boldsymbol{X}^{\text{imp},h} \,\big|\, \boldsymbol{\theta}\big)$$

in the last step.

7.1.6 *Practical Concerns*

For the implementation of the considered MCMC scheme, some further issues have to be considered. These are the choice of the update interval, the number of auxiliary time points, and the handling of path proposals which lie outside the admissible state space.

Choice of Update Interval

Selection of the update interval (t_a, t_b) in whose interior the imputed data is to be renewed may be of high relevance. Path updates on intervals containing large numbers of imputed data points cause major changes and may speed up convergence of the Markov chain. On the other hand, proposals for large data sets are more likely to be rejected. Furthermore, the modified bridge proposal is a good approximation to the path segment only if (t_a, t_b) is not too large.

Assume we have $S{+}1$ observed or imputed consecutive data points $\boldsymbol{Y}_0$, $\boldsymbol{Y}_1, \ldots, \boldsymbol{Y}_S$ and we wish to bound the number of updated data points for each iteration by $R \leq S-1$. As before, let the update interval (a, b) correspond to a proposal for $\{\boldsymbol{Y}_{a+1}, \ldots, \boldsymbol{Y}_{b-1}\}$. The term *update* refers to both accepted and rejected proposals here. An obvious procedure to draw (a, b) with $a, b \in \{0, \ldots, S\}$ and $2 \leq b - a \leq R + 1$ is the following:

Algorithm 7.1.

1. *Draw* $a \sim U(\{0, \ldots, S-2\})$.
2. *Draw* $b \sim U(\{a+2, \ldots, \min\{a+R+1, S\}\})$.

However, the sampling algorithm for (a, b) should ensure that all data points $\boldsymbol{Y}_1, \ldots, \boldsymbol{Y}_{S-1}$ have the same probability to be updated; Algorithm 7.1 discriminates data points $\boldsymbol{Y}_j$ with j close to 0 or S as there are fewer intervals (a, b) fullfilling the above requirements at the boundaries than in the centre of $(0, S)$. A more detailed reasoning is included in Sect. B.3 in the Appendix. A corrected algorithm is proposed in the following. Section B.3 provides the proof that with this algorithm the probability to be updated is the same for all $\boldsymbol{Y}_1, \ldots, \boldsymbol{Y}_{S-1}$.

Algorithm 7.2.

1. *Draw* $a^* \sim U(\{1-R, \ldots, S-2\})$.
2. *Draw* $b^* \sim U(\{a^*+2, \ldots, \min\{a^*+R+1, S+R-1\}\})$.
3. *Set* $a = \max\{a^*, 0\}$ *and* $b = \min\{b^*, S\}$.
4. *In case of* $b - a < 2$, *repeat the above steps.*

Alternatively, Elerian et al. (2001) suggest a blockwise update of the entire data $\boldsymbol{Y}_1, \ldots, \boldsymbol{Y}_{S-1}$ with proposals for adjacent blocks with Poisson distributed sizes for some fixed intensity parameter $\lambda \in \mathbb{R}_+$:

Algorithm 7.3.

1. *Set* $c_0 = 0$ *and* $j = 1$.
2. *While* $c_{j-1} < S$:
   a. *Draw* $Z \sim \text{Po}(\lambda)$ *and set* $c_j = \min\{c_{j-1} + Z, S\}$.
   b. *Increment* j.

The path is then successively updated on (t_{c_0}, t_{c_1}), (t_{c_1}, t_{c_2}) etc. The individual proposals are independently accepted or rejected.

The decision whether to employ Algorithm 7.2 or 7.3 is problem-specific. Algorithm 7.3 updates the sample path more rigorously but is therefore more time-consuming than Algorithm 7.2. The choice might hence depend on the amount of imputed data or the severeness of measurement error (cf. Sect. 7.2.2). The simulation study in Sect. 7.1.7 uses Algorithm 7.3 as the subsequent evaluation includes the calculation of inefficiency factors; this is meaningful only if in each iteration of the MCMC scheme all data points are investigated.

Sampling Strategy

The number m_i of subintervals between every two consecutive observations at times τ_i and τ_{i+1}, $i = 0, \ldots, M-1$, crucially influences the estimation results. A small number of intermediate time points degrades the accuracy of the Euler approximation (7.3) to the true posterior density and may hence cause a discretisation bias. Large numbers of auxiliary time points, on the other hand, are computationally costly. Hence, the numbers m_i of subintervals should be chosen both sufficiently large and sustainably small. In general, they will be identified empirically.

Eraker (2001) suggests to start with small m_i and to subsequently increase these numbers after convergence of the Markov chain has been achieved. This procedure is pursued until further increases of m_i have negligible impact on the estimation results.

However, too large amounts of imputed data can also deteriorate the behaviour of the whole procedure. Section 7.3 deals with the convergence of the constructed Markov chains as the m_i tend to infinity.

Validity of Path Proposals

The random walk proposal for the parameter update in Sect. 7.1.3 automatically generates proposals from the parameter space Θ. For the path update in Sect. 7.1.2, however, there is no guarantee that the path proposals maintain the boundaries of the state space. There are two general solutions to this problem:

The first possibility is to consider transformations of the process such that the transformed sample paths are unrestricted. An SDE describing the transformed process can be obtained using Itô's formula, which was provided in Sect. 3.2.10. For example, Elerian et al. (2001) consider the logarithm of the one-dimensional Cox-Ingersoll-Ross process, which has non-negative state space before transformation and is introduced in Sect. A.3 in the Appendix.

An alternative solution to these possibly complicated calculations is to include an appropriate indicator function in the acceptance probability of the path proposal such that invalid proposals are rejected.

7.1.7 Example: Ornstein-Uhlenbeck Process

In this section, the implementation of the above methodology is illustrated on the example of a specific diffusion. Consider the one-dimensional Ornstein-Uhlenbeck process $X = (X_t)_{t \geq 0}$ which is described by the SDE

$$\mathrm{d}X_t = \alpha(\beta - X_t)\mathrm{d}t + \sigma \mathrm{d}B_t \quad , X_0 = x_0, \tag{7.20}$$

for parameters $\beta \in \mathbb{R}$, $\alpha, \sigma^2 \in \mathbb{R}_+$ and initial value $x_0 \in \mathbb{R}$. The solution of this process is a Gaussian process, i.e. the exact transition density is available. A detailed description of the Ornstein-Uhlenbeck process is included in Sect. A.2 in the Appendix. The full conditional densities of the parameters are given in what follows. Complete derivations are provided in Sect. B.4 in the Appendix.

Be aware that in case of improper or partially improper prior distributions it is not guaranteed that the joint posterior density of all parameters is proper even if the full conditional densities are. Hence an analysis of the joint posterior should precede the application of the full conditional distributions in an MCMC algorithm. Section B.4 investigates in which cases the posterior density is proper. It turns out that for fixed $\alpha \in \mathbb{R}_+$ this is true even for flat priors for β and σ^2.

Exact Full Conditional Proposal

Assume we have observed or imputed data $Y_0, Y_1, \ldots, Y_m$ at time points $t_0, t_1, \ldots, t_m$. Since the transition density of the Ornstein-Uhlenbeck process is explicitly known, the full conditional densities of the model parameters α, β and σ^2 can immediately be written in an unnormalised form. For β and σ^2, these are of the following types: For flat priors

$$p(\beta) \propto 1 \quad \text{for } \beta \in \mathbb{R} \qquad \text{and} \qquad p(\sigma^2) \propto 1 \quad \text{for } \sigma^2 \in \mathbb{R}_+, \tag{7.21}$$

one has

$$\beta \,\big|\, \alpha, \sigma^2, Y_0, \ldots, Y_m \sim \mathcal{N}\left(\frac{\displaystyle\sum_{k=0}^{m-1} \frac{Y_{k+1} - Y_k \mathrm{e}^{-\alpha \Delta t_k}}{1 + \mathrm{e}^{-\alpha \Delta t_k}}}{\displaystyle\sum_{k=0}^{m-1} \frac{1 - \mathrm{e}^{-\alpha \Delta t_k}}{1 + \mathrm{e}^{-\alpha \Delta t_k}}}, \frac{\dfrac{\sigma^2}{2\alpha}}{\displaystyle\sum_{k=0}^{m-1} \frac{1 - \mathrm{e}^{-\alpha \Delta t_k}}{1 + \mathrm{e}^{-\alpha \Delta t_k}}} \right),$$

$$\sigma^2 \,\big|\, \alpha, \beta, Y_0, \ldots, Y_m \sim \mathrm{IG}\left(\tfrac{m}{2} - 1, \alpha \sum_{k=0}^{m-1} \frac{\big(Y_{k+1} - Y_k \mathrm{e}^{-\alpha \Delta t_k} - \beta\big(1 - \mathrm{e}^{-\alpha \Delta t_k}\big)\big)^2}{1 - \mathrm{e}^{-2\alpha \Delta t_k}} \right).$$

For conjugate priors

$$\beta \sim \mathcal{N}\big(\beta_0, \rho_\beta^2\big) \quad \text{and} \quad \sigma^2 \sim \mathrm{IG}\big(\kappa_0, \nu_0\big) \tag{7.22}$$

with hyperparameters $\beta_0 \in \mathbb{R}$ and $\rho_\beta, \kappa_0, \nu_0 \in \mathbb{R}_+$, one obtains

$$\beta \,\big|\, \alpha, \sigma^2, Y_0, \ldots, Y_m$$
$$\sim \mathcal{N}\left(\frac{\dfrac{\sigma^2 \beta_0}{2\alpha\rho_\beta^2} + \displaystyle\sum_{k=0}^{m-1} \frac{Y_{k+1} - Y_k \mathrm{e}^{-\alpha \Delta t_k}}{1 + \mathrm{e}^{-\alpha \Delta t_k}}}{\dfrac{\sigma^2}{2\alpha\rho_\beta^2} + \displaystyle\sum_{k=0}^{m-1} \frac{1 - \mathrm{e}^{-\alpha \Delta t_k}}{1 + \mathrm{e}^{-\alpha \Delta t_k}}}, \frac{\dfrac{\sigma^2}{2\alpha}}{\dfrac{\sigma^2}{2\alpha\rho_\beta^2} + \displaystyle\sum_{k=0}^{m-1} \frac{1 - \mathrm{e}^{-\alpha \Delta t_k}}{1 + \mathrm{e}^{-\alpha \Delta t_k}}} \right), \tag{7.23}$$

$$\sigma^2 \,\big|\, \alpha, \beta, Y_0, \ldots, Y_m$$
$$\sim \mathrm{IG}\left(\frac{m}{2} + \kappa_0,\ \nu_0 + \alpha \sum_{k=0}^{m-1} \frac{\big(Y_{k+1} - Y_k \mathrm{e}^{-\alpha \Delta t_k} - \beta\big(1 - \mathrm{e}^{-\alpha \Delta t_k}\big)\big)^2}{1 - \mathrm{e}^{-2\alpha \Delta t_k}} \right), \tag{7.24}$$

where IG denotes the inverse gamma distribution (see the notation tables on pp. xvii). For $\rho_\beta = \infty$, $\kappa_0 = -1$ and $\nu_0 = 0$, this corresponds to the results above for the flat priors. The full conditional density of α,

$$\pi\big(\alpha \,\big|\, \beta, \sigma^2, Y_0, \ldots, Y_m\big) \propto \frac{p(\alpha)\alpha^{m/2} \exp\left(-\dfrac{\alpha}{\sigma^2} \displaystyle\sum_{k=0}^{m-1} \frac{\big(Y_{k+1} - Y_k \mathrm{e}^{-\alpha \Delta t_k} - \beta\big(1 - \mathrm{e}^{-\alpha \Delta t_k}\big)\big)^2}{1 - \mathrm{e}^{-2\alpha \Delta t_k}}\right)}{\displaystyle\prod_{k=0}^{m-1} \sqrt{1 - \mathrm{e}^{-2\alpha \Delta t_k}}},$$

cannot be recognised to be of any standard distribution type. Naturally, as β and σ^2 are a priori independent in both (7.21) and (7.22), the above posterior distributions remain valid if a mix of flat and conjugate priors is chosen. Full derivations of the posterior densities are included in Sect. B.4 in the Appendix.

Approximate Full Conditional Proposal

In the general case, where the transition density of the diffusion process is not available, approximate full conditional densities may be employed instead. These are for the Ornstein-Uhlenbeck process as follows: For flat priors

$$p(\alpha) \propto 1 \;\; \text{for}\, \alpha \in \mathbb{R}_+, \quad p(\beta) \propto 1 \;\; \text{for}\, \beta \in \mathbb{R}, \quad p(\sigma^2) \propto 1; \;\text{for}\, \sigma^2 \in \mathbb{R}_+,$$

one has

$$\alpha \,\big|\, \beta, \sigma^2, Y_0, \ldots, Y_m \sim \mathcal{N}_{\text{trunc}}\left(\frac{\displaystyle\sum_{k=0}^{m-1} (Y_{k+1} - Y_k)(\beta - Y_k)}{\displaystyle\sum_{k=0}^{m-1} (\beta - Y_k)^2 \Delta t_k}, \frac{\sigma^2}{\displaystyle\sum_{k=0}^{m-1} (\beta - Y_k)^2 \Delta t_k}\right),$$

$$\beta \,\big|\, \alpha, \sigma^2, Y_0, \ldots, Y_m \sim \mathcal{N}\left(\frac{\dfrac{Y_m - Y_0}{\alpha} + \displaystyle\sum_{k=0}^{m-1} Y_k \Delta t_k}{t_m - t_0}, \frac{\sigma^2}{\alpha^2 (t_m - t_0)}\right),$$

$$\sigma^2 \,\big|\, \alpha, \beta, Y_0, \ldots, Y_m \sim \mathrm{IG}\left(\frac{m}{2} - 1, \frac{1}{2} \sum_{k=0}^{m-1} \frac{\big(Y_{k+1} - Y_k - \alpha(\beta - Y_k)\Delta t_k\big)^2}{\Delta t_k}\right),$$

where $\mathcal{N}_{\text{trunc}}$ denotes the normal distribution truncated at zero, which generates Gaussian random numbers on the positive real line (see also the notation tables on pp. xvii). The conjugate priors

$$\alpha \sim \mathcal{N}_{\text{trunc}}\big(\alpha_0, \rho_\alpha^2\big), \quad \beta \sim \mathcal{N}\big(\beta_0, \rho_\beta^2\big), \quad \sigma^2 \sim \text{IG}\big(\kappa_0, \nu_0\big)$$

lead to

$$\begin{aligned}
&\alpha \,\big|\, \beta, \sigma^2, Y_0, \ldots, Y_m \\
&\sim \mathcal{N}_{\text{trunc}}\left(\frac{\rho_\alpha^2 \sum_{k=0}^{m-1}(Y_{k+1}-Y_k)(\beta - Y_k) + \alpha_0\sigma^2}{\rho_\alpha^2 \sum_{k=0}^{m-1}(\beta - Y_k)^2 \Delta t_k + \sigma^2}, \frac{\sigma^2\rho_\alpha^2}{\rho_\alpha^2 \sum_{k=0}^{m-1}(\beta - Y_k)^2 \Delta t_k + \sigma^2}\right), \\
&\beta \,\big|\, \alpha, \sigma^2, Y_0, \ldots, Y_m \\
&\sim \mathcal{N}\left(\frac{\alpha^2\rho_\beta^2\left(\frac{Y_m - Y_0}{\alpha} + \sum_{k=0}^{m-1} Y_k \Delta t_k\right) + \sigma^2\beta_0}{\alpha^2\rho_\beta^2(t_m - t_0) + \sigma^2}, \frac{\sigma^2\rho_\beta^2}{\alpha^2\rho_\beta^2(t_m - t_0) + \sigma^2}\right), \qquad (7.25) \\
&\sigma^2 \,\big|\, \alpha, \beta, Y_0, \ldots, Y_m \\
&\sim \text{IG}\left(\frac{m}{2} + \kappa_0\,,\ \nu_0 + \frac{1}{2}\sum_{k=0}^{m-1}\frac{\big(Y_{k+1} - Y_k - \alpha(\beta - Y_k)\Delta t_k\big)^2}{\Delta t_k}\right). \qquad (7.26)
\end{aligned}$$

Setting $\rho_\alpha = \infty$, $\rho_\beta = \infty$, $\kappa_0 = -1$ and $\nu_0 = 0$ in these formulas yields the full conditional densities which were obtained using flat priors. Again, the calculations of these posterior densities are provided in Sect. B.4.

Random Walk Proposal

Assume that the current value of the parameter is $\boldsymbol{\theta} = (\alpha, \beta, \sigma^2)'$. Due to the range of admissible values for the parameter components, the following random walk proposals are apparent: Draw

$$\begin{aligned}
\log\alpha^* &\sim \mathcal{N}\big(\log\alpha, \gamma_\alpha^2\big) \\
\beta^* &\sim \mathcal{N}\big(\beta, \gamma_\beta^2\big) \qquad (7.27) \\
\log\sigma^{2*} &\sim \mathcal{N}\big(\log\sigma^2, \gamma_\sigma^2\big) \qquad (7.28)
\end{aligned}$$

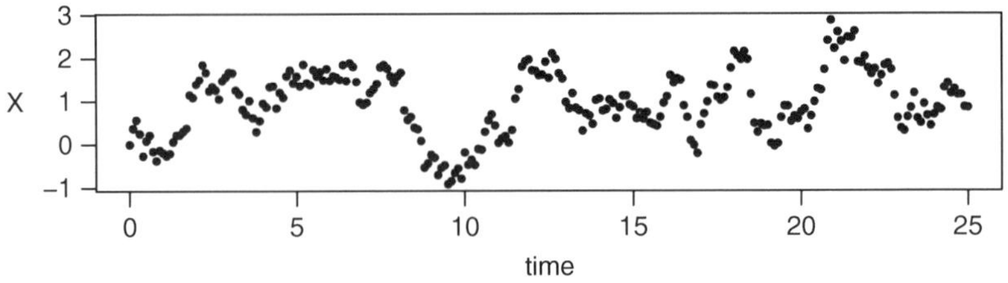

Fig. 7.5 Exactly sampled diffusion path at times $0, 0.1, 0.2, \ldots, 25$ of an Ornstein-Uhlenbeck process satisfying (7.20) with parameter $\boldsymbol{\theta} = (0.5, 0.9, 1)'$. The estimation results in this section condition on subsets of these observations

for some predefined positive constants γ_α, γ_β and γ_σ. The acceptance probability for a so-obtained proposal $\boldsymbol{\theta}^* = (\alpha^*, \beta^*, \sigma^{2*})'$ is

$$\zeta(\boldsymbol{\theta}^*, \boldsymbol{\theta}) = 1 \wedge \left(\prod_{k=0}^{m-1} \frac{\pi^{\text{Euler}}(\boldsymbol{Y}_{k+1} | \boldsymbol{Y}_k, \boldsymbol{\theta}^*)}{\pi^{\text{Euler}}(\boldsymbol{Y}_{k+1} | \boldsymbol{Y}_k, \boldsymbol{\theta})} \right) \cdot \frac{p(\boldsymbol{\theta}^*)}{p(\boldsymbol{\theta})} \cdot \frac{\alpha^* \sigma^{2*}}{\alpha \sigma^2}.$$

Simulation Study

In the following, we generate exact discrete realisations $\{x_{\tau_1}, \ldots, x_{\tau_M}\}$ of the Ornstein-Uhlenbeck process at times $\tau_1, \ldots, \tau_M$ given the parameter $\boldsymbol{\theta} = (\alpha, \beta, \sigma^2)' = (0.5, 0.9, 1)'$ and the initial value $x_0 = 0$. Given the observed data, we apply the estimating schemes described in this section in order to infer on $\boldsymbol{\theta}$. All functions have been implemented in R.

To be more precise, in all experiments the employed dataset is a subset of the discretely sampled diffusion path displayed in Fig. 7.5. Inter-observation intervals are chosen evenly spaced, i.e. $\tau_i = i/M$ for $i = 0, \ldots, M$. Each interval $[\tau_i, \tau_{i+1}]$ is then again partitioned into m equidistant intervals with boundaries $t_{i,j} = (i + j/m)/M$ for $i = 0, \ldots, M-1$ and $j = 0, \ldots, m$.

The parameter $\alpha = 0.5$ is considered known whilst β and σ^2 are supposed unknown. The synthetic data setting, however, allows for comparison of simulation results with the true parameter values $\beta = 0.9$ and $\sigma^2 = 1$.

For the path and parameter proposals, all considered approaches are studied. Their abbreviations and respective formulas are summarised in Table 7.1. For β and σ^2, the conjugate priors (7.22) with $\beta_0 = 0$, $\rho_\beta^2 = 1$ and $\kappa_0 = \nu_0 = 3$ are applied. The a priori expectations and variances of the parameters are thus $\mathbb{E}(\beta) = 0$, $\text{Var}(\beta) = 1$, $\mathbb{E}(\sigma^2) = 1.5$ and $\text{Var}(\sigma^2) = 2.25$.

The estimation procedure performs the following steps:

1. Initialise Y^{imp} by linear interpolation.
2. Draw initial values for β and σ^2 from (7.22) with $\beta_0 = 0$, $\rho_\beta^2 = 1$ and $\kappa_0 = \nu_0 = 3$.

Table 7.1 Abbreviations and formulas for parameter and path proposals used for the parameter estimation in Figs. 7.6–7.20 and Tables 7.2–7.5. The parameter priors are chosen as in (7.22). Hyperparameters are $\beta_0 = 0$, $\rho_\beta^2 = 1$ and $\kappa_0 = \nu_0 = 3$

Abbreviation	Path proposal	$Y_{(a,b)}^{\text{imp}}$	
E	Euler	(7.5)	
E2-MH	Double-sided Euler (Metropolis-Hastings)	(7.6)	
E2-MG	Double-sided Euler (Metropolis-within-Gibbs)	(7.6)	
MB	Modified bridge	(7.7)	
DB	Diffusion bridge	(7.9)	
G	Gaussian	(7.11)	
t	Student t	(7.13)	
Abbreviation	Parameter proposal	β	σ^2
eFC	Exact full conditionals	(7.23)	(7.24)
aFC	Approximate full conditionals	(7.25)	(7.26)
RW	Random walk	(7.27)	(7.28)

3. Repeat the following steps 10^5 times:

   Path update:

   (a) Choose an interval (t_a, t_b) using Algorithm 7.3 with $\lambda = 5$.
   (b) Draw a proposal $Y_{(a,b)}^{\text{imp}*}$ according to the investigated method; accept or reject.

   Parameter update:

   If full conditional proposals are applied:

   (a) Draw a proposal β^* (conditioned on the current σ^2) and accept.
   (b) Draw a proposal σ^{2*} (conditioned on the new β^*) and accept.

   If random walk proposals are applied:

   (a) Draw a proposal β^* with $\gamma_\beta = 0.5$.
   (b) Draw a proposal σ^{2*} with $\gamma_\sigma = 0.5$.
   (c) Accept both or none.

Results for $T = 25$, $M = 25$ and $m = 2$ are shown in Figs. 7.6–7.11. Figures 7.12–7.17 display results for $T = 25$, $M = 25$ and $m = 10$. These are summarised in Tables 7.2–7.5 and Fig. 7.19. A discussion follows in Sect. 7.1.8.

7.1.8 Discussion

In this section, a variety of path and parameter proposals were introduced as modules in the general MCMC scheme (7.2) to alternately estimate the model parameter and imputed sample path of a diffusion process. The different proposals were applied to infer on the parameters of an Ornstein-Uhlenbeck process whose solution is available in explicit form. The evaluation of each proposal technique is the objective of the following.

To begin with, consider the seven different path proposal schemes from Sect. 7.1.2; these are the Euler proposal, the double-sided Euler proposal (in a Metropolis-Hastings and a Metropolis-within-Gibbs version), the modified bridge proposal, the diffusion bridge proposal, the Gaussian proposal and the Student t proposal. The most important criterion to rate an estimation scheme is certainly to consider whether the parameter estimates approximately match the true values.

In short, the Euler proposal, double-sided Euler proposal (in both versions), modified bridge proposal and diffusion bridge proposal yield satisfying estimation results with respect to the obtained 95 % highest probability density intervals. The Gaussian and Student t proposals, on the other hand, fail on exactly this account as they obviously do not correctly estimate the parameters of the diffusion coefficient in the considered example. This is apparent in Fig. 7.19.

A further issue is the investigation of the acceptance rates for the path and parameter update which are listed in Tables 7.4 and 7.5. For $m = 10$, these are apprehensively low for the random walk parameter update in combination with the Gaussian and Student t proposals. The acceptance rates for the other path proposal schemes seem inconspicuous yet but should possibly be further evaluated for higher amounts of imputed data. In particular, the rates for the Euler proposal are expected to further decrease as m increases as this proposal distribution does not condition on the end point of the path segment.

There remains the question why the Gaussian and Student t proposals perform so poorly although the simulated paths in Figs. 7.2 and 7.3 on pp. 178 and 179 do not appear to substantially differ from those obtained with the modified or diffusion bridge proposal. Empirical investigations yield that within the MCMC procedure, the Gaussian and Student t proposals seem to be unable to reproduce the shape of a diffusion path. This is illustrated in Fig. 7.20 under consideration of one characteristic property of diffusion paths: the quadratic variation. This attribute was introduced in Sect. 3.2.6 for time-continuous data. In the present situation, we estimate the quadratic variation of the discrete skeleton $Y = \{Y_0, \ldots, Y_K\}$ of the Ornstein-Uhlenbeck process as

$$\widehat{\langle Y, Y \rangle}_{[0,T]} = \sum_{i=0}^{K-1} \big(Y_{i+1} - Y_i\big)^2. \tag{7.29}$$

If the grid of time points is sufficiently fine, one should obtain $\widehat{\langle Y, Y \rangle}_{[0,T]}/T \approx \sigma^2 = 1$ for $\sigma^2 = 1$. The left column of Fig. 7.20 shows trace plots of the quadratic variation of the diffusion paths obtained with the usual MCMC scheme considered in this chapter so far, where $T = 1$, $M = 10$ and $m = 10$. In a second experiment, the MCMC algorithm was modified such that the parameter $\boldsymbol{\theta}$ is fixed to its true value and only the imputed data is updated. The resulting trace plots of the quadratic variation are shown in the middle column. The last column shows the quadratic variation of a series of diffusion path proposals conditioned on the true parameter value without any accept/reject mechanism. Apparently, the Gaussian and Student t proposals do in fact propose diffusion paths that match the required quadratic variation. However, these paths seem to generally be rejected in the MCMC procedure even when they are conditioned on the true model parameter.

As these are empirical results, the Gaussian and Student t proposals may work better for different models like those considered in Elerian et al. (2001). In any case, the experiments in that paper are not comparable to the simulation study here with respect to the numbers of observations, which are 500 in Elerian et al. (2001) and 25 here. However, there are other reasons speaking against the Gaussian and Student t proposals: Chib and Shephard (2002) already point out the computational cost which is necessary to search for the mode of the Gaussian or t distribution for large numbers of missing data points. In fact, even in the relatively simple simulation study in Sect. 7.1.7, these two approaches turned out to be computationally much more costly than other proposal schemes. In the context of importance sampling, by the way, Eraker (2002) notices that the Gaussian proposal does generally not meet the regularity conditions which are required for the convergence of the related SMLE scheme in Sect. 6.3.3.

Now consider the remaining five path proposal schemes, that are the Euler proposal, double-sided Euler proposal (two versions), modified bridge proposal and diffusion bridge proposal. Figures 7.10, 7.11, 7.16 and 7.17 show autocorrelation plots for the parameters β and σ^2. Figure 7.18 displays the inefficiency factors of the serially correlated imputed data; see Sect. B.5 in the Appendix for details on inefficiency factors. The modified bridge and diffusion bridge proposals show best performance concerning the autocorrelation of both the parameter and the imputed data. Thereby, the modified bridge proposal seems to work slightly better. It hence emerges from the simulation study as the first choice of a path proposal scheme.

For the update of the model parameter, three different proposal schemes were considered in Sect. 7.1.3: the exact full conditional proposal, the approximate full conditional proposal, and the random walk proposal. The first one comes into question only if the transition density of the considered diffusion process is known in closed form; in practice, this is seldomly the case, and hence this proposal cannot generally be selected. Moreover, even if the transition density was tractable and the full conditional densities could be determined up to a normalising constant, this would be of practical use only if one was able to generate random variates from this density. In Sect. 7.1.7, the exact full conditional densities of the parameters β and σ^2 could be associated with a normal and inverse gamma distribution, respectively. The full conditional distribution of the parameter α, however, was not recognised to be of any standard distribution type.

The approximate full conditional density can be computed for all diffusion processes up to the normalising constant. The above comments on the practical usability, however, apply here as well: There is no benefit of the approximate full conditional density kernel unless a possibility to sample from it is at hand. Furthermore, the calculations on pp. 193–195 and in Sect. B.4 in the Appendix show that the derivation of both the exact and the approximate full conditional densities can be quite elaborate even for a fairly standard diffusion process.

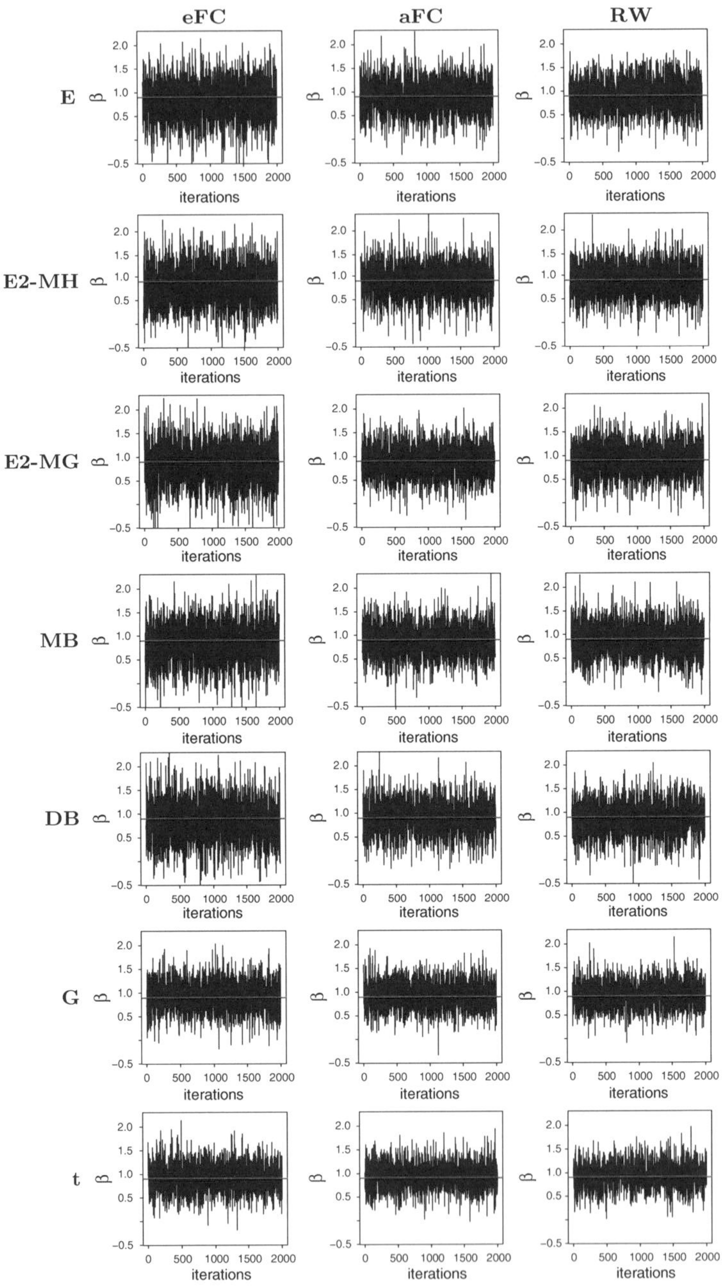

Fig. 7.6 Estimation of parameters of the Ornstein-Uhlenbeck process (7.20) as described on pp. 196–197. The MCMC scheme conditions on observed data at times $0, 1, \ldots, 25$ and introduces $m = 2$ subintervals in between every two observations. This figure shows the trace plots of β. The Markov chains have length 10^5 but have been thinned by factor 50. The true value for β equals 0.9 and is indicated by the *red horizontal line*. Abbreviations for the path and parameter proposals are listed in Table 7.1

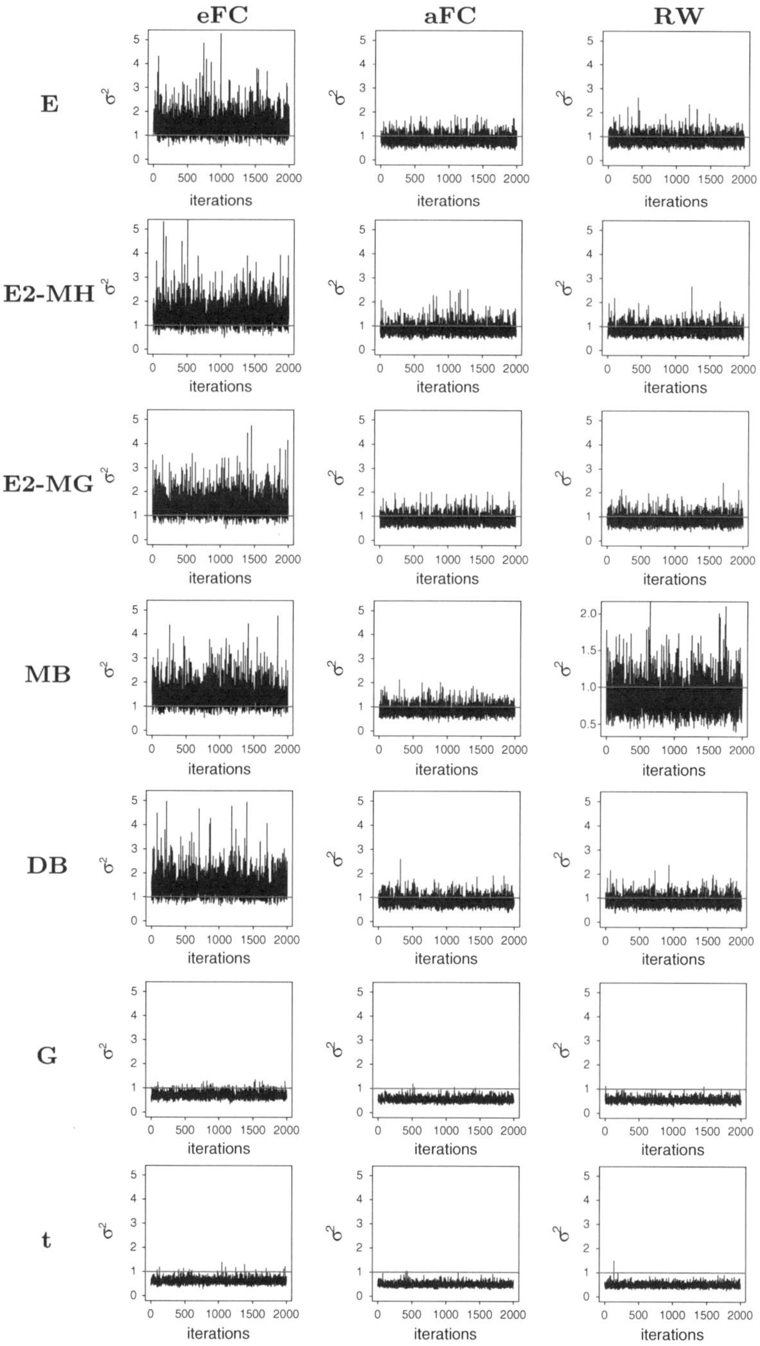

Fig. 7.7 Estimation results as described in Fig. 7.6. This figure shows the trace plots for σ^2. The true parameter value for σ^2 equals 1 and is indicated by the *red horizontal line*

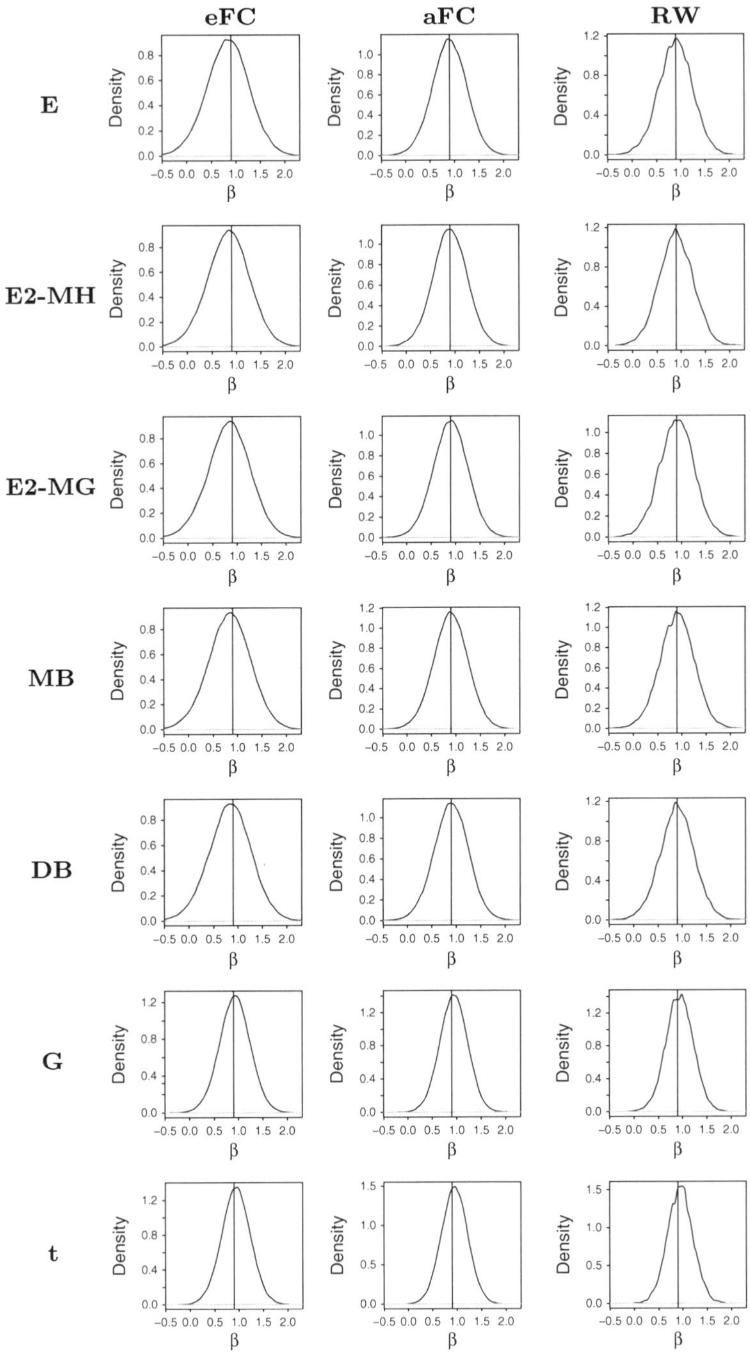

Fig. 7.8 Estimation of the posterior density of β based on the results from Fig. 7.6. Density estimation takes into account the full Markov chain, i.e. without thinning, after having discarded a 10 % burn-in phase. The true value of the parameter is indicated by the *vertical line*

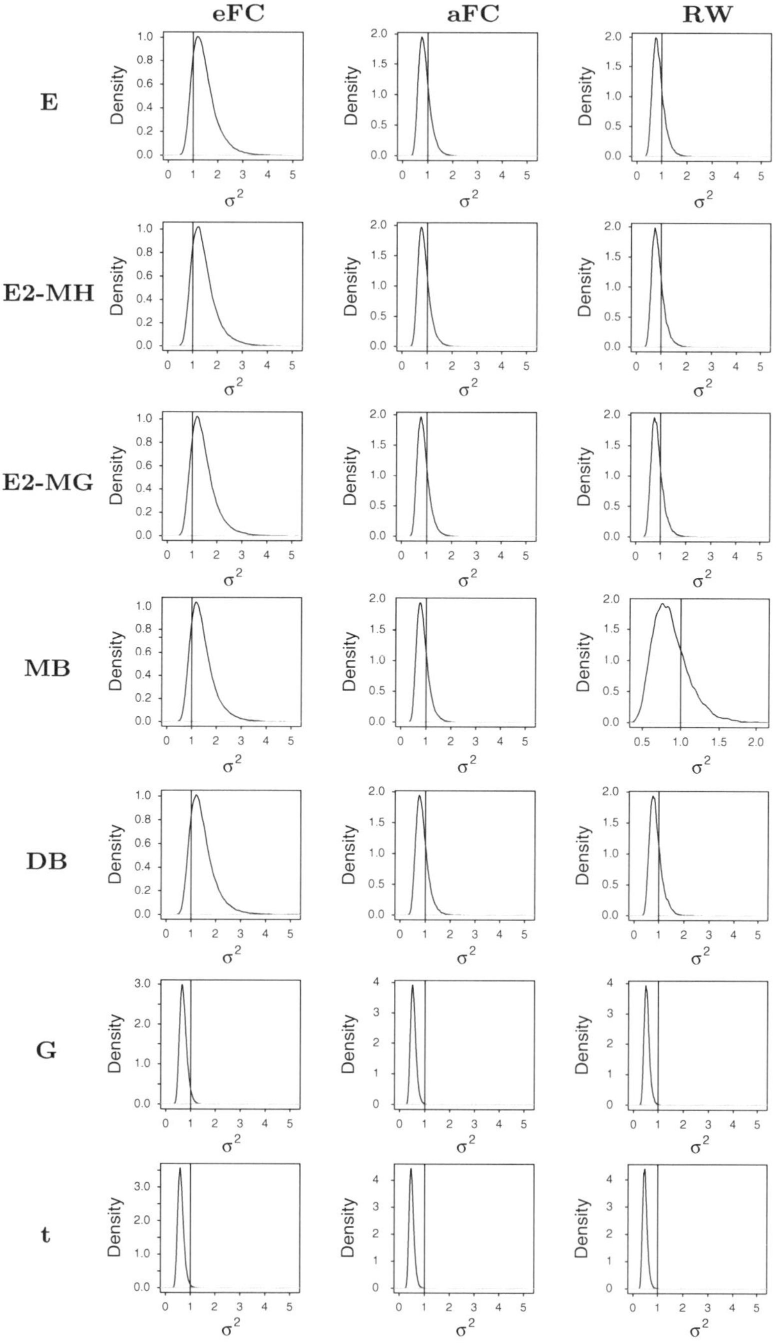

Fig. 7.9 Estimation of the posterior density of σ^2 based on the results from Fig. 7.7. Density estimation takes into account the full Markov chain, i.e. without thinning, after having discarded a 10 % burn-in phase. The true value of the parameter is indicated by the *vertical line*

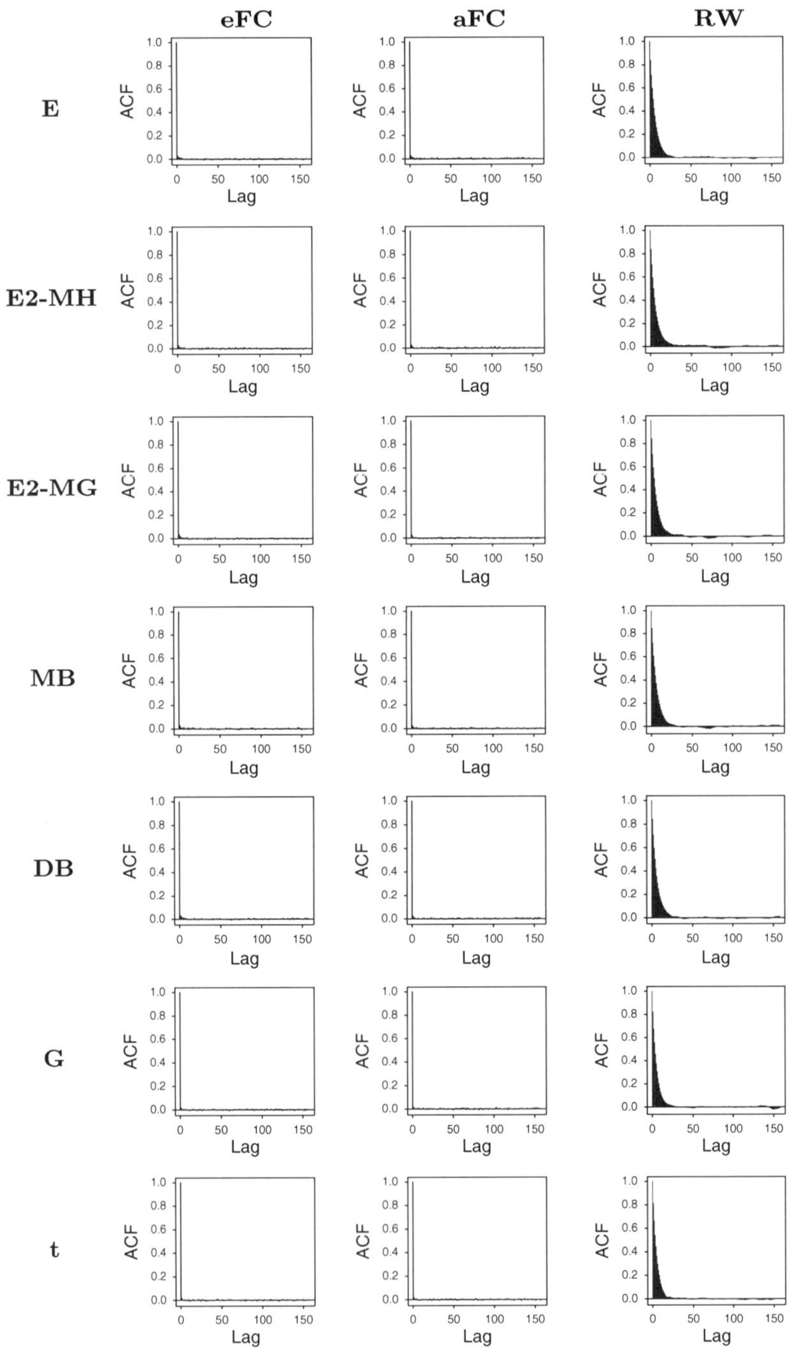

Fig. 7.10 Autocorrelation plots for β based on the results from Fig. 7.6. Calculation of the autocorrelation takes into account the full Markov chain, i.e. without thinning, after having discarded a 10 % burn-in phase

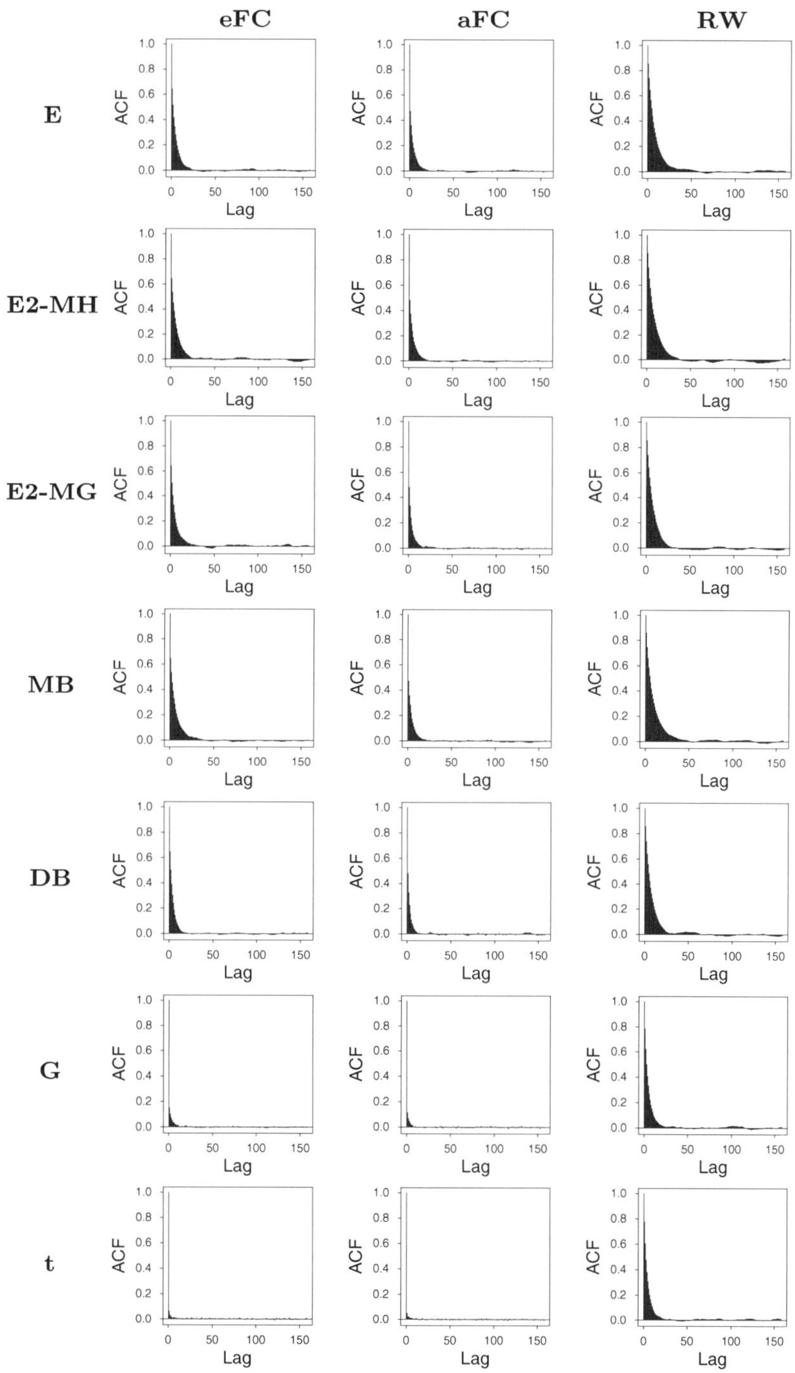

Fig. 7.11 Autocorrelation plots for σ^2 based on the results from Fig. 7.7. Calculation of the autocorrelation takes into account the full Markov chain, i.e. without thinning, after having discarded a 10 % burn-in phase

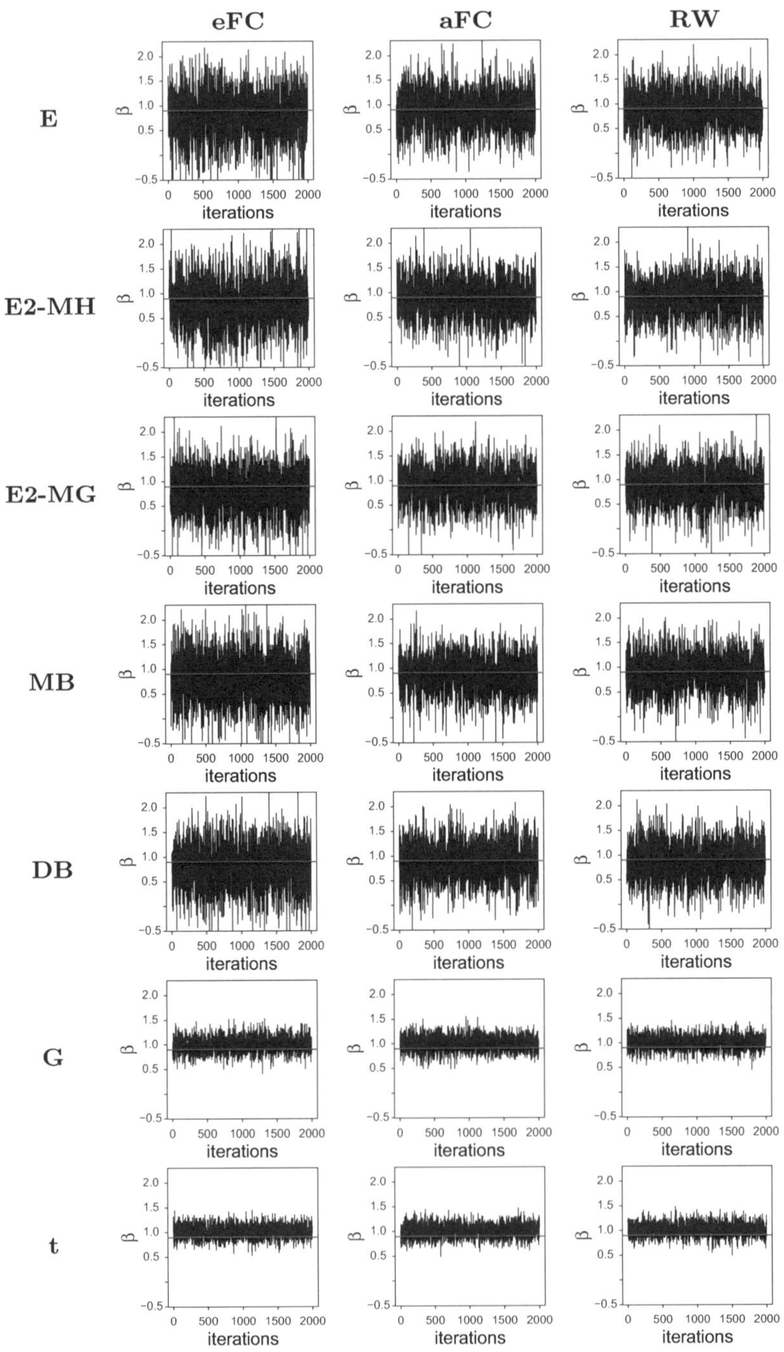

Fig. 7.12 Estimation of parameters of the Ornstein-Uhlenbeck process (7.20) as in Fig. 7.6, this time introducing $m = 10$ subintervals in between every two observations. This figure shows the trace plots of β. The Markov chains have length 10^5 but have been thinned by factor 50. The true value for β equals 0.9 and is indicated by the *red horizontal line*

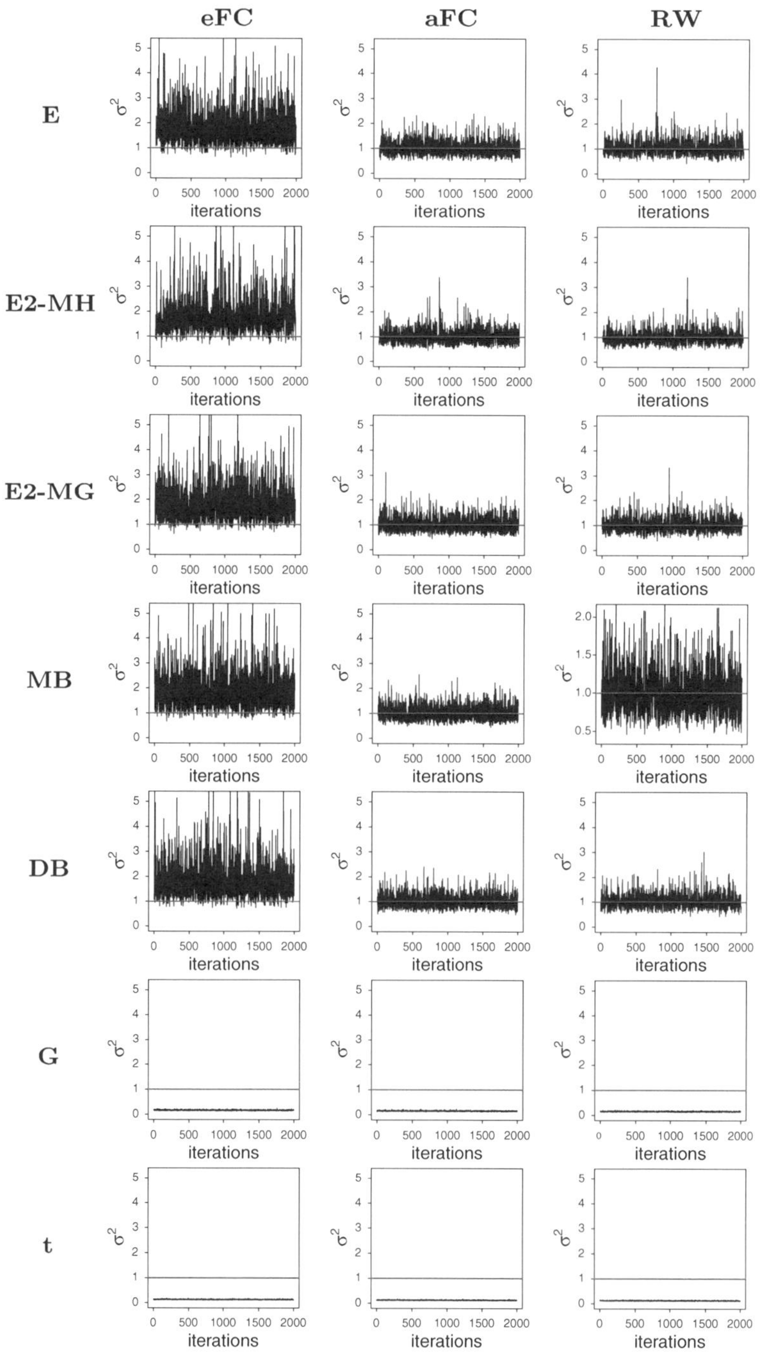

Fig. 7.13 Estimation results as described in Fig. 7.12. This figure shows the trace plots for σ^2. The true parameter value equals $\sigma^2 = 1$ and is indicated by the *red horizontal line*

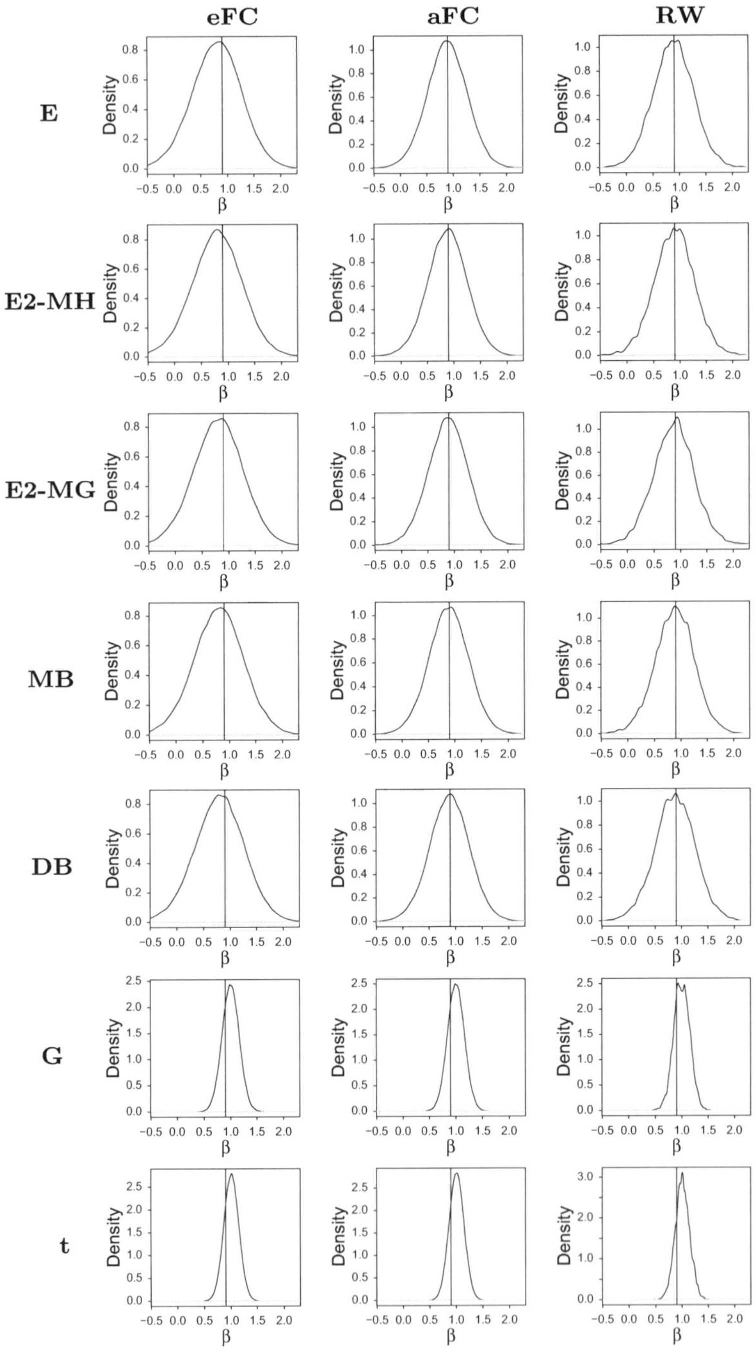

Fig. 7.14 Estimation of the posterior density of β based on the results from Fig. 7.12. Density estimation takes into account the full Markov chain, i.e. without thinning, after having discarded a 10 % burn-in phase. The true value of the parameter is indicated by the *vertical line*

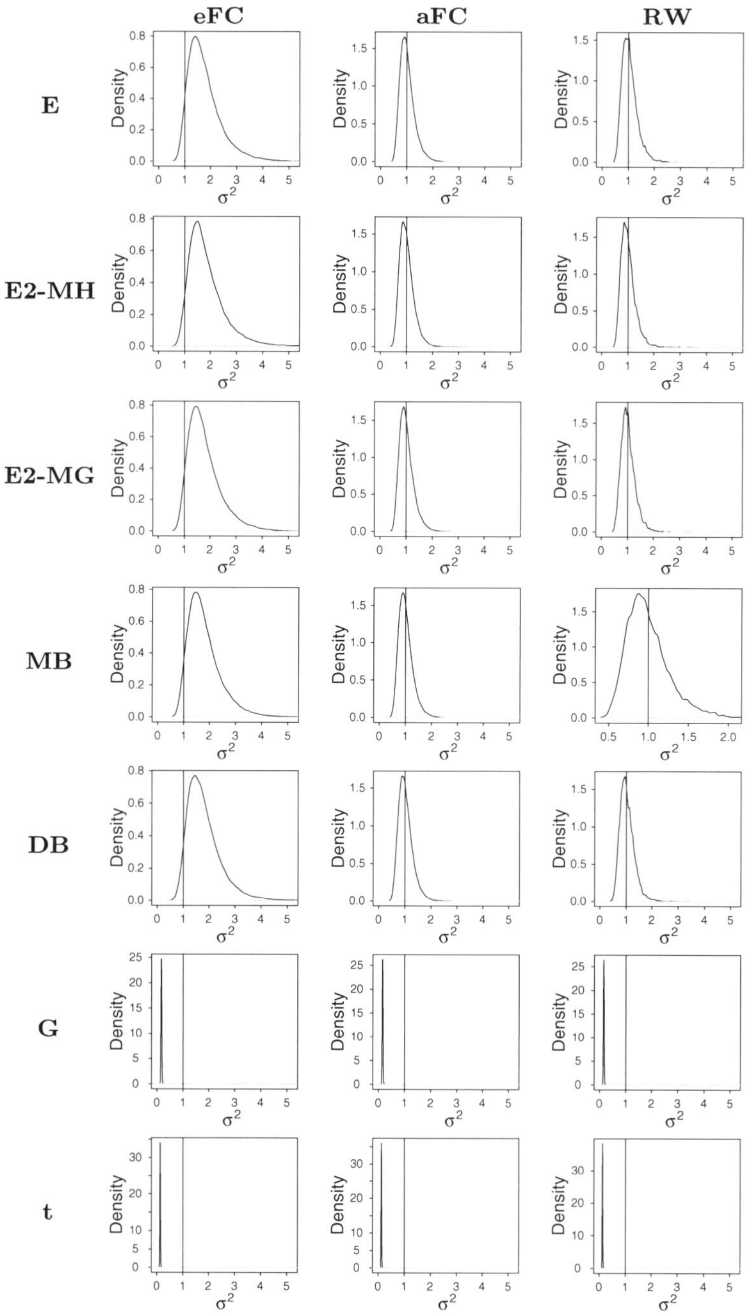

Fig. 7.15 Estimation of the posterior density of σ^2 based on the results from Fig. 7.13. Density estimation takes into account the full Markov chain, i.e. without thinning, after having discarded a 10 % burn-in phase. The true value of the parameter is indicated by the *vertical line*

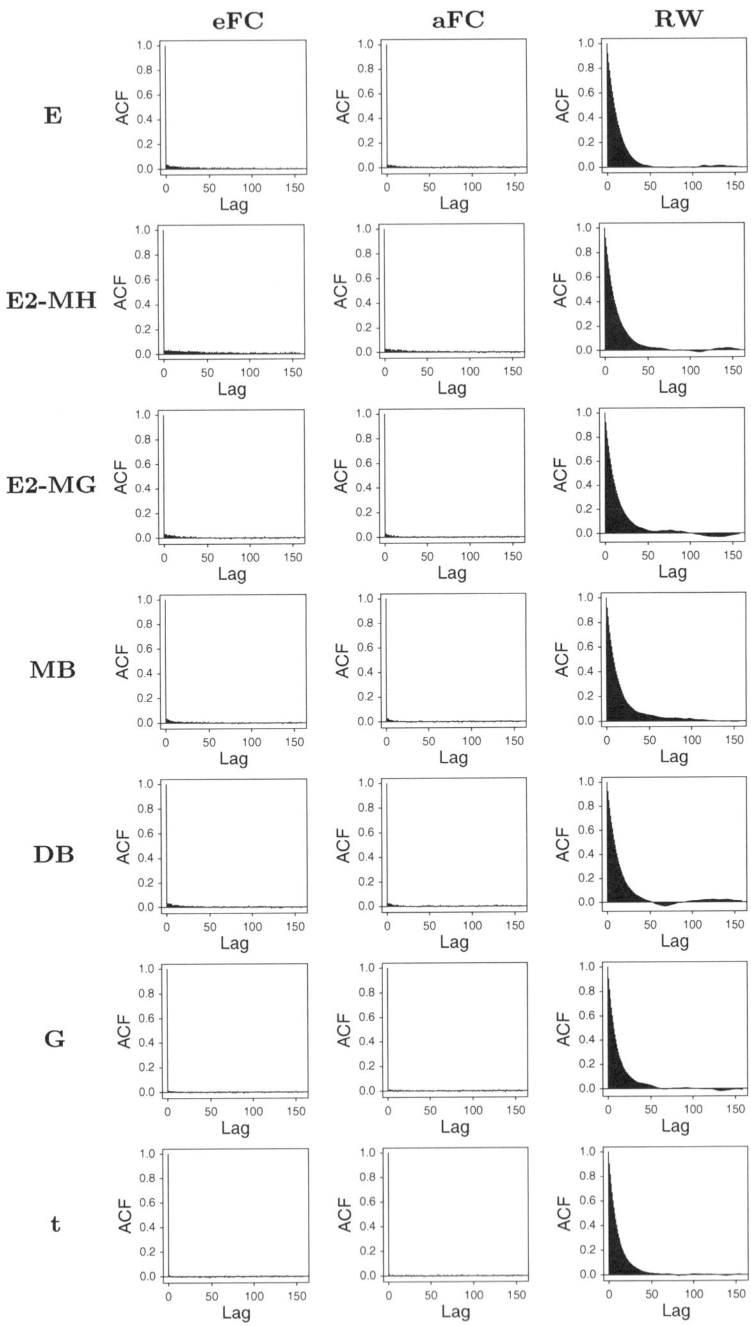

Fig. 7.16 Autocorrelation plots for β based on the results from Fig. 7.12. Calculation of the autocorrelation takes into account the full Markov chain, i.e. without thinning, after having discarded a 10 % burn-in phase

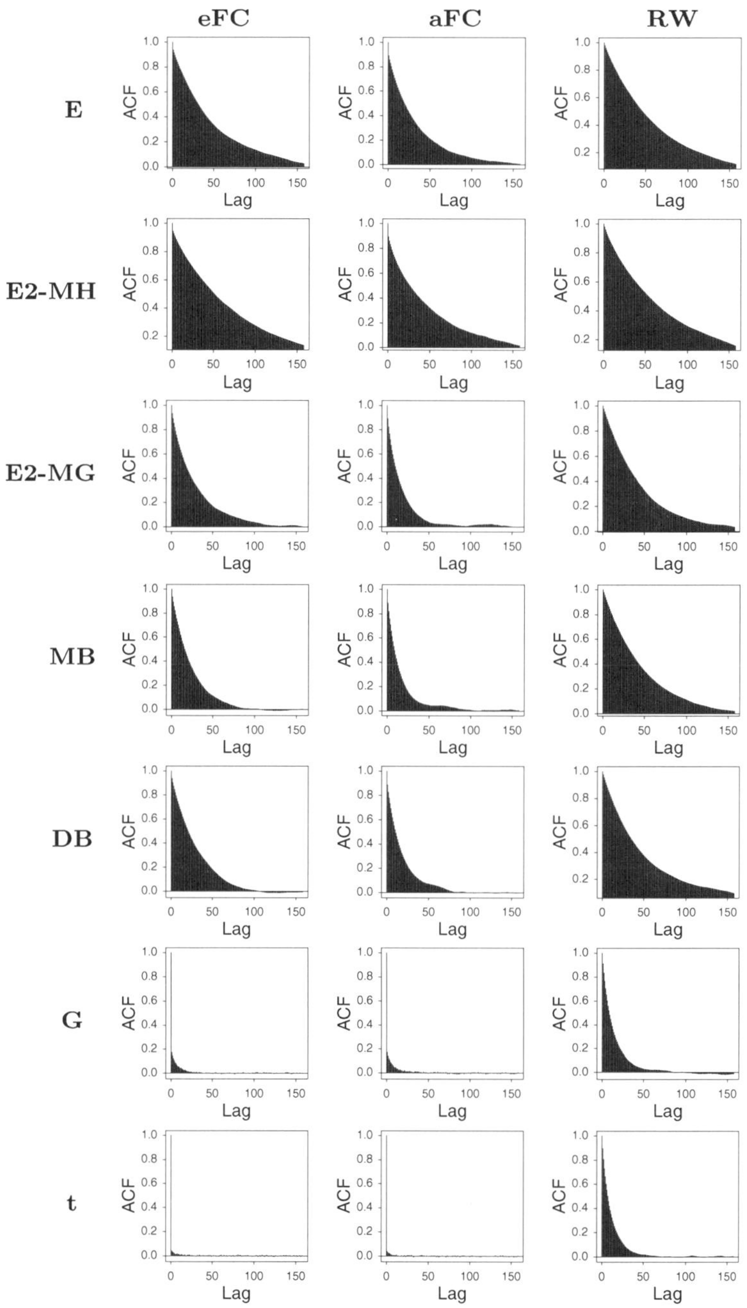

Fig. 7.17 Autocorrelation plots for σ^2 based on the results from Fig. 7.13. Calculation of the autocorrelation takes into account the full Markov chain, i.e. without thinning, after having discarded a 10 % burn-in phase

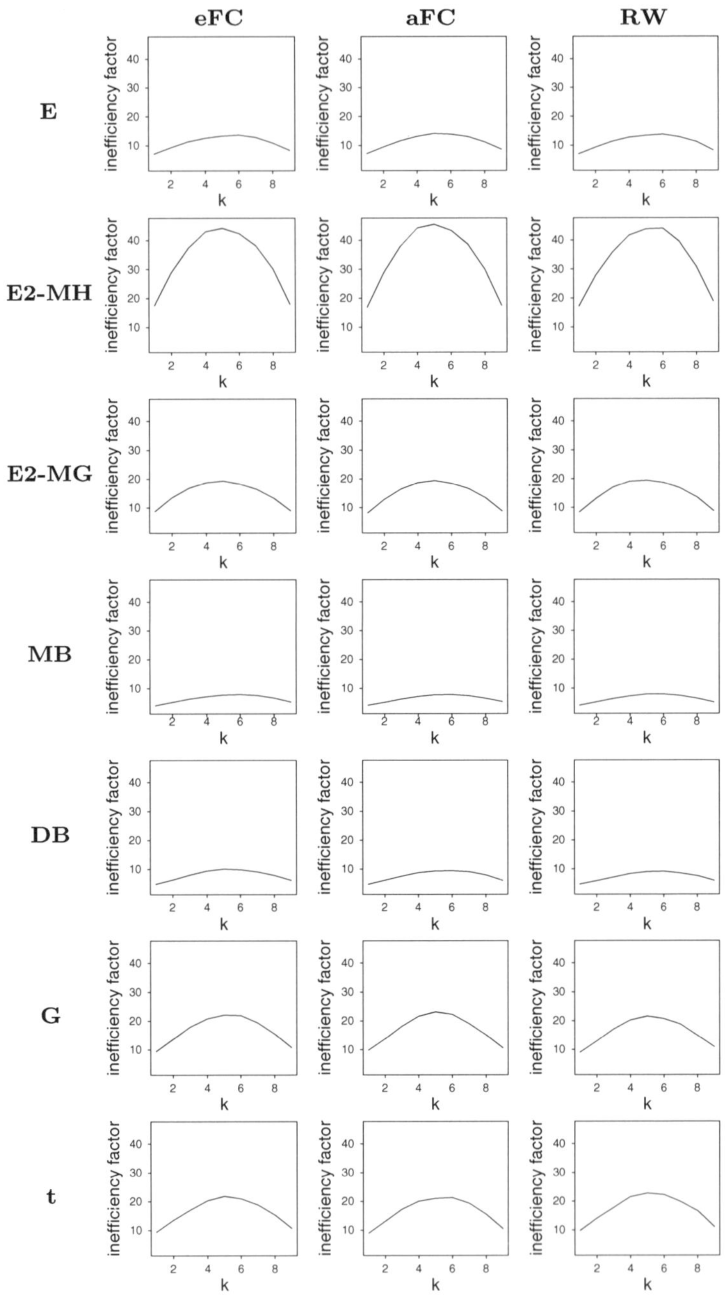

Fig. 7.18 Inefficiency factors for imputed data generated by the MCMC scheme as described in the simulation study on pp. 196–197. The estimation procedure conditions on observed data at times 0 and 1 and introduces $m = 10$ subintervals in between these two observations. This figure shows the inefficiency factor $\iota(Y_k)$ for $k = 1, \ldots, m-1$ as described in Sect. B.5, where $t_0 = 0$ and $t_m = 1$. The Markov chains have length 10^5 less a discarded burn-in phase of 10 %

Table 7.2 Estimation results as in Figs. 7.6 and 7.7 with $T = 25$, $M = 25$ and $m = 2$. This table displays the posterior means and posterior 95 %-hpd intervals after a 10 % burn-in phase. The latter are computed according to Chen and Shao (1999). The true values of the parameters are $\beta = 0.9$ and $\sigma^2 = 1$. The hpd intervals are also shown in Fig. 7.19 on p. 215

	eFC	aFC	RW
E	β : 0.84, (−0.01,1.71) σ^2: 1.45, (0.66,2.45)	β : 0.89, (0.20,1.59) σ^2: 0.87, (0.48,1.35)	β : 0.89, (0.20,1.58) σ^2: 0.87, (0.47,1.34)
E2-MH	β : 0.84, (−0.02,1.70) σ^2: 1.44, (0.65,2.44)	β : 0.90, (0.19,1.58) σ^2: 0.87, (0.47,1.35)	β : 0.89, (0.19,1.57) σ^2: 0.87, (0.47,1.36)
E2-MG	β : 0.84, (−0.02,1.71) σ^2: 1.44, (0.65,2.44)	β : 0.89, (0.22,1.61) σ^2: 0.87, (0.47,1.34)	β : 0.90, (0.22,1.61) σ^2: 0.87, (0.46,1.35)
MB	β : 0.84, (−0.03,1.69) σ^2: 1.43, (0.66,2.44)	β : 0.89, (0.20,1.59) σ^2: 0.87, (0.47,1.35)	β : 0.90, (0.20,1.58) σ^2: 0.88, (0.48,1.37)
DB	β : 0.84, (−0.02,1.71) σ^2: 1.45, (0.65,2.46)	β : 0.89, (0.19,1.58) σ^2: 0.87, (0.47,1.35)	β : 0.89, (0.19,1.58) σ^2: 0.87, (0.48,1.38)
G	β : 0.92, (0.29,1.55) σ^2: 0.70, (0.44,1.00)	β : 0.94, (0.38,1.49) σ^2: 0.55, (0.35,0.77)	β : 0.94, (0.38,1.48) σ^2: 0.55, (0.35,0.77)
t	β : 0.93, (0.35,1.53) σ^2: 0.62, (0.40,0.87)	β : 0.95, (0.41,1.48) σ^2: 0.50, (0.32,0.69)	β : 0.94, (0.43,1.47) σ^2: 0.50, (0.33,0.70)

Table 7.3 Estimation results as in Figs. 7.12 and 7.13 with $T = 25$, $M = 25$ and $m = 10$. This table displays the posterior means and posterior 95 %-hpd intervals after a 10 % burn-in phase. The true values of the parameters are $\beta = 0.9$ and $\sigma^2 = 1$. The hpd intervals are also shown in Fig. 7.19 on p. 215

	eFC	aFC	RW
E	β : 0.80, (−0.14,1.73) σ^2: 1.77, (0.77,3.08)	β : 0.88, (0.15,1.63) σ^2: 1.02, (0.53,1.58)	β : 0.88, (0.14,1.63) σ^2: 1.05, (0.56,1.65)
E2-MH	β : 0.79, (−0.16,1.74) σ^2: 1.86, (0.76,3.30)	β : 0.88, (0.14,1.62) σ^2: 1.01, (0.54,1.55)	β : 0.89, (0.11,1.61) σ^2: 1.00, (0.55,1.56)
E2-MG	β : 0.80, (−0.13,1.74) σ^2: 1.79, (0.78,3.10)	β : 0.88, (0.12,1.60) σ^2: 1.02, (0.55,1.58)	β : 0.87, (0.12,1.63) σ^2: 1.02, (0.53,1.55)
MB	β : 0.80, (−0.14,1.74) σ^2: 1.81, (0.78,3.12)	β : 0.88, (0.12,1.61) σ^2: 1.01, (0.56,1.58)	β : 0.88, (0.14,1.62) σ^2: 1.00, (0.52,1.55)
DB	β : 0.80, (−0.14,1.74) σ^2: 1.81, (0.78,3.13)	β : 0.88, (0.12,1.62) σ^2: 1.02, (0.55,1.57)	β : 0.87, (0.13,1.64) σ^2: 1.04, (0.57,1.58)
G	β : 1.00, (0.68,1.32) σ^2: 0.17, (0.14,0.20)	β : 1.00, (0.69,1.31) σ^2: 0.16, (0.13,0.19)	β : 1.00, (0.72,1.31) σ^2: 0.16, (0.13,0.19)
t	β : 1.01, (0.73,1.29) σ^2: 0.13, (0.11,0.16)	β : 1.01, (0.73,1.28) σ^2: 0.13, (0.10,0.15)	β : 1.01, (0.74,1.28) σ^2: 0.13, (0.11,0.15)

Table 7.4 Acceptance rates for the path update corresponding to the experiments in Figs. 7.6, 7.7, 7.12 and 7.13

	eFC (%)	aFC (%)	RW (%)
E	$m = 2$: 62	$m = 2$: 57	$m = 2$: 57
	$m = 10$: 46	$m = 10$: 44	$m = 10$: 45
E2-MH	$m = 2$: 88	$m = 2$: 88	$m = 2$: 88
	$m = 10$: 67	$m = 10$: 67	$m = 10$: 67
E2-MG	$m = 2$: 92	$m = 2$: 92	$m = 2$: 92
	$m = 10$: 98	$m = 10$: 98	$m = 10$: 98
MB	$m = 2$: 88	$m = 2$: 88	$m = 2$: 88
	$m = 10$: 97	$m = 10$: 97	$m = 10$: 97
DB	$m = 2$: 79	$m = 2$: 79	$m = 2$: 79
	$m = 10$: 72	$m = 10$: 72	$m = 10$: 72
G	$m = 2$: 53	$m = 2$: 53	$m = 2$: 53
	$m = 10$: 37	$m = 10$: 37	$m = 10$: 37
t	$m = 2$: 52	$m = 2$: 52	$m = 2$: 52
	$m = 10$: 36	$m = 10$: 36	$m = 10$: 36

Table 7.5 Acceptance rates for the random walk parameter update corresponding to the Markov chains displayed in Figs. 7.6, 7.7, 7.12 and 7.13. The acceptance rates for the exact and approximate full conditional proposals are 100 % due to the construction of the algorithm

	RW (%)
E	$m = 2$: 29
	$m = 10$: 16
E2-MH	$m = 2$: 29
	$m = 10$: 16
E2-MG	$m = 2$: 29
	$m = 10$: 16
MB	$m = 2$: 29
	$m = 10$: 16
DB	$m = 2$: 29
	$m = 10$: 16
G	$m = 2$: 25
	$m = 10$: 9
t	$m = 2$: 25
	$m = 10$: 8

The random walk proposal, in contrast, is always available, easy to implement and not problem-specific apart from the domain of the parameter.

The approximate full conditional proposal and the random walk proposal yield similar posterior means for the parameters in Tables 7.2 and 7.3 and highest probability density (hpd) intervals in Fig. 7.19. For all but the Gaussian and Student t path updates, these posterior means match the true parameter values quite well for $m = 10$. The exact full conditional proposal, on the other hand, underestimates β, overestimates σ^2 and produces fairly large hpd intervals unless it is combined with the Gaussian or Student t proposal. Therefore, the approximate full conditional and random walk proposals should be preferred for the parameter update.

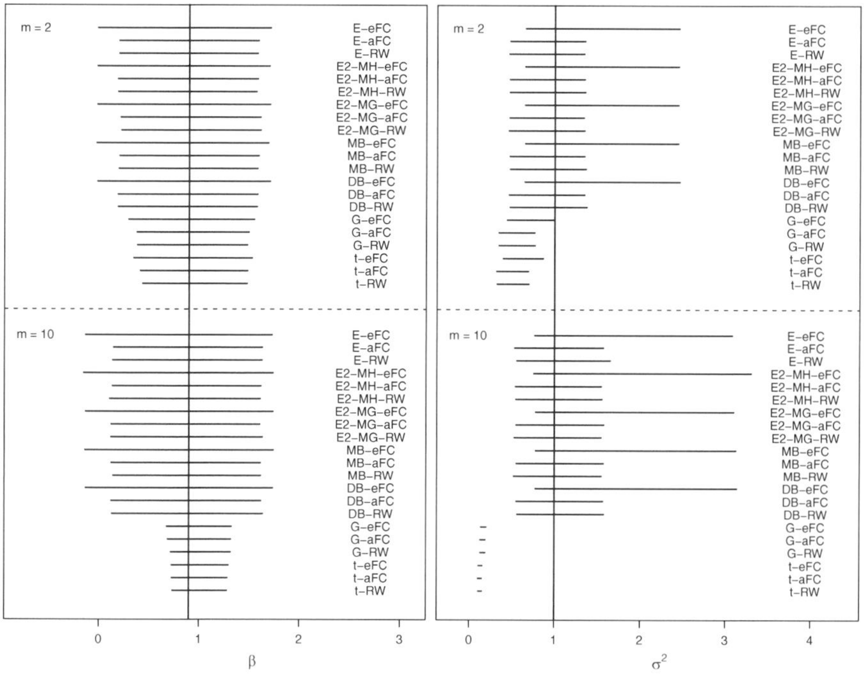

Fig. 7.19 95% hpd intervals for β (*left*) and σ^2 (*right*) as displayed in Tables 7.2 and 7.3

For the specific random walk variances γ_β^2 and γ_σ^2, the random walk proposal causes higher autocorrelation in the Markov chains for β and σ^2 in Figs. 7.10, 7.11, 7.16 and 7.17 than the full conditional proposals. Yet the universal and convenient employability of the random walk overweighs this last issue. Overall, the random walk is the favourite parameter proposal scheme.

To summarise, this section introduces the general methodology of Bayesian inference for diffusions via data augmentation. The implementation of the general procedure is extensively discussed by considering specific path and parameter proposal schemes and other practical concerns. The methodology is illustrated on a simulation study for the one-dimensional Ornstein-Uhlenbeck process. To that end, all algorithms have been implemented in R. As expected, for those update schemes which were classified appropriate techniques in this discussion, estimation results improve as the amount of imputed data increases from $m = 2$ to $m = 10$ intermediate subintervals. Overall, the modified brigde proposal for the missing data in combination with the random walk proposal for the parameter show best performance. Hence the following section concentrates on these two approaches and extends them to a more general framework than considered in this section.

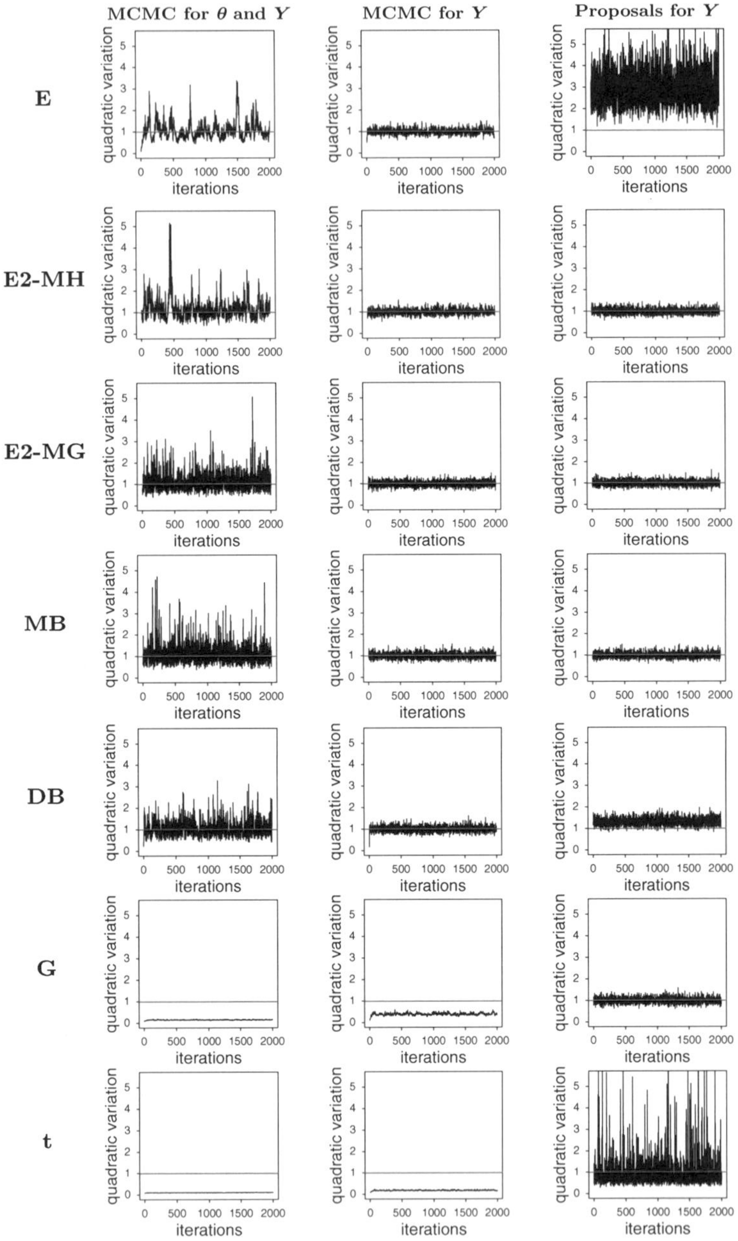

Fig. 7.20 Trace plots of quadratic variation (7.29) of discretely sampled diffusion paths obtained by three different experiments. *Left*: By application of the MCMC scheme as described in the simulation study for unknown β and σ^2 with $T = 1$, $M = 10$ and $m = 10$. *Middle*: By application of the same MCMC scheme, but for known β and σ^2. *Right*: By proposing paths conditional on the true value of $\boldsymbol{\theta}$ but without any accept/reject mechanism. All Markov chains have length 10^5 but have been thinned by factor 50. The true value of the quadratic variation equals 1 and is indicated by the *red horizontal line*

7.2 Extension to Latent Data and Observation with Error

The previous section introduced general concepts of parameter estimation for diffusions using data augmentation schemes. Observations of the diffusion paths at discrete time points were assumed both complete and without measurement error. In applications in life sciences, however, these conditions are seldomly fulfilled. Section 7.2.1 therefore extends the algorithms from Sect. 7.1 to latent data. Section 7.2.2 additionally adapts them to observation with error. Some of the calculations in this section have also been carried out for equidistant time steps by Golightly and Wilkinson (2006a, 2008), who partly arrive at different results though.

7.2.1 Latent Data

It often occurs in applications in life sciences that the state variable of a diffusion process $(\boldsymbol{X}_t)_{t\geq 0}$ is not fully observable and hence consists of an observed and an unobserved latent part: In infectious disease epidemiology, for example, one reports the numbers of infected individuals in a population but usually does not know the numbers of susceptibles. In chemical kinetics, one may observe the sum of the concentrations of two species but possibly cannot measure the single concentrations. Typical examples outside of life sciences are stochastic volatility models being composed of an observed asset price and a latent volatility.

This section extends the estimation schemes from Sect. 7.1, which alternately perform a path and a parameter update, to a latent data framework. To that end, the parameter update can be adopted from Sect. 7.1.3 without change. The path update, however, needs to be modified for the latent data framework. This section hence deals with path proposals in the presence of incomplete observations.

As before, assume that there are observations of the state of the process available at times $0 = \tau_0 < \tau_1 < \ldots < \tau_M = T$. In order to reduce the lengths of inter-observation time intervals, impute $m_i - 1$ auxiliary time points in between every two observation times τ_i and τ_{i+1} for $i = 0, \ldots, M-1$. Then, in all, there are $K+1$ observation and auxiliary times, where $K = m_0 + \ldots + m_{M-1}$. Figure 7.21 shows a discretised two-dimensional diffusion path which consists of a one-dimensional observable part V and a one-dimensional latent part L, where auxiliary data has been imputed. The indices in the notation are as in Fig. 7.4 on p. 189.

To simplify notation, label all observed and auxiliary time points in ascending order by t_k, $k = 0, \ldots, K$. Define $\mathfrak{O} = \{k \in \{0, \ldots, K\} \,|\, t_k \in \{\tau_0, \ldots, \tau_M\}\}$, that is the set of indices of observation times. Like in the previous section, let $\boldsymbol{Y}_k = \boldsymbol{X}_{t_k}$ for all $k = 0, \ldots, K$. For $k \notin \mathfrak{O}$, $\boldsymbol{Y}_k$ is completely unobserved and needs to be fully investigated in the path update. For $k \in \mathfrak{O}$, $\boldsymbol{Y}_k$ is partially observed, i.e. its components can be rearranged in a way such that $\boldsymbol{Y}_k = (\boldsymbol{V}_k', \boldsymbol{L}_k')' \in \mathbb{R}^d$ for observed $\boldsymbol{V}_k \in \mathbb{R}^{d_1}$ and latent $\boldsymbol{L}_k \in \mathbb{R}^{d_2}$, where

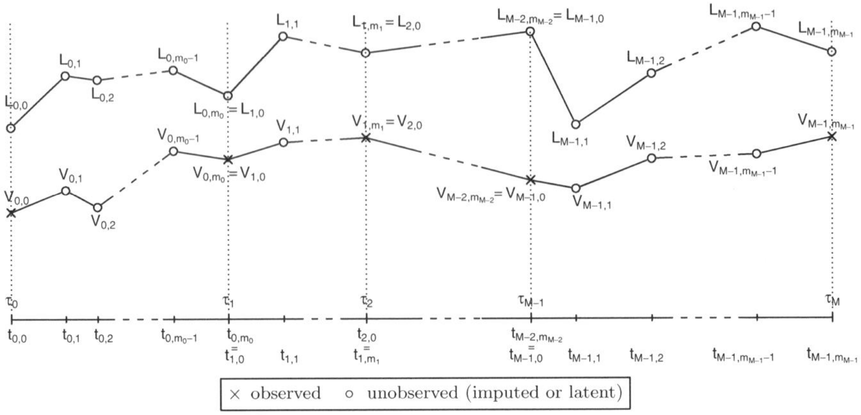

Fig. 7.21 Illustration of a two-dimensional path with an observable component V and a latent component L. As in Fig. 7.4 on p. 189, the observed part of the path consists of the initial value at time τ_0 and M discrete observations at times $\tau_1, \ldots, \tau_M$. Observed data is labelled with *crosses*, imputed data with *circles*. The latent components are imputed at all observation and auxiliary times and hence also labelled with *circles*

$d = d_1 + d_2$. In that case, the path update changes $\boldsymbol{L}_k$ but leaves $\boldsymbol{V}_k$ unaltered. For simplicity, suppose that the decomposition of $\boldsymbol{Y}_k$ into $\boldsymbol{V}_k$ and $\boldsymbol{L}_k$ is the same for all $k \in \mathfrak{O}$, although this assumption is not necessary for the path proposal schemes considered in the following.

In what follows, we first investigate the general path update procedure and afterwards provide the required proposal distributions. Adapted acceptance probabilities are presented on pp. 223.

Choose an update interval (t_a, t_b) such that $|\{a+1, \ldots, b-1\} \cap \mathfrak{O}| \leq 1$, i.e. there is not more than one observation time in the interior of (t_a, t_b). The following situations may now occur:

1. One has $|\{a+1, \ldots, b-1\} \cap \mathfrak{O}| = 0$, i.e. there is no observation time in the interior of (t_a, t_b). In this case path proposals are obtained as in Sect. 7.1.2.
2. One has $|\{a+1, \ldots, b-1\} \cap \mathfrak{O}| = 1$, i.e. there is exactly one observation time t_r in the interior of (t_a, t_b). As in the framework without latent data in Sect. 7.1.2, there are various possibilities to propose the path segment between t_a and t_b. The discussion in Sect. 7.1.8 showed that satisfactory results can be achieved by application of the modified bridge proposal. Hence this approach is extended to the latent data case in the following. There are two strategies for this extension:

   a. First, propose $\boldsymbol{L}_r^* \,|\, \boldsymbol{Y}_a, \boldsymbol{V}_r, \boldsymbol{Y}_b, \boldsymbol{\theta}$ as in (7.31) below, that is the latent vector at the intermediate observation time t_r. Then, generate two conditionally independent proposals on (t_a, t_r) and (t_r, t_b) conditioned on $\boldsymbol{Y}_a, \boldsymbol{V}_r, \boldsymbol{L}_r^*, \boldsymbol{Y}_b, \boldsymbol{\theta}$ as in item 1.
   This approach is illustrated in Fig. 7.22.

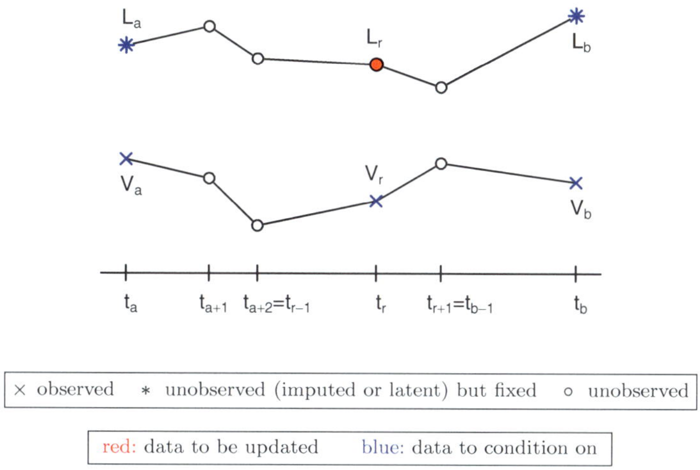

Fig. 7.22 Illustration of update strategy (2a) on p. 218 for a two-dimensional process: the objective is the update of a discretised path segment on the time interval (t_a, t_b) in presence of an intermediate observation time t_r. The path consists of a one-dimensional observable component V (*lower curve*) and a one-dimensional latent component L (*upper curve*). Under the assumption that $a \neq 0$ and $b \neq K$, the end points L_a and L_b are kept fixed. *Crosses* indicate observed values, *stars* label unobserved but fixed values, and *circles* stand for unobserved values that are still to be considered in the path update. This figure shows the first step of strategy (2a): Update L_r (*red*) conditional on V_a, L_a, V_r, V_b and L_b (*blue*). In the second step, which is not shown here, L_r is considered fixed and the path is updated on (t_a, t_r) and (t_r, t_b) as considered in Sect. 7.1.2

b. Update the path segment from the left to the right. More precisely:
   - For $k = a, \ldots, r-2$, propose $\boldsymbol{Y}_{k+1}^* \,|\, \boldsymbol{Y}_k^*, \boldsymbol{V}_r, \boldsymbol{Y}_b, \boldsymbol{\theta}$ as in (7.35) below, where $\boldsymbol{Y}_a^* = \boldsymbol{Y}_a$. Alternatively—and computationally less costly—, propose $\boldsymbol{Y}_{k+1}^* \,|\, \boldsymbol{Y}_k^*, \boldsymbol{V}_r, \boldsymbol{\theta}$ as in (7.33).
   - Propose $\boldsymbol{L}_r^* \,|\, \boldsymbol{Y}_{r-1}^*, \boldsymbol{V}_r, \boldsymbol{Y}_b, \boldsymbol{\theta}$ as in (7.31) below.
   - For $k = r, \ldots, b-2$, propose $\boldsymbol{Y}_{k+1}^* \,|\, \boldsymbol{Y}_k^*, \boldsymbol{Y}_b, \boldsymbol{\theta}$ as in item 1, where $\boldsymbol{Y}_r^*$ is composed of $\boldsymbol{V}_r$ and $\boldsymbol{L}_r^*$.

   This procedure is shown in Fig. 7.23.

Special situations occur when $a = 0$ or $b = K$: Usually, the imputed data is updated merely on the interior of (t_a, t_b) such that $\boldsymbol{Y}_a$ and $\boldsymbol{Y}_b$ remain unaltered. If, however, $\boldsymbol{Y}_0$ or $\boldsymbol{Y}_K$ are only partially observed, their update has to be included in the update of adjoining path segments. Under the assumption that $|\{a+1, \ldots, b-1\} \cap \mathfrak{O}| = 0$, this involves drawing from $\mathfrak{L}(\boldsymbol{L}_0 | \boldsymbol{V}_0, \boldsymbol{Y}_b, \boldsymbol{\theta})$, $\mathfrak{L}(\boldsymbol{L}_K | \boldsymbol{Y}_{K-1}, \boldsymbol{V}_K, \boldsymbol{\theta})$ and $\mathfrak{L}(\boldsymbol{Y}_{k+1} | \boldsymbol{Y}_k, \boldsymbol{V}_K, \boldsymbol{\theta})$, where $\mathfrak{L}$ denotes the distribution of a random variable. The first two distributions are provided in (7.37) and (7.39) below. The third one corresponds to (7.33) with r replaced by K.

Exact sampling of the above mentioned conditional distributions is generally not possible, but they can be approximated under consideration of the Euler scheme

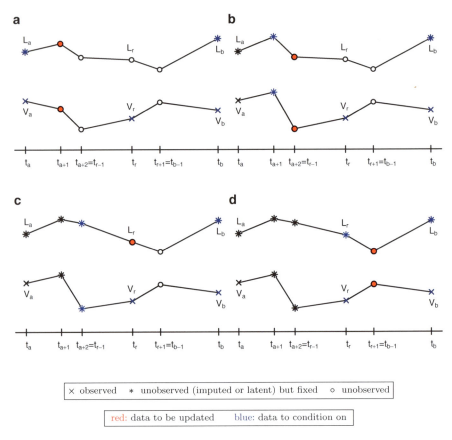

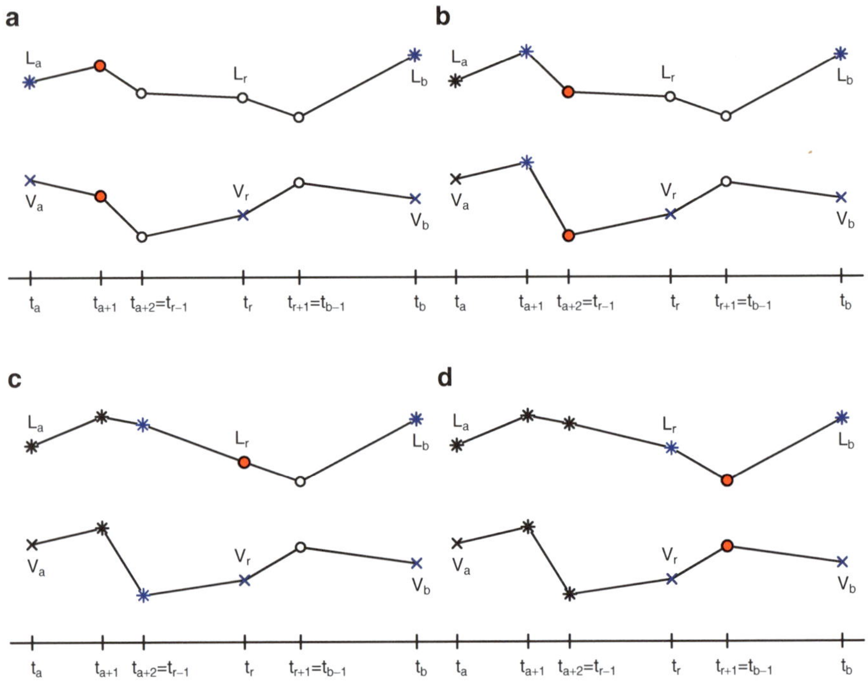

× observed ∗ unobserved (imputed or latent) but fixed ∘ unobserved

red: data to be updated blue: data to condition on

Fig. 7.23 Illustration of update strategy (2b) on p. 218 for a two-dimensional process: The objective is the update of a discretised path segment on the time interval (t_a, t_b) in presence of an intermediate observation time t_r. The path consists of a one-dimensional observable component V and a one-dimensional latent component L. Under the assumption that $a \neq 0$ and $b \neq K$, the end points L_a and L_b are kept fixed. *Crosses* indicate observed values, *stars* label unobserved but fixed values, and *circles* stand for unobserved values that are still to be considered in the path update. This figure shows the single steps of strategy (2b): (**a**) Update the path from the left to the right, i.e. start by investigation of the data at time t_{a+1}. As t_{a+1} is an auxiliary time point, both V_{a+1} and L_{a+1} (*red coloured*) are to be updated. The proposals are conditioned on V_a, L_a, V_r, V_b and L_b (*blue*). (**b**) Continue with the update of the data at time t_{a+2} (*red*). The just updated V_{a+1} and L_{a+1} are now considered fixed, hence condition on these and V_r, V_b, L_b (*blue*). (**c**) As V_r is observed, update only the latent component L_r (*red*) conditioned on V_{r-1}, L_{r-1}, V_r, V_b and L_b (*blue*). (**d**) Last, update V_{b-1} and L_{b-1} (*red*) conditioned on its direct left and right neighbours (*blue*)

and some further simplifications. Hence, the distributions that have not yet been investigated in Sect. 7.1.2 are now approximated in order to provide appropriate proposal distributions for the diffusion paths. Most results are based on standard multivariate normal theory. For the sake of brevity, only the outcomes are shown here. Full derivations are given in Sect. B.6 in the Appendix.

For shorter notation, abbreviate $\boldsymbol{\mu}_k = \boldsymbol{\mu}(\boldsymbol{Y}_k, \boldsymbol{\theta})$ and $\boldsymbol{\Sigma}_k = \boldsymbol{\Sigma}(\boldsymbol{Y}_k, \boldsymbol{\theta})$ for all k in the following. Furthermore, decompose $\boldsymbol{\mu}$ and $\boldsymbol{\Sigma}$ into

$$\boldsymbol{\mu} = \begin{pmatrix} \boldsymbol{\mu}^v \\ \boldsymbol{\mu}^l \end{pmatrix} \qquad \text{and} \qquad \boldsymbol{\Sigma} = \begin{pmatrix} \boldsymbol{\Sigma}^{vv} & \boldsymbol{\Sigma}^{vl} \\ \boldsymbol{\Sigma}^{lv} & \boldsymbol{\Sigma}^{ll} \end{pmatrix}$$

such that $\boldsymbol{\mu}^v \in \mathbb{R}^{d_1}$, $\boldsymbol{\mu}^l \in \mathbb{R}^{d_2}$, $\boldsymbol{\Sigma}^{vv} \in \mathbb{R}^{d_1 \times d_1}$ and $\boldsymbol{\Sigma}^{ll} \in \mathbb{R}^{d_2 \times d_2}$.

Approximation of $\mathfrak{L}(L_r \mid Y_k, V_r, Y_b, \theta)$ for $k < r$

Let $a \leq k < r$. The derivations in Appendix B.6 yield that the joint distribution of $\boldsymbol{V}_r$ and $\boldsymbol{L}_r$ conditioned on $\boldsymbol{Y}_k$, $\boldsymbol{Y}_b$ and $\boldsymbol{\theta}$ can be approximated as

$$\begin{pmatrix} \boldsymbol{V}_r \\ \boldsymbol{L}_r \end{pmatrix} \Big| \boldsymbol{Y}_k, \boldsymbol{Y}_b, \boldsymbol{\theta} \sim \mathcal{N}\left(\begin{pmatrix} \boldsymbol{V}_k + \dfrac{\boldsymbol{V}_b - \boldsymbol{V}_k}{\Delta_{kb}} \Delta_{kr} \\ \boldsymbol{L}_k + \dfrac{\boldsymbol{L}_b - \boldsymbol{L}_k}{\Delta_{kb}} \Delta_{kr} \end{pmatrix}, \frac{\Delta_{rb}\Delta_{kr}}{\Delta_{kb}} \begin{pmatrix} \boldsymbol{\Sigma}_k^{vv} & \boldsymbol{\Sigma}_k^{vl} \\ \boldsymbol{\Sigma}_k^{lv} & \boldsymbol{\Sigma}_k^{ll} \end{pmatrix} \right), \tag{7.30}$$

where $\Delta_{kr} = t_r - t_k$, $\Delta_{rb} = t_b - t_r$ and $\Delta_{kb} = t_b - t_k$. Further conditioning on $\boldsymbol{V}_r$ yields

$$\boldsymbol{L}_r \mid \boldsymbol{Y}_k, \boldsymbol{V}_r, \boldsymbol{Y}_b, \boldsymbol{\theta} \sim \mathcal{N}(\boldsymbol{\eta}_k, \boldsymbol{\Lambda}_k) \tag{7.31}$$

with

$$\boldsymbol{\eta}_k = \boldsymbol{L}_k + \frac{\boldsymbol{L}_b - \boldsymbol{L}_k}{\Delta_{kb}} \Delta_{kr} + \boldsymbol{\Sigma}_k^{lv} \left(\boldsymbol{\Sigma}_k^{vv}\right)^{-1} \left(\boldsymbol{V}_r - \boldsymbol{V}_k - \frac{\boldsymbol{V}_b - \boldsymbol{V}_k}{\Delta_{kb}} \Delta_{kr} \right)$$

and

$$\boldsymbol{\Lambda}_k = \frac{\Delta_{rb}\Delta_{kr}}{\Delta_{kb}} \left(\boldsymbol{\Sigma}_k^{ll} - \boldsymbol{\Sigma}_k^{lv} \left(\boldsymbol{\Sigma}_k^{vv}\right)^{-1} \boldsymbol{\Sigma}_k^{vl} \right).$$

Approximation of $\mathfrak{L}(Y_{k+1} \mid Y_k, V_r, \theta)$ for $k < r - 1$

Let $a \leq k < r - 1$. Appendix B.6 shows that one has approximately

$$\begin{pmatrix} \boldsymbol{Y}_{k+1} \\ \boldsymbol{V}_r \end{pmatrix} \Big| \boldsymbol{Y}_k, \boldsymbol{\theta} \sim \mathcal{N}\left(\begin{pmatrix} \boldsymbol{Y}_k + \boldsymbol{\mu}_k \Delta t_k \\ \boldsymbol{V}_k + \boldsymbol{\mu}_k^v \Delta_{kr} \end{pmatrix}, \begin{pmatrix} \boldsymbol{\Sigma}_k \Delta t_k & \boldsymbol{D}_k' \Delta t_k \\ \boldsymbol{D}_k \Delta t_k & \boldsymbol{\Sigma}_k^{vv} \Delta_{kr} \end{pmatrix} \right), \tag{7.32}$$

where $\Delta t_k = t_{k+1} - t_k$, $\Delta_{kr-} = t_r - t_{k+1}$, $\Delta_{kr} = \Delta_{kr-} + \Delta t_k = t_r - t_k$ and $\boldsymbol{D}_k = (\boldsymbol{\Sigma}_k^{vv}, \boldsymbol{\Sigma}_k^{vl})$. This implies

$$\boldsymbol{Y}_{k+1} \mid \boldsymbol{Y}_k, \boldsymbol{V}_r, \boldsymbol{\theta} \sim \mathcal{N}(\boldsymbol{\rho}_k, \boldsymbol{\Gamma}_k) \tag{7.33}$$

with

$$\boldsymbol{\rho}_k = \begin{pmatrix} \boldsymbol{V}_k + \dfrac{\boldsymbol{V}_r - \boldsymbol{V}_k}{\Delta_{kr}} \Delta t_k \\ \boldsymbol{L}_k + \boldsymbol{\mu}_k^l \Delta t_k + \boldsymbol{\Sigma}_k^{lv} (\boldsymbol{\Sigma}_k^{vv})^{-1} \left(\dfrac{\boldsymbol{V}_r - \boldsymbol{V}_k}{\Delta_{kr}} - \boldsymbol{\mu}_k^v \right) \Delta t_k \end{pmatrix}$$

and

$$\boldsymbol{\Gamma}_k = \begin{pmatrix} \boldsymbol{\Sigma}_k^{vv} \Delta_{kr-} & \boldsymbol{\Sigma}_k^{vl} \Delta_{kr-} \\ \boldsymbol{\Sigma}_k^{lv} \Delta_{kr-} & \boldsymbol{\Sigma}_k^{ll} \Delta_{kr} - \boldsymbol{\Sigma}_k^{lv} (\boldsymbol{\Sigma}_k^{vv})^{-1} \boldsymbol{\Sigma}_k^{vl} \Delta t_k \end{pmatrix} \frac{\Delta t_k}{\Delta_{kr}} .$$

Approximation of $\mathfrak{L}\big(\boldsymbol{Y}_{k+1} \,|\, \boldsymbol{Y}_k, \boldsymbol{V}_r, \boldsymbol{Y}_b, \boldsymbol{\theta}\big)$ for $k < r-1$

The next extension is to further condition (7.33) on $\boldsymbol{Y}_b$ in order not to lose this end point information. As demonstrated in Appendix B.6, the conditional joint distribution of $\boldsymbol{Y}_{k+1}$, $\boldsymbol{V}_r$ and $\boldsymbol{Y}_b$ reads

$$\begin{pmatrix} \boldsymbol{Y}_{k+1} \\ \boldsymbol{V}_r \\ \boldsymbol{Y}_b \end{pmatrix} \Bigg| \boldsymbol{Y}_k, \boldsymbol{\theta} \sim \mathcal{N}\left(\begin{pmatrix} \boldsymbol{Y}_k + \boldsymbol{\mu}_k \Delta t_k \\ \boldsymbol{V}_k + \boldsymbol{\mu}_k^v \Delta_{kr} \\ \boldsymbol{Y}_k + \boldsymbol{\mu}_k \Delta_{kb} \end{pmatrix}, \begin{pmatrix} \boldsymbol{\Sigma}_k \Delta t_k & \boldsymbol{D}_k' \Delta t_k & \boldsymbol{\Sigma}_k \Delta t_k \\ \boldsymbol{D}_k \Delta t_k & \boldsymbol{\Sigma}_k^{vv} \Delta_{kr} & \boldsymbol{D}_k \Delta_{kr} \\ \boldsymbol{\Sigma}_k \Delta t_k & \boldsymbol{D}_k' \Delta_{kr} & \boldsymbol{\Sigma}_k \Delta_{kb} \end{pmatrix} \right) \tag{7.34}$$

with $\Delta_{kb} = t_b - t_k$ and the notation introduced so far. It follows that

$$\boldsymbol{Y}_{k+1} \,|\, \boldsymbol{Y}_k, \boldsymbol{V}_r, \boldsymbol{Y}_b, \boldsymbol{\theta} \sim \mathcal{N}(\boldsymbol{\xi}_k, \boldsymbol{\Psi}_k) \tag{7.35}$$

with

$$\boldsymbol{\xi}_k = \boldsymbol{Y}_k + \boldsymbol{\mu}_k \Delta t_k + (\boldsymbol{D}_k' \Delta t_k, \boldsymbol{\Sigma}_k \Delta t_k) \begin{pmatrix} \boldsymbol{\Sigma}_k^{vv} \Delta_{kr} & \boldsymbol{D}_k \Delta_{kr} \\ \boldsymbol{D}_k' \Delta_{kr} & \boldsymbol{\Sigma}_k \Delta_{kb} \end{pmatrix}^{-1} \begin{pmatrix} \boldsymbol{V}_r - \boldsymbol{V}_k - \boldsymbol{\mu}_k^v \Delta_{kr} \\ \boldsymbol{Y}_b - \boldsymbol{Y}_k - \boldsymbol{\mu}_k \Delta_{kb} \end{pmatrix}$$

and

$$\boldsymbol{\Psi}_k = \boldsymbol{\Sigma}_k \Delta t_k - \big(\boldsymbol{D}_k' \Delta t_k, \boldsymbol{\Sigma}_k \Delta t_k\big) \begin{pmatrix} \boldsymbol{\Sigma}_k^{vv} \Delta_{kr} & \boldsymbol{D}_k \Delta_{kr} \\ \boldsymbol{D}_k' \Delta_{kr} & \boldsymbol{\Sigma}_k \Delta_{kb} \end{pmatrix}^{-1} \begin{pmatrix} \boldsymbol{D}_k \Delta t_k \\ \boldsymbol{\Sigma}_k \Delta t_k \end{pmatrix} .$$

Simulation from this distribution is computationally more elaborate than drawing from (7.33) as (7.35) involves the inversion of larger matrices.

Approximation of $\mathfrak{L}\big(\boldsymbol{L}_0 \,|\, \boldsymbol{V}_0, \boldsymbol{Y}_b, \boldsymbol{\theta}\big)$

Consideration of

$$\boldsymbol{Y}_0 \,|\, \boldsymbol{Y}_b, \boldsymbol{\theta} \sim \mathcal{N}\big(\boldsymbol{Y}_b - \boldsymbol{\mu}_b \Delta_{0b}, \boldsymbol{\Sigma}_b \Delta_{0b}\big) \tag{7.36}$$

with $\Delta_{0b} = t_b - t_0$ immediately yields

$$\boldsymbol{L}_0 \,|\, \boldsymbol{V}_0, \boldsymbol{Y}_b, \boldsymbol{\theta} \sim \mathcal{N}(\boldsymbol{\chi}, \boldsymbol{\Xi}) \tag{7.37}$$

with

$$\boldsymbol{\chi} = \boldsymbol{L}_b - \boldsymbol{\mu}_b^l \Delta_{0b} + \boldsymbol{\Sigma}_b^{lv} (\boldsymbol{\Sigma}_b^{vv})^{-1} \big(\boldsymbol{V}_0 - \boldsymbol{V}_b + \boldsymbol{\mu}_b^v \Delta_{0b}\big)$$

and

$$\boldsymbol{\Xi} = \big(\boldsymbol{\Sigma}_b^{ll} - \boldsymbol{\Sigma}_b^{lv} (\boldsymbol{\Sigma}_b^{vv})^{-1} \boldsymbol{\Sigma}_b^{vl}\big) \Delta_{0b}.$$

Approximation of $\mathfrak{L}\big(\boldsymbol{L}_K \,|\, \boldsymbol{Y}_{K-1}, \boldsymbol{V}_K, \boldsymbol{\theta}\big)$

Analogously,

$$\boldsymbol{Y}_K \,|\, \boldsymbol{Y}_{K-1}, \boldsymbol{\theta} \sim \mathcal{N}\big(\boldsymbol{Y}_{K-1} + \boldsymbol{\mu}_{K-1} \Delta t_{K-1}, \boldsymbol{\Sigma}_{K-1} \Delta t_{K-1}\big) \tag{7.38}$$

implies

$$\boldsymbol{L}_K \,|\, \boldsymbol{Y}_{K-1}, \boldsymbol{V}_K, \boldsymbol{\theta} \sim \mathcal{N}(\boldsymbol{\kappa}, \boldsymbol{\Pi}) \tag{7.39}$$

with

$$\boldsymbol{\kappa} = \boldsymbol{L}_{K-1} + \boldsymbol{\mu}_{K-1}^l \Delta t_{K-1} + \boldsymbol{\Sigma}_{K-1}^{lv} (\boldsymbol{\Sigma}_{K-1}^{vv})^{-1} \big(\boldsymbol{V}_K - \boldsymbol{V}_{K-1} - \boldsymbol{\mu}_{K-1}^v \Delta t_{K-1}\big)$$

and

$$\boldsymbol{\Pi} = \big(\boldsymbol{\Sigma}_{K-1}^{ll} - \boldsymbol{\Sigma}_{K-1}^{lv} (\boldsymbol{\Sigma}_{K-1}^{vv})^{-1} \boldsymbol{\Sigma}_{K-1}^{vl}\big) \Delta t_{K-1}.$$

Conclusion

Now that all required proposal distributions are at hand, we are able to write down the adapted acceptance probability for the proposed imputed data opposed to the current imputed data. We do this for the more complicated update strategy (2b) on p. 218. The acceptance probability for strategy (2a) can then easily be obtained.

First assume $|\{a+1, \ldots, b-1\} \cap \mathfrak{O}| = 1$ and $a \neq 0$, $b \neq K$, i.e. there is one observation time $t_r \in (t_a, t_b)$. Then the acceptance probability reads

$$\begin{aligned}
&\zeta\big(\{\boldsymbol{Y}_{(a,r)}^{\text{imp}*}, \boldsymbol{L}_r^*, \boldsymbol{Y}_{(r,b)}^{\text{imp}*}\}, \{\boldsymbol{Y}_{(a,r)}^{\text{imp}}, \boldsymbol{L}_r, \boldsymbol{Y}_{(r,b)}^{\text{imp}}\}\big) \\
&\quad = 1 \wedge \frac{\pi\big(\boldsymbol{Y}_{(a,r)}^{\text{imp}*}, \boldsymbol{L}_r^*, \boldsymbol{Y}_{(r,b)}^{\text{imp}*} \,|\, \boldsymbol{Y}_a, \boldsymbol{V}_r, \boldsymbol{Y}_b, \boldsymbol{\theta}\big)}{\pi\big(\boldsymbol{Y}_{(a,r)}^{\text{imp}}, \boldsymbol{L}_r, \boldsymbol{Y}_{(r,b)}^{\text{imp}} \,|\, \boldsymbol{Y}_a, \boldsymbol{V}_r, \boldsymbol{Y}_b, \boldsymbol{\theta}\big)} \cdot \frac{q\big(\boldsymbol{Y}_{(a,r)}^{\text{imp}}, \boldsymbol{L}_r, \boldsymbol{Y}_{(r,b)}^{\text{imp}} \,|\, \boldsymbol{Y}_a, \boldsymbol{V}_r, \boldsymbol{Y}_b, \boldsymbol{\theta}\big)}{q\big(\boldsymbol{Y}_{(a,r)}^{\text{imp}*}, \boldsymbol{L}_r^*, \boldsymbol{Y}_{(r,b)}^{\text{imp}*} \,|\, \boldsymbol{Y}_a, \boldsymbol{V}_r, \boldsymbol{Y}_b, \boldsymbol{\theta}\big)}.
\end{aligned}$$

The components of this acceptance probability are

$$\pi\big(\boldsymbol{Y}_{(a,r)}^{\text{imp}*}, \boldsymbol{L}_r^*, \boldsymbol{Y}_{(r,b)}^{\text{imp}*} \,|\, \boldsymbol{Y}_a, \boldsymbol{V}_r, \boldsymbol{Y}_b, \boldsymbol{\theta}\big) \propto \prod_{k=a}^{b-1} \pi^{\text{Euler}}\big(\boldsymbol{Y}_{k+1}^* \,|\, \boldsymbol{Y}_k^*, \boldsymbol{\theta}\big)$$

and

$$q(\boldsymbol{Y}^{\text{imp}*}_{(a,r)}, \boldsymbol{L}^*_r, \boldsymbol{Y}^{\text{imp}*}_{(r,b)} \mid \boldsymbol{Y}_a, \boldsymbol{V}_r, \boldsymbol{Y}_b, \boldsymbol{\theta})$$
$$= \left(\prod_{k=a}^{r-2} q(\boldsymbol{Y}^*_{k+1} | \boldsymbol{Y}^*_k, \boldsymbol{V}_r, \boldsymbol{Y}_b, \boldsymbol{\theta})\right) \cdot q(\boldsymbol{L}^*_r | \boldsymbol{Y}^*_{r-1}, \boldsymbol{V}_r, \boldsymbol{Y}_b, \boldsymbol{\theta}) \cdot \left(\prod_{k=r}^{b-2} q(\boldsymbol{Y}^*_{k+1} \mid \boldsymbol{Y}^*_k, \boldsymbol{Y}_b, \boldsymbol{\theta})\right),$$

where $\boldsymbol{Y}^*_a = \boldsymbol{Y}_a$, $\boldsymbol{Y}^*_b = \boldsymbol{Y}_b$ and $\boldsymbol{Y}^*_r = (\boldsymbol{V}'_r, \boldsymbol{L}^{*\prime}_r)'$. For $|\{a+1,\ldots,b-1\} \cap \mathfrak{O}| = 0$ and $a = 0$, $b \neq K$, the acceptance probability is

$$\zeta\big(\{\boldsymbol{L}^*_0, \boldsymbol{Y}^{\text{imp}*}_{(0,b)}\}, \{\boldsymbol{L}_0, \boldsymbol{Y}^{\text{imp}}_{(0,b)}\}\big)$$
$$= 1 \wedge \frac{\pi\big(\boldsymbol{L}^*_0, \boldsymbol{Y}^{\text{imp}*}_{(0,b)} \mid \boldsymbol{V}_0, \boldsymbol{Y}_b, \boldsymbol{\theta}\big)}{\pi\big(\boldsymbol{L}_0, \boldsymbol{Y}^{\text{imp}}_{(0,b)} \mid \boldsymbol{V}_0, \boldsymbol{Y}_b, \boldsymbol{\theta}\big)} \cdot \frac{q\big(\boldsymbol{L}_0, \boldsymbol{Y}^{\text{imp}}_{(0,b)} \mid \boldsymbol{V}_0, \boldsymbol{Y}_b, \boldsymbol{\theta}\big)}{q\big(\boldsymbol{L}^*_0, \boldsymbol{Y}^{\text{imp}*}_{(0,b)} \mid \boldsymbol{V}_0, \boldsymbol{Y}_b, \boldsymbol{\theta}\big)}$$

with

$$\pi\big(\boldsymbol{L}^*_0, \boldsymbol{Y}^{\text{imp}*}_{(0,b)} \mid \boldsymbol{V}_0, \boldsymbol{Y}_b, \boldsymbol{\theta}\big) \propto \left(\prod_{k=a}^{b-1} \pi^{\text{Euler}}\big(\boldsymbol{Y}^*_{k+1} \mid \boldsymbol{Y}^*_k, \boldsymbol{\theta}\big)\right) \pi\big(\boldsymbol{Y}^*_0 \mid \boldsymbol{\theta}\big)$$

and

$$q\big(\boldsymbol{L}^*_0, \boldsymbol{Y}^{\text{imp}*}_{(0,b)} \mid \boldsymbol{V}_0, \boldsymbol{Y}_b, \boldsymbol{\theta}\big) = q\big(\boldsymbol{L}^*_0 \mid \boldsymbol{V}_0, \boldsymbol{Y}_b, \boldsymbol{\theta}\big) \left(\prod_{k=0}^{b-2} q\big(\boldsymbol{Y}^*_{k+1} \mid \boldsymbol{Y}^*_k, \boldsymbol{Y}_b, \boldsymbol{\theta}\big)\right),$$

where $\boldsymbol{Y}^*_0 = (\boldsymbol{V}'_0, \boldsymbol{L}^{*\prime}_0)'$ and $\pi(\boldsymbol{Y}^*_0|\boldsymbol{\theta}) \propto \pi(\boldsymbol{L}^*_0|\boldsymbol{V}_0, \boldsymbol{\theta})$ is some model-specific density for the initial value. Similarly, for $|\{a+1,\ldots,b-1\} \cap \mathfrak{O}| = 0$ and $a \neq 0$, $b = K$ one has

$$\zeta\big(\{\boldsymbol{Y}^{\text{imp}*}_{(a,K)}, \boldsymbol{L}^*_K\}, \{\boldsymbol{Y}^{\text{imp}}_{(a,K)}, \boldsymbol{L}_K\}\big)$$
$$= 1 \wedge \frac{\pi\big(\boldsymbol{Y}^{\text{imp}*}_{(a,K)}, \boldsymbol{L}^*_K \mid \boldsymbol{Y}_a, \boldsymbol{V}_K, \boldsymbol{\theta}\big)}{\pi\big(\boldsymbol{Y}^{\text{imp}}_{(a,K)}, \boldsymbol{L}_K \mid \boldsymbol{Y}_a, \boldsymbol{V}_K, \boldsymbol{\theta}\big)} \cdot \frac{q\big(\boldsymbol{Y}^{\text{imp}}_{(a,K)}, \boldsymbol{L}_K \mid \boldsymbol{Y}_a, \boldsymbol{V}_K, \boldsymbol{\theta}\big)}{q\big(\boldsymbol{Y}^{\text{imp}*}_{(a,K)}, \boldsymbol{L}^*_K \mid \boldsymbol{Y}_a, \boldsymbol{V}_K, \boldsymbol{\theta}\big)}$$

with

$$\pi\big(\boldsymbol{Y}^{\text{imp}*}_{(a,K)}, \boldsymbol{L}^*_K \mid \boldsymbol{Y}_a, \boldsymbol{V}_K, \boldsymbol{\theta}\big) \propto \prod_{k=a}^{K-1} \pi^{\text{Euler}}\big(\boldsymbol{Y}^*_{k+1} \mid \boldsymbol{Y}^*_k, \boldsymbol{\theta}\big)$$

and

$$q\big(\boldsymbol{Y}^{\text{imp}*}_{(a,K)}, \boldsymbol{L}^*_K \mid \boldsymbol{Y}_a, \boldsymbol{V}_K, \boldsymbol{\theta}\big) = \left(\prod_{k=a}^{K-2} q\big(\boldsymbol{Y}^*_{k+1} \mid \boldsymbol{Y}^*_k, \boldsymbol{V}_K, \boldsymbol{\theta}\big)\right) q\big(\boldsymbol{L}^*_K \mid \boldsymbol{Y}^*_{K-1}, \boldsymbol{V}_K, \boldsymbol{\theta}\big)$$

with $\boldsymbol{Y}^*_a = \boldsymbol{Y}_a$. This concludes the extension of the MCMC scheme (7.2) to a latent data framework.

7.2.2 Observation with Error

Another issue that is of importance in practice is that observations are often measured with error, i.e. one has for all t_k

$$\boldsymbol{v}_k = \boldsymbol{V}_k + \boldsymbol{\varepsilon}_k, \qquad \boldsymbol{\varepsilon}_k \sim \mathcal{N}(\boldsymbol{0}, \boldsymbol{\Upsilon}_k), \tag{7.40}$$

where $\boldsymbol{V}_k \in \mathbb{R}^{d_1}$ is the observable part of $\boldsymbol{X}_{t_k}$, $\boldsymbol{v}_k$ is the measurement of $\boldsymbol{V}_k$, and $\boldsymbol{\varepsilon}_k$ is the observation error with mean zero and positive definite covariance matrix $\boldsymbol{\Upsilon}_k$. Observation errors are considered independent for unequal observation times. This is a setting that also underlies Kalman filters (e.g. Maybeck 1979). The $\boldsymbol{\Upsilon}_k$ are either assumed known from empirical data, or their estimation is included in the inference procedure for the diffusion path and the model parameter. In the latter case, let $\boldsymbol{\theta}$ stand for the collection of all parameters to estimate including the $\boldsymbol{\Upsilon}_k$. This section adapts the MCMC scheme from Sect. 7.2.1 to also handle observation errors in addition to latent data.

As before, suppose there are $K+1$ observation and auxiliary times $t_0 < t_1 < \ldots < t_K$, and let $\mathfrak{O} = \{k \in \{0,\ldots,K\} \,|\, t_k \in \{\tau_0,\ldots,\tau_M\}\}$ be the set of indices of observation times. As observations are assumed to be subject to measurement error, the vectors $\boldsymbol{V}_k$ have to be updated also for $k \in \mathfrak{O}$ now.

The posterior density of the entire diffusion path $\{\boldsymbol{Y}_k\}_{k-0,\ldots,K}$ conditional on the observations $\{\boldsymbol{v}_k\}_{k\in\mathfrak{O}}$ and the parameter $\boldsymbol{\theta}$ then equals

$$\begin{aligned}
&\pi\left(\{\boldsymbol{Y}_k\}_{k=0,\ldots,K} \,\middle|\, \{\boldsymbol{v}_k\}_{k\in\mathfrak{O}}, \boldsymbol{\theta}\right) \\
&\quad \propto \pi\big(\{\boldsymbol{v}_k\}_{k\in\mathfrak{O}} \,\big|\, \{\boldsymbol{Y}_k\}_{k=0,\ldots,K}, \boldsymbol{\theta}\big)\pi\big(\{\boldsymbol{Y}_k\}_{k=0,\ldots,K} \,\big|\, \boldsymbol{\theta}\big) \\
&\quad = \left(\prod_{k\in\mathfrak{O}} \phi\big(\boldsymbol{v}_k \,\big|\, \boldsymbol{V}_k, \boldsymbol{\Upsilon}_k\big)\right)\left(\prod_{k=0}^{K-1} \pi\big(\boldsymbol{Y}_{k+1} \,\big|\, \boldsymbol{Y}_k, \boldsymbol{\theta}\big)\right)\pi\big(\boldsymbol{Y}_0 \,\big|\, \boldsymbol{\theta}\big).
\end{aligned}$$

The posterior density of the parameter $\boldsymbol{\theta}$ conditioned on both the estimated and observed path is

$$\begin{aligned}
&\pi\big(\boldsymbol{\theta} \,\big|\, \{\boldsymbol{Y}_k\}_{k=0,\ldots,K}, \{\boldsymbol{v}_k\}_{k\in\mathfrak{O}}\big) \\
&\quad \propto \left(\prod_{k\in\mathfrak{O}} \phi\big(\boldsymbol{v}_k \,\big|\, \boldsymbol{V}_k, \boldsymbol{\Upsilon}_k\big)\right)\left(\prod_{k=0}^{K-1} \pi\big(\boldsymbol{Y}_{k+1} \,\big|\, \boldsymbol{Y}_k, \boldsymbol{\theta}\big)\right)\pi\big(\boldsymbol{Y}_0 \,\big|\, \boldsymbol{\theta}\big)p(\boldsymbol{\theta}).
\end{aligned}$$

The path proposal distributions from Sect. 7.2.1 have to be adjusted to the new setting. In particular, the observable parts $\boldsymbol{V}_k$ need to be updated for $k \in \mathfrak{O}$ in consideration of (7.40). The new path update algorithm is as follows.

Choose an update interval (t_a, t_b) such that $|\{a+1,\ldots,b-1\} \cap \mathfrak{O}| \leq 1$, i.e. there is not more than one observation time in the interior of (t_a, t_b). The following situations may occur:

1. One has $|\{a+1,\ldots,b-1\} \cap \mathfrak{O}| = 0$, i.e. there is no observation time in the interior of (t_a, t_b). In this case path proposals are again obtained as in Sect. 7.1.2.

2. One has $|\{a+1,\dots,b-1\} \cap \mathfrak{O}| = 1$, i.e. there is exactly one observation time t_r in the interior of (t_a, t_b). The two strategies from Sect. 7.2.1 now read as follows.

   a. Propose $\boldsymbol{Y}_r^* \,|\, \boldsymbol{Y}_a, \boldsymbol{v}_r, \boldsymbol{Y}_b, \boldsymbol{\theta}$ as in (7.41) below. Then, generate two conditionally independent proposals on (t_a, t_r) and (t_r, t_b) conditioned on $\boldsymbol{Y}_a, \boldsymbol{Y}_r^*, \boldsymbol{Y}_b, \boldsymbol{\theta}$ as in item 1.
   b. Update the path segment from the left to the right. More precisely:

      - For $k = a, \dots, r-2$, propose $\boldsymbol{Y}_{k+1}^* \,|\, \boldsymbol{Y}_k^*, \boldsymbol{v}_r, \boldsymbol{Y}_b, \boldsymbol{\theta}$ as in (7.43) below, where $\boldsymbol{Y}_a^* = \boldsymbol{Y}_a$. Alternatively, propose $\boldsymbol{Y}_{k+1}^* \,|\, \boldsymbol{Y}_k^*, \boldsymbol{v}_r, \boldsymbol{\theta}$ as in (7.42).
      - Propose $\boldsymbol{Y}_r^* \,|\, \boldsymbol{Y}_{r-1}^*, \boldsymbol{v}_r, \boldsymbol{Y}_b, \boldsymbol{\theta}$ as in (7.41) below.
      - For $k = r, \dots, b-2$, propose $\boldsymbol{Y}_{k+1}^* \,|\, \boldsymbol{Y}_k^*, \boldsymbol{Y}_b, \boldsymbol{\theta}$ as in item 1.

The special cases $a = 0$ and $b = K$ involve drawing from $\mathfrak{L}(\boldsymbol{Y}_0 | \boldsymbol{v}_0, \boldsymbol{Y}_b, \boldsymbol{\theta})$, $\mathfrak{L}(\boldsymbol{Y}_K | \boldsymbol{Y}_{K-1}, \boldsymbol{v}_K, \boldsymbol{\theta})$ and $\mathfrak{L}(\boldsymbol{Y}_{k+1} | \boldsymbol{Y}_k, \boldsymbol{v}_K, \boldsymbol{\theta})$ under the assumption that $|\{a+1,\dots,b-1\} \cap \mathfrak{O}| = 0$. The first two distributions are provided in (7.44) and (7.45). The third one corresponds to (7.42) with r replaced by K.

The following shows the required approximate proposal distributions. The notation is adopted from Sect. 7.2.1.

Approximation of $\mathfrak{L}(\boldsymbol{Y}_r \mid \boldsymbol{Y}_k, \boldsymbol{v}_r, \boldsymbol{Y}_b, \boldsymbol{\theta})$ for $k < r$

With (7.30), one obtains

$$\pi\left(\boldsymbol{Y}_r \,\middle|\, \boldsymbol{Y}_k, \boldsymbol{v}_r, \boldsymbol{Y}_b, \boldsymbol{\theta}\right) \propto \phi\left(\boldsymbol{v}_r \,\middle|\, \boldsymbol{V}_r, \boldsymbol{\Upsilon}_r\right) \phi\left(\boldsymbol{Y}_r \,\middle|\, \boldsymbol{Y}_k + \frac{\boldsymbol{Y}_b - \boldsymbol{Y}_k}{\Delta_{kb}} \Delta_{kr}, \frac{\Delta_{rb}\Delta_{kr}}{\Delta_{kb}} \boldsymbol{\Sigma}_k\right),$$

which results in

$$\boldsymbol{Y}_r \,|\, \boldsymbol{Y}_k, \boldsymbol{v}_r, \boldsymbol{Y}_b, \boldsymbol{\theta} \sim \mathcal{N}(\tilde{\boldsymbol{\eta}}_k, \tilde{\boldsymbol{\Lambda}}_k) \tag{7.41}$$

with

$$\tilde{\boldsymbol{\eta}}_k = \tilde{\boldsymbol{\Lambda}}_k \left(\begin{pmatrix} \boldsymbol{\Upsilon}_r^{-1} \boldsymbol{v}_r \\ \boldsymbol{0} \end{pmatrix} + \frac{\Delta_{kb}}{\Delta_{rb}\Delta_{kr}} \boldsymbol{\Sigma}_k^{-1} \left(\boldsymbol{Y}_k + \frac{\boldsymbol{Y}_b - \boldsymbol{Y}_k}{\Delta_{kb}} \Delta_{kr} \right) \right)$$

and

$$\tilde{\boldsymbol{\Lambda}}_k = \left(\begin{pmatrix} \boldsymbol{\Upsilon}_r^{-1} & \boldsymbol{0} \\ \boldsymbol{0} & \boldsymbol{0} \end{pmatrix} + \frac{\Delta_{kb}}{\Delta_{rb}\Delta_{kr}} \boldsymbol{\Sigma}_k^{-1} \right)^{-1}.$$

Approximation of $\mathfrak{L}(\boldsymbol{Y}_{k+1} \mid \boldsymbol{Y}_k, \boldsymbol{v}_r, \boldsymbol{\theta})$ for $k < r - 1$

Use (7.32) and $\boldsymbol{v}_r = \boldsymbol{V}_r + \boldsymbol{\varepsilon}_r$ to obtain

$$\begin{pmatrix} \boldsymbol{Y}_{k+1} \\ \boldsymbol{v}_r \end{pmatrix} \Bigg| \boldsymbol{Y}_k, \boldsymbol{\theta} \sim \mathcal{N}\left(\begin{pmatrix} \boldsymbol{Y}_k + \boldsymbol{\mu}_k \Delta t_k \\ \boldsymbol{V}_k + \boldsymbol{\mu}_k^v \Delta_{kr} \end{pmatrix}, \begin{pmatrix} \boldsymbol{\Sigma}_k \Delta t_k & \boldsymbol{D}_k' \Delta t_k \\ \boldsymbol{D}_k \Delta t_k & \boldsymbol{\Sigma}_k^{vv} \Delta_{kr} + \boldsymbol{\Upsilon}_r \end{pmatrix} \right).$$

That yields

$$\boldsymbol{Y}_{k+1} \,|\, \boldsymbol{Y}_k, \boldsymbol{v}_r, \boldsymbol{\theta} \sim \mathcal{N}(\tilde{\boldsymbol{\rho}}_k, \tilde{\boldsymbol{\Gamma}}_k) \tag{7.42}$$

with

$$\tilde{\boldsymbol{\rho}}_k = \boldsymbol{Y}_k + \boldsymbol{\mu}_k \Delta t_k + \boldsymbol{D}_k' \big(\boldsymbol{\Sigma}_k^{vv} \Delta_{kr} + \boldsymbol{\Upsilon}_r\big)^{-1} \big(\boldsymbol{v}_r - \boldsymbol{V}_k - \boldsymbol{\mu}_k^v \Delta_{kr}\big) \Delta t_k$$

and

$$\tilde{\boldsymbol{\Gamma}}_k = \Big(\boldsymbol{\Sigma}_k - \boldsymbol{D}_k' \big(\boldsymbol{\Sigma}_k^{vv} \Delta_{kr} + \boldsymbol{\Upsilon}_r\big)^{-1} \boldsymbol{D}_k \Delta t_k\Big) \Delta t_k \,.$$

This formula corresponds to (7.33) in the case of no observation error.

Approximation of $\mathfrak{L}\big(Y_{k+1} \,|\, Y_k, v_r, Y_b, \theta\big)$ for $k < r - 1$

Analogously, with (7.34),

$$\begin{pmatrix} \boldsymbol{Y}_{k+1} \\ \boldsymbol{v}_r \\ \boldsymbol{Y}_b \end{pmatrix} \Bigg|\, \boldsymbol{Y}_k, \boldsymbol{\theta} \sim \mathcal{N}\left(\begin{pmatrix} \boldsymbol{Y}_k + \boldsymbol{\mu}_k \Delta t_k \\ \boldsymbol{V}_k + \boldsymbol{\mu}_k^v \Delta_{kr} \\ \boldsymbol{Y}_k + \boldsymbol{\mu}_k \Delta_{kb} \end{pmatrix}, \begin{pmatrix} \boldsymbol{\Sigma}_k \Delta t_k & \boldsymbol{D}_k' \Delta t_k & \boldsymbol{\Sigma}_k \Delta t_k \\ \boldsymbol{D}_k \Delta t_k & \boldsymbol{\Sigma}_k^{vv} \Delta_{kr} + \boldsymbol{\Upsilon}_r & \boldsymbol{D}_k \Delta_{kr} \\ \boldsymbol{\Sigma}_k \Delta t_k & \boldsymbol{D}_k' \Delta_{kr} & \boldsymbol{\Sigma}_k \Delta_{kb} \end{pmatrix} \right),$$

and hence

$$\boldsymbol{Y}_{k+1} \,|\, \boldsymbol{Y}_k, \boldsymbol{v}_r, \boldsymbol{Y}_b, \boldsymbol{\theta} \sim \mathcal{N}(\tilde{\boldsymbol{\xi}}_k, \tilde{\boldsymbol{\Psi}}_k) \tag{7.43}$$

with

$$\tilde{\boldsymbol{\xi}}_k = \boldsymbol{Y}_k + \boldsymbol{\mu}_k \Delta t_k + (\boldsymbol{D}_k', \boldsymbol{\Sigma}_k) \begin{pmatrix} \boldsymbol{\Sigma}_k^{vv} \Delta_{kr} + \boldsymbol{\Upsilon}_r & \boldsymbol{D}_k \Delta_{kr} \\ \boldsymbol{D}_k' \Delta_{kr} & \boldsymbol{\Sigma}_k \Delta_{kb} \end{pmatrix}^{-1} \begin{pmatrix} \boldsymbol{v}_r - \boldsymbol{V}_k - \boldsymbol{\mu}_k^v \Delta_{kr} \\ \boldsymbol{Y}_b - \boldsymbol{Y}_k - \boldsymbol{\mu}_k \Delta_{kb} \end{pmatrix} \Delta t_k$$

and

$$\tilde{\boldsymbol{\Psi}}_k = \left(\boldsymbol{\Sigma}_k - (\boldsymbol{D}_k', \boldsymbol{\Sigma}_k) \begin{pmatrix} \boldsymbol{\Sigma}_k^{vv} \Delta_{kr} + \boldsymbol{\Upsilon}_r & \boldsymbol{D}_k \Delta_{kr} \\ \boldsymbol{D}_k' \Delta_{kr} & \boldsymbol{\Sigma}_k \Delta_{kb} \end{pmatrix}^{-1} \begin{pmatrix} \boldsymbol{D}_k \\ \boldsymbol{\Sigma}_k \end{pmatrix} \Delta t_k \right) \Delta t_k.$$

Approximation of $\mathfrak{L}\big(Y_0 \,|\, v_0, Y_b, \theta\big)$

Equation (7.36) implies

$$\pi\big(\boldsymbol{Y}_0 \,\big|\, \boldsymbol{v}_0, \boldsymbol{Y}_b, \boldsymbol{\theta}\big) \propto \phi\big(\boldsymbol{v}_0 \,\big|\, \boldsymbol{V}_0, \boldsymbol{\Upsilon}_0\big) \phi\Big(\boldsymbol{Y}_0 \,\Big|\, \boldsymbol{Y}_b - \boldsymbol{\mu}_b \Delta_{0b}\,,\, \boldsymbol{\Sigma}_b \Delta_{0b}\Big),$$

which leads to

$$\boldsymbol{Y}_0 \,|\, \boldsymbol{v}_0, \boldsymbol{Y}_b, \boldsymbol{\theta} \sim \mathcal{N}(\tilde{\boldsymbol{\chi}}, \tilde{\boldsymbol{\Xi}}) \tag{7.44}$$

with

$$\tilde{\chi} = \tilde{\Xi}\left(\begin{pmatrix}\boldsymbol{\Upsilon}_0^{-1}\boldsymbol{v}_0\\ \mathbf{0}\end{pmatrix} + \boldsymbol{\Sigma}_b^{-1}\,\frac{\boldsymbol{Y}_b - \boldsymbol{\mu}_b\Delta_{0b}}{\Delta_{0b}}\right)$$

and

$$\tilde{\Xi} = \left(\begin{pmatrix}\boldsymbol{\Upsilon}_0^{-1} & \mathbf{0}\\ \mathbf{0} & \mathbf{0}\end{pmatrix} + \frac{\boldsymbol{\Sigma}_b^{-1}}{\Delta_{0b}}\right)^{-1}.$$

Approximation of $\mathfrak{L}(\boldsymbol{Y}_K \mid \boldsymbol{Y}_{K-1}, \boldsymbol{v}_K, \boldsymbol{\theta})$

Finally, with (7.38), one gets

$$\boldsymbol{Y}_K \mid \boldsymbol{Y}_{K-1}, \boldsymbol{v}_K, \boldsymbol{\theta} \sim \mathcal{N}(\tilde{\boldsymbol{\kappa}}, \tilde{\boldsymbol{\Pi}}) \tag{7.45}$$

with

$$\tilde{\boldsymbol{\kappa}} = \tilde{\boldsymbol{\Pi}}\left(\begin{pmatrix}\boldsymbol{\Upsilon}_K^{-1}\boldsymbol{v}_K\\ \mathbf{0}\end{pmatrix} + \boldsymbol{\Sigma}_{K-1}^{-1}\,\frac{\boldsymbol{Y}_{K-1} + \boldsymbol{\mu}_{K-1}\Delta t_{K-1}}{\Delta t_{K-1}}\right)$$

and

$$\tilde{\boldsymbol{\Pi}} = \left(\begin{pmatrix}\boldsymbol{\Upsilon}_K^{-1} & \mathbf{0}\\ \mathbf{0} & \mathbf{0}\end{pmatrix} + \frac{\boldsymbol{\Sigma}_{K-1}^{-1}}{\Delta t_{K-1}}\right)^{-1}.$$

Conclusion

Altogether, the path proposal density for partial observations with error according to the update strategy (2b) on p. 226 equals for $a \neq 0$, $b \neq K$ and $|\{a+1,\ldots,b-1\} \cap \mathfrak{O}| = 1$

$$q(\boldsymbol{Y}_{(a,b)}^{\text{imp}*} \mid \boldsymbol{Y}_a, \boldsymbol{v}_r, \boldsymbol{Y}_b, \boldsymbol{\theta}) = \left(\prod_{k=a}^{r-2} q(\boldsymbol{Y}_{k+1}^* \mid \boldsymbol{Y}_k^*, \boldsymbol{v}_r, \boldsymbol{Y}_b, \boldsymbol{\theta})\right)$$

$$\cdot\, q(\boldsymbol{Y}_r^* \mid \boldsymbol{Y}_{r-1}^*, \boldsymbol{v}_r, \boldsymbol{Y}_b, \boldsymbol{\theta}) \cdot \left(\prod_{k=r}^{b-2} q(\boldsymbol{Y}_{k+1}^* \mid \boldsymbol{Y}_k^*, \boldsymbol{Y}_b, \boldsymbol{\theta})\right),$$

where $\boldsymbol{Y}_a^* = \boldsymbol{Y}_a$. For $a = 0$, $b \neq K$ and $|\{a+1,\ldots,b-1\} \cap \mathfrak{O}| = 0$, the proposal density is

$$q(\boldsymbol{Y}_{[0,b)}^{\text{imp}*} \mid \boldsymbol{v}_0, \boldsymbol{Y}_b, \boldsymbol{\theta}) = q(\boldsymbol{Y}_0^* \mid \boldsymbol{v}_0, \boldsymbol{Y}_b, \boldsymbol{\theta})\left(\prod_{k=0}^{b-2} q(\boldsymbol{Y}_{k+1}^* \mid \boldsymbol{Y}_k^*, \boldsymbol{Y}_b, \boldsymbol{\theta})\right).$$

Similarly, for $b = K$, $a \neq 0$ and $|\{a+1, \ldots, b-1\} \cap \mathfrak{O}| = 0$, one has

$$q\big(\boldsymbol{Y}^{\text{imp}*}_{(a,K]} \,\big|\, \boldsymbol{Y}_a, \boldsymbol{v}_K, \boldsymbol{\theta}\big) = \left(\prod_{k=a}^{K-2} q\big(\boldsymbol{Y}^*_{k+1} \,\big|\, \boldsymbol{Y}^*_k, \boldsymbol{v}_K, \boldsymbol{\theta}\big)\right) q\big(\boldsymbol{Y}^*_K \,\big|\, \boldsymbol{Y}^*_{K-1}, \boldsymbol{v}_K, \boldsymbol{\theta}\big)$$

with $\boldsymbol{Y}^*_a = \boldsymbol{Y}_a$.

This concludes the extension of the MCMC concepts from Sect. 7.1 to latent data and observation with error.

7.3 Convergence Problems

Now return to the simulation study from Sect. 7.1.7, i.e. consider again the situation where discrete-time observations are complete and without measurement error. In these experiments, relatively low amounts of auxiliary data were imputed; time intervals between every two observations were divided into $m = 2$ and $m = 10$ subintervals only. Standard computers can easily deal with much higher numbers. Figure 7.24 thus shows the trace plots of β and σ^2 for $M = 25$ observations and $m \in \{10, 100, 1000\}$. This time, only the modified bridge proposal for the imputed data and the random walk proposal for the parameter are applied as these turned out to perform best in the discussion in Sect. 7.1.8. By increasing the number m of subintervals, one hopes to further improve the results for $m = 10$ in Figs. 7.12 and 7.13.

However, Fig. 7.24 shows astonishing behaviour of the MCMC output: Instead of reducing the estimation bias and delivering steadily improving parameter estimates, mixing of the Markov chain for σ^2 becomes substantially worse as m increases. Acceptance rates for $\boldsymbol{\theta}$ decrease from 16 % ($m = 10$) to 5 % ($m = 100$) and 2 % ($m = 1{,}000$). Inference for β, on the other hand, appears relatively unaffected.

This section aims to investigate the above phenomenon. Its understanding is essential for the remaining chapter. It was first analysed by Roberts and Stramer (2001) and has been addressed by a number of researchers since then. The outcomes of this section are the basis for Sect. 7.4 which provides improvements of the MCMC procedure investigated so far.

Without loss of generality, consider a time-homogeneous diffusion process $\boldsymbol{X}$ on a time interval $[0, T]$, where the initial value $\boldsymbol{X}_0 = \boldsymbol{x}_0$ is known and the final value $\boldsymbol{X}_T = \boldsymbol{x}$ is completely observed, i.e. $\boldsymbol{X}^{\text{obs}} = \{\boldsymbol{x}_0, \boldsymbol{x}\}$. The remaining path segment $\boldsymbol{X}_{(0,T)} = (\boldsymbol{X}_t)_{t \in (0,T)}$ is unobserved. Section 7.1.1 explains why the restriction to this setting is sufficient and can easily be generalised to more observations.

In order to get to the bottom of the problem of this section, assume that the missing path segment can be imputed continuously. That means, $\boldsymbol{X}^{\text{imp}} = \boldsymbol{X}_{(0,T)}$ is an infinite-dimensional object rather than a countable collection of discrete data points as considered in the previous sections. One will not face this situation in practice; however, it corresponds to increasing the number m of subintervals of $[0, T]$ to infinity in the discrete framework.

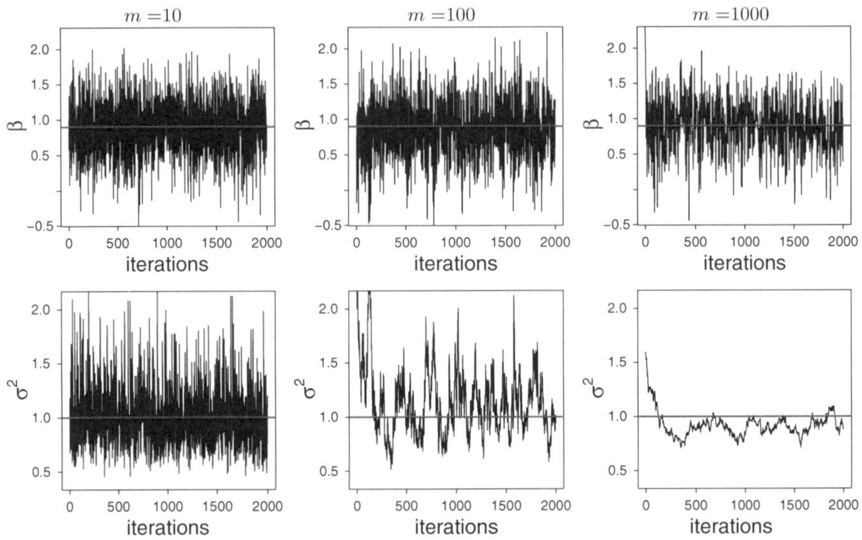

Fig. 7.24 Estimation of parameters of the Ornstein-Uhlenbeck process (7.20) as described on pp. 196–197. The path and parameter are updated via the modified bridge and random walk proposals. The MCMC scheme conditions on data points at times $0, 1, \ldots, 25$ (i.e. $M = 25$) which are observed without error and introduces $m \in \{10, 100, 1{,}000\}$ subintervals in between every two observations. This figure shows the trace plots of β and σ^2. The Markov chains have length 10^5 but have been thinned by factor 50. The true values $\beta = 0.9$ and $\sigma^2 = 1$ are indicated by the *red horizontal lines*. Update intervals are sampled with Algorithm 7.3

The key to the explanation for the diminishing convergence as m grows larger is the quadratic variation identity for diffusion processes,

$$\langle \boldsymbol{X}, \boldsymbol{X} \rangle_{[0,T]} = \lim_{\delta(\mathcal{Z}_n)\downarrow 0} \sum_{i=1}^{h(n)} \left(\boldsymbol{X}_{t_i^{(n)}} - \boldsymbol{X}_{t_{i-1}^{(n)}}\right)\left(\boldsymbol{X}_{t_i^{(n)}} - \boldsymbol{X}_{t_{i-1}^{(n)}}\right)' = \int_0^T \boldsymbol{\Sigma}(\boldsymbol{X}_\tau, \boldsymbol{\theta})\mathrm{d}\tau, \tag{7.46}$$

which was introduced in Sect. 3.2.6. In this equation, $\delta(\mathcal{Z}_n)$ denotes the fineness of a partition $\mathcal{Z}_n = (0 = t_0^{(n)} < t_1^{(n)} < \ldots < t_{h(n)}^{(n)} = T)$ of $[0, T]$ into $h(n)$ subintervals for arbitrary n and h. Equation (7.46) holds in probability and, for sufficiently smooth $\boldsymbol{\Sigma}$, almost surely. Papaspiliopoulos et al. (2003) label very similar properties which tie the data and the parameters an *ergodicity constraint*.

Roberts and Stramer (2001) point out that this equality implies that the quadratic variation $\langle \boldsymbol{X}, \boldsymbol{X} \rangle_{[0,T]}$ of the path and the diffusion matrix $\boldsymbol{\Sigma}$ are unavoidably linked together: As soon as the full path $\boldsymbol{X}_{[0,T]} = \boldsymbol{X}^{\text{obs}} \cup \boldsymbol{X}^{\text{imp}}$ is known, the diffusion matrix can be calculated rather than estimated via (7.46); and the other way round, a fixed diffusion matrix determines a path with the appropriate quadratic variation. See Polson and Roberts (1994) for a detailed elaboration. In Sect. 6.1.1, this connection was emphasised as a convenient property as it theoretically allows feasible identification of those model parameters that are uniquely determined by the value of $\boldsymbol{\Sigma}$. In the context of the present chapter, however, this characteristic limits the performance of the MCMC scheme. The crucial difference is that in Sect. 6.1.1 the diffusion path was assumed continuously *observed*. Here, it is considered continuously *imputed*.

Mathematically, the problem can also be formulated as follows: Let $\mathbb{P}_{\boldsymbol{\theta}}$ denote the probability measure induced by the diffusion process fulfilling the SDE

$$\mathrm{d}\boldsymbol{X}_t = \boldsymbol{\mu}(\boldsymbol{X}_t, \boldsymbol{\theta})\mathrm{d}t + \boldsymbol{\sigma}(\boldsymbol{X}_t, \boldsymbol{\theta})\mathrm{d}\boldsymbol{B}_t\,, \quad \boldsymbol{X}_{t_0} = \boldsymbol{x}_0,$$

for fixed parameter $\boldsymbol{\theta} \in \Theta$. Denote by $\mathbb{W}_{\boldsymbol{\theta}}$ the measure for the respective driftless version,

$$\mathrm{d}\tilde{\boldsymbol{X}}_t = \boldsymbol{\sigma}(\tilde{\boldsymbol{X}}_t, \boldsymbol{\theta})\mathrm{d}\boldsymbol{B}_t\,, \quad \tilde{\boldsymbol{X}}_{t_0} = \boldsymbol{x}_0.$$

Conditioned on the path $\boldsymbol{X}_{[0,T]}$, the distribution of the diffusion matrix is just a point mass at the value implicated by the quadratic variation link (7.46). For $\boldsymbol{\Sigma}(\cdot, \boldsymbol{\theta}) \neq \boldsymbol{\Sigma}(\cdot, \boldsymbol{\theta}^*)$, where $\boldsymbol{\theta}, \boldsymbol{\theta}^* \in \Theta$, the measures $\mathbb{W}_{\boldsymbol{\theta}}$ and $\mathbb{W}_{\boldsymbol{\theta}^*}$ are thus mutually singular, i.e. they have disjoint support. This is denoted by $\mathbb{W}_{\boldsymbol{\theta}} \perp \mathbb{W}_{\boldsymbol{\theta}^*}$. The two measures $\mathbb{P}_{\boldsymbol{\theta}}$ and $\mathbb{W}_{\boldsymbol{\theta}}$ are equivalent according to Girsanov's theorem in Sect. 3.2.12, that means they have identical null sets. Hence, $\mathbb{P}_{\boldsymbol{\theta}} \perp \mathbb{P}_{\boldsymbol{\theta}^*}$ as well. For the likelihood of $\boldsymbol{\theta}$ with respect to Lebesgue measure $\mathbb{L}$,

$$L(\boldsymbol{\theta}; \boldsymbol{X}) = \frac{\mathrm{d}\mathbb{P}_{\boldsymbol{\theta}}}{\mathrm{d}\mathbb{L}}(\boldsymbol{X}),$$

this implies

$$\forall\, \boldsymbol{X} \in \mathcal{X} \quad \forall\, \boldsymbol{\theta}, \boldsymbol{\theta}^* \in \Theta$$
$$\Big(\boldsymbol{\Sigma}(\cdot, \boldsymbol{\theta}) \neq \boldsymbol{\Sigma}(\cdot, \boldsymbol{\theta}^*) \;\Rightarrow\; L(\boldsymbol{\theta}; \boldsymbol{X}) = 0 \,\vee\, L(\boldsymbol{\theta}^*; \boldsymbol{X}) = 0 \quad a.s.\Big)$$

and

$$\forall\, \boldsymbol{X}, \boldsymbol{X}^* \in \mathcal{X} \quad \forall\, \boldsymbol{\theta} \in \Theta$$
$$\Big(\langle \boldsymbol{X}, \boldsymbol{X} \rangle \neq \langle \boldsymbol{X}^*, \boldsymbol{X}^* \rangle \;\Rightarrow\; L(\boldsymbol{\theta}; \boldsymbol{X}) = 0 \,\vee\, L(\boldsymbol{\theta}; \boldsymbol{X}^*) = 0 \quad a.s.\Big).$$

Now consider the updating scheme (7.2) on p. 173, where both the path update and the parameter update are performed using the Metropolis-Hastings algorithm. These update steps use acceptance probabilities including the factors $\pi(\boldsymbol{X}^*|\boldsymbol{\theta})/\pi(\boldsymbol{X}|\boldsymbol{\theta}) = L(\boldsymbol{\theta}; \boldsymbol{X}^*)/L(\boldsymbol{\theta}; \boldsymbol{X})$ (path update) and $\pi(\boldsymbol{X}|\boldsymbol{\theta}^*)/\pi(\boldsymbol{X}|\boldsymbol{\theta}) = L(\boldsymbol{\theta}^*; \boldsymbol{X})/L(\boldsymbol{\theta}; \boldsymbol{X})$ (parameter update), where the asterisk tags the proposals. Presumably, the previous state $(\boldsymbol{X}, \boldsymbol{\theta})$ of the Markov chain is consistent and has positive likelihood. Then, unless $\boldsymbol{\Sigma}(\cdot, \boldsymbol{\theta}) = \boldsymbol{\Sigma}(\cdot, \boldsymbol{\theta}^*)$, the numerators in both acceptance probabilities involve the factor 0, i.e. the proposals $\boldsymbol{X}^*$ and $\boldsymbol{\theta}^*$ will be rejected. The update scheme is degenerate.

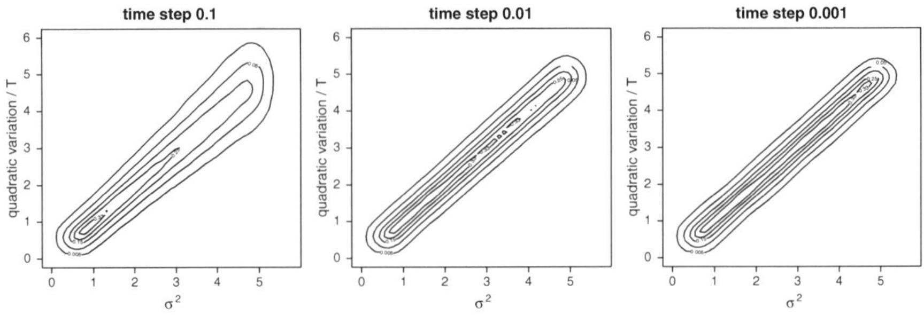

Fig. 7.25 Kernel density estimates of joint densities of σ^2 and the quadratic variation per time $\widehat{\langle Y,Y\rangle}_{[0,25]}/25$, where Y is an exactly sampled discrete skeleton of an Ornstein-Uhlenbeck process on the time interval $[0,25]$ satisfying (7.20) on p. 192. The quadratic variation is calculated as in Eq. (7.29) on p. 198. Density estimation is based on skeletons for 25,000 uniformly sampled $\sigma^2 \sim U([0.5,5.0])$ with $\alpha = 0.5$ and $\beta = 0.9$ fixed. The skeletons are simulated on an equidistant time grid with step lengths 0.1 (*left*), 0.01 (*middle*) and 0.001 (*right*)

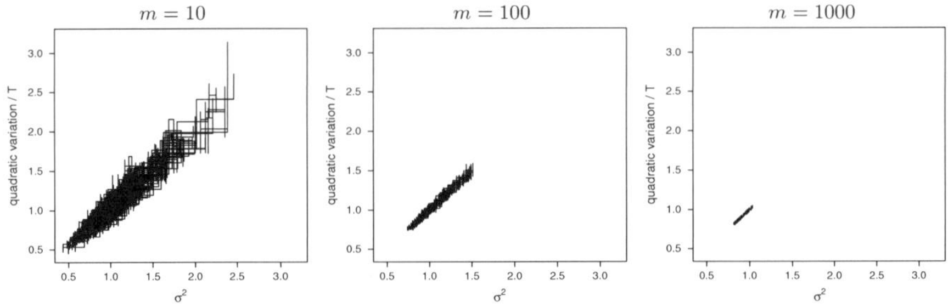

Fig. 7.26 Joint trace plots of σ^2 and $\widehat{\langle Y,Y\rangle}_{[0,25]}/25$ corresponding to iterations 80,000–85,000 (without thinning) of the Markov chains shown in Fig. 7.24. The quadratic variation is calculated as in Eq. (7.29) on p. 198

In practice, we do not come into contact with either continuous observation or continuous imputation. However, the above considerations imply that the algorithm slows down as more and more data points are imputed and is even degenerate in the limit. This explains the decreasing acceptance rates mentioned at the beginning of this section. Even worse, the Markov chain runs a risk of appearing to converge when in fact it is trapped in a consistent combination of $\boldsymbol{X}^{\mathrm{imp}}$ and $\boldsymbol{\theta}$.

Figure 7.25 shows empirical joint densities for σ^2 and the quadratic variation per time, $\widehat{\langle Y,Y\rangle}_{[0,25]}/25$, of a discrete path skeleton Y of an Ornstein-Uhlenbeck process with diffusion coefficient σ. Due to (7.46), one expects $\widehat{\langle Y,Y\rangle}_{[0,25]} \approx 25\sigma^2$. The quadratic variation is estimated as in (7.29) on p. 198. The time steps 0.1, 0.01 and 0.001 in the three graphics correspond to $M \cdot m = 250, 2{,}500, 25{,}000$ observed and auxiliary data points on the time interval $[0,25]$, respectively; that matches the situations in Fig. 7.24. Figure 7.26 shows joint trace plots of σ^2 and $\widehat{\langle Y,Y\rangle}_{[0,25]}/25$ for the MCMC experiments considered in Fig. 7.24.

Altogether, one faces the dilemma that on the one hand it is essential to increase the amount of imputed data in order to reduce the estimation bias, but on the other hand this action results in arbitrarily slow mixing of the Markov chains. This difficulty is well-observed in Fig. 7.24. The algorithm can even arrive in a deadlocked situation where both the imputed data and the model parameter remain almost unaltered.

The following section reviews and develops novel improvements on the MCMC algorithm in order to establish convergence that is not constrained by disturbing dependence structures.

7.4 Improvements of Convergence

The previous section described the bad mixing behaviour of the MCMC scheme considered in Sects. 7.1 and 7.2 which originates from the close connection between the quadratic variation of a diffusion path and the parameters determined by the diffusion matrix. Since the discovery of this cause by Roberts and Stramer (2001), several authors have attempted to modify the basic MCMC scheme (7.2) in such a way that it is not degenerate in the limit $m \to \infty$.

This section reviews some of these approaches, in particular a change of factorisation of the dominating measure in Sect. 7.4.1, time change transformations in Sect. 7.4.2 and particle filters in Sect. 7.4.3. The first two approaches fall into the class of reparameterisations that cause a priori independence between the parameter and the missing data; these are called *(partially) non-centred parameterisations* (Papaspiliopoulos et al. 2003). The third approach modifies the MCMC scheme in such a way that updates of the path and the parameter happen simultaneously rather than alternately.

The just mentioned approaches work well for general one-dimensional diffusion processes but are, however, not applicable or not appropriate for many multi-dimensional diffusion processes in life sciences such as those considered in Chaps. 8 and 9. Details are given in the respective sections.

Hence, Sect. 7.4.4 develops a novel concept for infinite-dimensional state spaces which is applicable to general multi-dimensional diffusions under fairly general regularity conditions. We adopt the name *innovation scheme* for this method due to similar but different approaches which are pointed out in Sect. 7.4.4. The convergence of the innovation scheme is proven, and its computational efficiency is demonstrated in a simulation study.

7.4.1 Changing the Factorisation of the Dominating Measure

The approach described in this section is based on a parallel, drawn by Roberts and Stramer (2001), between a suitable reparameterisation of a diffusion process and the change of factorisation of the dominating measure. The latter should not depend on the parameter to estimate.

Consider a one-dimensional diffusion X satisfying the SDE

$$\mathrm{d}X_t = \mu(X_t)\mathrm{d}t + \sigma \mathrm{d}B_t, \quad X_0 = x_0, \tag{7.47}$$

where $\sigma \in \mathbb{R}_+$ is the volatility parameter to be estimated and the drift function μ may depend on a parameter $\boldsymbol{\theta}$ not containing σ. In the current context, $\boldsymbol{\theta}$ is of lower priority and hence not included in the notation. Denote by $\mathbb{P}_\sigma$ the law induced by (7.47), and let $\mathbb{W}_\sigma$ be the law of Brownian motion with volatility parameter σ and initial value x_0 (recall Sect. 3.1.1 for the definition). $\mathbb{P}_\sigma$ and $\mathbb{W}_\sigma$ are equivalent measures and their Radon-Nikodym derivative $\mathrm{d}\mathbb{P}_\sigma/\mathrm{d}\mathbb{W}_\sigma(X) = G(X;\mu,\sigma)$ is available. However, as noticed in Sect. 7.3, $\mathbb{W}_\sigma$ is not suitable as dominating measure due to the mutual singularity $\mathbb{W}_\sigma \perp \mathbb{W}_{\sigma^*}$ for $\sigma \neq \sigma^*$.

The reparameterisation suggested by Roberts and Stramer is motivated by the following construction of Brownian motion $B_{[0,T]}$ with volatility parameter σ and initial value b_0: First, draw the end point $b \sim \mathbb{W}_\sigma$ conditional on b_0. Next, simulate a Brownian $(0,0,T,0)$-bridge $\tilde{B}_{(0,T)}$ with volatility parameter 1 as defined in Sect. 3.1.1. Imagine that this was possible in continuous time. Last, transform this bridge to appropriate volatility and boundary points as in step 3 in Sect. 3.3.3 and utilise it as the path segment $B_{(0,T)}$. This procedure is a purely theoretical consideration; in practice, simulation of Brownian motion at discrete time points would be performed as in step 1 in Sect. 3.3.3. However, the above construction illustrates that the measure $\mathbb{W}_\sigma$ can be factorised as

$$\mathbb{W}_\sigma\big(B_{[0,T)}, b\big) = \big(\mathbb{B}_\sigma^{(0,b_0,T,b)} \otimes \mathbb{W}_\sigma\big)\big(h^{-1}(\tilde{B}_{[0,T)};\sigma,b_0,b),b\big),$$

where $\tilde{B} = h(B;\sigma,b_0,b)$, and $\mathbb{B}$ denotes the law of a Brownian bridge with the volatility parameter as subscript and the boundary specification as superscript.

The idea is now to decompose $\mathbb{P}_\sigma$ in a corresponding manner. To that end, consider the two transformations

$$\begin{aligned} \dot{X}_t &= h_1(X_t;\sigma) &&= \frac{1}{\sigma} X_t \\ \tilde{X}_t &= h_2(\dot{X}_t;\dot{x}_0,\dot{x}) &&= \dot{X}_t - \frac{(T-t)\dot{x}_0 + t\dot{x}}{T} \end{aligned}$$

for $t \in [0,T]$. The function h_1 transforms X to a diffusion process which is the solution of

$$\mathrm{d}\dot{X}_t = \frac{\mu(\sigma \dot{X}_t)}{\sigma}\mathrm{d}t + \mathrm{d}B_t, \quad \dot{X}_0 = \frac{x_0}{\sigma}, \tag{7.48}$$

according to Itô's lemma in Sect. 3.2.10. Note that this process has unit diffusion and does therefore not experience the difficulties investigated in Sect. 7.3 concerning the quadratic variation. However, the original process X cannot simply be replaced by $\dot{X}$ as our inference conditions on the observed end point x. Knowledge of $\dot{x} = h_1(x;\sigma) = x/\sigma$ then again requires knowledge of σ. Hence, the second function h_2 transforms the unit volatility process $\dot{X}_{[0,T]}$ such that it starts and ends at zero. The concatenation of h_1 and h_2 carries out the same transformation as the function h in the Brownian motion construction above:

$$\tilde{B} = h\big(B;\sigma,b_0,b\big) = h_2\big(h_1(B;\sigma);h_1(b_0;\sigma),h_1(b;\sigma)\big).$$

This implies

$$\frac{\mathrm{d}\mathbb{P}_\sigma}{\mathrm{d}\mathbb{W}_\sigma}\big(X_{[0,T)}\,\big|\,x\big) = \frac{\mathrm{d}\mathbb{H}}{\mathrm{d}\mathbb{B}_1^{(0,0,T,0)}}\big(\tilde{X}_{[0,T)}\,\big|\,\dot{x}\big),$$

where $\mathbb{H}$ denotes the law of $\tilde{X}$. That means that conditional on the final point x, the dominating measure can be written independently of σ. Moreover,

$$\frac{\mathrm{d}\mathbb{P}_\sigma}{\mathrm{d}\mathbb{W}_\sigma}\big(X_{[0,T)},x\big) = \frac{\mathrm{d}\mathbb{Q}}{\mathrm{d}\mathbb{B}_1^{(0,\dot{x}_0,T,\dot{x})}}\big(\dot{X}_{[0,T)},\dot{x}\big) = G(\dot{X};\dot{\mu},1)$$

with $\mathbb{Q}$ being the law of $\dot{X}$, and $\dot{\mu}$ being defined as the drift function in (7.48). This means that the likelihood of $\dot{X}$ is of known form.

Based on this reparameterisation one can now develop MCMC algorithms which achieve improved convergence results. See Roberts and Stramer (2001) for details. As the transformation is invertible, back-transformation to the original diffusion path is straightforward.

Unfortunately, this method cannot be applied to general multi-dimensional diffusion processes. Although some extensions are possible (Roberts and Stramer 2001; Kalogeropoulos 2007; Kalogeropoulos et al. 2011), an appropriate transformation requires reparameterising to unit diffusion coefficient. It was already noted in Sect. 3.2.11 that in the multi-dimensional case such a transform does generally not exist.

7.4.2 Time Change Transformations

An alternative approach to reparameterise a diffusion process such that there is a dominating measure which does not depend on the volatility parameters is via time change transformations as suggested by Kalogeropoulos et al. (2010). This method has been developed for several but not all possibly multi-dimensional diffusion processes. The central tool in this procedure is the following *time change formula* which can be obtained from a more general theorem in Øksendal (2003, Chap. 8.5); see also Klebaner (2005, Chap. 7): Let $\boldsymbol{X} = (\boldsymbol{X}_t)_{t\in[0,T]}$ be a diffusion process fulfilling the SDE

$$\mathrm{d}\boldsymbol{X}_t = \boldsymbol{\mu}(\boldsymbol{X}_t,\boldsymbol{\theta})\mathrm{d}t + \boldsymbol{\sigma}(\boldsymbol{X}_t,\boldsymbol{\theta})\mathrm{d}\boldsymbol{B}_t, \quad \boldsymbol{X}_0 = \boldsymbol{x}_0,$$

and

$$h(t) = \int_0^t c(\tau)\mathrm{d}\tau \qquad \text{for } t \in [0,T]$$

be a *time change* with positive *time change rate* $c : [0,T] \to \mathbb{R}_+$. Note that the function h is strictly increasing and hence invertible. Define another diffusion process $\boldsymbol{Z} = (\boldsymbol{Z}_s)_{s\in[0,h(T)]}$ on a new time scale such that $s = h(t)$ and $\boldsymbol{Z}_s = \boldsymbol{X}_t = \boldsymbol{X}_{h^{-1}(s)}$ for all $t \in [0,T]$ and $s \in [0,h(T)]$. Then $\boldsymbol{Z}$ satisfies the SDE

$$\mathrm{d}\boldsymbol{Z}_s = \frac{\boldsymbol{\mu}(\boldsymbol{Z}_s,\boldsymbol{\theta})}{c(h^{-1}(s))}\,ds + \frac{\boldsymbol{\sigma}(\boldsymbol{Z}_s,\boldsymbol{\theta})}{\sqrt{c(h^{-1}(s))}}\,\mathrm{d}\boldsymbol{B}_s, \quad \boldsymbol{Z}_0 = \boldsymbol{x}_0.$$

The following illustrates the idea by Kalogeropoulos et al. (2010) to utilise the time change transformation for our purposes on the example of a one-dimensional diffusion process satisfying the SDE

$$\mathrm{d}X_t = \mu(X_t,\boldsymbol{\theta})\mathrm{d}t + \sigma\mathrm{d}B_t, \quad X_0 = x_0,$$

where $t \in [0,1]$. In this representation, the diffusion coefficient $\sigma \in \mathbb{R}_+$ is considered as one component of $\boldsymbol{\theta}$. Let $\mathbb{P}^X$ be the measure induced by the process X and $\mathbb{W}^X$ the measure of a respective driftless version $dM_t = \sigma\mathrm{d}B_t$. Suppose that $X_1 = x$. Then, similarly to the considerations in the previous section, we can write $\mathbb{W}^X(X_{[0,1]}) = \mathbb{W}^X_{1,x}(X_{[0,1)})\mathbb{W}^X(x)$ with $\mathbb{W}^X_{1,x}$ being the measure $\mathbb{W}^X$ further conditioned on the end point x at time 1. Then

$$\frac{\mathrm{d}\mathbb{P}^X}{\mathrm{d}(\mathbb{W}^X_{1,x}\otimes\mathbb{L})}(X_{[0,1)},x) = G(X_{[0,1)},\boldsymbol{\theta})f_\sigma(x),$$

where the function G is obtained via Girsanov's formula from Sect. 3.2.12, and f_σ is the Lebesgue density of the end point X_1 under $\mathbb{W}^X$. In this expression, the dominating measure $\mathbb{W}^X_{1,x}\otimes\mathbb{L}$ clearly depends on σ as $\mathbb{W}^X_{1,x}$ is the law of a Brownian bridge with volatility parameter σ. Hence consider the following time change transformation which reparameterises the SDE to unit diffusion coefficient: Let

$$s = h_1(t) = \sigma^2 t$$

and consider the process

$$U_s = \begin{cases} X_{h_1^{-1}(s)} & \text{for } 0 \le s \le \sigma^2 \\ M_{h_1^{-1}(s)} & \text{for } s > \sigma^2. \end{cases}$$

With the above time change formula we obtain

$$dU_s = \begin{cases} \dfrac{\mu(U_s,\boldsymbol{\theta})}{\sigma^2}\,ds + \mathrm{d}B_s & \text{for } 0 \le s \le \sigma^2 \\ \mathrm{d}B_s & \text{for } s > \sigma^2. \end{cases}$$

Clearly, the process U has unit diffusion, and $U_{\sigma^2} = X_1 = x$. Let $\mathbb{P}^U$ be the probability measure induced by U and $\mathbb{W}^U$ the driftless counterpart. Then

$$\frac{\mathrm{d}\mathbb{P}^U}{\mathrm{d}(\mathbb{W}^U_{\sigma^2,x}\otimes\mathbb{L})}\left(U_{[0,\sigma^2)},x\right)=G(U_{[0,\sigma^2)},\boldsymbol{\theta})f_\sigma(x),$$

where $\mathbb{W}^U_{\sigma^2,x}$ is the law of the driftless version of U conditioned on $U_{\sigma^2}=x$. Although the parameter σ has been eliminated from the diffusion coefficient, it is still included in the time point σ^2 where U reaches the state x. That means that again the dominating measure $\mathbb{W}^U_{\sigma^2,x}\otimes\mathbb{L}$ in the above expression depends on σ. Therefore introduce a second time change

$$u=h_2(s)=\frac{s}{\sigma^2(\sigma^2-s)}$$

for $s\in[0,\sigma^2)$ and apply this in the transformation

$$Z_u=\frac{1}{\sigma^2-s}\left(U_s-\left(1-\frac{s}{\sigma^2}\right)x_0-\frac{s}{\sigma^2}x\right)\quad\text{for } u\in[0,\infty).$$

In the following let $x_0=x=0$. Then

$$Z_u=\frac{1+u\sigma^2}{\sigma^2}U_{h_2^{-1}(u)}.$$

For better understanding of the derivation of an SDE for Z, introduce an intermediate process $\tilde{Z}$ such that $\tilde{Z}_u=U_{h_2^{-1}(s)}$ and $Z_u=(1+u\sigma^2)\tilde{Z}_u/\sigma^2$ for all $u\in[0,\infty)$ and $s\in[0,\sigma^2)$. Then, with

$$\frac{\partial h_2(s)}{\partial s}=\frac{1}{(\sigma^2-s)^2},\ \text{i.e.}\quad\frac{\partial h_2}{\partial s}\left(h_2^{-1}(u)\right)=\frac{\partial h_2}{\partial s}\left(\frac{u\sigma^4}{1+u\sigma^2}\right)=\frac{(1+u\sigma^2)^2}{\sigma^4},$$

the time change formula yields

$$\mathrm{d}\tilde{Z}_u=\frac{\sigma^4}{(1+u\sigma^2)^2}\frac{\mu(\tilde{Z}_u,\boldsymbol{\theta})}{\sigma^2}du+\frac{\sigma^2}{1+u\sigma^2}\mathrm{d}B_u.$$

Next, Itô's formula from Sect. 3.2.10 leads to

$$\mathrm{d}Z_u=\tilde{Z}_u du+\frac{1+u\sigma^2}{\sigma^2}\mathrm{d}\tilde{Z}_u=\frac{\mu\left(\dfrac{\sigma^2Z_u}{1+u\sigma^2},\boldsymbol{\theta}\right)+\sigma^2Z_u}{1+u\sigma^2}du+\mathrm{d}B_u.$$

Note that $U_{\sigma^2}=0$ implies $Z_\infty=0$. Hence, if $\mathbb{P}^Z$ denotes the law of Z and $\mathbb{W}^Z_{\infty,0}$ is the law of a unit diffusion process starting in state zero at time $u=0$ and reaching state zero at time $u=\infty$, then

$$\frac{\mathrm{d}\mathbb{P}^Z}{\mathrm{d}(\mathbb{W}^Z_{\infty,0}\otimes\mathbb{L})}\left(Z_{[0,\infty)},0\right)=G(Z_{[0,\infty)},\boldsymbol{\theta})f_\sigma(0).$$

Kalogeropoulos et al. (2010) prove that $\mathbb{W}^Z_{\infty,0}$ is standard Wiener measure. In other words, the dominating measure $\mathbb{W}^Z_{\infty,0} \otimes \mathbb{L}$ does not depend on σ. One can hence perform inference for Z using the MCMC schemes from Sects. 7.1 and 7.2 without the risk of bad mixing. As all transformations above are invertible, estimation results for Z can easily be transferred to the original diffusion process X.

The above concept of time change transformations is generalised by Kalogeropoulos et al. (2010) to certain higher-dimensional stochastic volatility models. The extension to general multi-dimensional diffusion processes with state-dependent diffusion coefficients, however, is still the subject of ongoing research. The approach described in this section hence cannot be applied to the large-dimensional applications from life sciences which are considered in Chaps. 8 and 9.

7.4.3 Particle Filters

Another idea to overcome disturbing dependencies between the parameter and the quadratic variation of the diffusion path is the use of particle filters, where the path and the parameter are updated simultaneously rather than alternately. The proposed path and parameter are then consistent at any time. The principle of particle filters for diffusions is described in the following.

Suppose there are M partial observations $\boldsymbol{v}_1, \ldots, \boldsymbol{v}_M$ of the diffusion in addition to the initial value $\boldsymbol{v}_0$ at times $\tau_0 < \tau_1 < \ldots < \tau_M$. These observations may be subject to measurement error. A particle filter successively performs inference for the parameter and the diffusion path concentrating on the time interval $[\tau_k, \tau_{k+1}]$ for $k = 0, \ldots, M-1$. Certainly, estimation results for different path segments shall not be independent; when focusing on $[\tau_k, \tau_{k+1}]$ for fixed k, findings for $[\tau_0, \tau_k]$ are taken into account by conditioning on a set of particles $\{\boldsymbol{X}^{(i)}_{\tau_k}, \boldsymbol{\theta}^{(i)}\}_{i=1,\ldots,I}$ for some large $I \in \mathbb{N}$. These particles are considered as draws from $\pi(\boldsymbol{X}_{\tau_k}, \boldsymbol{\theta}|\boldsymbol{v}_0, \ldots, \boldsymbol{v}_k)$. For $k = 0$, they are generated from some initial distribution. In case of complete observations without measurement error, one has $\boldsymbol{X}^{(i)}_{\tau_k} = \boldsymbol{v}_k$ for all i.

Joint inference for $\boldsymbol{X}_{[\tau_k, \tau_{k+1}]}$ and $\boldsymbol{\theta}$ is now accomplished as follows: Based on the particles $\{\boldsymbol{X}^{(i)}_{\tau_k}, \boldsymbol{\theta}^{(i)}\}_{i=1,\ldots,I}$, a discrete probability function $\hat{\pi}_k$ is obtained as an estimate of the density $\pi(\boldsymbol{X}_{\tau_k}, \boldsymbol{\theta}|\boldsymbol{v}_0, \ldots, \boldsymbol{v}_k)$. This could for example be the empirical probability function putting equal weight on all particles. Next, a new Markov chain $\{\boldsymbol{X}^{(i)}_{[\tau_k, \tau_{k+1}]}, \boldsymbol{\theta}^{(i)}\}_{i=1,\ldots,I}$ is constructed conditional on the observations $\boldsymbol{v}_0, \ldots, \boldsymbol{v}_{k+1}$. Previous estimation results are incorporated by using $\hat{\pi}_k$ in

$$\begin{aligned}
&\pi\big(\boldsymbol{X}_{[\tau_k, \tau_{k+1}]}, \boldsymbol{\theta}\big|\boldsymbol{v}_0, \ldots, \boldsymbol{v}_{k+1}\big) \\
&\quad = \pi\big(\boldsymbol{X}_{(\tau_k, \tau_{k+1}]}\big|\boldsymbol{X}_{\tau_k}, \boldsymbol{\theta}, \boldsymbol{v}_0, \ldots, \boldsymbol{v}_{k+1}\big)\pi\big(\boldsymbol{X}_{\tau_k}, \boldsymbol{\theta}\big|\boldsymbol{v}_0, \ldots, \boldsymbol{v}_k\big).
\end{aligned}$$

Discarding $\boldsymbol{X}^{(i)}_{[\tau_k, \tau_{k+1})}$ for all i yields a Markov chain $\{\boldsymbol{X}^{(i)}_{\tau_{k+1}}, \boldsymbol{\theta}^{(i)}\}_{i=1,\ldots,I}$ which can be regarded as a set of draws from $\pi(\boldsymbol{X}_{\tau_{k+1}}, \boldsymbol{\theta}|\boldsymbol{v}_0, \ldots, \boldsymbol{v}_{k+1})$ because of

$$\pi\big(\boldsymbol{X}_{\tau_{k+1}}, \boldsymbol{\theta} \,\big|\, \boldsymbol{v}_0, \ldots, \boldsymbol{v}_{k+1}\big) = \int\limits_{\mathcal{X}^\infty} \pi\big(\boldsymbol{X}_{[\tau_k, \tau_{k+1}]}, \boldsymbol{\theta} \,\big|\, \boldsymbol{v}_0, \ldots, \boldsymbol{v}_{k+1}\big) \mathrm{d}\boldsymbol{X}_{[\tau_k, \tau_{k+1})}.$$

This is the set of particles used for inference on the subsequent interval $[\tau_{k+1}, \tau_{k+2}]$.

Golightly and Wilkinson (2006a,b) implement such particle filters by using MCMC techniques based on a discretisation of the path segment $\boldsymbol{X}_{[\tau_k, \tau_{k+1}]}$ and its transition density as in Sects. 7.1 and 7.2. In that case, the set of particles can for each k be obtained as a Markov chain after thinning and discarding a burn-in phase. Fearnhead et al. (2008) propose particle filters for diffusions based on the Exact Algorithm from Sect. 6.5, thus not requiring any time-discretisations. Filtering for (jump-)diffusions has also been applied e.g. by Del Moral et al. (2001), Chib et al. (2004) and Johannes et al. (2006).

The crucial point why a particle filter theoretically solves the convergence problems discussed in Sect. 7.3 is that the parameter $\boldsymbol{\theta}$ and the path segment $\boldsymbol{X}_{[\tau_k, \tau_{k+1}]}$ are always generated in a way such that they are consistent. In particular, first a parameter $\boldsymbol{\theta}^*$ and state $\boldsymbol{X}^*_{\tau_k}$ are drawn from $\hat{\pi}_k(\boldsymbol{X}_{\tau_k}, \boldsymbol{\theta} | \boldsymbol{v}_0, \ldots, \boldsymbol{v}_k)$, and the remaining path segment $\boldsymbol{X}^*_{(\tau_k, \tau_{k+1}]}$ conditions on these. In the MCMC context, then either both $(\boldsymbol{\theta}^*, \boldsymbol{X}^*_{\tau_k})$ and $\boldsymbol{X}^*_{(\tau_k, \tau_{k+1}]}$ are accepted or none.

As a convenient by-product, the particle filter enables *online estimation*, i.e. it does not have to discard previous estimation results when new observations become available at times larger than τ_M. That means, a Monte Carlo sampler does not have to be restarted but simply continues the procedure conditional on the new observations. Online estimation is especially in demand in real-time analysis, i.e. in applications where instantaneous action is required and results of time-consuming estimation procedures cannot be awaited. Examples are the monitoring of the spread of an infectious disease or modelling asset prices at the stock market.

However, whilst fixing one problem, the use of particle filters brings up other difficulties in practice: One issue is that poor approximations to the particles $\{\boldsymbol{X}^{(i)}_{\tau_k}, \boldsymbol{\theta}^{(i)}\}_{i=1,\ldots,I}$ propagate poor approximations to subsequent sets of particles. Second, the use of MCMC methods in combination with particle filters as in Golightly and Wilkinson (2006a,b) are generally exposed to fairly low acceptance probabilities and hence slow mixing of the Markov chains. Therefore the methodology is not appropriate for the inherently computer-intensive data augmentation of large-dimensional processes that may occur in life sciences; see for example the application in Chap. 8.

7.4.4 Innovation Scheme on Infinite-Dimensional State Spaces

This section now develops a novel and widely applicable update scheme which works for any multi-dimensional diffusion process under fairly general regularity conditions. In particular, no special form of the diffusion coefficient such as a unit diffusion matrix is required. The MCMC method is computationally efficient and experiences satisfying acceptance rates for any amount of imputed data. Most importantly, it does not break down as the amount of imputed data grows to infinity.

In the form presented here, the method is newly investigated in this book. There are however related approaches in the literature as described forthcoming. Adopting the notation from corresponding references, the introduced method will be referred to as *innovation scheme* in the following.

The idea of the innovation scheme can be motivated by means of the parameter update for a diffusion process $\boldsymbol{X}$ on the time interval $[0, T]$ as follows: As in Sect. 7.3, restrict the following considerations to the case where the initial value $\boldsymbol{X}_0 = \boldsymbol{x}_0$ and the final value $\boldsymbol{X}_T = \boldsymbol{x}$ are known and the remaining path segment $\boldsymbol{X}_{(0,T)}$ is unknown. Once more, denote the measure of the target diffusion by $\mathbb{P}_{\boldsymbol{\theta}}$, that is the measure induced by a diffusion satisfying

$$\mathrm{d}\boldsymbol{X}_t = \boldsymbol{\mu}(\boldsymbol{X}_t, \boldsymbol{\theta})\mathrm{d}t + \boldsymbol{\sigma}(\boldsymbol{X}_t, \boldsymbol{\theta})\mathrm{d}\boldsymbol{B}_t, \quad \boldsymbol{X}_0 = \boldsymbol{x}_0.$$

Assume that $\boldsymbol{\sigma}$ is invertible. Then, given $\boldsymbol{X} = (\boldsymbol{X}_t)_{t\in[0,T]} \sim \mathbb{P}_{\boldsymbol{\theta}}$, the process $\boldsymbol{B} = (\boldsymbol{B}_t)_{t\in[0,T]}$ with

$$\mathrm{d}\boldsymbol{B}_t = \boldsymbol{\sigma}^{-1}(\boldsymbol{X}_t, \boldsymbol{\theta})\big(\mathrm{d}\boldsymbol{X}_t - \boldsymbol{\mu}(\boldsymbol{X}_t, \boldsymbol{\theta})\mathrm{d}t\big), \quad \boldsymbol{B}_0 = \boldsymbol{0}, \tag{7.49}$$

is d-dimensional standard Brownian motion. In particular, $\boldsymbol{B}$ has unit volatility and hence possesses a property which is desirable in the context of the general data imputation scheme considered in this chapter.

The above equations mean that there is a deterministic link between the target process $\boldsymbol{X}$ and the parameter-free Brownian motion process $\boldsymbol{B}$. This relationship however conditions on the parameter. Define a function h such that $\boldsymbol{X}_t = h(\boldsymbol{B}_t, \boldsymbol{\theta})$ for all $t \in [0, T]$ and given $\boldsymbol{\theta}$. This function is invertible in its first argument, i.e. there is another function h^{-1} such that $h^{-1}(\boldsymbol{X}_t, \boldsymbol{\theta}) = \boldsymbol{B}_t$. The connection between $\boldsymbol{X}$ and $\boldsymbol{B}$ can be exploited in the parameter update of the MCMC scheme (7.2) by conditioning the acceptance or rejection decision on $\boldsymbol{B}$ instead of $\boldsymbol{X}$. In particular, one updates $\boldsymbol{\theta}|\{\boldsymbol{B}^{\mathrm{imp}}, \boldsymbol{x}_0, \boldsymbol{x}\}$ rather than $\boldsymbol{\theta}|\{\boldsymbol{X}^{\mathrm{imp}}, \boldsymbol{x}_0, \boldsymbol{x}\}$, where $\boldsymbol{B}^{\mathrm{imp}} = h^{-1}(\boldsymbol{X}^{\mathrm{imp}}, \boldsymbol{\theta})$.

For given $\boldsymbol{\theta}$, $\boldsymbol{x}_0$, $\boldsymbol{x}$ and $\boldsymbol{X}^{\mathrm{imp}}$, the parameter update could then look as follows:

1. Draw $\boldsymbol{\theta}^* \sim q(\boldsymbol{\theta}^*|\boldsymbol{\theta})$.
2. Compute $\boldsymbol{B}^{\mathrm{imp}} = h^{-1}(\boldsymbol{X}^{\mathrm{imp}}, \boldsymbol{\theta})$.
3. Accept $\boldsymbol{\theta}^*$ with probability

$$\zeta(\boldsymbol{\theta}^*, \boldsymbol{\theta}) = 1 \wedge \frac{\pi(\boldsymbol{\theta}^* \mid \boldsymbol{B}^{\mathrm{imp}}, \boldsymbol{x}_0, \boldsymbol{x})q(\boldsymbol{\theta} \mid \boldsymbol{\theta}^*)}{\pi(\boldsymbol{\theta} \mid \boldsymbol{B}^{\mathrm{imp}}, \boldsymbol{x}_0, \boldsymbol{x})q(\boldsymbol{\theta}^* \mid \boldsymbol{\theta})},$$

   otherwise keep $\boldsymbol{\theta}$.
4. If $\boldsymbol{\theta}^*$ was accepted, replace $\boldsymbol{X}^{\mathrm{imp}}$ by $\boldsymbol{X}^{\mathrm{imp}*} = h(\boldsymbol{B}^{\mathrm{imp}}, \boldsymbol{\theta}^*)$.

Note that the order of steps 1 and 2 could be exchanged. The path correction in step 4 is new in comparison to the update schemes considered in Sects. 7.1 and 7.2. Due to this step, the just presented algorithm overcomes the degeneracy problems explored in Sect. 7.3, because the parameters $\boldsymbol{\theta}$ and $\boldsymbol{\theta}^*$ are consistent with $\boldsymbol{X}^{\text{imp}} = h(\boldsymbol{B}^{\text{imp}}, \boldsymbol{\theta})$ and $\boldsymbol{X}^{\text{imp}*} = h(\boldsymbol{B}^{\text{imp}}, \boldsymbol{\theta}^*)$, respectively.

The algorithm however experiences a different drawback: The acceptance of $\boldsymbol{\theta}^*$ implies the just mentioned path correction from $\boldsymbol{X}^{\text{imp}}$ to $\boldsymbol{X}^{\text{imp}*} = h(\boldsymbol{B}^{\text{imp}}, \boldsymbol{\theta}^*)$. The observed end point $\boldsymbol{x}$, however, remains the same. As generally $h(h^{-1}(\boldsymbol{x}, \boldsymbol{\theta}), \boldsymbol{\theta}^*) \neq \boldsymbol{x}$ for $\boldsymbol{\theta} \neq \boldsymbol{\theta}^*$, there is no guarantee that $\boldsymbol{X}^{\text{imp}*}$ satisfyingly bridges the gap between $\boldsymbol{x}_0$ and $\boldsymbol{x}$. One may trust in the Metropolis-Hastings algorithm rejecting all unlikely proposals in a discrete-time framework. A more reliable and desired tool is however an efficient algorithm for which convergence in the continuous-time setting is proven.

In the following we hence introduce a similar but different update mechanism and prove its convergence in a continuous-time framework.

Related Work

The idea to base the parameter update on a parameter-independent Brownian motion process has already been mentioned by Chib et al. (2004). They consider it in a framework where there is not necessarily an end point condition for the imputed diffusion process. The above accentuated difficulty does hence not appear. Although the modified update scheme is applied in a simulation study, the authors do not give details for the calculation of the Brownian motion construct.

Golightly and Wilkinson (2008, 2010) seize the general concept of Chib et al. and apply it to the parameter update as it is also investigated in this book. They however do not consider a continuous-time framework as done here but exclusively concentrate on discrete-time skeletons. In particular, Eq. (7.49) is replaced by

$$\mathring{\boldsymbol{B}}_{k+1} = \mathring{\boldsymbol{B}}_k + \boldsymbol{\sigma}^{-1}(\boldsymbol{Y}_k, \boldsymbol{\theta})\Big(\boldsymbol{Y}_{k+1} - \boldsymbol{Y}_k - \boldsymbol{\mu}(\boldsymbol{Y}_k, \boldsymbol{\theta})\Delta t_k\Big) \tag{7.50}$$

for appropriate indices k, $\Delta t_k = t_{k+1} - t_k$ and $\mathring{\boldsymbol{B}}_0 = \boldsymbol{0}$. To emphasise this difference, the notation $\mathring{\boldsymbol{B}}_k = \boldsymbol{B}_{t_k}$ and $\boldsymbol{Y}_k = \boldsymbol{X}_{t_k}$ is used here for observation and auxiliary times t_k. The back-transformation happens via

$$\boldsymbol{Y}_{k+1} = \boldsymbol{Y}_k + \boldsymbol{\sigma}(\boldsymbol{Y}_k, \boldsymbol{\theta})\big(\mathring{\boldsymbol{B}}_{k+1} - \mathring{\boldsymbol{B}}_k\big) + \boldsymbol{\mu}(\boldsymbol{Y}_k, \boldsymbol{\theta})\Delta t_k. \tag{7.51}$$

As this construction does not satisfyingly handle possible end point conditions, Golightly and Wilkinson also consider other deterministic links between $\mathring{\boldsymbol{B}}$ and $\boldsymbol{Y}$. To that end, define a function f such that $\boldsymbol{Y}_k = f(\mathring{\boldsymbol{B}}_k, \boldsymbol{\theta})$ and $\mathring{\boldsymbol{B}}_k = f^{-1}(\boldsymbol{Y}_k, \boldsymbol{\theta})$ for $k = 1, \ldots, m$ and $t_m = T$. When conditioning the parameter update on this transformation, the acceptance probability for the parameter becomes by change of variables

$$\begin{aligned}
&\zeta(\boldsymbol{\theta}^*, \boldsymbol{\theta}) \\
&= 1 \wedge \frac{\pi(\boldsymbol{\theta}^* \,|\, \mathring{\boldsymbol{B}}_0, \ldots, \mathring{\boldsymbol{B}}_m) q(\boldsymbol{\theta} \,|\, \boldsymbol{\theta}^*)}{\pi(\boldsymbol{\theta} \,|\, \mathring{\boldsymbol{B}}_0, \ldots, \mathring{\boldsymbol{B}}_m) q(\boldsymbol{\theta}^* \,|\, \boldsymbol{\theta})} \\
&= 1 \wedge \left(\prod_{k=0}^{m-1} \frac{\pi(\mathring{\boldsymbol{B}}_{k+1} | \mathring{\boldsymbol{B}}_k, \boldsymbol{\theta}^*)}{\pi(\mathring{\boldsymbol{B}}_{k+1} | \mathring{\boldsymbol{B}}_k, \boldsymbol{\theta})} \right) \cdot \frac{p(\boldsymbol{\theta}^*)}{p(\boldsymbol{\theta})} \cdot \frac{q(\boldsymbol{\theta} \,|\, \boldsymbol{\theta}^*)}{q(\boldsymbol{\theta}^* | \boldsymbol{\theta})} \\
&= 1 \wedge \left(\prod_{k=0}^{m-1} \frac{|J(f(\mathring{\boldsymbol{B}}_{k+1}, \boldsymbol{\theta}^*))|}{|J(f(\mathring{\boldsymbol{B}}_{k+1}, \boldsymbol{\theta}))|} \right) \left(\prod_{k=0}^{m-1} \frac{\pi(\boldsymbol{Y}^*_{k+1} | \boldsymbol{Y}^*_k, \boldsymbol{\theta}^*)}{\pi(\boldsymbol{Y}_{k+1} | \boldsymbol{Y}_k, \boldsymbol{\theta})} \right) \cdot \frac{p(\boldsymbol{\theta}^*)}{p(\boldsymbol{\theta})} \cdot \frac{q(\boldsymbol{\theta} \,|\, \boldsymbol{\theta}^*)}{q(\boldsymbol{\theta}^* | \boldsymbol{\theta})},
\end{aligned}$$

where $\boldsymbol{Y}^*_0 = \boldsymbol{Y}_0$, $\boldsymbol{Y}^*_k = f(\mathring{\boldsymbol{B}}_k, \boldsymbol{\theta}^*)$ for $k = 1, \ldots, m$, and

$$J\big(f(\mathring{\boldsymbol{B}}_{k+1}, \boldsymbol{\theta})\big) = \left| \frac{\partial f(\mathring{\boldsymbol{B}}_{k+1}, \boldsymbol{\theta})}{\partial \mathring{\boldsymbol{B}}_{k+1}} \right|$$

is the Jacobian determinant of f. In this acceptance probability, the numerators and denominators differ in both parameter and sample path, i.e. a critical situation as described in Sect. 7.3 should not occur. There, however, remains to be proven that the acceptance probability behaves nicely as part of the MCMC algorithm as the time step between two consecutive imputed data points tends to zero.

The above approach has been explicitly designed for discrete path skeletons. Golightly and Wilkinson (2010) emphasise that $\mathring{\boldsymbol{B}}$ can in principal be any deterministic transformation of $\boldsymbol{Y}$ and that f does actually not have to be related to the original diffusion process. They point out that certain transformations such as the modified bridge from p. 181 are however advantageous with respect to the end point condition of an imputed path segment.

In contrast to that, we in the following consider a specific transformation of the original infinite-dimensional diffusion process. We employ this transformation in both the parameter and the path update and show that the resulting MCMC scheme works when applied to continuously imputed path segments. This proceeding also supplies further insight on the method by Golightly and Wilkinson. See the conclusion on pp. 261 for corresponding remarks.

Contribution of This Book

In the remainder of this section, we present an MCMC mechanism for infinite-dimensional imputed path segments and show that it does not experience the degeneracy problems pointed out in Sect. 7.3. To that end, we investigate both the parameter update and the path update in continuous time. We derive explicit formulas for the involved acceptance probabilities so that these can be used in practice. The performance of the new approach is illustrated afterwards in a simulation study.

Table 7.6 Overview of introduced probability measures. All measures assume $\boldsymbol{X}_0 = \boldsymbol{x}_0$, and $\boldsymbol{B}$ is d-dimensional standard Brownian motion

$\mathbb{P}_{\boldsymbol{\theta}}$	$: \mathrm{d}\boldsymbol{X}_t = \boldsymbol{\sigma}(\boldsymbol{X}_t,\boldsymbol{\theta})\,\mathrm{d}\boldsymbol{B}_t + \boldsymbol{\mu}(\boldsymbol{X}_t,\boldsymbol{\theta})\,\mathrm{d}t$	
$\tilde{\mathbb{P}}_{\boldsymbol{\theta}}$	$: \mathrm{d}\boldsymbol{X}_t = \boldsymbol{\sigma}(\boldsymbol{X}_t,\boldsymbol{\theta})\,\mathrm{d}\boldsymbol{B}_t + \boldsymbol{\mu}(\boldsymbol{X}_t,\boldsymbol{\theta})\,\mathrm{d}t$	$, \boldsymbol{X}_T = \boldsymbol{x}$
$\mathbb{D}_{\mathbf{0},\boldsymbol{\theta}}$	$: \mathrm{d}\boldsymbol{X}_t = \boldsymbol{\sigma}(\boldsymbol{X}_t,\boldsymbol{\theta})\,\mathrm{d}\boldsymbol{B}_t + \frac{\boldsymbol{x}-\boldsymbol{X}_t}{T-t}\,\mathrm{d}t$	
$\mathbb{D}_{\boldsymbol{\mu},\boldsymbol{\theta}}$	$: \mathrm{d}\boldsymbol{X}_t = \boldsymbol{\sigma}(\boldsymbol{X}_t,\boldsymbol{\theta})\,\mathrm{d}\boldsymbol{B}_t + \left(\boldsymbol{\mu}(\boldsymbol{X}_t,\boldsymbol{\theta}) + \frac{\boldsymbol{x}-\boldsymbol{X}_t}{T-t}\right)\mathrm{d}t$	
$\mathbb{W}_{\boldsymbol{\theta}}$	$: \mathrm{d}\boldsymbol{X}_t = \boldsymbol{\sigma}(\boldsymbol{X}_t,\boldsymbol{\theta})\,\mathrm{d}\boldsymbol{B}_t$	
$\tilde{\mathbb{W}}_{\boldsymbol{\theta}}$	$: \mathrm{d}\boldsymbol{X}_t = \boldsymbol{\sigma}(\boldsymbol{X}_t,\boldsymbol{\theta})\,\mathrm{d}\boldsymbol{B}_t$	$, \boldsymbol{X}_T = \boldsymbol{x}$
$\mathbb{W}$	$: \mathrm{d}\boldsymbol{X}_t = \mathrm{d}\boldsymbol{B}_t$	
$\tilde{\mathbb{W}}$	$: \mathrm{d}\boldsymbol{X}_t = \mathrm{d}\boldsymbol{B}_t$	$, \boldsymbol{X}_T = \boldsymbol{x}$
$\mathbb{Z}_{\boldsymbol{\theta}}$	$: \mathrm{d}\boldsymbol{Z}_t = \boldsymbol{\sigma}^{-1}(\boldsymbol{X}_t,\boldsymbol{\theta})\left(\mathrm{d}\boldsymbol{X}_t - \frac{\boldsymbol{x}-\boldsymbol{X}_t}{T-t}\,\mathrm{d}t\right)$	$, \boldsymbol{Z}_0 = \mathbf{0}, \quad \boldsymbol{X}_t \sim \tilde{\mathbb{P}}_{\boldsymbol{\theta}}$

The general concept of the proposed update mechanism is as follows: As before, let $\mathbb{P}_{\boldsymbol{\theta}}$ denote the law induced by a diffusion satisfying the SDE

$$\mathrm{d}\boldsymbol{X}_t = \boldsymbol{\mu}(\boldsymbol{X}_t,\boldsymbol{\theta})\mathrm{d}t + \boldsymbol{\sigma}(\boldsymbol{X}_t,\boldsymbol{\theta})\mathrm{d}\boldsymbol{B}_t, \quad \boldsymbol{X}_0 = \boldsymbol{x}_0,$$

and let $\tilde{\mathbb{P}}_{\boldsymbol{\theta}}$ be the law of the same process but further conditioned on the end point $\boldsymbol{X}_T = \boldsymbol{x}$. In this section, a number of probability measures is introduced. For better lucidity, these are summarised in Table 7.6.

Once more, assume that the diffusion coefficient $\boldsymbol{\sigma}$ is invertible. For $\boldsymbol{X} \sim \tilde{\mathbb{P}}_{\boldsymbol{\theta}}$, define a process $\boldsymbol{Z} = (\boldsymbol{Z}_t)_{t\in[0,T]}$ through

$$\begin{aligned}\mathrm{d}\boldsymbol{Z}_t &= \boldsymbol{\sigma}^{-1}(\boldsymbol{X}_t,\boldsymbol{\theta})\left(\mathrm{d}\boldsymbol{X}_t - \frac{\boldsymbol{x}-\boldsymbol{X}_t}{T-t}\,\mathrm{d}t\right), \qquad \boldsymbol{Z}_0 = \mathbf{0}, \qquad (7.52)\\ &= \mathrm{d}\boldsymbol{B}_t + \boldsymbol{\sigma}^{-1}(\boldsymbol{X}_t,\boldsymbol{\theta})\left(\boldsymbol{\mu}(\boldsymbol{X}_t,\boldsymbol{\theta}) - \frac{\boldsymbol{x}-\boldsymbol{X}_t}{T-t}\right)\mathrm{d}t.\end{aligned}$$

Let $\mathbb{Z}_{\boldsymbol{\theta}}$ denote the law of $\boldsymbol{Z}$, and define a function g which is invertible in its first argument such that $\boldsymbol{X} = g(\boldsymbol{Z},\boldsymbol{\theta})$ and $\boldsymbol{Z} = g^{-1}(\boldsymbol{X},\boldsymbol{\theta})$. In terms of the process $\boldsymbol{Z}$, the respective SDE for this back-transformation reads

$$\mathrm{d}\boldsymbol{X}_t = \boldsymbol{\sigma}(\boldsymbol{X}_t,\boldsymbol{\theta})\mathrm{d}\boldsymbol{Z}_t + \frac{\boldsymbol{x}-\boldsymbol{X}_t}{T-t}\,\mathrm{d}t, \quad \boldsymbol{X}_0 = \boldsymbol{x}_0.$$

The initial value $\boldsymbol{x}_0$ and the final point $\boldsymbol{x}$ of $\boldsymbol{X}$ are considered fixed and are hence not included in the notation $\tilde{\mathbb{P}}_{\boldsymbol{\theta}}$, $\mathbb{Z}_{\boldsymbol{\theta}}$ and g.

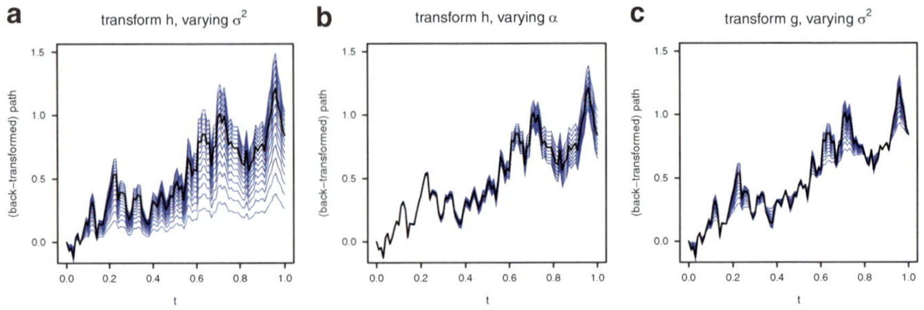

Fig. 7.27 Back-transformation of a sample path: The *thick black line* shows an exact discrete-time realisation of a one-dimensional Ornstein-Uhlenbeck process X satisfying the SDE $\mathrm{d}X_t = -\alpha X_t \mathrm{d}t + \sigma \mathrm{d}B_t$ with $X_0 = 0$ and $\boldsymbol{\theta} = (\alpha, \sigma^2)' = (0.5, 1)'$. The time grid consists of 100 equidistant time points in the interval $[0, 1]$. The path is transformed to $B = h^{-1}(X, \boldsymbol{\theta})$ and $Z = g^{-1}(X, \boldsymbol{\theta})$ defined through (7.49) and (7.52). Formulas for the discrete-time setting are given in (7.50), (7.51), (7.68) and (7.69). The *thin blue lines* show the back-transformations with respect to different parameters. Using the transform function h, the back-transformations do not hit the original end point of the diffusion path. (**a**) Back-transformations $h(B, \boldsymbol{\theta}^*)$ for $\boldsymbol{\theta}^* = (0.5, \sigma^{2*})'$ with $\sigma^{2*} \in \{0.1, 0.2, \ldots, 1.5\}$. (**b**) Back-transformations $h(B, \boldsymbol{\theta}^*)$ for $\boldsymbol{\theta}^* = (\alpha^*, 1)'$ with $\alpha^* \in \{0.1, 0.2, \ldots, 1\}$. (**c**) Back-transformations $g(Z, \boldsymbol{\theta}^*)$ for $\boldsymbol{\theta}^* = (0.5, \sigma^{2*})'$ with $\sigma^{2*} \in \{0.1, 0.2, \ldots, 1.5\}$. For $\boldsymbol{\theta}^* = (\alpha^*, 1)'$ with arbitrary α^*, the back-transformation equals the original process, i.e. $X = g(Z, \boldsymbol{\theta}^*)$

Like the process $\boldsymbol{B}$ defined in (7.49) above, $\boldsymbol{Z}$ has unit diffusion. Moreover, the construction of $\mathbb{Z}_{\boldsymbol{\theta}}$ explicitly involves the end point $\boldsymbol{x}$ of $\boldsymbol{X}$. It ensures that $g(g^{-1}(\boldsymbol{x}, \boldsymbol{\theta}), \boldsymbol{\theta}^*) = \boldsymbol{x}$ even for $\boldsymbol{\theta} \neq \boldsymbol{\theta}^*$. This can be seen from the following informal argument: Because of $\boldsymbol{X}_T = \boldsymbol{x}$, the time-discretisation of (7.52) at time T for a small time step ε,

$$\boldsymbol{Z}_T - \boldsymbol{Z}_{T-\varepsilon} = \boldsymbol{\sigma}^{-1}(\boldsymbol{X}_{T-\varepsilon}, \boldsymbol{\theta}) \left(\boldsymbol{X}_T - \boldsymbol{X}_{T-\varepsilon} - \frac{\boldsymbol{x} - \boldsymbol{X}_{T-\varepsilon}}{\varepsilon} \varepsilon \right),$$

implies that $\boldsymbol{Z}_T = \boldsymbol{Z}_{T-\varepsilon}$. As a consequence, the back-transformation at time T,

$$\boldsymbol{X}_T^* - \boldsymbol{X}_{T-\varepsilon}^* = \boldsymbol{\sigma}(\boldsymbol{X}_{T-\varepsilon}^*, \boldsymbol{\theta}^*)(\boldsymbol{Z}_T - \boldsymbol{Z}_{T-\varepsilon}) + \frac{\boldsymbol{x} - \boldsymbol{X}_{T-\varepsilon}^*}{\varepsilon} \varepsilon,$$

yields $\boldsymbol{X}_T^* = \boldsymbol{x}$ also for $\boldsymbol{\theta} \neq \boldsymbol{\theta}^*$. A formal reasoning is postponed to the proof of Lemma 7.3 on p. 248.

Figure 7.27 displays back-transformations of a one-dimensional diffusion path X based on the processes $B = h^{-1}(X, \boldsymbol{\theta})$ and $Z = g^{-1}(X, \theta)$ defined through the SDEs (7.49) and (7.52). Note that for all diffusion processes $\boldsymbol{X}$ one has $\boldsymbol{X} = g(g^{-1}(\boldsymbol{X}, \boldsymbol{\theta}), \boldsymbol{\theta}^*)$ if $\boldsymbol{\theta}^*$ is such that $\boldsymbol{\sigma}(\cdot, \boldsymbol{\theta}) = \boldsymbol{\sigma}(\cdot, \boldsymbol{\theta}^*)$ even for $\boldsymbol{\theta} \neq \boldsymbol{\theta}^*$. This is a meaningful characteristic of the transformation as the degeneracy issues from Sect. 7.3 involve only those components of $\boldsymbol{\theta}$ which enter the diffusion coefficient.

The process $\boldsymbol{Z}$ is not Brownian motion, but the corresponding measure $\mathbb{Z}_{\boldsymbol{\theta}}$ is absolutely continuous with respect to Wiener measure as will be shown in Lemma 7.2 on p. 248. As $\boldsymbol{Z}^{\text{imp}}$ has unit diffusion, it qualifies to take over the role of $\boldsymbol{B}^{\text{imp}} = h^{-1}(\boldsymbol{X}^{\text{imp}}, \boldsymbol{\theta})$ in the idea presented at the very beginning of the present Sect. 7.4.4. Indeed, the construction of $\boldsymbol{Z}$ can be seen as an attempt to mimic Brownian motion. We hence adopt the notation from Chib et al. (2004) and call $\boldsymbol{Z}$ an *innovation process*. Moreover, an update algorithm based on $\boldsymbol{Z}^{\text{imp}} = g^{-1}(\boldsymbol{X}^{\text{imp}}, \boldsymbol{\theta})$ will be referred to as *innovation scheme* in the following. The idea is to update $\boldsymbol{\theta}|\boldsymbol{Z}^{\text{imp}}, \boldsymbol{x}_0, \boldsymbol{x}$ and $\boldsymbol{Z}^{\text{imp}}|\boldsymbol{\theta}, \boldsymbol{x}_0, \boldsymbol{x}$ instead of $\boldsymbol{\theta}|\boldsymbol{X}^{\text{imp}}, \boldsymbol{x}_0, \boldsymbol{x}$ and $\boldsymbol{X}^{\text{imp}}|\boldsymbol{\theta}, \boldsymbol{x}_0, \boldsymbol{x}$, respectively. We suggest that the algorithms for the parameter and path updates then look as follows:

Algorithm 7.4 (Parameter Update). *Given $\boldsymbol{\theta}$, $\boldsymbol{x}_0$, $\boldsymbol{x}$ and $\boldsymbol{X}^{\text{imp}}$, perform the following steps:*

1. *Draw $\boldsymbol{\theta}^* \sim q(\boldsymbol{\theta}^*|\boldsymbol{\theta})$.*
2. *Compute $\boldsymbol{Z}^{\text{imp}} = g^{-1}(\boldsymbol{X}^{\text{imp}}, \boldsymbol{\theta})$.*
3. *Accept $\boldsymbol{\theta}^*$ with probability*

$$\zeta(\boldsymbol{\theta}^*, \boldsymbol{\theta}) = 1 \wedge \frac{\pi(\boldsymbol{\theta}^* \mid \boldsymbol{Z}^{\text{imp}}, \boldsymbol{x}_0, \boldsymbol{x})q(\boldsymbol{\theta} \mid \boldsymbol{\theta}^*)}{\pi(\boldsymbol{\theta} \mid \boldsymbol{Z}^{\text{imp}}, \boldsymbol{x}_0, \boldsymbol{x})q(\boldsymbol{\theta}^* \mid \boldsymbol{\theta})}, \tag{7.53}$$

   otherwise keep $\boldsymbol{\theta}$.
4. *If $\boldsymbol{\theta}^*$ was accepted, replace $\boldsymbol{X}^{\text{imp}}$ by $\boldsymbol{X}^{\text{imp}*} = g(\boldsymbol{Z}^{\text{imp}}, \boldsymbol{\theta}^*)$.*

Note that in step 1, the proposal density for the parameter is chosen such that it does neither depend on the observed nor on the imputed data.

Algorithm 7.5 (Path Update). *Given $\boldsymbol{\theta}$, $\boldsymbol{x}_0$, $\boldsymbol{x}$ and $\boldsymbol{X}^{\text{imp}}$, perform the following steps:*

1. *Compute $\boldsymbol{Z}^{\text{imp}} = g^{-1}(\boldsymbol{X}^{\text{imp}}, \boldsymbol{\theta})$.*
2. *Draw $\boldsymbol{Z}^{\text{imp}*} \sim q(\boldsymbol{Z}^{\text{imp}*}|\boldsymbol{Z}^{\text{imp}}, \boldsymbol{x}_0, \boldsymbol{x}, \boldsymbol{\theta})$ such that it has unit diffusion.*
3. *Accept $\boldsymbol{X}^{\text{imp}*} = g(\boldsymbol{Z}^{\text{imp}*}, \boldsymbol{\theta})$ with probability*

$$\zeta(\boldsymbol{Z}^{\text{imp}*}, \boldsymbol{Z}^{\text{imp}}) = 1 \wedge \frac{\pi(\boldsymbol{Z}^{\text{imp}*}|\boldsymbol{x}_0, \boldsymbol{x}, \boldsymbol{\theta})q(\boldsymbol{Z}^{\text{imp}}|\boldsymbol{Z}^{\text{imp}*}, \boldsymbol{x}_0, \boldsymbol{x}, \boldsymbol{\theta})}{\pi(\boldsymbol{Z}^{\text{imp}}|\boldsymbol{x}_0, \boldsymbol{x}, \boldsymbol{\theta})q(\boldsymbol{Z}^{\text{imp}*}|\boldsymbol{Z}^{\text{imp}}, \boldsymbol{x}_0, \boldsymbol{x}, \boldsymbol{\theta})}, \tag{7.54}$$

   otherwise keep $\boldsymbol{X}^{\text{imp}}$.

A concluding correction of the parameter corresponding to step 4 in Algorithm 7.4 is not necessary in the path update as the quadratic variation of both $\boldsymbol{X}^{\text{imp}}$ and $\boldsymbol{X}^{\text{imp}*}$ should be consistent with $\boldsymbol{\theta}$. Be aware that $\boldsymbol{X}^{\text{imp}*}$ has different definitions in the two algorithms: In the parameter update, it is constructed as $\boldsymbol{X}^{\text{imp}*} = g(\boldsymbol{Z}^{\text{imp}}, \boldsymbol{\theta}^*)$, and in the path update as $\boldsymbol{X}^{\text{imp}*} = g(\boldsymbol{Z}^{\text{imp}*}, \boldsymbol{\theta})$.

In the remainder of this section, we will show for both the parameter and the path update

(a) That the algorithms converge and
(b) That explicit formulas for the acceptance probabilities can be derived.

Assumptions

For the purposes of this section, we assume that the drift function $\boldsymbol{\mu}$ and the diffusion coefficient $\boldsymbol{\sigma}$ are bounded and that $\boldsymbol{\sigma}$ is invertible with bounded inverse $\boldsymbol{\sigma}^{-1}$. We generally consider time-homogeneous diffusions in this section, but in the following some results will also be proven for time-dependent drift and diffusion coefficient. In that case, $\boldsymbol{\mu}$ and $\boldsymbol{\sigma}$ are not only supposed to be twice continuously differentiable with respect to the state variable but also continuously differentiable with respect to time. These derivatives are required to be bounded as well.

Parameter Update

The acceptance probabilities (7.53) and (7.54) have been formulated using a rather informal generic notation where π denotes a collection of Lebesgue densities; compare with Sect. 7.1.1. In the following, we will distinguish between the densities of $\boldsymbol{X}^{\text{imp}}$ and $\boldsymbol{Z}^{\text{imp}} = g^{-1}(\boldsymbol{X}^{\text{imp}}, \boldsymbol{\theta})$ by writing

$$\pi(\boldsymbol{X}^{\text{imp}} \,|\, \boldsymbol{x}_0, \boldsymbol{x}, \boldsymbol{\theta}) = \frac{\mathrm{d}\tilde{\mathbb{P}}_{\boldsymbol{\theta}}}{\mathrm{d}\mathbb{L}}\left(\boldsymbol{X}^{\text{imp}}\right) \quad \text{and} \quad \pi(\boldsymbol{Z}^{\text{imp}} \,|\, \boldsymbol{x}_0, \boldsymbol{x}, \boldsymbol{\theta}) = \frac{\mathrm{d}\mathbb{Z}_{\boldsymbol{\theta}}}{\mathrm{d}\mathbb{L}}\left(\boldsymbol{Z}^{\text{imp}}\right),$$

where $\mathbb{L}$ is Lebesgue measure. Define $f_{\boldsymbol{\theta}}$ and $\tilde{f}_{\boldsymbol{\theta}}$ to be the Lebesgue densities under $\mathbb{P}_{\boldsymbol{\theta}}$ and $\tilde{\mathbb{P}}_{\boldsymbol{\theta}}$, respectively, i.e.

$$\mathrm{d}\mathbb{P}_{\boldsymbol{\theta}} = f_{\boldsymbol{\theta}}\,\mathrm{d}\mathbb{L} \qquad \text{and} \qquad \mathrm{d}\tilde{\mathbb{P}}_{\boldsymbol{\theta}} = \tilde{f}_{\boldsymbol{\theta}}\,\mathrm{d}\mathbb{L}.$$

Then the acceptance probability (7.53) for the parameter becomes

$$\zeta(\boldsymbol{\theta}^*, \boldsymbol{\theta}) = 1 \wedge \frac{\mathrm{d}\mathbb{Z}_{\boldsymbol{\theta}^*}(\boldsymbol{Z}^{\text{imp}})\, f_{\boldsymbol{\theta}^*}(\boldsymbol{x})\, p(\boldsymbol{\theta}^*)\, q(\boldsymbol{\theta} \,|\, \boldsymbol{\theta}^*)}{\mathrm{d}\mathbb{Z}_{\boldsymbol{\theta}}(\boldsymbol{Z}^{\text{imp}})\ f_{\boldsymbol{\theta}}(\boldsymbol{x})\ p(\boldsymbol{\theta})\ q(\boldsymbol{\theta}^* | \,\boldsymbol{\theta}\,)}\,. \tag{7.55}$$

Two objectives are considered in the following: First, show that this acceptance probability behaves nicely as part of the MCMC algorithm. Second, obtain explicit expressions for (7.55) such that it is of practical use. We assume that p and q are known, sufficiently regular and can be evaluated.

Corollary 7.1. *The quotient* $(\mathrm{d}\mathbb{Z}_{\boldsymbol{\theta}^*}/\mathrm{d}\mathbb{Z}_{\boldsymbol{\theta}})(\boldsymbol{Z}^{\text{imp}})$ *does not degenerate as described on p. 231; both its numerator and denominator are finite.*

Proof. Introduce the probability measure $\mathbb{D}_{\mathbf{0},\boldsymbol{\theta}}$ which is induced by a diffusion fulfilling the SDE

$$\mathrm{d}\boldsymbol{X}_t = \boldsymbol{\sigma}(\boldsymbol{X}_t,\boldsymbol{\theta})\,\mathrm{d}\boldsymbol{B}_t + \frac{\boldsymbol{x}-\boldsymbol{X}_t}{T-t}\,\mathrm{d}t, \quad \boldsymbol{X}_0 = \boldsymbol{x}_0. \tag{7.56}$$

For an overview, Table 7.6 on p. 243 lists all measures defined in the context of the innovation scheme. $\mathbb{D}_{\mathbf{0},\boldsymbol{\theta}}$ defines a diffusion process which almost surely reaches the state $\boldsymbol{x}$ at time T (Delyon and Hu 2006, Lemma 4). For $\boldsymbol{\Sigma} = \boldsymbol{\sigma}\boldsymbol{\sigma}'$ not depending on the state variable, one has $\mathbb{D}_{\mathbf{0},\boldsymbol{\theta}} = \tilde{\mathbb{W}}_{\boldsymbol{\theta}}$. In particular, for $\boldsymbol{\Sigma} \equiv \boldsymbol{I}$ the measure $\mathbb{D}_{\mathbf{0},\boldsymbol{\theta}}$ reduces to $\tilde{\mathbb{W}}$, that is the law of a d-dimensional Brownian $(0,\boldsymbol{x}_0,T,\boldsymbol{x})$-bridge (e.g. Karatzas and Shreve 1991, Sect. 5.6).

The function g which connects $\boldsymbol{X}^{\text{imp}}$ and $\boldsymbol{Z}^{\text{imp}}$ is chosen such that the relationship between $\tilde{\mathbb{P}}_{\boldsymbol{\theta}}$ and $\mathbb{Z}_{\boldsymbol{\theta}}$ is the same as the link between $\mathbb{D}_{\mathbf{0},\boldsymbol{\theta}}$ and Wiener measure $\mathbb{W}$, i.e.

$$\boldsymbol{Z} \sim \mathbb{Z}_{\boldsymbol{\theta}} \qquad \Leftrightarrow \qquad g(\boldsymbol{Z},\boldsymbol{\theta}) \sim \tilde{\mathbb{P}}_{\boldsymbol{\theta}} \tag{7.57}$$

and

$$\boldsymbol{B} \sim \mathbb{W} \qquad \Leftrightarrow \qquad g(\boldsymbol{B},\boldsymbol{\theta}) \sim \mathbb{D}_{\mathbf{0},\boldsymbol{\theta}}. \tag{7.58}$$

Consequently, the change of variables theorem yields

$$\frac{\mathrm{d}\mathbb{Z}_{\boldsymbol{\theta}}}{\mathrm{d}\mathbb{L}}\left(\boldsymbol{Z}^{\text{imp}}\right) = \left|J\left(g(\boldsymbol{Z}^{\text{imp}},\boldsymbol{\theta})\right)\right| \frac{\mathrm{d}\tilde{\mathbb{P}}_{\boldsymbol{\theta}}}{\mathrm{d}\mathbb{L}}\left(g(\boldsymbol{Z}^{\text{imp}},\boldsymbol{\theta})\right) \tag{7.59}$$

and

$$\frac{\mathrm{d}\mathbb{W}}{\mathrm{d}\mathbb{L}}\left(\boldsymbol{Z}^{\text{imp}}\right) = \left|J\left(g(\boldsymbol{Z}^{\text{imp}},\boldsymbol{\theta})\right)\right| \frac{\mathrm{d}\mathbb{D}_{\mathbf{0},\boldsymbol{\theta}}}{\mathrm{d}\mathbb{L}}\left(g(\boldsymbol{Z}^{\text{imp}},\boldsymbol{\theta})\right), \tag{7.60}$$

where

$$J\left(g(\boldsymbol{Z}^{\text{imp}},\boldsymbol{\theta})\right) = \left|\frac{\partial g(\boldsymbol{Z}^{\text{imp}},\boldsymbol{\theta})}{\partial \boldsymbol{Z}^{\text{imp}}}\right|$$

is the Jacobian determinant of g. Then

$$\begin{aligned}\frac{\mathrm{d}\mathbb{Z}_{\boldsymbol{\theta}^*}(\boldsymbol{Z}^{\text{imp}})}{\mathrm{d}\mathbb{Z}_{\boldsymbol{\theta}}(\boldsymbol{Z}^{\text{imp}})} \\ = \frac{\dfrac{\mathrm{d}\mathbb{Z}_{\boldsymbol{\theta}^*}}{\mathrm{d}\mathbb{W}}(\boldsymbol{Z}^{\text{imp}})}{\dfrac{\mathrm{d}\mathbb{Z}_{\boldsymbol{\theta}}}{\mathrm{d}\mathbb{W}}(\boldsymbol{Z}^{\text{imp}})} = \frac{\dfrac{\mathrm{d}\tilde{\mathbb{P}}_{\boldsymbol{\theta}^*}}{\mathrm{d}\mathbb{D}_{\mathbf{0},\boldsymbol{\theta}^*}}\left(g(\boldsymbol{Z}^{\text{imp}},\boldsymbol{\theta}^*)\right)}{\dfrac{\mathrm{d}\tilde{\mathbb{P}}_{\boldsymbol{\theta}}}{\mathrm{d}\mathbb{D}_{\mathbf{0},\boldsymbol{\theta}}}\left(g(\boldsymbol{Z}^{\text{imp}},\boldsymbol{\theta})\right)} = \frac{\dfrac{\mathrm{d}\tilde{\mathbb{P}}_{\boldsymbol{\theta}^*}}{\mathrm{d}\mathbb{D}_{\mathbf{0},\boldsymbol{\theta}^*}}(\boldsymbol{X}^{\text{imp}*})}{\dfrac{\mathrm{d}\tilde{\mathbb{P}}_{\boldsymbol{\theta}}}{\mathrm{d}\mathbb{D}_{\mathbf{0},\boldsymbol{\theta}}}(\boldsymbol{X}^{\text{imp}})}\end{aligned} \tag{7.61}$$

In the last expression, the numerator and denominator differ in both parameter and imputed data. A situation as described on p. 231, where always either the numerator or denominator of the acceptance probability is zero, does therefore not occur. Under the assumptions on p. 246 regarding $\boldsymbol{\mu}$ and $\boldsymbol{\sigma}$, Delyon and Hu (2006, Theorem 6) prove that $\tilde{\mathbb{P}}_{\boldsymbol{\theta}}$ is absolutely continuous with respect to $\mathbb{D}_{0,\boldsymbol{\theta}}$ for all $\boldsymbol{\theta}$. Hence, both the numerator and denominator of the last fraction in (7.61) are finite. □

We are now in the position to show the following two propositions which were already mentioned earlier in this section.

Lemma 7.2. *$\mathbb{Z}_{\boldsymbol{\theta}}$ is absolutely continuous with respect to* $\mathbb{W}$.

Proof. In the proof of Corollary 7.1 it was pointed out that $\tilde{\mathbb{P}}_{\boldsymbol{\theta}}$ is absolutely continuous with respect to $\mathbb{D}_{0,\boldsymbol{\theta}}$, where the latter is induced by the solution of (7.56). As these two measures are linked with the measures $\mathbb{Z}_{\boldsymbol{\theta}}$ and $\mathbb{W}$ in the same deterministic way—see (7.57) and (7.58)—, this also proves that $\mathbb{Z}_{\boldsymbol{\theta}}$ is absolutely continuous with respect to $\mathbb{W}$. □

Lemma 7.3. *The back-transformation $\boldsymbol{X}^* = g(g^{-1}(\boldsymbol{X},\boldsymbol{\theta}),\boldsymbol{\theta}^*)$ for $\boldsymbol{X} \sim \tilde{\mathbb{P}}_{\boldsymbol{\theta}}$ hits the required end point $\boldsymbol{X}^*_T = \boldsymbol{x}$.*

Proof. Let $\boldsymbol{Z} \sim \mathbb{Z}_{\boldsymbol{\theta}}$ and $\boldsymbol{B} \sim \mathbb{W}$. As $\mathbb{Z}_{\boldsymbol{\theta}} \ll \mathbb{W}$, the process $\boldsymbol{X}^* = g(\boldsymbol{Z},\boldsymbol{\theta}^*)$ induces a law $\mathbb{Q}_{\boldsymbol{\theta},\boldsymbol{\theta}^*}$ which is absolutely continuous with respect to the law $\mathbb{D}_{0,\boldsymbol{\theta}^*}$ of $g(\boldsymbol{B},\boldsymbol{\theta}^*)$. More precisely, $\mathbb{D}_{0,\boldsymbol{\theta}^*}(\boldsymbol{X}^*) = 0$ implies $\mathbb{Q}_{\boldsymbol{\theta},\boldsymbol{\theta}^*}(\boldsymbol{X}^*)=0$. Under $\mathbb{D}_{0,\boldsymbol{\theta}^*}$, the diffusion process almost surely hits the desired end point $\boldsymbol{x}$. Consequently, this must be true also under $\mathbb{Q}_{\boldsymbol{\theta},\boldsymbol{\theta}^*}$. Hence $g(g^{-1}(\boldsymbol{x},\boldsymbol{\theta}),\boldsymbol{\theta}^*) = \boldsymbol{x}$ as was to be shown. □

Be aware that the imputed data $\boldsymbol{X}^{\mathrm{imp}}$ consists of all values $\boldsymbol{X}_t$ for $t \in (0,T)$. It is crucial that $\boldsymbol{X}_T$ does not belong to $\boldsymbol{X}^{\mathrm{imp}}$. The starting value $\boldsymbol{x}_0$ is formally not included in $\boldsymbol{X}^{\mathrm{imp}}$ either. This value is however inherent in all measures considered in this section (compare with Table 7.6) and does not depend on $\boldsymbol{\theta}$. It is hence reasonable to incorporate the initial value in the integrals on the following pages. For convenience, let $\boldsymbol{X}^{\mathrm{imp}} = (\boldsymbol{X}_t)_{t\in(0,T-\varepsilon]}$ for a small but positive constant ε. This is also abbreviated as $\boldsymbol{X}_{(0,T-\varepsilon]}$.

The utilisation of the acceptance probability (7.55) in an MCMC algorithm requires an explicit formula such that it can be evaluated in practice. Hence consider the following corollary.

Corollary 7.4. *An explicit expression for $(\mathrm{d}\mathbb{Z}_{\boldsymbol{\theta}^*}/\mathrm{d}\mathbb{Z}_{\boldsymbol{\theta}})(\boldsymbol{Z}^{\mathrm{imp}}) \cdot (f_{\boldsymbol{\theta}^*}/f_{\boldsymbol{\theta}})(\boldsymbol{x})$ as part of the acceptance probability* (7.55) *is available.*

Proof. Consider the relationship between $\tilde{\mathbb{P}}_{\boldsymbol{\theta}}$ and $\mathbb{P}_{\boldsymbol{\theta}}$. These two measures differ by the end point condition of $\tilde{\mathbb{P}}_{\boldsymbol{\theta}}$. Heuristically, one has

$$\begin{aligned}\frac{\mathrm{d}\tilde{\mathbb{P}}_{\boldsymbol{\theta}}}{\mathrm{d}\mathbb{L}}\left(\boldsymbol{X}^{\mathrm{imp}}\right) = \tilde{f}_{\boldsymbol{\theta}}(\boldsymbol{X}^{\mathrm{imp}}) &= \frac{f_{\boldsymbol{\theta}}(\boldsymbol{x}|\boldsymbol{X}^{\mathrm{imp}})f_{\boldsymbol{\theta}}(\boldsymbol{X}^{\mathrm{imp}})}{f_{\boldsymbol{\theta}}(\boldsymbol{x})} \\ &= \frac{f_{\boldsymbol{\theta}}(\boldsymbol{x}|\boldsymbol{X}_{T-\varepsilon})}{f_{\boldsymbol{\theta}}(\boldsymbol{x})}\frac{\mathrm{d}\mathbb{P}_{\boldsymbol{\theta}}}{\mathrm{d}\mathbb{L}}\left(\boldsymbol{X}^{\mathrm{imp}}\right).\end{aligned} \tag{7.62}$$

Hence

$$\frac{\mathrm{d}\mathbb{Z}_{\boldsymbol{\theta}^*}(\boldsymbol{Z}^{\mathrm{imp}})\, f_{\boldsymbol{\theta}^*}(\boldsymbol{x})}{\mathrm{d}\mathbb{Z}_{\boldsymbol{\theta}}(\boldsymbol{Z}^{\mathrm{imp}})\, f_{\boldsymbol{\theta}}(\boldsymbol{x})}$$

$$= \frac{\left(\dfrac{\mathrm{d}\tilde{\mathbb{P}}_{\boldsymbol{\theta}^*}}{\mathrm{d}\mathbb{P}_{\boldsymbol{\theta}^*}} \dfrac{\mathrm{d}\mathbb{P}_{\boldsymbol{\theta}^*}}{\mathrm{d}\mathbb{D}_{\mathbf{0},\boldsymbol{\theta}^*}}\right)(\boldsymbol{X}^{\mathrm{imp}*})\, f_{\boldsymbol{\theta}^*}(\boldsymbol{x})}{\left(\dfrac{\mathrm{d}\tilde{\mathbb{P}}_{\boldsymbol{\theta}}}{\mathrm{d}\mathbb{P}_{\boldsymbol{\theta}}} \dfrac{\mathrm{d}\mathbb{P}_{\boldsymbol{\theta}}}{\mathrm{d}\mathbb{D}_{\mathbf{0},\boldsymbol{\theta}}}\right)(\boldsymbol{X}^{\mathrm{imp}})\, f_{\boldsymbol{\theta}}(\boldsymbol{x})} = \frac{\dfrac{\mathrm{d}\mathbb{P}_{\boldsymbol{\theta}^*}}{\mathrm{d}\mathbb{D}_{\mathbf{0},\boldsymbol{\theta}^*}}(\boldsymbol{X}^{\mathrm{imp}*})\, f_{\boldsymbol{\theta}^*}(\boldsymbol{x}|\boldsymbol{X}^*_{T-\varepsilon})}{\dfrac{\mathrm{d}\mathbb{P}_{\boldsymbol{\theta}}}{\mathrm{d}\mathbb{D}_{\mathbf{0},\boldsymbol{\theta}}}(\boldsymbol{X}^{\mathrm{imp}})\, f_{\boldsymbol{\theta}}(\boldsymbol{x}|\boldsymbol{X}_{T-\varepsilon})}.$$

There is no analytically explicit form for $f_{\boldsymbol{\theta}}(\boldsymbol{x}|\boldsymbol{X}_{T-\varepsilon})$; that is the Lebesgue density for the transition from $\boldsymbol{X}_{T-\varepsilon}$ to $\boldsymbol{x}$ within time ε, where $\boldsymbol{X} \sim \mathbb{P}_{\boldsymbol{\theta}}$. However, for small ε, an approximation via e.g. the Euler scheme should be possible. For the calculation of $\mathrm{d}\mathbb{P}_{\boldsymbol{\theta}}/\mathrm{d}\mathbb{D}_{\mathbf{0},\boldsymbol{\theta}} = (\mathrm{d}\mathbb{P}_{\boldsymbol{\theta}}/\mathrm{d}\mathbb{W}_{\boldsymbol{\theta}})(\mathrm{d}\mathbb{W}_{\boldsymbol{\theta}}/\mathrm{d}\mathbb{D}_{\mathbf{0},\boldsymbol{\theta}})$, Girsanov's formula from Sect. 3.2.12 seems appropriate. The drift term $(\boldsymbol{x} - \boldsymbol{X}_t)/(T-t)$ under $\mathbb{D}_{\mathbf{0},\boldsymbol{\theta}}$, however, explodes as $t \to T$ and hence Novikov's condition in Sect. 3.2.12 may not be fulfilled. Nevertheless, Delyon and Hu (2006, Theorem 1) prove a generalisation of Girsanov's formula which holds under weaker conditions and which is applicable in the present case. With this theorem, one obtains the same result as under uncritical application of (3.25); that is

$$\log\left(\frac{\mathrm{d}\mathbb{P}_{\boldsymbol{\theta}}}{\mathrm{d}\mathbb{D}_{\mathbf{0},\boldsymbol{\theta}}}(\boldsymbol{X}^{\mathrm{imp}})\right) = \log\left(\frac{\mathrm{d}\mathbb{P}_{\boldsymbol{\theta}}}{\mathrm{d}\mathbb{W}_{\boldsymbol{\theta}}}(\boldsymbol{X}^{\mathrm{imp}})\right) + \log\left(\frac{\mathrm{d}\mathbb{W}_{\boldsymbol{\theta}}}{\mathrm{d}\mathbb{D}_{\mathbf{0},\boldsymbol{\theta}}}(\boldsymbol{X}^{\mathrm{imp}})\right)$$

$$= \int_0^{T-\varepsilon} \left(\boldsymbol{\mu}(\boldsymbol{X}_t, \boldsymbol{\theta}) - \frac{\boldsymbol{x} - \boldsymbol{X}_t}{T-t}\right)' \boldsymbol{\Sigma}^{-1}(\boldsymbol{X}_t, \boldsymbol{\theta})\, \mathrm{d}\boldsymbol{X}_t \tag{7.63}$$

$$-\frac{1}{2}\int_0^{T-\varepsilon} \left(\boldsymbol{\mu}'(\boldsymbol{X}_t, \boldsymbol{\theta})\boldsymbol{\Sigma}^{-1}(\boldsymbol{X}_t, \boldsymbol{\theta})\boldsymbol{\mu}(\boldsymbol{X}_t, \boldsymbol{\theta}) - \frac{(\boldsymbol{x} - \boldsymbol{X}_t)'\boldsymbol{\Sigma}^{-1}(\boldsymbol{X}_t, \boldsymbol{\theta})(\boldsymbol{x} - \boldsymbol{X}_t)}{(T-t)^2}\right)\mathrm{d}t.$$

All integrands in this expression are explicitly known as functions of $\boldsymbol{X}$ and $\boldsymbol{\theta}$. □

Corollary 7.1 proves that both the numerator and denominator of (7.61) are finite. Since all other components of (7.55) are supposed to be sufficiently regular, this property carries forward to the entire quotient in the acceptance probability (7.55). A further desirable property would be that the numerator and denominator are even bounded. Otherwise, for a bounded proposal density q, the Markov chain generated by the Metropolis-Hastings algorithm with acceptance probability (7.55) may dwell too long in single states and show bad mixing behaviour. For instance, Tierney (1994) and Mengersen and Tweedie (1996) show that an independence sampler with target density π and proposal density q is uniformly ergodic if there exists a constant $c > 0$ such that $q(y)/\pi(y) \geq c$ for all y in a possibly multi-dimensional real state space. Otherwise, the algorithm is not even geometrically ergodic. For details, see the original papers or the book by Roberts and Tweedie (2012). Analogous statements for the framework of this section are, however, beyond the scope of this book.

The proof of Corollary 7.4 provides an explicit formula for the critical part of the acceptance probability (7.55); because of the term $(T-t)^2$ in the denominator of the last term of (7.63), it is however not evident whether the exponential function of this expression, together with the factor $f_{\boldsymbol{\theta}}(\boldsymbol{x}|\boldsymbol{X}_{T-\varepsilon})$, is bounded. In the following we hence rearrange the above terms in an appropriate way. To that end, we follow the line of the proof of Theorem 5 in Delyon and Hu (2006) who derive expressions for $\mathrm{d}\tilde{\mathbb{P}}_{\boldsymbol{\theta}}/\mathrm{d}\mathbb{D}_{\boldsymbol{\mu},\boldsymbol{\theta}}$ and $\mathrm{d}\tilde{\mathbb{P}}_{\boldsymbol{\theta}}/\mathrm{d}\mathbb{D}_{\boldsymbol{0},\boldsymbol{\theta}}$, where $\mathbb{D}_{\boldsymbol{\mu},\boldsymbol{\theta}}$ is defined in Table 7.6 on p. 243 and employed in the path update below.

Corollary 7.5. *One has*

$$\frac{\mathrm{d}\tilde{\mathbb{P}}_{\boldsymbol{\theta}}}{\mathrm{d}\mathbb{D}_{\boldsymbol{\mu},\boldsymbol{\theta}}}\big(\boldsymbol{X}_{(0,T-\varepsilon]}\big) = \exp\left(-\int_0^{T-\varepsilon} \frac{D_1(\boldsymbol{X}_t,\boldsymbol{\theta})+D_2(\boldsymbol{X}_t,\boldsymbol{\theta})+D_3(\boldsymbol{X}_t,\boldsymbol{\theta})}{2(T-t)}\right)$$
$$\cdot \left(\frac{T}{\varepsilon}\right)^{-\frac{d(d-1)}{2}} \frac{\phi\big(\boldsymbol{x}\,|\,\boldsymbol{x}_0, T\boldsymbol{\Sigma}(\boldsymbol{x}_0,\boldsymbol{\theta})\big)}{f_{\boldsymbol{\theta}}(\boldsymbol{x})} \cdot \frac{|\boldsymbol{\Sigma}(\boldsymbol{x}_0,\boldsymbol{\theta})|^{\frac{1}{2}}}{|\boldsymbol{\Sigma}(\boldsymbol{X}_{T-\varepsilon},\boldsymbol{\theta})|^{\frac{1}{2}}} \tag{7.64}$$

and

$$\frac{\mathrm{d}\tilde{\mathbb{P}}_{\boldsymbol{\theta}}}{\mathrm{d}\mathbb{D}_{\boldsymbol{0},\boldsymbol{\theta}}}\big(\boldsymbol{X}_{(0,T-\varepsilon]}\big) = \exp\left(-\int_0^{T-\varepsilon} \frac{D_2(\boldsymbol{X}_t,\boldsymbol{\theta})+D_3(\boldsymbol{X}_t,\boldsymbol{\theta})}{2(T-t)}\right)$$
$$\cdot \exp\left(\int_0^{T-\varepsilon} \boldsymbol{\mu}'(\boldsymbol{X}_t,\boldsymbol{\theta})\boldsymbol{\Sigma}^{-1}(\boldsymbol{X}_t,\boldsymbol{\theta})\mathrm{d}\boldsymbol{X}_t - \frac{1}{2}\int_0^{T-\varepsilon} \boldsymbol{\mu}'(\boldsymbol{X}_t,\boldsymbol{\theta})\boldsymbol{\Sigma}^{-1}(\boldsymbol{X}_t,\boldsymbol{\theta})\boldsymbol{\mu}(\boldsymbol{X}_t,\boldsymbol{\theta})\mathrm{d}t\right) \tag{7.65}$$
$$\cdot \frac{\phi\big(\boldsymbol{x}\,|\,\boldsymbol{x}_0, T\boldsymbol{\Sigma}(\boldsymbol{x}_0,\boldsymbol{\theta})\big)}{f_{\boldsymbol{\theta}}(\boldsymbol{x})} \frac{|\boldsymbol{\Sigma}(\boldsymbol{x}_0,\boldsymbol{\theta})|^{\frac{1}{2}}}{|\boldsymbol{\Sigma}(\boldsymbol{X}_{T-\varepsilon},\boldsymbol{\theta})|^{\frac{1}{2}}} \left(\frac{T}{\varepsilon}\right)^{-\frac{d(d-1)}{2}},$$

where

$$D_1(\boldsymbol{X}_t,\boldsymbol{\theta}) = -2(\boldsymbol{x}-\boldsymbol{X}_t)'\boldsymbol{\Sigma}^{-1}(\boldsymbol{X}_t,\boldsymbol{\theta})\boldsymbol{\mu}(\boldsymbol{X}_t,\boldsymbol{\theta})\mathrm{d}t$$

$$D_2(\boldsymbol{X}_t,\boldsymbol{\theta}) = (\boldsymbol{x}-\boldsymbol{X}_t)'\big(\mathrm{d}\boldsymbol{\Sigma}^{-1}(\boldsymbol{X}_t,\boldsymbol{\theta})\big)(\boldsymbol{x}-\boldsymbol{X}_t)$$

$$D_3(\boldsymbol{X}_t, \boldsymbol{\theta}) = -\sum_{i=1}^{d}\sum_{j=1}^{d} \frac{(\boldsymbol{x} - \boldsymbol{X}_t)'\left(\frac{\partial \boldsymbol{\Sigma}^{-1}(\boldsymbol{X}_t, \boldsymbol{\theta})}{\partial x^{(j)}}\boldsymbol{e}_i + \frac{\partial \boldsymbol{\Sigma}^{-1}(\boldsymbol{X}_t, \boldsymbol{\theta})}{\partial x^{(i)}}\boldsymbol{e}_j\right)}{T - t} \mathrm{d}X_t^{(i)}\mathrm{d}X_t^{(j)}.$$

In these formulas, $\phi(\boldsymbol{y}|\boldsymbol{\nu}, \boldsymbol{\Lambda})$ *is the multivariate Gaussian density with mean* $\boldsymbol{\nu}$ *and covariance matrix* $\boldsymbol{\Lambda}$ *evaluated at* $\boldsymbol{y}$*, and* $|\boldsymbol{A}|$ *denotes the determinant of a square matrix* $\boldsymbol{A}$*. Furthermore,* $\boldsymbol{e}_i$ *is the ith unit vector of dimension d,* $\mathrm{d}X_t^{(i)}$ *is the ith component of* $\mathrm{d}\boldsymbol{X}_t$*, and* $\partial/\partial x^{(i)}$ *denotes differentiation with respect to the ith component of the state variable.*

Proof. The calculations are carried out using the heuristic approach (7.62). Due to space restrictions, they are moved to Sect. B.7 in the Appendix. □

Remark 7.2. Under the regularity conditions from p. 246, Delyon and Hu derive very similar expressions for $\mathrm{d}\tilde{\mathbb{P}}_{\boldsymbol{\theta}}/\mathrm{d}\mathbb{D}_{\boldsymbol{\mu},\boldsymbol{\theta}}$ and $\mathrm{d}\tilde{\mathbb{P}}_{\boldsymbol{\theta}}/\mathrm{d}\mathbb{D}_{\boldsymbol{0},\boldsymbol{\theta}}$. However, their results are obtained in a different context than here; applied to $\boldsymbol{X}_{[0,T]}$, the formulas are provided up to proportionality constants which do not depend on $\boldsymbol{X}_{(0,T)}$ but on $\boldsymbol{\theta}$, $\boldsymbol{x}_0$ and $\boldsymbol{x}$. In particular, Delyon and Hu (2006, Theorems 5 and 6) show that

$$\frac{\mathrm{d}\tilde{\mathbb{P}}_{\boldsymbol{\theta}}}{\mathrm{d}\mathbb{D}_{\boldsymbol{\mu},\boldsymbol{\theta}}}\left(\boldsymbol{X}_{[0,T]}\right) \propto \exp\left(-\int_0^T \frac{D_1(\boldsymbol{X}_t, \boldsymbol{\theta}) + D_2(\boldsymbol{X}_t, \boldsymbol{\theta}) + D_3(\boldsymbol{X}_t, \boldsymbol{\theta})}{2(T-t)}\right)$$

and

$$\frac{\mathrm{d}\tilde{\mathbb{P}}_{\boldsymbol{\theta}}}{\mathrm{d}\mathbb{D}_{\boldsymbol{0},\boldsymbol{\theta}}}\left(\boldsymbol{X}_{[0,T]}\right) \propto \exp\left(-\int_0^T \frac{D_2(\boldsymbol{X}_t, \boldsymbol{\theta}) + D_3(\boldsymbol{X}_t, \boldsymbol{\theta})}{2(T-t)}\right)$$

$$\cdot \exp\left(\int_0^T \boldsymbol{\mu}'(\boldsymbol{X}_t, \boldsymbol{\theta})\boldsymbol{\Sigma}^{-1}(\boldsymbol{X}_t, \boldsymbol{\theta})\mathrm{d}\boldsymbol{X}_t - \frac{1}{2}\int_0^T \boldsymbol{\mu}'(\boldsymbol{X}_t, \boldsymbol{\theta})\boldsymbol{\Sigma}^{-1}(\boldsymbol{X}_t, \boldsymbol{\theta})\boldsymbol{\mu}(\boldsymbol{X}_t, \boldsymbol{\theta})\mathrm{d}t\right),$$

where proportionality constants contain $\boldsymbol{\theta}$, $\boldsymbol{x}_0$ and $\boldsymbol{x}$. In this section, however, we want to apply the Radon-Nikodym derivatives to $\boldsymbol{X}_{(0,T-\varepsilon]}$ instead of $\boldsymbol{X}_{[0,T]}$, which leads to additional changes of the above formulas. In order to obtain all components of the derivatives which are relevant in the present context, the derivations including all constants were performed in this book.

The results in Corollary 7.5 require that ε is chosen arbitrarily small such that for the transition from $\boldsymbol{X}_{T-\varepsilon}$ to $\boldsymbol{x}$ one can simply assume a Gaussian increment

$$\boldsymbol{X}_T|\boldsymbol{X}_{T-\varepsilon}, \boldsymbol{\theta} \sim \mathcal{N}\big(\boldsymbol{X}_{T-\varepsilon}, \varepsilon\boldsymbol{\Sigma}(\boldsymbol{X}_{T-\varepsilon}, \boldsymbol{\theta})\big).$$

Otherwise, the multiplicative correction term

$$\frac{f_{\boldsymbol{\theta}}(\boldsymbol{x} \mid \boldsymbol{X}_{T-\varepsilon})}{\phi(\boldsymbol{x} \mid \boldsymbol{X}_{T-\varepsilon}, \varepsilon\boldsymbol{\Sigma}(\boldsymbol{X}_{T-\varepsilon}, \boldsymbol{\theta}))}$$

$$\approx \frac{\phi(\boldsymbol{x} \mid \boldsymbol{X}_{T-\varepsilon} + \varepsilon\boldsymbol{\mu}(\boldsymbol{X}_{T-\varepsilon}, \boldsymbol{\theta}), \varepsilon\boldsymbol{\Sigma}(\boldsymbol{X}_{T-\varepsilon}, \boldsymbol{\theta}))}{\phi(\boldsymbol{x} \mid \boldsymbol{X}_{T-\varepsilon}, \varepsilon\boldsymbol{\Sigma}(\boldsymbol{X}_{T-\varepsilon}, \boldsymbol{\theta}))}$$

$$= \exp\Big(\boldsymbol{\mu}'(\boldsymbol{X}_{T-\varepsilon}, \boldsymbol{\theta})\boldsymbol{\Sigma}^{-1}(\boldsymbol{X}_{T-\varepsilon}, \boldsymbol{\theta})\big(\boldsymbol{x} - \boldsymbol{X}_{T-\varepsilon} - \frac{\varepsilon}{2}\,\boldsymbol{\mu}(\boldsymbol{X}_{T-\varepsilon}, \boldsymbol{\theta})\big)\Big) \quad (7.66)$$

should be included. Respective results as in Corollary 7.5 also hold if $\boldsymbol{\mu}$ and $\boldsymbol{\sigma}$ are time-dependent. These are likewise derived and provided in Sect. B.7.

Overall, we arrive at the following concluding theorem.

Theorem 7.6. *The parameter update can be performed by application of Algorithm 7.4 on p. 245 with acceptance probability*

$$\zeta(\boldsymbol{\theta}^*, \boldsymbol{\theta}) = 1 \wedge \frac{\frac{d\tilde{\mathbb{P}}_{\boldsymbol{\theta}^*}}{d\mathbb{D}_{\mathbf{0},\boldsymbol{\theta}^*}}\big(g(\boldsymbol{Z}^{\mathrm{imp}}, \boldsymbol{\theta}^*)\big)\, f_{\boldsymbol{\theta}^*}(\boldsymbol{x})\, p(\boldsymbol{\theta}^*)\, q(\boldsymbol{\theta} \mid \boldsymbol{\theta}^*)}{\frac{d\tilde{\mathbb{P}}_{\boldsymbol{\theta}}}{d\mathbb{D}_{\mathbf{0},\boldsymbol{\theta}}}\big(g(\boldsymbol{Z}^{\mathrm{imp}}, \boldsymbol{\theta})\big)\, f_{\boldsymbol{\theta}}(\boldsymbol{x})\, p(\boldsymbol{\theta})\; q(\boldsymbol{\theta}^* \mid \boldsymbol{\theta})}. \quad (7.67)$$

An explicit formula for this probability is available due to Corollary 7.5. The acceptance probability is regular in the sense that both the numerator and denominator of the quotient in (7.67) *are bounded. A degenerate situation as described on p. 231 cannot occur.*

Proof. Formula (7.67) is straightforward using (7.55) and (7.61). All constituents of ζ are known due to the derivative (7.65) obtained in Corollary 7.5. In particular, the unknown functions $f_{\boldsymbol{\theta}}(\boldsymbol{x})$ and $f_{\boldsymbol{\theta}^*}(\boldsymbol{x})$ cancel with respective corresponding parts. All integrals in (7.65) are well-defined (Delyon and Hu 2006, Lemma 4), and the quotient of determinants is bounded as $\boldsymbol{\Sigma}$ and $\boldsymbol{\Sigma}^{-1}$ are bounded. The terms $(T/\varepsilon)^{-d(d-1)/2}$ in (7.65) cancel when plugged in into the acceptance probability (7.67). It follows from Corollary 7.1 that the acceptance probability is non-degenerate. □

The implementation of the parameter update is described in the following paragraph, and its performance is shown in a simulation study on pp. 260.

Remark 7.3. Equation (7.65) shows that the acceptance probability (7.67) is sufficiently regular. An explicit form of this acceptance probability, however, would also have been available without (7.65) but under consideration of (7.63). These different representations lead to identical functions in the continuous case. They however differ once discretised as will be seen on p. 255.

Implementation of Parameter Update

The above considerations showed that the parameter update algorithm proposed in this section converges in a continuous-time framework. In practice, however, discretisations of the diffusion paths are considered. Suppose one has a path segment with fixed starting value $\boldsymbol{X}_0 = \boldsymbol{x}_0$ at time $t_0 = 0$ and observed end point $\boldsymbol{X}_T = \boldsymbol{x}$ at time $t_m = T$. Assume that data $\boldsymbol{X}_{t_1}, \ldots, \boldsymbol{X}_{t_{m-1}}$ at times $0 < t_1 < \ldots < t_{m-1} < T$ is imputed. During the update mechanism, the variable $\boldsymbol{X}_t$ will be transformed to a variable $\boldsymbol{Z}_t$. For shorter notation, let $\boldsymbol{X}_{t_k} = \boldsymbol{Y}_k$ and $\boldsymbol{Z}_{t_k} = \mathring{\boldsymbol{Z}}_k$ for all $k = 0, \ldots, m$. As argued on p. 248, include the starting value $\boldsymbol{x}_0$ in the imputed data. Hence define $\boldsymbol{Y}^{\text{imp}} = \{\boldsymbol{Y}_0, \boldsymbol{Y}_1, \ldots, \boldsymbol{Y}_{m-1}\}$ and $\boldsymbol{Y}^{\text{imp}*} = \{\boldsymbol{Y}_0^*, \boldsymbol{Y}_1^*, \ldots, \boldsymbol{Y}_{m-1}^*\}$. Then Algorithm 7.4 adapted to the discretised data reads as follows.

Algorithm 7.6 (Parameter Update for Discretised Data). *Given $\boldsymbol{\theta}$, $\boldsymbol{Y}^{\text{imp}}$ and $\boldsymbol{Y}_m = \boldsymbol{x}$, perform the following steps:*

1. *Draw $\boldsymbol{\theta}^* \sim q(\boldsymbol{\theta}^* | \boldsymbol{\theta})$.*
2. *Successively compute for $k = 0, \ldots, m-2$*

$$\mathring{\boldsymbol{Z}}_{k+1} = \mathring{\boldsymbol{Z}}_k + \boldsymbol{\sigma}^{-1}(\boldsymbol{Y}_k, \boldsymbol{\theta}) \left(\boldsymbol{Y}_{k+1} - \boldsymbol{Y}_k - \frac{\boldsymbol{x} - \boldsymbol{Y}_k}{T - t_k} \Delta t_k \right), \tag{7.68}$$

where $\mathring{\boldsymbol{Z}}_0 = \boldsymbol{0}$, $\boldsymbol{Y}_0 = \boldsymbol{x}_0$ and $\Delta t_k = t_{k+1} - t_k$. Furthermore, obtain the back-transformation with respect to the proposed parameter $\boldsymbol{\theta}^*$,

$$\boldsymbol{Y}_{k+1}^* = \boldsymbol{Y}_k^* + \boldsymbol{\sigma}(\boldsymbol{Y}_k^*, \boldsymbol{\theta}^*) \big(\mathring{\boldsymbol{Z}}_{k+1} - \mathring{\boldsymbol{Z}}_k \big) + \frac{\boldsymbol{x} - \boldsymbol{Y}_k^*}{T - t_k} \Delta t_k \tag{7.69}$$

for $k = 0, \ldots, m-2$, where $\boldsymbol{Y}_0^* = \boldsymbol{Y}_0$.

3. *Accept $\boldsymbol{\theta}^*$ with probability*

$$\zeta(\boldsymbol{\theta}^*, \boldsymbol{\theta}) = 1 \wedge \frac{H(\boldsymbol{Y}^{\text{imp}*}, \boldsymbol{\theta}^*) p(\boldsymbol{\theta}^*) q(\boldsymbol{\theta} \,|\, \boldsymbol{\theta}^*)}{H(\boldsymbol{Y}^{\text{imp}}, \boldsymbol{\theta}) p(\boldsymbol{\theta}) q(\boldsymbol{\theta}^* \,|\, \boldsymbol{\theta})},$$

where

$$\log H(\boldsymbol{Y}^{\text{imp}}, \boldsymbol{\theta}) = -\frac{1}{2} \sum_{k=0}^{m-2} \frac{(\boldsymbol{x} - \boldsymbol{Y}_k)' \big(\boldsymbol{\Sigma}^{-1}(\boldsymbol{Y}_{k+1}, \boldsymbol{\theta}) - \boldsymbol{\Sigma}^{-1}(\boldsymbol{Y}_k, \boldsymbol{\theta}) \big) (\boldsymbol{x} - \boldsymbol{Y}_k)}{T - t_k} \tag{7.70}$$

$$-\frac{1}{2}\sum_{k=0}^{m-2}\sum_{i,j=1}^{d}\frac{(\boldsymbol{x}-\boldsymbol{Y}_k)'\left(\frac{\partial\boldsymbol{\Sigma}^{-1}(\boldsymbol{Y}_k,\boldsymbol{\theta})}{\partial y^{(j)}}\boldsymbol{e}_i+\frac{\partial\boldsymbol{\Sigma}^{-1}(\boldsymbol{Y}_k,\boldsymbol{\theta})}{\partial y^{(i)}}\boldsymbol{e}_j\right)}{T-t_k}\Delta Y_k^{(i)}\Delta Y_k^{(j)} \quad (7.71)$$

$$+\sum_{k=0}^{m-1}\boldsymbol{\mu}'(\boldsymbol{Y}_k,\boldsymbol{\theta})\boldsymbol{\Sigma}^{-1}(\boldsymbol{Y}_k,\boldsymbol{\theta})\left(\Delta\boldsymbol{Y}_k-\frac{1}{2}\boldsymbol{\mu}(\boldsymbol{Y}_k,\boldsymbol{\theta})\Delta t_k\right) \quad (7.72)$$

$$+\log\phi(\boldsymbol{x}\,|\,\boldsymbol{x}_0,T\boldsymbol{\Sigma}(\boldsymbol{x}_0,\boldsymbol{\theta}))+\frac{1}{2}\left(\log|\boldsymbol{\Sigma}(\boldsymbol{x}_0,\boldsymbol{\theta})|-\log|\boldsymbol{\Sigma}(\boldsymbol{Y}_{m-1},\boldsymbol{\theta})|\right) \quad (7.73)$$

with $\boldsymbol{x}_0 = \boldsymbol{Y}_0$ *and* $\Delta\boldsymbol{Y}_k = \boldsymbol{Y}_{k+1} - \boldsymbol{Y}_k$ *with components* $\Delta Y_k^{(i)}$. *Otherwise keep* $\boldsymbol{\theta}$.

4. *If* $\boldsymbol{\theta}^*$ *was accepted, replace* $\boldsymbol{Y}^{\text{imp}}$ *by* $\boldsymbol{Y}^{\text{imp}*}$.

The function H equals the time-discretisation of Eq. (7.65) times $f_{\boldsymbol{\theta}}(\boldsymbol{x})$ without multiplicative constants but including the correction term (7.66). The latter corresponds to $k = m-1$ in line (7.72). Note that lines (7.70) and (7.71) disappear when $\boldsymbol{\Sigma}$ does not depend on the state variable. The same holds for the second summand in line (7.73).

It was required on p. 246 that $\boldsymbol{\Sigma}^{-1}$ is differentiable. In case the derivatives of $\boldsymbol{\Sigma}^{-1}$ are not analytically available, one can approximate them through difference quotients, i.e. one uses in line (7.71)

$$\left(\frac{\partial\boldsymbol{\Sigma}^{-1}(\boldsymbol{Y}_k,\boldsymbol{\theta})}{\partial y^{(j)}}\boldsymbol{e}_i+\frac{\partial\boldsymbol{\Sigma}^{-1}(\boldsymbol{Y}_k,\boldsymbol{\theta})}{\partial y^{(i)}}\boldsymbol{e}_j\right)\Delta Y_k^{(i)}\Delta Y_k^{(j)}$$

$$\approx\left(\boldsymbol{\Sigma}^{-1}(\boldsymbol{Y}_{k,[j]},\boldsymbol{\theta})-\boldsymbol{\Sigma}^{-1}(\boldsymbol{Y}_k,\boldsymbol{\theta})\right)\boldsymbol{e}_i\Delta Y_k^{(i)}$$

$$+\left(\boldsymbol{\Sigma}^{-1}(\boldsymbol{Y}_{k,[i]},\boldsymbol{\theta})-\boldsymbol{\Sigma}^{-1}(\boldsymbol{Y}_k,\boldsymbol{\theta})\right)\boldsymbol{e}_j\Delta Y_k^{(j)},$$

where the components of $\boldsymbol{Y}_{k,[j]}$ are defined as $Y_{k,[j]}^{(i)} = Y_k^{(i)}$ for $i \neq j$ and $Y_{k,[j]}^{(j)} = Y_{k+1}^{(j)}$.

Because of $\mathrm{d}X_t^{(i)}\mathrm{d}X_t^{(j)} = \Sigma_{ij}(\boldsymbol{X}_t,\boldsymbol{\theta})\mathrm{d}t$ as shown by Eq. (B.28) on p. 405, the term $\Delta Y_k^{(i)}\Delta Y_k^{(j)}$ in line (7.71) could furthermore be replaced by $\Sigma_{ij}(\boldsymbol{Y}_k,\boldsymbol{\theta})\Delta t_k$, where Σ_{ij} is the component in the ith row and jth column of $\boldsymbol{\Sigma}$.

An alternative representation of the acceptance probability in Algorithm 7.6 follows the discretisation of $(\mathrm{d}\mathbb{P}_{\boldsymbol{\theta}}/\mathrm{d}\mathbb{D}_{\boldsymbol{0},\boldsymbol{\theta}})(\boldsymbol{X}^{\text{imp}})f_{\boldsymbol{\theta}}(\boldsymbol{x}|\boldsymbol{X}_{T-\varepsilon})$ according to Eq. (7.63); compare with the remark on p. 252. Then

$$
\begin{aligned}
&\log H(\boldsymbol{Y}^{\mathrm{imp}}, \boldsymbol{\theta}) \\
&= \sum_{k=0}^{m-2} \left(\boldsymbol{\mu}(\boldsymbol{Y}_k, \boldsymbol{\theta}) - \frac{\boldsymbol{x} - \boldsymbol{Y}_k}{T - t_k} \right)' \boldsymbol{\Sigma}^{-1}(\boldsymbol{Y}_k, \boldsymbol{\theta}) \Delta \boldsymbol{Y}_k \qquad (7.74) \\
&-\frac{1}{2} \sum_{k=0}^{m-2} \left(\boldsymbol{\mu}'(\boldsymbol{Y}_k, \boldsymbol{\theta}) \boldsymbol{\Sigma}^{-1}(\boldsymbol{Y}_k, \boldsymbol{\theta}) \boldsymbol{\mu}(\boldsymbol{Y}_k, \boldsymbol{\theta}) - \frac{(\boldsymbol{x} - \boldsymbol{Y}_k)' \boldsymbol{\Sigma}^{-1}(\boldsymbol{Y}_k, \boldsymbol{\theta})(\boldsymbol{x} - \boldsymbol{Y}_k)}{(T - t_k)^2} \right) \Delta t_k \\
&+ \log \phi\big(\boldsymbol{x} \,\big|\, \boldsymbol{Y}_{m-1} + \Delta t_{m-1} \boldsymbol{\mu}(\boldsymbol{Y}_{m-1}, \boldsymbol{\theta}), \Delta t_{m-1} \boldsymbol{\Sigma}(\boldsymbol{Y}_{m-1}, \boldsymbol{\theta})\big).
\end{aligned}
$$

If the diffusion coefficient does not depend on the state of the process, this formula yields the same acceptance probability as the one in Algorithm 7.6.

Path Update

We now turn to the path update, i.e. the imputation of the missing data. Algorithm 7.5 on p. 245 already proposed how to utilise the one-to-one relationship between the target process $\boldsymbol{X}$ and the unit diffusion process $\boldsymbol{Z}$ in that context. At the end, the following elaborations provide the mathematical proof that the modified bridge proposal from p. 181, the diffusion bridge proposal from p. 182 and the proposal by Delyon and Hu from p. 184 work in the continuous-time framework, i.e. for an infinite amount of imputed data.

Recall the suggested Algorithm 7.5 which generates a proposal $\boldsymbol{Z}^{\mathrm{imp}*}$ with unit diffusion and transforms this to a candidate $\boldsymbol{X}^{\mathrm{imp}*} = g(\boldsymbol{Z}^{\mathrm{imp}*}, \boldsymbol{\theta})$ as an alternative choice to the current data $\boldsymbol{X}^{\mathrm{imp}} = g(\boldsymbol{Z}^{\mathrm{imp}}, \boldsymbol{\theta})$. We require that the target measure $\mathbb{Z}_{\boldsymbol{\theta}}$ of the innovation process is absolutely continuous with respect to the proposal measure for $\boldsymbol{Z}^{\mathrm{imp}*}$, i.e. the target measure $\tilde{\mathbb{P}}_{\boldsymbol{\theta}}$ of the diffusion process is absolutely continuous with respect to the proposal measure for $\boldsymbol{X}^{\mathrm{imp}*} = g(\boldsymbol{Z}^{\mathrm{imp}*}, \boldsymbol{\theta})$. Only in that case all possible paths are (theoretically) proposed and the acceptance probability is non-degenerate. An obvious choice is to propose $\boldsymbol{Z}^{\mathrm{imp}*} \sim \mathbb{W}$, i.e. to let $q = \mathrm{d}\mathbb{W}/\mathrm{d}\mathbb{L}$ be the Lebesgue density under Wiener measure.

Corollary 7.7. *The path update can be performed by application of Algorithm 7.5 on p. 245, where the innovation process $\boldsymbol{Z}^{\mathrm{imp}*}$ is proposed from $\mathbb{W}$ and accepted with probability*

$$
\begin{aligned}
&\zeta(\boldsymbol{Z}^{\mathrm{imp}*}, \boldsymbol{Z}^{\mathrm{imp}}) \\
&= 1 \wedge \left(\frac{\mathrm{d}\tilde{\mathbb{P}}_{\boldsymbol{\theta}}}{\mathrm{d}\mathbb{D}_{\boldsymbol{0},\boldsymbol{\theta}}} \big(g(\boldsymbol{Z}^{\mathrm{imp}*}, \boldsymbol{\theta})\big) \right) \Big/ \left(\frac{\mathrm{d}\tilde{\mathbb{P}}_{\boldsymbol{\theta}}}{\mathrm{d}\mathbb{D}_{\boldsymbol{0},\boldsymbol{\theta}}} \big(g(\boldsymbol{Z}^{\mathrm{imp}}, \boldsymbol{\theta})\big) \right). \qquad (7.75)
\end{aligned}
$$

This algorithm is non-degenerate.

Proof. In terms of the probability measures introduced in this section, the acceptance probability (7.54) from p. 245 reads

$$\varsigma(\boldsymbol{Z}^{\mathrm{imp}*}, \boldsymbol{Z}^{\mathrm{imp}}) = 1 \wedge \frac{\dfrac{\mathrm{d}\mathbb{Z}_{\boldsymbol{\theta}}}{\mathrm{d}\mathbb{L}}(\boldsymbol{Z}^{\mathrm{imp}*})\, q(\boldsymbol{Z}^{\mathrm{imp}}|\boldsymbol{Z}^{\mathrm{imp}*}, \boldsymbol{x}_0, \boldsymbol{x}, \boldsymbol{\theta})}{\dfrac{\mathrm{d}\mathbb{Z}_{\boldsymbol{\theta}}}{\mathrm{d}\mathbb{L}}(\boldsymbol{Z}^{\mathrm{imp}})\, q(\boldsymbol{Z}^{\mathrm{imp}*}|\boldsymbol{Z}^{\mathrm{imp}}, \boldsymbol{x}_0, \boldsymbol{x}, \boldsymbol{\theta})}. \tag{7.76}$$

Change of variables as in (7.59) and (7.60) yields

$$\frac{\dfrac{\mathrm{d}\mathbb{Z}_{\boldsymbol{\theta}}}{\mathrm{d}\mathbb{L}}(\boldsymbol{Z}^{\mathrm{imp}*})}{\dfrac{\mathrm{d}\mathbb{Z}_{\boldsymbol{\theta}}}{\mathrm{d}\mathbb{L}}(\boldsymbol{Z}^{\mathrm{imp}})} = \frac{\left(\dfrac{\mathrm{d}\mathbb{Z}_{\boldsymbol{\theta}}}{\mathrm{d}\mathbb{W}}\dfrac{\mathrm{d}\mathbb{W}}{\mathrm{d}\mathbb{L}}\right)(\boldsymbol{Z}^{\mathrm{imp}*})}{\left(\dfrac{\mathrm{d}\mathbb{Z}_{\boldsymbol{\theta}}}{\mathrm{d}\mathbb{W}}\dfrac{\mathrm{d}\mathbb{W}}{\mathrm{d}\mathbb{L}}\right)(\boldsymbol{Z}^{\mathrm{imp}})} = \frac{\dfrac{\mathrm{d}\tilde{\mathbb{P}}_{\boldsymbol{\theta}}}{\mathrm{d}\mathbb{D}_{\boldsymbol{0},\boldsymbol{\theta}}}\left(g(\boldsymbol{Z}^{\mathrm{imp}*}, \boldsymbol{\theta})\right)\dfrac{\mathrm{d}\mathbb{W}}{\mathrm{d}\mathbb{L}}(\boldsymbol{Z}^{\mathrm{imp}*})}{\dfrac{\mathrm{d}\tilde{\mathbb{P}}_{\boldsymbol{\theta}}}{\mathrm{d}\mathbb{D}_{\boldsymbol{0},\boldsymbol{\theta}}}\left(g(\boldsymbol{Z}^{\mathrm{imp}}, \boldsymbol{\theta})\right)\dfrac{\mathrm{d}\mathbb{W}}{\mathrm{d}\mathbb{L}}(\boldsymbol{Z}^{\mathrm{imp}})}.$$

Plugging in $q = \mathrm{d}\mathbb{W}/\mathrm{d}\mathbb{L}$ into (7.76) yields (7.75). This acceptance probability is non-degenerate as due to the construction in Algorithms 7.4 and 7.5 the quadratic variation of both $\boldsymbol{X}^{\mathrm{imp}} = g(\boldsymbol{Z}^{\mathrm{imp}}, \boldsymbol{\theta})$ and $\boldsymbol{X}^{\mathrm{imp}*} = g(\boldsymbol{Z}^{\mathrm{imp}*}, \boldsymbol{\theta})$ is consistent with $\boldsymbol{\theta}$. An explicit formula for (7.75) is available using Eq. (7.65) on p. 250. □

These considerations show that for suitable proposal measures the detour around the innovation process $\boldsymbol{Z}$ is actually not necessary in the path update: Instead of sampling $\boldsymbol{Z}^{\mathrm{imp}*}$ from a measure $\mathbb{Q}^Z_{\boldsymbol{\theta}}$ and then deterministically calculating $\boldsymbol{X}^{\mathrm{imp}*} = g(\boldsymbol{Z}^{\mathrm{imp}*}, \boldsymbol{\theta})$, one can directly obtain $\boldsymbol{X}^{\mathrm{imp}*}$ from the resulting measure $\mathbb{Q}^X_{\boldsymbol{\theta}}$—if this measure is known and simulation from it is possible. The only requirement is that $\mathbb{Z}_{\boldsymbol{\theta}} \ll \mathbb{Q}^Z_{\boldsymbol{\theta}}$, i.e. $\tilde{\mathbb{P}}_{\boldsymbol{\theta}} \ll \mathbb{Q}^X_{\boldsymbol{\theta}}$.

Two appropriate choices for $\mathbb{Q}^X_{\boldsymbol{\theta}}$ are $\mathbb{D}_{\boldsymbol{0},\boldsymbol{\theta}}$ and $\mathbb{D}_{\boldsymbol{\mu},\boldsymbol{\theta}}$; recall the definitions from Table 7.6 on p. 243. As already mentioned before, these measures fulfil $\tilde{\mathbb{P}}_{\boldsymbol{\theta}} \ll \mathbb{D}_{\boldsymbol{0},\boldsymbol{\theta}}$ and $\tilde{\mathbb{P}}_{\boldsymbol{\theta}} \ll \mathbb{D}_{\boldsymbol{\mu},\boldsymbol{\theta}}$ (Delyon and Hu 2006). Approximate simulation is possible via e.g. the Euler scheme.

Theorem 7.8. *The path update can be performed by application of Algorithm 7.5 with acceptance probabilities*

$$\varsigma(\boldsymbol{X}^{\mathrm{imp}*}, \boldsymbol{X}^{\mathrm{imp}}) = 1 \wedge \left(\frac{\mathrm{d}\tilde{\mathbb{P}}_{\boldsymbol{\theta}}}{\mathrm{d}\mathbb{D}_{\boldsymbol{0},\boldsymbol{\theta}}}(\boldsymbol{X}^{\mathrm{imp}*})\right) \Big/ \left(\frac{\mathrm{d}\tilde{\mathbb{P}}_{\boldsymbol{\theta}}}{\mathrm{d}\mathbb{D}_{\boldsymbol{0},\boldsymbol{\theta}}}(\boldsymbol{X}^{\mathrm{imp}})\right) \tag{7.77}$$

for proposals $\boldsymbol{X}^{\mathrm{imp}*} \sim \mathbb{D}_{\boldsymbol{0},\boldsymbol{\theta}}$ *and*

$$\varsigma(\boldsymbol{X}^{\mathrm{imp}*}, \boldsymbol{X}^{\mathrm{imp}}) = 1 \wedge \left(\frac{\mathrm{d}\tilde{\mathbb{P}}_{\boldsymbol{\theta}}}{\mathrm{d}\mathbb{D}_{\boldsymbol{\mu},\boldsymbol{\theta}}}(\boldsymbol{X}^{\mathrm{imp}*})\right) \Big/ \left(\frac{\mathrm{d}\tilde{\mathbb{P}}_{\boldsymbol{\theta}}}{\mathrm{d}\mathbb{D}_{\boldsymbol{\mu},\boldsymbol{\theta}}}(\boldsymbol{X}^{\mathrm{imp}})\right) \tag{7.78}$$

for proposals $\boldsymbol{X}^{\mathrm{imp}*} \sim \mathbb{D}_{\boldsymbol{\mu},\boldsymbol{\theta}}$. *For both choices, the algorithm is non-degenerate.*

Proof. Both equations are obvious if one considers the path update directly for $\boldsymbol{X}^{\mathrm{imp}}$ without regarding $\boldsymbol{Z}^{\mathrm{imp}}$. The acceptance probability (7.77) is naturally also the same as (7.75). Section B.8 in the Appendix briefly shows the according derivation for (7.78). Explicit expressions for (7.77) and (7.78) are available with (7.64) and (7.65), in which all integrals are well-defined (Delyon and Hu 2006, Lemma 4). The reasoning of the regularity of the above acceptance probabilities follows the line of the proof of Theorem 7.6. □

The proposal measures $\mathbb{D}_{\mathbf{0},\boldsymbol{\theta}}$ and $\mathbb{D}_{\boldsymbol{\mu},\boldsymbol{\theta}}$ have already been considered in Sect. 7.1 as the diffusion bridge proposal (p. 182) and a proposal due to Delyon and Hu (p. 184). The proposal measure $\mathbb{D}_{\mathbf{0},\boldsymbol{\theta}}$ is also covered by the limit of the modified bridge proposal (p. 181) as the amount of imputed data tends to infinity. This section hence proves that these proposals from the discrete-time framework also work in continuous time.

The following paragraph describes the practical implementation of the path update with the two options $\mathbb{D}_{\mathbf{0},\boldsymbol{\theta}}$ and $\mathbb{D}_{\boldsymbol{\mu},\boldsymbol{\theta}}$ as proposal measures. Afterwards, the entire modified MCMC algorithm is applied in a simulation study on pp. 260.

Implementation of Path Update

As in the implementation of the parameter update, consider a discrete path skeleton consisting of observed and imputed data $\boldsymbol{x}_0 = \boldsymbol{X}_{t_0}, \boldsymbol{X}_{t_1}, \ldots, \boldsymbol{X}_{t_{m-1}}, \boldsymbol{X}_{t_m} = \boldsymbol{x}$ at time points $0 = t_0 < t_1 < \ldots < t_{m-1} < t_m = T$. Let $\boldsymbol{X}_{t_k} = \boldsymbol{Y}_k$ for all $k = 0, \ldots, m$ and define $\boldsymbol{Y}^{\mathrm{imp}} = \{\boldsymbol{Y}_0, \boldsymbol{Y}_1, \ldots, \boldsymbol{Y}_{m-1}\}$ and $\boldsymbol{Y}^{\mathrm{imp}*} = \{\boldsymbol{Y}_0^*, \boldsymbol{Y}_1^*, \ldots, \boldsymbol{Y}_{m-1}^*\}$.

For the path proposal measure $\mathbb{D}_{\mathbf{0},\boldsymbol{\theta}}$, the path algorithm 7.5 adapted to the discretised data reads as follows.

Algorithm 7.7 (Path Update for Discretised Data I). *Given $\boldsymbol{\theta}$, $\boldsymbol{Y}^{\mathrm{imp}}$ and $\boldsymbol{Y}_m = \boldsymbol{x}$, perform the following steps:*

1. *Draw an approximate discrete skeleton $\boldsymbol{Y}^{\mathrm{imp}*} \sim \mathbb{D}_{\mathbf{0},\boldsymbol{\theta}}$, i.e. successively simulate*

$$\boldsymbol{Y}_{k+1}^* = \boldsymbol{Y}_k^* + \frac{\boldsymbol{x} - \boldsymbol{Y}_k^*}{T - t_k} \Delta t_k + \boldsymbol{\sigma}(\boldsymbol{Y}_k^*, \boldsymbol{\theta}) \mathcal{N}(\mathbf{0}, \Delta t_k \boldsymbol{I})$$

   for $k = 0, \ldots, m-2$, where $\boldsymbol{Y}_0^* = \boldsymbol{Y}_0 = \boldsymbol{x}_0$ and $\Delta t_k = t_{k+1} - t_k$.
2. *Accept $\boldsymbol{Y}^{\mathrm{imp}*}$ with probability*

$$\zeta(\boldsymbol{Y}^{\mathrm{imp}*}, \boldsymbol{Y}^{\mathrm{imp}}) = 1 \wedge \frac{\tilde{H}(\boldsymbol{Y}^{\mathrm{imp}*}, \boldsymbol{\theta})}{\tilde{H}(\boldsymbol{Y}^{\mathrm{imp}}, \boldsymbol{\theta})},$$

   where

$$\log \tilde{H}(\boldsymbol{Y}^{\mathrm{imp}}, \boldsymbol{\theta}) = -\frac{1}{2} \sum_{k=0}^{m-2} \frac{(\boldsymbol{x} - \boldsymbol{Y}_k)' \big(\boldsymbol{\Sigma}^{-1}(\boldsymbol{Y}_{k+1}, \boldsymbol{\theta}) - \boldsymbol{\Sigma}^{-1}(\boldsymbol{Y}_k, \boldsymbol{\theta})\big)(\boldsymbol{x} - \boldsymbol{Y}_k)}{T - t_k} \tag{7.79}$$

$$-\frac{1}{2}\sum_{k=0}^{m-2}\sum_{i,j=1}^{d}\frac{(\boldsymbol{x}-\boldsymbol{Y}_k)'\left(\frac{\partial\boldsymbol{\Sigma}^{-1}(\boldsymbol{Y}_k,\boldsymbol{\theta})}{\partial y^{(j)}}\boldsymbol{e}_i+\frac{\partial\boldsymbol{\Sigma}^{-1}(\boldsymbol{Y}_k,\boldsymbol{\theta})}{\partial y^{(i)}}\boldsymbol{e}_j\right)}{T-t_k}\Delta Y_k^{(i)}\Delta Y_k^{(j)} \tag{7.80}$$

$$+\sum_{k=0}^{m-1}\boldsymbol{\mu}'(\boldsymbol{Y}_k,\boldsymbol{\theta})\boldsymbol{\Sigma}^{-1}(\boldsymbol{Y}_k,\boldsymbol{\theta})\left(\Delta\boldsymbol{Y}_k-\frac{1}{2}\boldsymbol{\mu}(\boldsymbol{Y}_k,\boldsymbol{\theta})\Delta t_k\right)-\frac{1}{2}\log|\boldsymbol{\Sigma}(\boldsymbol{Y}_{m-1},\boldsymbol{\theta})| \tag{7.81}$$

with $\Delta\boldsymbol{Y}_k=(\Delta Y_k^{(1)},\ldots,\Delta Y_k^{(d)})'=\boldsymbol{Y}_{k+1}-\boldsymbol{Y}_k$. *Otherwise keep* $\boldsymbol{Y}^{\mathrm{imp}}$.

The function $\tilde{H}$ is the discretisation of Eq. (7.65) times $f_{\boldsymbol{\theta}}(\boldsymbol{x})$ without constants but again incorporating the correction (7.66). As in Algorithm 7.6, this correction term corresponds to $k = m-1$ in line (7.81), and lines (7.79) and (7.80) disappear when $\boldsymbol{\Sigma}$ does not depend on the state variable. The previous remarks on possibly required approximations of the derivatives of $\boldsymbol{\Sigma}^{-1}$ naturally apply here as well.

For the path proposal measure $\mathbb{D}_{\boldsymbol{\mu},\boldsymbol{\theta}}$, the algorithm includes the discretisation of Eq. (7.64) and hence reads as follows.

Algorithm 7.8 (Path Update for Discretised Data II). *Given* $\boldsymbol{\theta}$, $\boldsymbol{Y}^{\mathrm{imp}}$ *and* $\boldsymbol{Y}_m=\boldsymbol{x}$, *perform the following steps:*

1. *Draw an approximate discrete skeleton* $\boldsymbol{Y}^{\mathrm{imp}*}\sim\mathbb{D}_{\boldsymbol{\mu},\boldsymbol{\theta}}$, *i.e. successively simulate*

$$\boldsymbol{Y}_{k+1}^*=\boldsymbol{Y}_k^*+\left(\boldsymbol{\mu}(\boldsymbol{Y}_k^*,\boldsymbol{\theta})+\frac{\boldsymbol{x}-\boldsymbol{Y}_k^*}{T-t_k}\right)\Delta t_k+\boldsymbol{\sigma}(\boldsymbol{Y}_k^*,\boldsymbol{\theta})\mathcal{N}\big(\boldsymbol{0},\Delta t_k\boldsymbol{I}\big)$$

   for $k=0,\ldots,m-2$, *where* $\boldsymbol{Y}_0^*=\boldsymbol{Y}_0=\boldsymbol{x}_0$ *and* $\Delta t_k=t_{k+1}-t_k$.
2. *Accept* $\boldsymbol{Y}^{\mathrm{imp}*}$ *with probability*

$$\zeta(\boldsymbol{Y}^{\mathrm{imp}*},\boldsymbol{Y}^{\mathrm{imp}})=1\wedge\frac{\bar{H}(\boldsymbol{Y}^{\mathrm{imp}*},\boldsymbol{\theta})}{\bar{H}(\boldsymbol{Y}^{\mathrm{imp}},\boldsymbol{\theta})},$$

   where

$$\begin{aligned}
&\log\bar{H}(\boldsymbol{Y}^{\mathrm{imp}},\boldsymbol{\theta})\\
&=\sum_{k=0}^{m-2}\frac{(\boldsymbol{x}-\boldsymbol{Y}_k)'\boldsymbol{\Sigma}^{-1}(\boldsymbol{Y}_k,\boldsymbol{\theta})\boldsymbol{\mu}(\boldsymbol{Y}_k,\boldsymbol{\theta})}{T-t_k}\Delta t_k\\
&-\frac{1}{2}\sum_{k=0}^{m-2}\frac{(\boldsymbol{x}-\boldsymbol{Y}_k)'\big(\boldsymbol{\Sigma}^{-1}(\boldsymbol{Y}_{k+1},\boldsymbol{\theta})-\boldsymbol{\Sigma}^{-1}(\boldsymbol{Y}_k,\boldsymbol{\theta})\big)(\boldsymbol{x}-\boldsymbol{Y}_k)}{T-t_k}\\
&-\frac{1}{2}\sum_{k=0}^{m-2}\sum_{i,j=1}^{d}\frac{(\boldsymbol{x}-\boldsymbol{Y}_k)'\left(\frac{\partial\boldsymbol{\Sigma}^{-1}(\boldsymbol{Y}_k,\boldsymbol{\theta})}{\partial y^{(j)}}\boldsymbol{e}_i+\frac{\partial\boldsymbol{\Sigma}^{-1}(\boldsymbol{Y}_k,\boldsymbol{\theta})}{\partial y^{(i)}}\boldsymbol{e}_j\right)}{T-t_k}\Delta Y_k^{(i)}\Delta Y_k^{(j)}
\end{aligned}$$

$$-\frac{1}{2}\log|\boldsymbol{\Sigma}(\boldsymbol{Y}_{m-1},\boldsymbol{\theta})|+\boldsymbol{\mu}'(\boldsymbol{Y}_{m-1},\boldsymbol{\theta})\boldsymbol{\Sigma}^{-1}(\boldsymbol{Y}_{m-1},\boldsymbol{\theta})\Big(\boldsymbol{x}-\boldsymbol{Y}_{m-1}-\frac{\Delta t_{m-1}}{2}\,\boldsymbol{\mu}(\boldsymbol{Y}_{m-1},\boldsymbol{\theta})\Big)$$

with $\Delta\boldsymbol{Y}_k=(\Delta Y_k^{(1)},\ldots,\Delta Y_k^{(d)})'=\boldsymbol{Y}_{k+1}-\boldsymbol{Y}_k$. *Otherwise keep* $\boldsymbol{Y}^{\text{imp}}$.

The performance of these two algorithms is shown and compared with one another in a simulation study on pp. 260.

Generalisation to Several Observation Times, Latent Data and Observation Error

The previously described methodology can easily be generalised to several observation times, latent data settings and observations with error. The first is briefly described in the following.

In the above considerations, in order to ease notation, the initial and final times 0 and T and the initial and final states $\boldsymbol{x}_0$ and $\boldsymbol{x}$ were not included in the symbol $\tilde{\mathbb{P}}_{\boldsymbol{\theta}}$ of the conditioned measure of the target diffusion satisfying

$$\mathrm{d}\boldsymbol{X}_t=\boldsymbol{\mu}(\boldsymbol{X}_t,\boldsymbol{\theta})\,\mathrm{d}t+\boldsymbol{\sigma}(\boldsymbol{X}_t,\boldsymbol{\theta})\,\mathrm{d}\boldsymbol{B}_t\quad,\boldsymbol{X}_T=\boldsymbol{x}.$$

In case of several observations at possibly non-equidistant time points, this specification is however required. Hence use the notation $\tilde{\mathbb{P}}_{\boldsymbol{\theta}}^{(0,\boldsymbol{x}_0,T,\boldsymbol{x})}$ for the measure induced by the above SDE. Furthermore, let $\mathbb{P}_{\boldsymbol{\theta}}^{(0,\boldsymbol{x}_0,T)}$ be the respective unconditioned target measure and $f_{\boldsymbol{\theta}}^{(0,\boldsymbol{x}_0,T)}$ the Lebesgue density of the end point $\boldsymbol{X}_T$ under $\mathbb{P}_{\boldsymbol{\theta}}^{(0,\boldsymbol{x}_0,T)}$.

Now suppose that for the target diffusion there are the fixed initial value $\boldsymbol{x}_{\tau_0}$ and M observations $\boldsymbol{x}_{\tau_1},\ldots,\boldsymbol{x}_{\tau_M}$ at times $\tau_0<\tau_1<\ldots<\tau_M$ available. For $i=0,\ldots,M-1$, impute auxiliary data $\boldsymbol{X}_i^{\text{imp}}$ in the time interval $[\tau_i,\tau_{i+1}]$. The posterior density of the parameter $\boldsymbol{\theta}$ with respect to Lebesgue measure then equals

$$\pi\big(\boldsymbol{\theta}\,\big|\,\boldsymbol{X}_{[\tau_0,\tau_M]}\big)\propto\left(\prod_{i=0}^{M-1}\frac{\mathrm{d}\tilde{\mathbb{P}}_{\boldsymbol{\theta}}^{(\tau_i,\boldsymbol{x}_{\tau_i},\tau_{i+1},\boldsymbol{x}_{\tau_{i+1}})}}{\mathrm{d}\mathbb{L}}\big(\boldsymbol{X}_i^{\text{imp}}\big)f_{\boldsymbol{\theta}}^{(\tau_i,\boldsymbol{x}_{\tau_i},\tau_{i+1})}(\boldsymbol{x}_{\tau_{i+1}})\right)p(\boldsymbol{\theta}).$$

The likelihood of $\boldsymbol{\theta}$ is most conveniently written as

$$\pi\big(\boldsymbol{X}_{[\tau_0,\tau_M]}\,\big|\,\boldsymbol{\theta}\big)\propto\prod_{i=0}^{M-1}\frac{\mathrm{d}\mathbb{P}_{\boldsymbol{\theta}}^{(\tau_i,\boldsymbol{x}_{\tau_i},\tau_{i+1})}}{\mathrm{d}\mathbb{L}}\big(\boldsymbol{X}_{(\tau_i,\tau_{i+1}]}\big).$$

With this, the previously described algorithms for the path and parameter update are easily generalised to several observation times. An extension of the parameter update to latent data and observation errors as considered in Sect. 7.2 is straightforward as well.

Simulation Study

The simulation study in Sect. 7.1.7 demonstrated the performance of the standard MCMC algorithms from Sect. 7.1 on the example of a one-dimensional Ornstein-Uhlenbeck process $X = (X_t)_{t\geq 0}$ satisfying the SDE

$$\mathrm{d}X_t = \alpha(\beta - X_t)\mathrm{d}t + \sigma \mathrm{d}B_t \quad , X_0 = x_0, \tag{7.82}$$

for parameters $\beta \in \mathbb{R}$, $\alpha, \sigma^2 \in \mathbb{R}_+$ and initial value $x_0 = 0$. Based on an exactly simulated realisation with $\boldsymbol{\theta} = (\alpha, \beta, \sigma^2)' = (0.5, 0.9, 1.0)'$, estimation was carried out for β and σ^2 with α considered known. The simulated sample path is displayed in Fig. 7.5 on p. 196.

The following simulation study revives the same example in order to evaluate the proficiencies of the innovation scheme. Results are compared with the outcomes for those schemes that worked best in Sect. 7.1.7; these are the modified bridge proposal for the path update and the random walk proposal for the parameter update. Again, all methods are implemented in R.

In all approaches in the present simulation study, the parameters β and σ^2 have again a priori distributions

$$\beta \sim \mathcal{N}(0,1) \quad \text{and} \quad \sigma^2 \sim \mathrm{IG}(3,3).$$

Given the current values β and σ^2, new parameters β^* and σ^{2*} are proposed via a random walk

$$\beta^* \sim \mathcal{N}(\beta, 0.025) \quad \text{and} \quad \log \sigma^{2*} \sim \mathcal{N}(\log \sigma^2, 0.025).$$

The three competing MCMC schemes in the simulation study differ from each other with respect to the path update and the acceptance mechanism of the parameter update as follows.

- *Standard Algorithm:* The acceptance probability for the proposed parameter $\boldsymbol{\theta}^* = (\beta^*, \sigma^{2*})'$ is as in Eq. (7.15) on p. 185. The diffusion path is updated via the modified bridge proposal as described on p. 181.
- *Innovation Scheme I:* The proposed parameter $\boldsymbol{\theta}^*$ is accepted or rejected according to Algorithm 7.6 on p. 253. The path update employs Algorithm 7.7 on p. 257.
- *Innovation Scheme II:* The proposed parameter $\boldsymbol{\theta}^*$ is accepted or rejected according to Algorithm 7.6 on p. 253. The path update employs Algorithm 7.8 on p. 258.

The update intervals are chosen with Algorithm 7.3 on p. 191 with mean length $\lambda = 5$. Different values for λ have been investigated as well but have not led to different conclusions.

Figures 7.28–7.41 and Tables 7.7 and 7.8 show the performance of the above three schemes for an increasing amount of imputed data and 10^5 iterations.

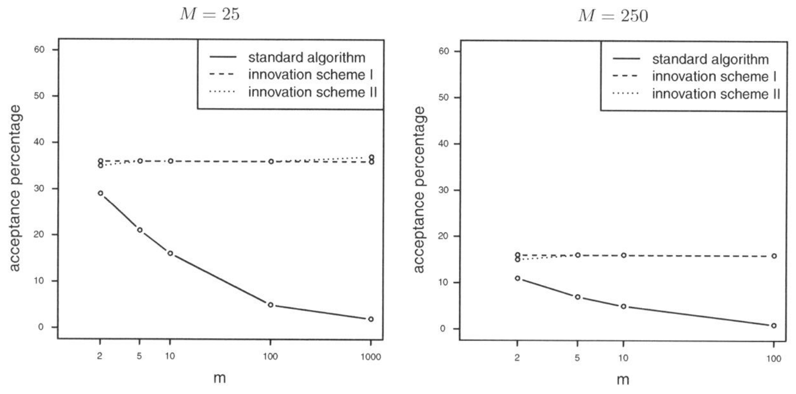

Fig. 7.28 Acceptance rates for the parameter update corresponding to the simulated Markov chains displayed in Figs. 7.29 and 7.30 ($M = 25$, *left graphic*) and Figs. 7.35 and 7.36 ($M = 250$, *right graphic*). The m-values are plotted on a log scale to the base 10

In particular, Figs. 7.29–7.34 display trace plots, posterior density estimates and autocorrelation plots for the MCMC procedure when observations of the diffusion path are available at times $0, 1, \ldots, 25$. With the notation from Sect. 7.1, this corresponds to the maximum time $T = 25$ and $M = 25$ observations. Figures 7.35–7.40 show these outcomes when the MCMC scheme conditions on observations at times $0, 0.1, \ldots, 25$, i.e. $T = 25$ and $M = 250$. All data is assumed to be measured without error. Figure 7.28 displays the acceptance rates of the parameter update in all experiments. Tables 7.7 and 7.8 summarise the posterior means and 95 %-hpd intervals for β and σ^2. The hpd intervals are also shown in Fig. 7.41.

The simulation results clearly demonstrate that the standard algorithm struggles when large amounts of data are imputed: The trace plots show poor mixing for $m \in \{100, 1{,}000\}$, high autocorrelation and crucially decreasing acceptance rates in the parameter update. In those cases, the standard scheme experiences severe difficulties to satisfyingly estimate the diffusion coefficient. The performance of the innovation scheme, in contrast, remains equally satisfactory for all values of m.

Further empirical investigations, which are not shown here, yield similar results for the MCMC scheme when the diffusion path is updated according to the modified bridge proposal on p. 181 or diffusion bridge proposal on p. 182 as long as the parameter update follows Algorithm 7.6.

Conclusion

To summarise, this section seizes the general idea from Chib et al. (2004) to base the MCMC algorithms considered in this chapter on an innovation process $\boldsymbol{Z}$ rather than on the original diffusion process $\boldsymbol{X}$. The innovation process is constructed such that it has unit diffusion and hence does not obstruct the convergence of the algorithms.

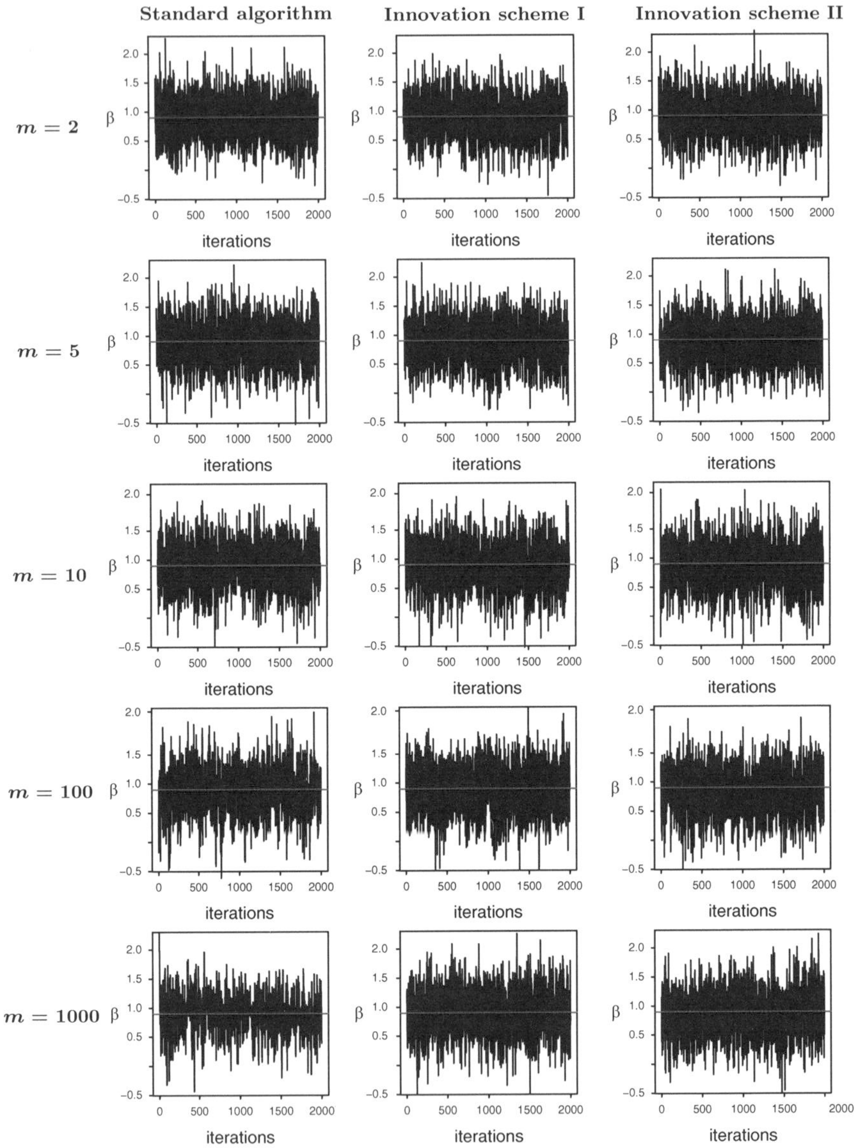

Fig. 7.29 Estimation of parameters of the Ornstein-Uhlenbeck process (7.82) as described on pp. 260. The MCMC scheme conditions on observed data at times $0, 1, \ldots, 25$ and introduces m subintervals in between every two observations. This figure shows the trace plots of β. The realisations of the Markov chains have length 10^5 but have been thinned by factor 50. The true value for β equals 0.9 and is indicated by the *red horizontal line*

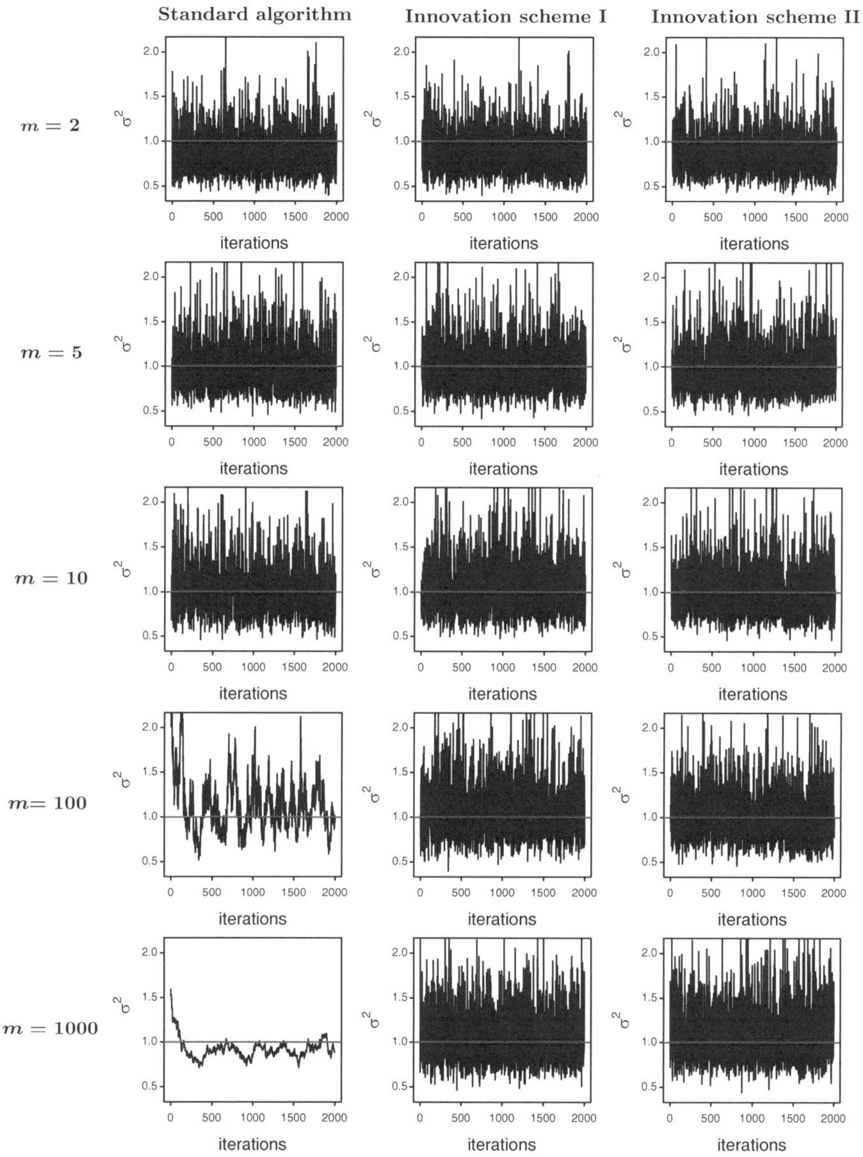

Fig. 7.30 Estimation results as described in Fig. 7.29. This figure shows the trace plots for σ^2. The true parameter value for σ^2 equals 1 and is indicated by the *red horizontal line*

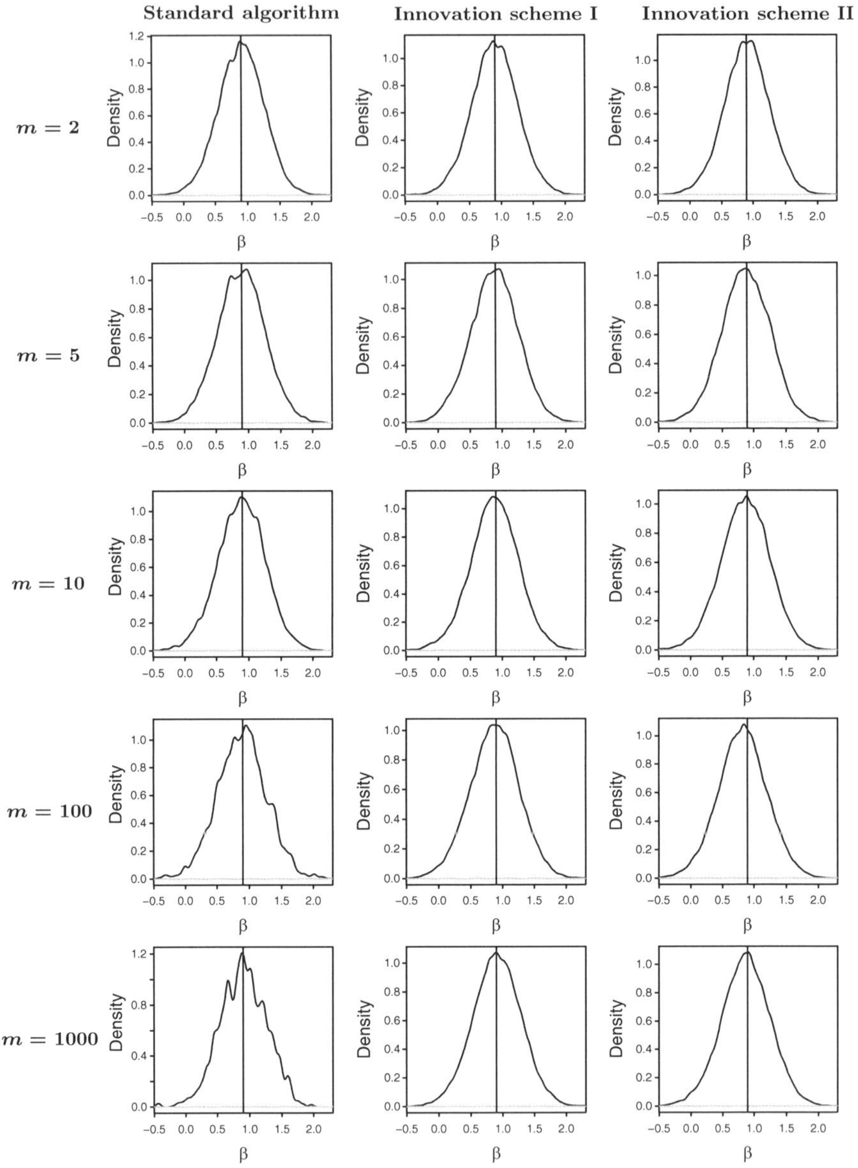

Fig. 7.31 Estimation of the posterior density of β based on the results from Fig. 7.29. Density estimation takes into account the full Markov chain, i.e. without thinning, after having discarded a 10 % burn-in phase. The true value of the parameter is indicated by the *vertical line*

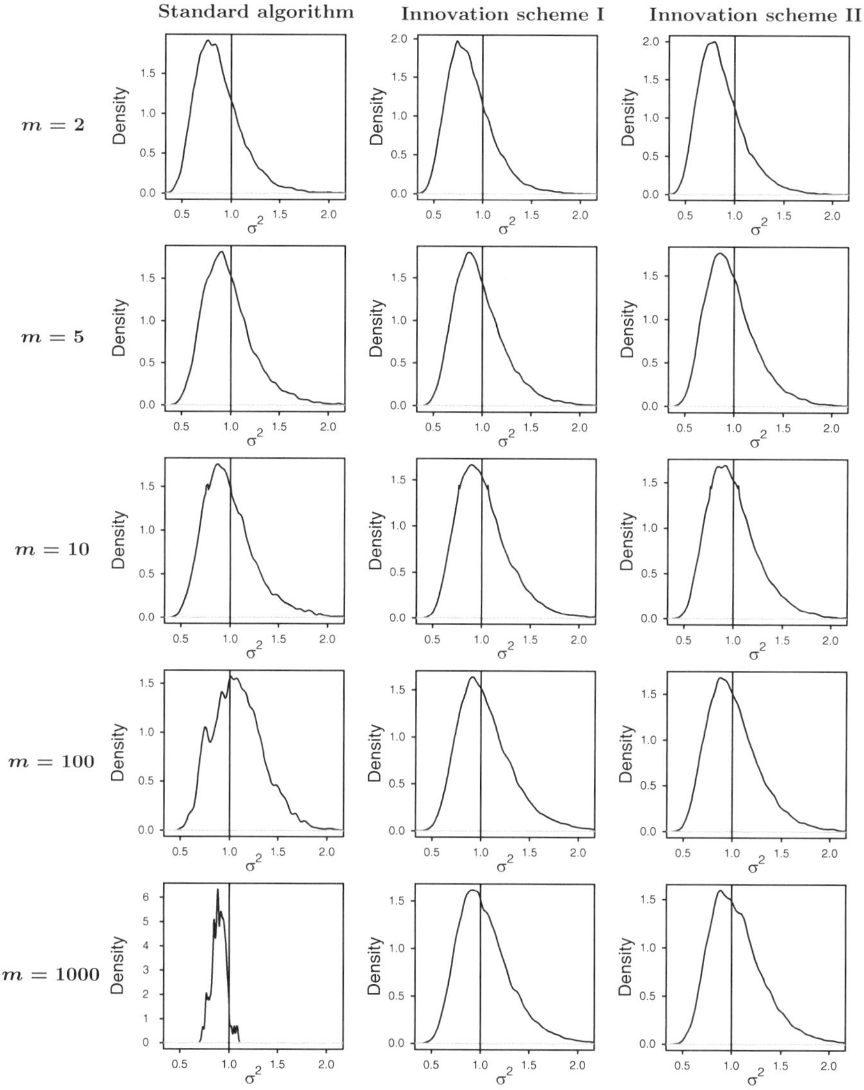

Fig. 7.32 Estimation of the posterior density of σ^2 based on the results from Fig. 7.30. Density estimation takes into account the full Markov chain, i.e. without thinning, after having discarded a 10 % burn-in phase. The true value of the parameter is indicated by *the vertical line*

This idea has already been investigated by Chib et al. (2004) for diffusion paths that are not conditioned on an end point and by Golightly and Wilkinson (2008, 2010) for discrete path skeletons, i.e. on finite-dimensional state spaces. This book considers conditioned diffusion paths on infinite-dimensional state spaces.

The important improvement in our approach is that we first assess the methodology in a continuous-time framework and then discretise resulting formulas for

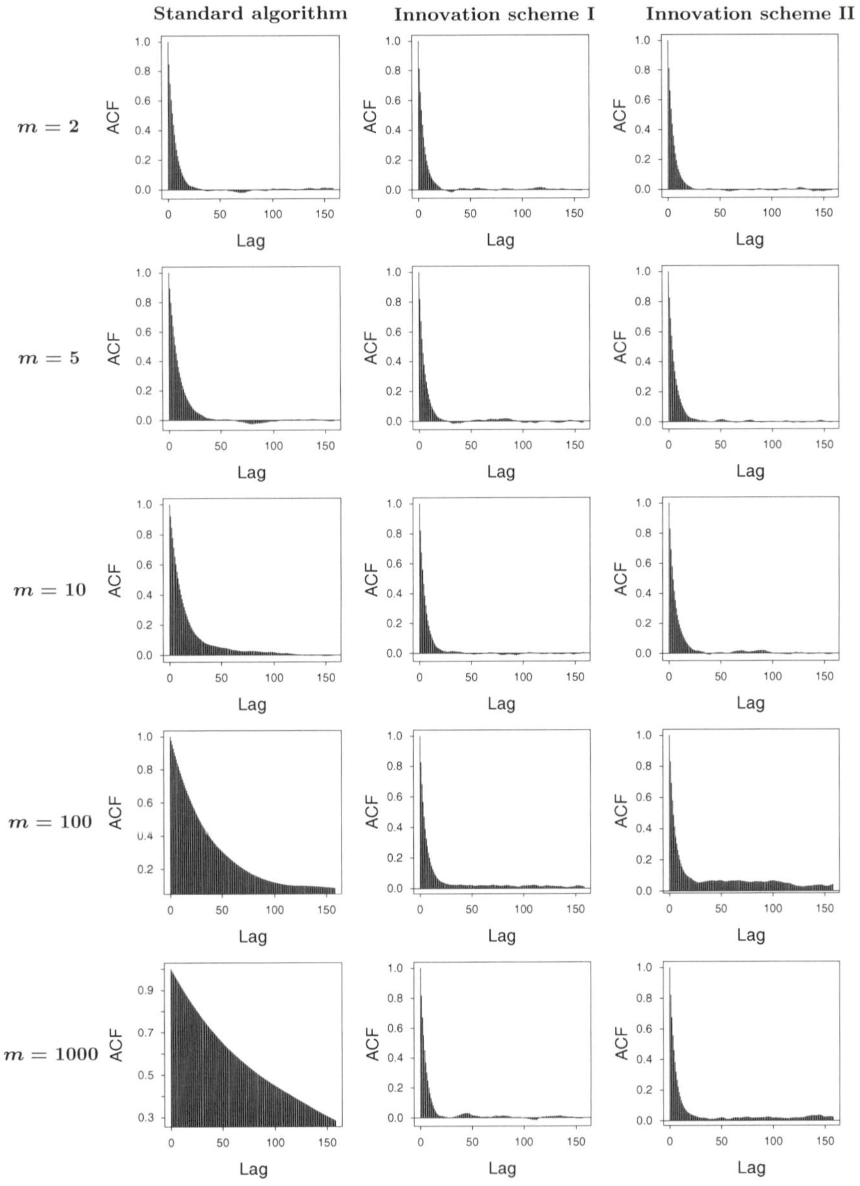

Fig. 7.33 Autocorrelation plots for β based on the results from Fig. 7.29. Calculation of the autocorrelation takes into account the full Markov chain, i.e. without thinning, after having discarded a 10 % burn-in phase

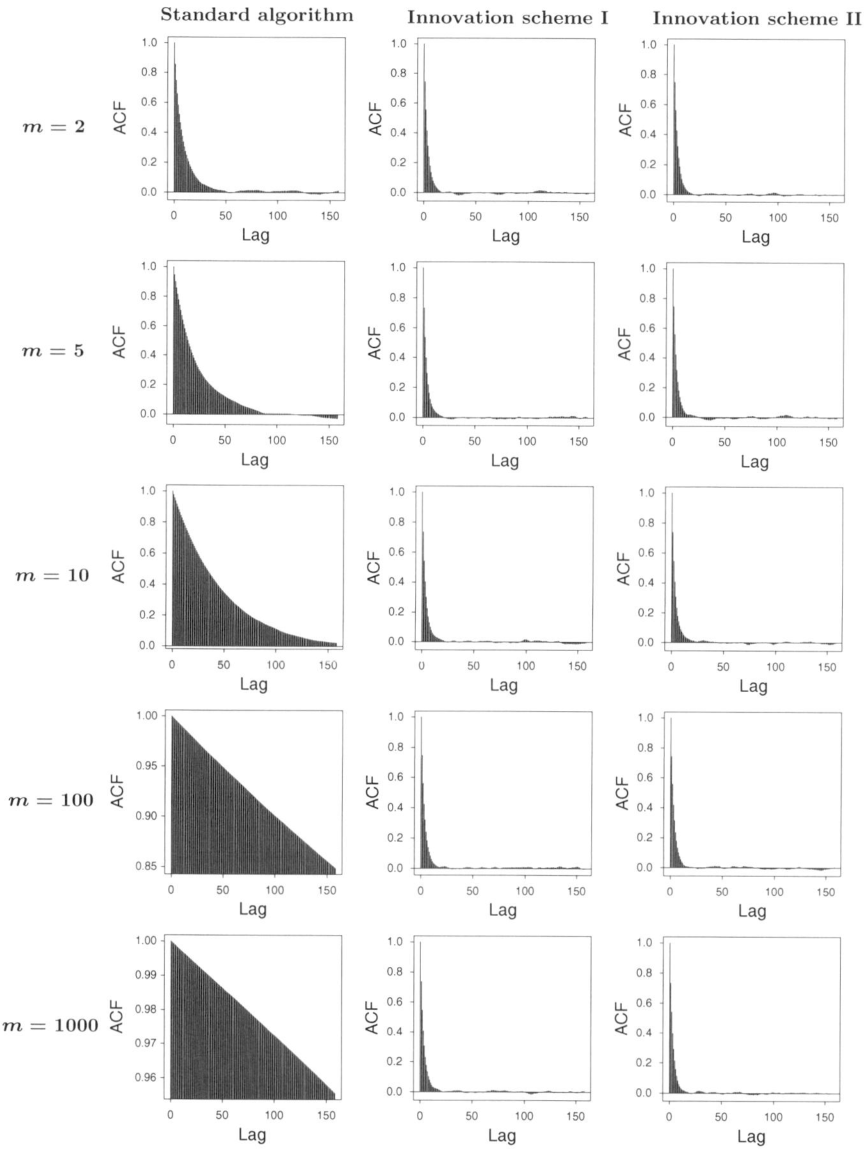

Fig. 7.34 Autocorrelation plots for σ^2 based on the results from Fig. 7.30. Calculation of the autocorrelation takes into account the full Markov chain, i.e. without thinning, after having discarded a 10 % burn-in phase

practical use. This is a more reliable method than starting from a discrete-time framework and then investigating its behaviour for decreasing time step. The former general concept is hence also favoured by Roberts and Stramer (2001) for MCMC sampling and by Papaspiliopoulos and Roberts (2012) in the context of importance sampling.

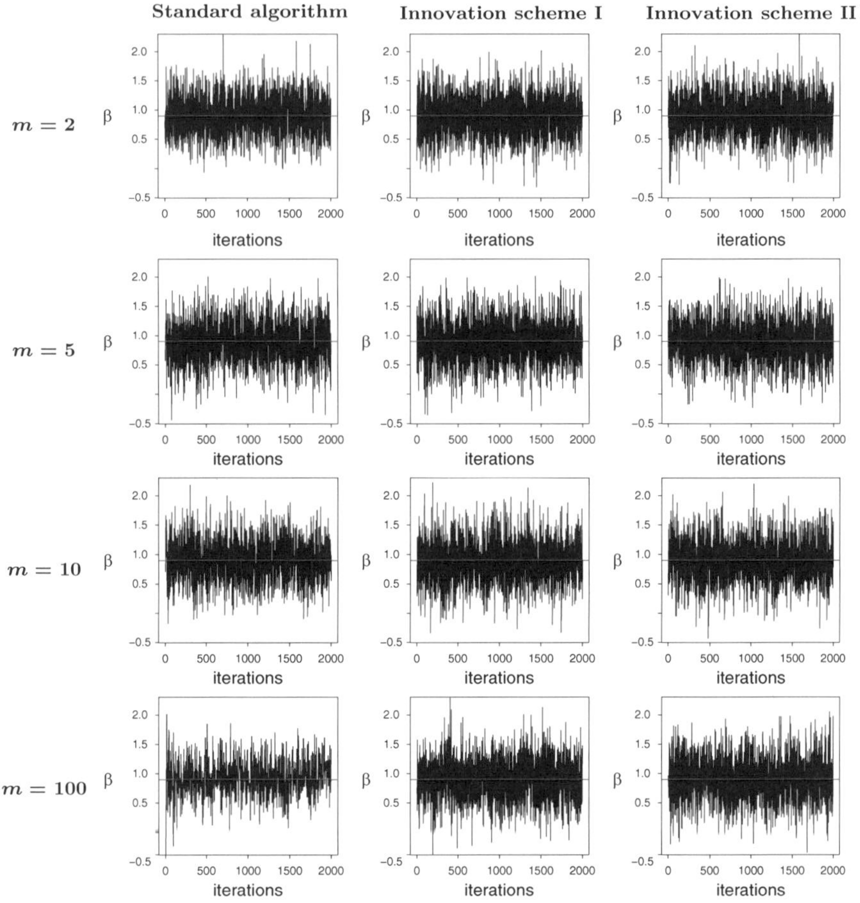

Fig. 7.35 Estimation of parameters of the Ornstein-Uhlenbeck process (7.82) as in Fig. 7.29, this time with the MCMC scheme conditioning on observed data at times $0, 0.1, \ldots, 25$. The procedure introduces m subintervals in between every two observations. This figure shows the trace plots of β. The Markov chains have length 10^5 but have been thinned by factor 50. The true value for β equals 0.9 and is indicated by the *red horizontal line*

This section proves that the newly proposed innovation scheme for conditioned diffusion paths on infinite-dimensional state spaces is non-degenerate. In short, our algorithm for the parameter update works because of the following two reasons: First of all, the construction in Algorithm 7.4 ensures consistency within $\{\boldsymbol{X}, \boldsymbol{\theta}\}$ and $\{\boldsymbol{X}^*, \boldsymbol{\theta}^*\}$. Second, the innovation process $\boldsymbol{Z}$ induces a law $\mathbb{Z}_{\boldsymbol{\theta}}$ which is absolutely continuous with respect to Wiener measure $\mathbb{W}$ for all values of $\boldsymbol{\theta}$. Moreover, there exists a law $\mathbb{D}_{0,\boldsymbol{\theta}}$ which dominates both the law $\tilde{\mathbb{P}}_{\boldsymbol{\theta}}$ of $\boldsymbol{X}$ and the law of the back-transformed process $\boldsymbol{X}^* = g(\boldsymbol{Z}, \boldsymbol{\theta}^*)$. The justification of the path update is based on similar arguments.

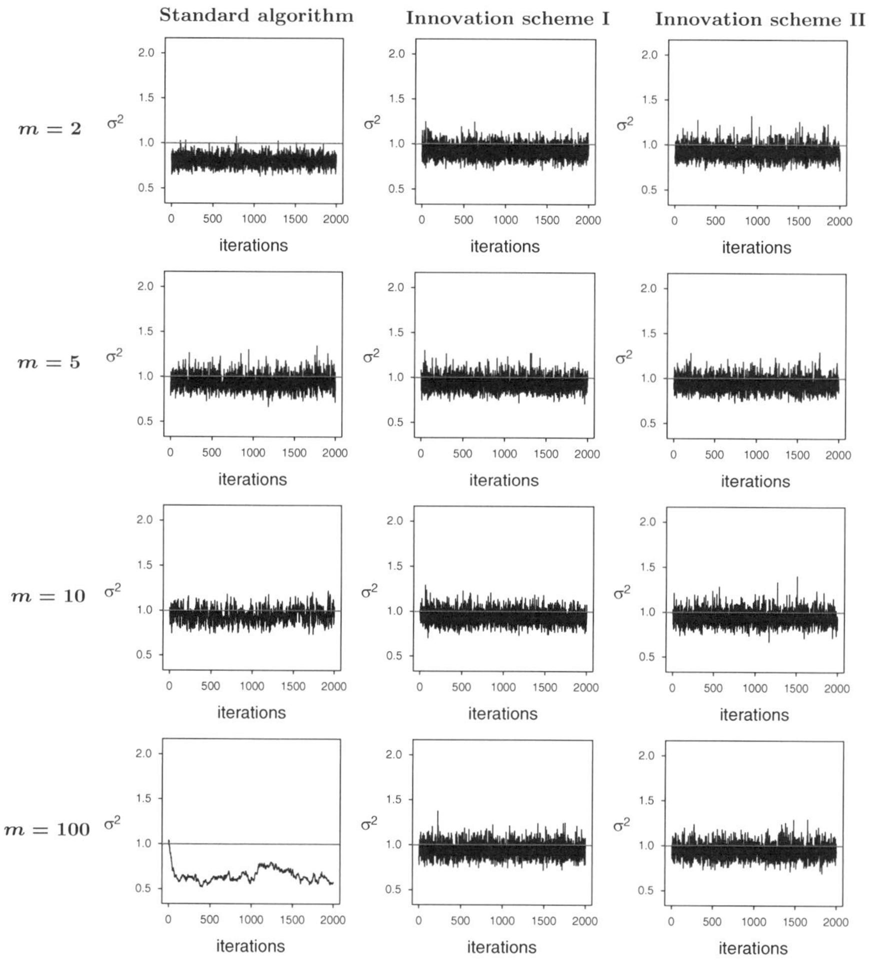

Fig. 7.36 Estimation results as described in Fig. 7.35. This figure shows the trace plots for σ^2. The true parameter value equals $\sigma^2 = 1$ and is indicated by the *red horizontal line*

For decreasing time steps, the modified bridge construct from p. 181 tends to the Euler approximation of the SDE inducing $\mathbb{D}_{0,\theta}$. Hence, our considerations eventually provide the proof that the parameter update proposed by Golightly and Wilkinson (2008), described on p. 241, works also in a continuous-time framework under the assumptions of this section when using the modified bridge transformation. This proposition is however not true for general deterministic links between the original process and the innovation process. Different constructs or a violation of the regularity conditions should carefully be investigated in a continuous-time setting.

In order to be able to apply the innovation scheme in practice, explicit formulas for all involved acceptance probabilities are obtained. In particular, this comprises the derivation of the Radon-Nikodym derivatives $d\tilde{\mathbb{P}}_\theta/d\mathbb{D}_{0,\theta}$ and $d\tilde{\mathbb{P}}_\theta/d\mathbb{D}_{\mu,\theta}$

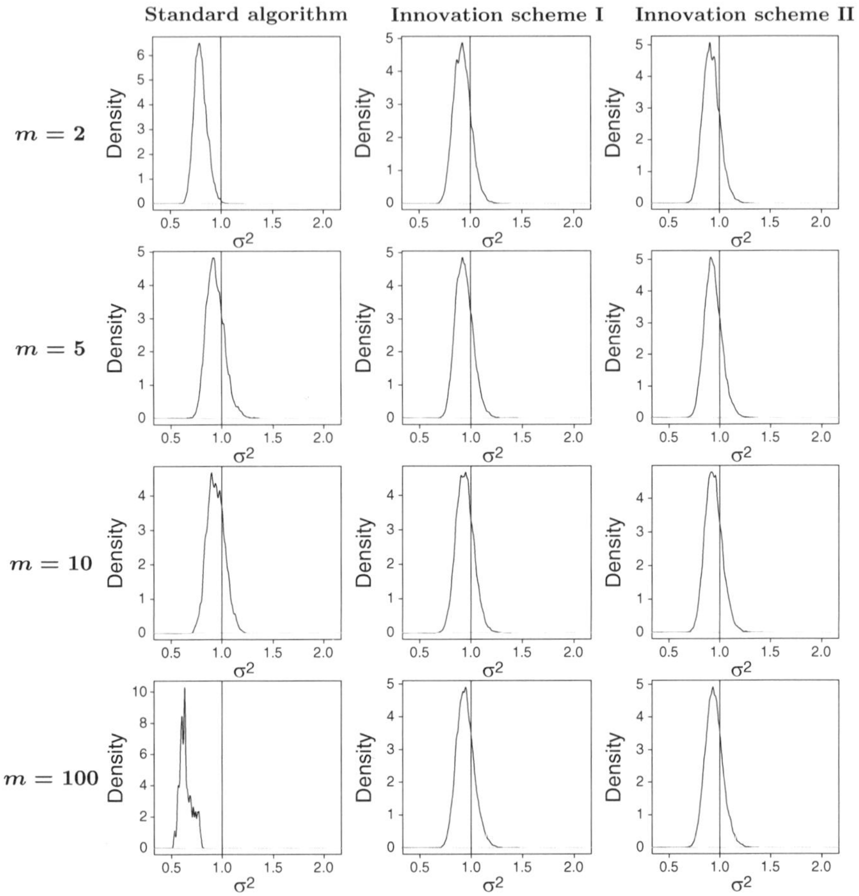

Fig. 7.37 Estimation of the posterior density of β based on the results from Fig. 7.35. Density estimation takes into account the full Markov chain, i.e. without thinning, after having discarded a 10 % burn-in phase. The true value of the parameter is indicated by the *vertical line*

including all factors which depend on the model parameter and the imputed data. These derivatives have also been utilised by Delyon and Hu (2006) and Papaspiliopoulos and Roberts (2012), though in a different representation than in this book, in the context of the simulation and importance sampling for conditioned diffusions. These applications, however, require different knowledge about factors which are proportional with respect to the parameter. Papaspiliopoulos and Roberts, for example, evaluate the proportionality constants as shown in (B.30) and (B.31) on p. 408 through Monte Carlo estimation. This measure would be impracticable in the context of this chapter.

Applied to discrete-time skeletons of diffusion paths, the discretised innovation scheme overcomes the convergence problems explained in Sect. 7.3 as the amount of imputed data increases. It is hence possible to raise the number of imputed data

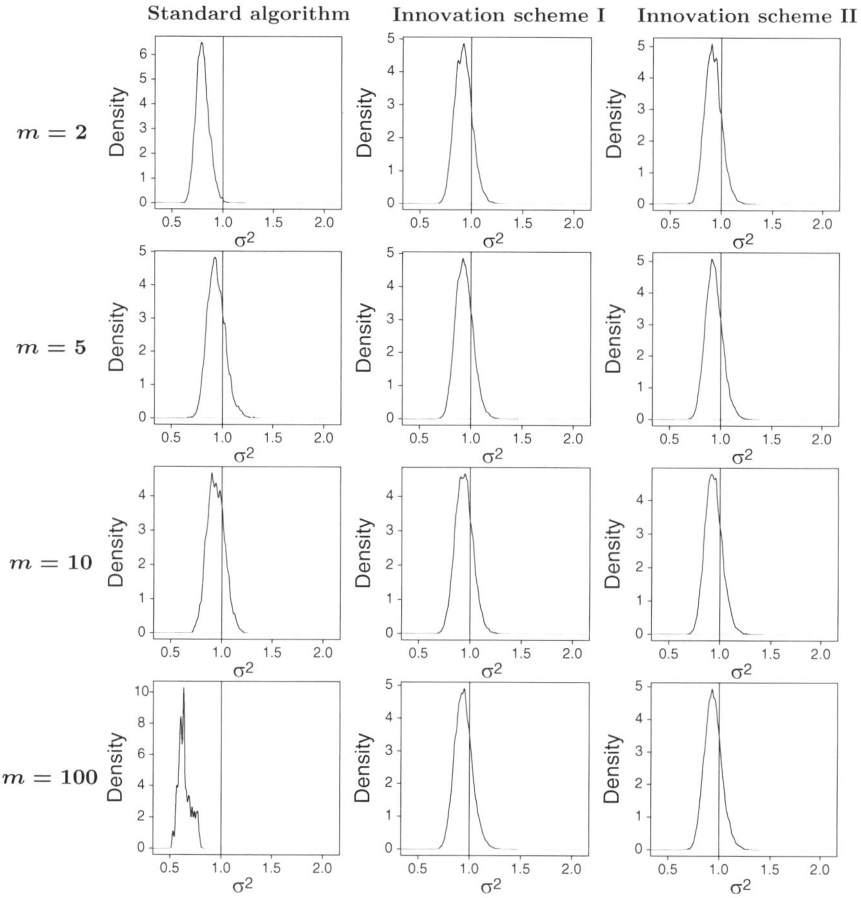

Fig. 7.38 Estimation of the posterior density of σ^2 based on the results from Fig. 7.36. Density estimation takes into account the full Markov chain, i.e. without thinning, after having discarded a 10 % burn-in phase. The true value of the parameter is indicated by the *vertical line*

points in between every two observations in order to reduce an estimation bias. All algorithms have been implemented in R. A simulation study illustrates that the innovation scheme initially improves as the number of imputed data points grows larger and then remains stable. It clearly outperforms standard schemes with respect to its mixing behaviour, serial correlation and acceptance rates.

To conclude, the innovation scheme on infinite-dimensional state spaces presented in this book provides an efficient and widely applicable MCMC mechanism which is appropriate for the parameter estimation also of large-dimensional diffusion processes. As it overcomes disturbing dependence structures which are inherent in most diffusion processes, convergence is guaranteed and practitioners are not restrained to bounded amounts of imputed data. The innovation scheme is applied in Chaps. 8 and 9 to multi-dimensional applications in life sciences.

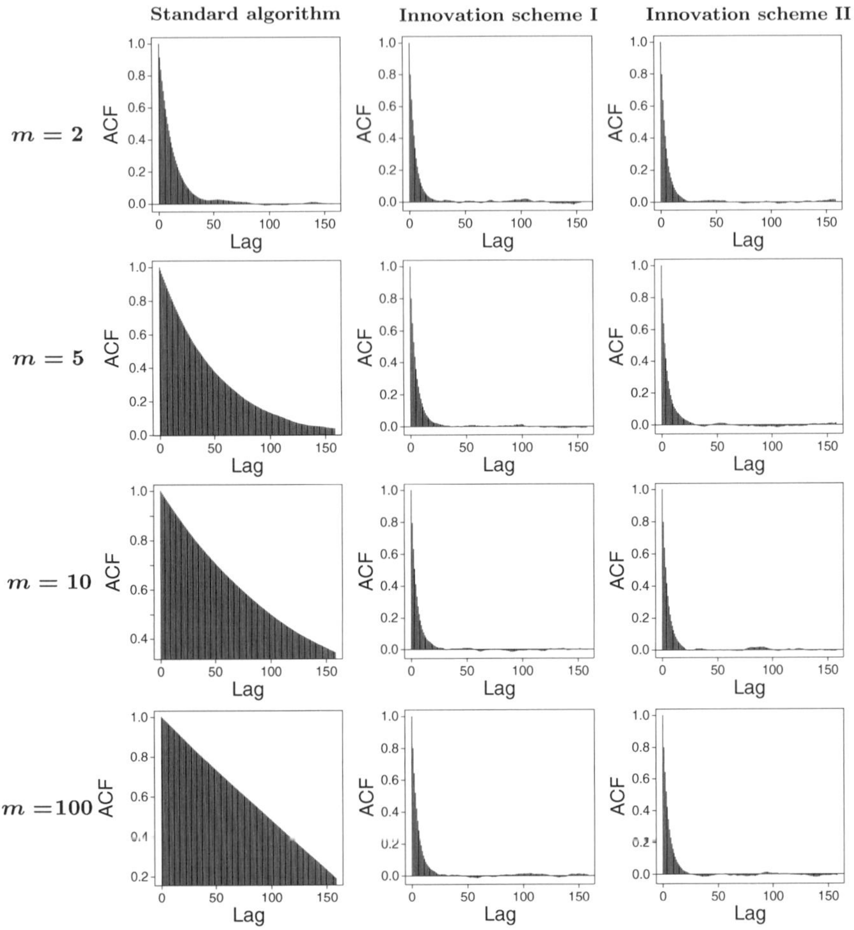

Fig. 7.39 Autocorrelation plots for β based on the results from Fig. 7.35. Calculation of the autocorrelation takes into account the full Markov chain, i.e. without thinning, after having discarded a 10 % burn-in phase

7.5 Discussion and Conclusion

This chapter introduces and comprehensively delves into the concept of Bayesian inference for diffusion processes via the imputation of auxiliary data. The introduction of this additional data reduces the distance between every two consecutive time points where observed or imputed data is available. This way it enables the approximation of the transition density of the diffusion process via the Euler scheme. It is then possible to construct an MCMC algorithm which alternately updates the imputed data and the model parameter. The resulting Markov chain can be utilised to infer on the parameter.

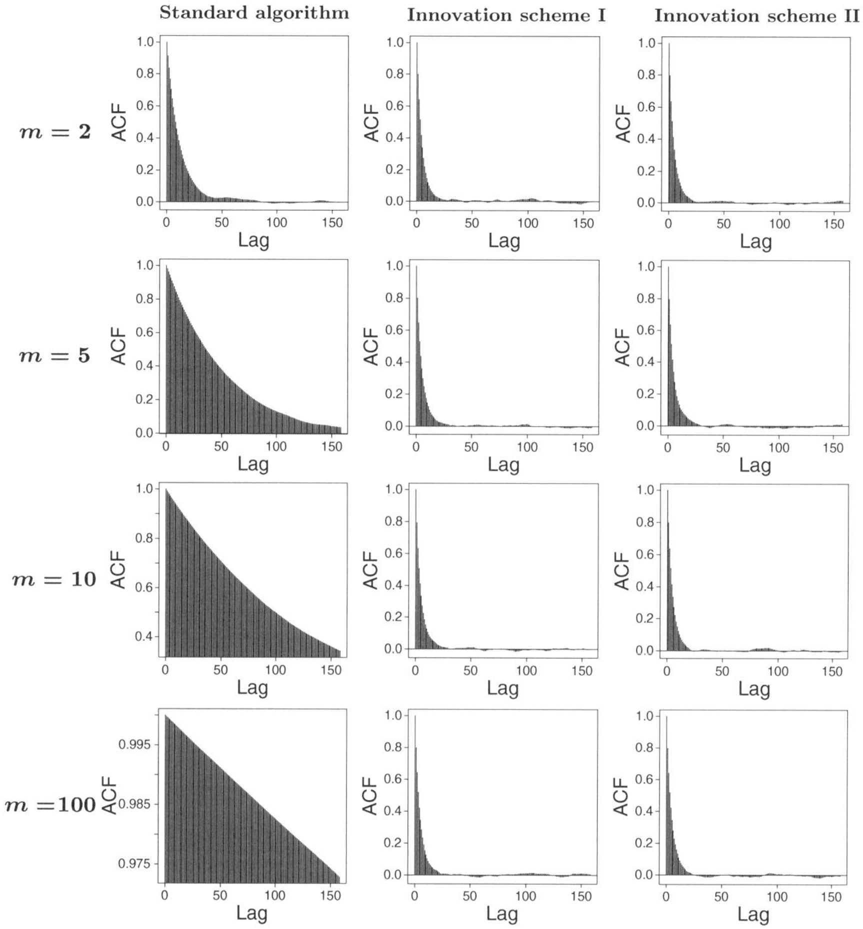

Fig. 7.40 Autocorrelation plots for σ^2 based on the results from Fig. 7.36. Calculation of the autocorrelation takes into account the full Markov chain, i.e. without thinning, after having discarded a 10 % burn-in phase

Section 7.1 comprehensively reviews general concepts for the update of the diffusion path and the update of the parameter and addresses related practical issues. In short, the path update involves proposing a new path segment which bridges the gap between given initial and final states. Naive proposal schemes ignore the end-point information and are hence inefficient. Improved techniques condition on the end point and respectively tie down the proposal distribution. For the parameter update, problem-specific full conditional densities or more general random walk proposals are employed. Empirical and analytical investigations show that for the path update the modified bridge proposal and for the parameter update the random walk proposal perform best.

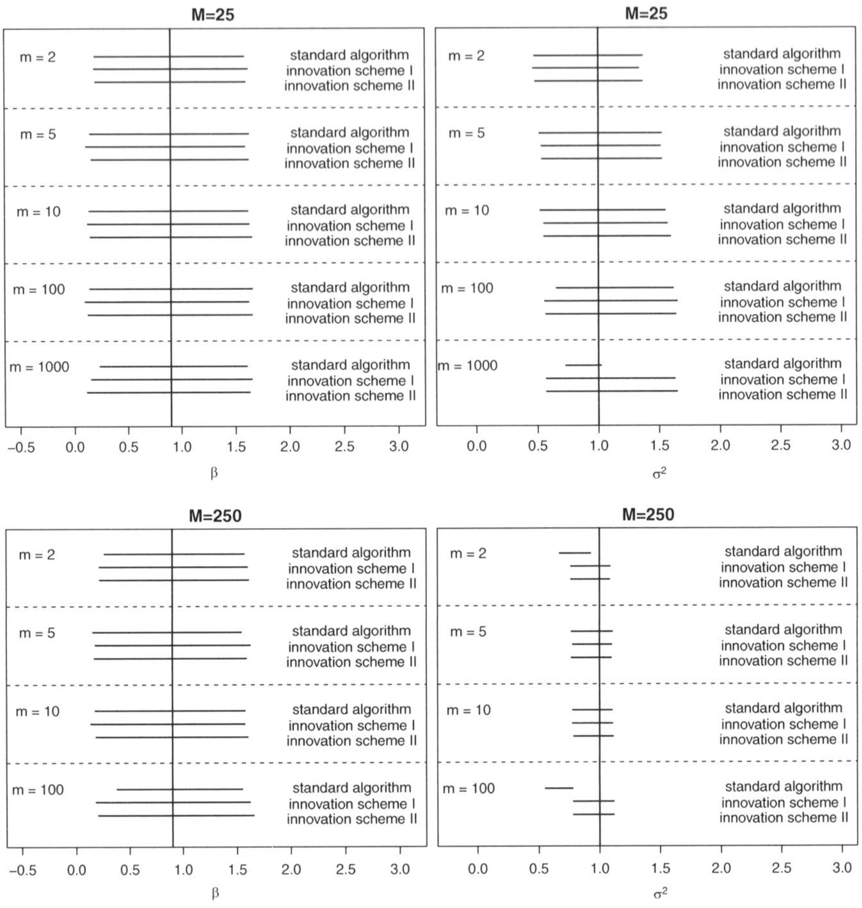

Fig. 7.41 95% highest probability density intervals for β (*left*) and σ^2 (*right*) as displayed in Tables 7.7 (*top*) and 7.8 (*bottom*)

The considerations in Sect. 7.1 are based on the assumption of complete observations without measurement error. As these requirements are typically not met in applications in life sciences, Sect. 7.2 adapts the previously introduced update schemes to the case of non-observed latent states and observations with error. Formulas required for the practical implementation are provided in that section.

Certainly, the MCMC schemes from Sects. 7.1 and 7.2 can also be used without introducing auxiliary time points. This has been done by Eraker et al. (2003), Eraker (2004), Asgharian and Bengtsson (2006) and Jacquier et al. (2007) for jump-diffusion processes in order to model prices at financial markets. In that context, dense data is available, hence no augmentation is necessary. The MCMC procedure then estimates the model parameters and latent variables. Kim et al. (1998) likewise apply the scheme to infer on parameters and latent variables for stochastic volatility models without imputing data; they however also touch the introduction of missing values.

Table 7.7 Estimation results as in Figs. 7.29 and 7.30, i.e. for $T = 25$ and $M = 25$. This table displays the posterior means and posterior 95 %-hpd intervals after a 10 % burn-in phase. The latter are computed according to Chen and Shao (1999). The true values of the parameters are $\beta = 0.9$ and $\sigma^2 = 1$. The hpd intervals are also shown in Fig. 7.41

	Standard algorithm	Innovation scheme I	Innovation scheme II
$m = 2$	β : 0.90, (0.20,1.58)	β : 0.90, (0.19,1.61)	β : 0.89, (0.20,1.59)
	σ^2: 0.88, (0.48,1.37)	σ^2: 0.87, (0.47,1.34)	σ^2: 0.87, (0.48,1.36)
$m = 5$	β : 0.89, (0.15,1.62)	β : 0.88, (0.11,1.59)	β : 0.89, (0.16,1.62)
	σ^2: 0.98, (0.52,1.52)	σ^2: 0.98, (0.53,1.51)	σ^2: 0.98, (0.54,1.52)
$m = 10$	β : 0.88, (0.14,1.62)	β : 0.88, (0.13,1.63)	β : 0.89, (0.15,1.65)
	σ^2: 1.00, (0.52,1.55)	σ^2: 1.02, (0.56,1.57)	σ^2: 1.02, (0.55,1.59)
$m = 100$	β : 0.88, (0.15,1.66)	β : 0.88, (0.10,1.62)	β : 0.88, (0.13,1.65)
	σ^2: 1.09, (0.66,1.61)	σ^2: 1.06, (0.56,1.65)	σ^2: 1.05, (0.57,1.64)
$m = 1{,}000$	β : 0.90, (0.24,1.60)	β : 0.91, (0.16,1.65)	β : 0.86, (0.12,1.63)
	σ^2: 0.90, (0.73,1.02)	σ^2: 1.05, (0.57,1.63)	σ^2: 1.07, (0.57,1.65)

Table 7.8 Estimation results as in Figs. 7.35 and 7.36, i.e. for $T = 25$ and $M = 250$. This table displays the posterior means and posterior 95 %-hpd intervals after a 10 % burn-in phase. The true values of the parameters are $\beta = 0.9$ and $\sigma^2 = 1$. The hpd intervals are also shown in Fig. 7.41

	Standard algorithm	Innovation scheme I	Innovation scheme II
$m = 2$	β : 0.91, (0.27,1.57)	β : 0.89, (0.22,1.60)	β : 0.90, (0.23,1.61)
	σ^2: 0.80, (0.67,0.93)	σ^2: 0.92, (0.77,1.09)	σ^2: 0.92, (0.77,1.08)
$m = 5$	β : 0.89, (0.16,1.54)	β : 0.90, (0.18,1.62)	β : 0.89, (0.17,1.59)
	σ^2: 0.94, (0.77,1.11)	σ^2: 0.94, (0.78,1.10)	σ^2: 0.94, (0.77,1.10)
$m = 10$	β : 0.91, (0.18,1.57)	β : 0.88, (0.14,1.57)	β : 0.89, (0.19,1.60)
	σ^2: 0.94, (0.78,1.10)	σ^2: 0.94, (0.78,1.11)	σ^2: 0.94, (0.79,1.11)
$m = 100$	β : 0.92, (0.38,1.55)	β : 0.90, (0.19,1.62)	β : 0.90, (0.21,1.65)
	σ^2: 0.64, (0.55,0.78)	σ^2: 0.94, (0.79,1.12)	σ^2: 0.94, (0.79,1.12)

Improvements of the update procedures in Sects. 7.1 and 7.2 may for example be obtained by replacing the Euler scheme, which is used for the approximation of the transition density based on the augmented dataset, by any higher order numerical scheme such as those presented in Sect. 3.3.2. For instance, Elerian (1998) considers the application of the Milstein scheme. This measure, however, does not solve a general conceptual convergence problem that appears as the amount of imputed data increases: For a time-continuously imputed diffusion path, the imputed data and the diffusion matrix are deterministically linked. Resulting difficulties for the update mechanisms from Sects. 7.1 and 7.2 are described in detail in Sect. 7.3. In practice, data is certainly never imputed continuously. However, this corresponds to the limiting case of steadily enhanced amounts of imputed data. The behaviour of the MCMC algorithm on an infinite-dimensional state space is hence an appropriate indicator for the behaviour of the MCMC scheme in the case where finite but increasing amounts of data are imputed. The consequence is poor mixing of the Markov chains, especially for those parameters determined by the diffusion matrix.

Some authors mention that such convergence issues are not a relevant problem in their practical applications when low numbers of auxiliary time points are introduced (see e.g. Eraker 2001). A crucial amount of imputed data may however easily be reached for low-frequency datasets or multi-dimensional diffusions. Moreover, in real data applications, where the true values of the parameters are unknown, it may be difficult to determine the threshold value for the number of imputed data points beyond which estimates deteriorate. In any case, it is desirable to have a reliable tool which guarantees that it does not break down as the amount of imputed data grows.

Thus, starting from the convergence problems of naive MCMC algorithms pointed out in Sects. 7.3 and 7.4 reviews and develops update techniques which are neatly modified such that they circumvent the sources of poor convergence. These methods are a change of factorisation of the dominating measure of the diffusion process, time change transformations, particle filters and the innovation scheme on infinite-dimensional state spaces. As the utilisation of the former three methods is inappropriate for the applications in Chaps. 8 and 9, this book concentrates on the innovation scheme. This method has been utilised before for unconditioned diffusion paths by Chib et al. (2004) and on finite-dimensional state spaces by Golightly and Wilkinson (2008, 2010). Its application to conditioned diffusions on infinite-dimensional state spaces, as contributed by this book, is however notably different.

In particular, Sect. 7.4.4 designs the innovation scheme for conditioned diffusions on infinite-dimensional state spaces and provides the mathematical proof that the so-constructed MCMC scheme converges. Consequently, this algorithm is non-degenerate also for arbitrarily large but finite sets of observed and imputed data. For practical usability, explicit formulas for all involved acceptance probabilities are derived. The modified parameter and path updates are described in algorithmic form including the time-discretisations of these acceptance probabilities. All algorithms are implemented and employed in a simulation study which certifies moderate computing times and verifies that the innovation scheme does not break down as the number of auxiliary data points increases. The enhanced innovation scheme hence outperforms the techniques introduced in Sect. 7.1.

To conclude, this chapter offers a detailed and critical inspection of Bayesian inference methods for diffusion processes based on data augmentation. The considered techniques are suitable for large and irregularly spaced observation intervals, multivariate diffusions with possibly latent components and for observations with error. Throughout the chapter, importance was attached to an understandable presentation of the update schemes and the convergence problems that arise in standard algorithms when more and more data is imputed. For the first time, this book surveys improved update schemes in Sect. 7.4 which aim to overcome the previously described convergence difficulties. These methods are all appropriate wherever their assumptions are true or where the considered diffusion process is low-dimensional, respectively. They however cannot be used for fairly complex and partly large-dimensional applications as investigated in Chaps. 8 and 9. In these cases, the enhanced innovation scheme is required. Its convergence has been proven and its practical implementation formulated in this chapter.

References

Asgharian H, Bengtsson C (2006) Jump spillover in international equity markets. J Financ Econom 4:167–203
Cano J, Kessler M, Salmerón D (2006) Approximation of the posterior density for diffusion processes. Stat Probab Lett 76:39–44
Chen MH, Shao QM (1999) Monte Carlo estimation of Bayesian credible and hpd intervals. J Comput Graph Stat 8:69–92
Chib S, Shephard N (2002) Numerical techniques for maximum likelihood estimation of continuous-time diffusion processes: comment. J Bus Econom Stat 20(3):325–27
Chib S, Pitt M, Shephard N (2004) Likelihood based inference for diffusion driven state space models. Working paper, Nuffield College, University of Oxford
Del Moral P, Jacod J, Protter P (2001) The Monte-Carlo method for filtering with discrete-time observations. Working paper at Cornell University Oper Res and Industrial Engineering
Delyon B, Hu Y (2006) Simulation of conditioned diffusion and application to parameter estimation. Stoch Process Appl 116:1660–1675
Dempster A, Laird N, Rubin D (1977) Maximum likelihood from incomplete data via the EM algorithm. J R Stat Soc Ser B 39:1–38
Durham G, Gallant A (2002) Numerical techniques for maximum likelihood estimation of continuous-time diffusion processes (with comments). J Bus Econom Stat 20:297–316
Elerian O (1998) A note on the existence of a closed form conditional transition density for the Milstein scheme. Working paper, Nuffield College, University of Oxford
Elerian O, Chib S, Shephard N (2001) Likelihood inference for discretely observed nonlinear diffusions. Econometrica 69:959–993
Eraker B (2001) MCMC analysis of diffusion models with application to finance. J Bus Econom Stat 19:177–191
Eraker B (2002) Comment on 'numerical techniques for maximum likelihood estimation of continuous-time diffusion processes'. J Bus Econom Stat 20:327–329
Eraker B (2004) Do stock prices and volatility jump? Reconciling evidence from spot and option prices. J Financ 59:1367–1404
Eraker B, Johannes M, Polson N (2003) The impact of jumps in volatility and returns. J Financ 58:1269–1300
Fearnhead P (2008) Computational methods for complex stochastic systems: a review of some alternatives to MCMC. Stat Comput 18:151–171
Fearnhead P, Papaspiliopoulos O, Roberts G (2008) Particle filters for partially-observed diffusions. J R Stat Soc Ser B 70:755–777
Gamerman D, Lopes H (2006) Markov chain Monte Carlo: stochastic simulation for Bayesian inference, 2nd edn. Chapman & Hall, Boca Raton/London/New York
Gilks W, Richardson S, Spiegelhalter D (eds) (1996) Markov chain Monte Carlo in practice. Chapman & Hall, London
Golightly A, Wilkinson D (2005) Bayesian inference for stochastic kinetic models using a diffusion approximation. Biometrics 61:781–788
Golightly A, Wilkinson D (2006a) Bayesian sequential inference for nonlinear multivariate diffusions. Stat Comput 16:323–338
Golightly A, Wilkinson D (2006b) Bayesian sequential inference for stochastic kinetic biochemical network models. J Comput Biol 13:838–851
Golightly A, Wilkinson D (2008) Bayesian inference for nonlinear multivariate diffusion models observed with error. Comput Stat Data Anal 52:1674–1693
Golightly A, Wilkinson D (2010) Markov chain Monte Carlo algorithms for SDE parameter estimation. In: Lawrence N, Girolami M, Rattray M, Sanguinetti G (eds) Introduction to learning and inference for computational systems biology. MIT Press, Cambridge/London, pp 253–275

Hurn A, Jeisman J, Lindsay K (2007) Seeing the wood for the trees: a critical evaluation of methods to estimate the parameters of stochastic differential equations. J Financ Econom 5:390–455

Jacquier E, Johannes M, Polson N (2007) MCMC maximum likelihood for latent state models. J Econom 137:615–640

Johannes M, Polson N, Stroud J (2006) Optimal filtering of jump diffusions: extracting latent states from asset prices. Rev Financ Stud 22:2759–2799

Jones C (1998) A simple Bayesian method for the analysis of diffusion processes. Working Paper, University of Pennsylvania

Kalogeropoulos K (2007) Likelihood-based inference for a class of multidimensional diffusions with unobserved paths. J Stat Plann Inference 137:3092–3102

Kalogeropoulos K, Roberts G, Dellaportas P (2010) Inference for stochastic volatility models using time change transformations. Ann Stat 38:784–807

Kalogeropoulos K, Dellaportas P, Roberts G (2011) Likelihood based inference for correlated diffusions. Can J Stat 39:52–72

Karatzas I, Shreve S (1991) Brownian motion and stochastic calculus, 2nd edn. Graduate texts in mathematics. Springer, New York

Kim S, Shephard N, Chib S (1998) Stochastic volatility: likelihood inference and comparison with ARCH models. Rev Econ Stud 65:361–393

Klebaner F (2005) Introduction to stochastic calculus with applications, 2nd edn. Imperial College Press, London

Maybeck P (1979) Stochastic models, estimation, and control, vol 1. Academic, New York/San Francisco/London

Mengersen K, Tweedie R (1996) Rates of convergence of the Hastings and Metropolis algorithms. Ann Stat 24:101–121

Øksendal B (2003) Stochastic differential equations. An introduction with applications, 6th edn. Springer, Berlin/Heidelberg

Papaspiliopoulos O, Roberts G (2012) Importance sampling techniques for estimation of diffusion models. In: Kessler M, Lindner A, Sorensen M (eds) Statistical methods for stochastic differential equations. Monographs on statistics and applied probability. Chapman & Hall (to appear), London, pp 311–337

Papaspiliopoulos O, Roberts G, Sköld M (2003) Non-centered parameterisations for hierarchical models and data augmentation (with discussion). In: Bernardo J, Bayarri M, Berger J, Dawid A, Heckerman D, Smith A, West M (eds) Bayesian statistics 7. Lecture notes in computer science, vol 4699. Oxford University Press, Oxford, pp 307–326

Pedersen A (1995) A new approach to maximum likelihood estimation for stochastic differential equations based on discrete observations. Scand J Stat 22:55–71

Polson N, Roberts G (1994) Bayes factors for discrete observations from diffusion processes. Biometrika 81:11–26

Robert C, Casella G (2004) Monte Carlo statistical methods, 2nd edn. Springer, New York

Roberts G, Rosenthal J (1997) Geometric ergodicity and hybrid Markov chains. Electron Commun Probab 2:13–25

Roberts G, Stramer O (2001) On inference for partially observed nonlinear diffusion models using the Metropolis-Hastings algorithm. Biometrika 88:603–621

Roberts G, Tweedie R (2012) Understanding MCMC. Springer (to appear), Berlin

Suda D (2009) Importance sampling on discretely-observed diffusions. In: Poster at BISP6, Bressanone/Brixen

Tanner M, Wong W (1987) The calculation of posterior distributions by data augmentation (with comments). J Am Stat Assoc 82:528–546

Tierney L (1994) Markov chains for exploring posterior distributions. Ann Stat 22:1701–1728

Part III
Applications

Chapter 8
Application I: Spread of Influenza

Influenza is a contagious disease caused by the influenza virus that affects mammals and birds. Human influenza morbidity and mortality is a major concern of public health institutions. According to recent assessments, the annual number of infected people lies between 5 and 15 % of the worldwide population, with 250,000–500,000 deaths every year (e.g. Russell et al. 2008).

This chapter deals with the statistical estimation of parameters in models for the spread of human influenza. To that end, the standard and multitype SIR models, which were introduced in Chap. 5, are applied. Out of the various mathematical representations considered in that chapter, the diffusion processes are chosen as the most appropriate ones here. Statistical inference is accomplished by means of the innovation scheme developed in Chap. 7.

To start with, a simulation study with synthetic datasets is carried out in Sect. 8.1. This gives an idea about the performance of the innovation scheme when applied to datasets of certain sizes and levels of completeness. In Sect. 8.2, the standard SIR model is applied to a dataset on an influenza outbreak in a British boarding school in 1978. Finally, in Sect. 8.3, the spatial spread of influenza in Germany is considered. To that end, the multitype SIR model is utilised with clusters corresponding to different geographic regions. Model parameters are estimated for a dataset on influenza occurrences in the season 2009/10. This study aims to be an initial analysis which can be extended in different directions in further investigations. Section 8.4 concludes and gives an outlook on such future work.

8.1 Simulation Study

This simulation study investigates three synthetic datasets: One dataset for the standard SIR model and two datasets for the multitype SIR model with $n = 3$ and $n = 10$ clusters, respectively. Both models were introduced in Sects. 5.1 and 5.2. The most relevant properties of the resulting diffusion approximations are

C. Fuchs, *Inference for Diffusion Processes: With Applications in Life Sciences*,
DOI 10.1007/978-3-642-25969-2_8,

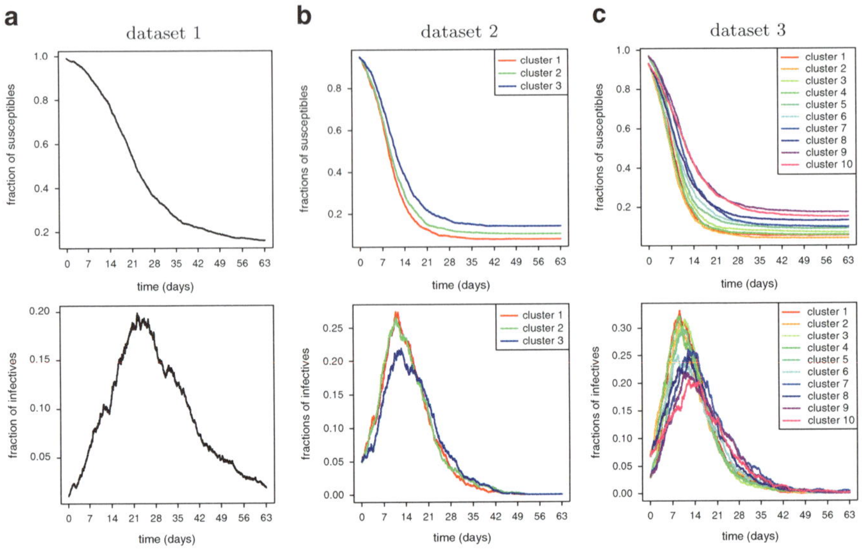

Fig. 8.1 Synthetic datasets used in the simulation study in this section. The *top row* shows the fractions of susceptibles, the *bottom row* the fractions of infected individuals. Simulations have been obtained by application of the Euler scheme from Sect. 3.3.2 with time step 0.025 and settings as described in the main text. Observations are assumed to be available at equidistant time steps of length 7 such that there are ten observations in the time interval $[0, 63]$, (**a**) dataset 1, (**b**) dataset 2, (**c**) dataset 3

summarised in Sects. 5.1.4 and 5.2.4 on pp. 109,121, respectively. In particular, the stochastic differential equations are given in (5.19) and (5.32) in these summaries. The notation in this chapter is adopted from Chap. 5.

8.1.1 Data

The sample paths from the three just mentioned datasets are shown in Fig. 8.1. They are generated by application of the Euler scheme from Sect. 3.3.2 with time step 0.025 on the time interval [0,63]. In order to reflect the observation interval in a real data situation, observations are provided only at time points $0, 7, 14, \ldots, 63$ in the simulation study. The model parameters and other variables are chosen as follows:

- *Dataset 1 (standard SIR model):* The population size equals $N = 1{,}000$ with initial state $\boldsymbol{x}_0 = (s_0, i_0)' = (0.99, 0.01)'$ at time zero. The model parameter, consisting of the contact rate α and the reciprocal length of the infectious period β, is chosen as $\boldsymbol{\theta} = (\alpha, \beta)' = (0.325, 0.15)'$. For consistent notation with datasets 2 and 3, the standard SIR model is also referred to as a multitype

SIR model with $n = 1$ cluster in the next section. The parameter α is then also denoted as α_1.

- *Dataset 2 (multitype SIR model with $n = 3$ clusters):* There are three clusters with identical population sizes $N_j = 1{,}000$, $j \in \{1, 2, 3\}$. The state variable of the diffusion process is $\boldsymbol{x} = (s_1, s_2, s_3, i_1, i_2, i_3)'$ with initial value $\boldsymbol{x}_0 = (0.95, 0.95, 0.95, 0.05, 0.05, 0.05)'$ at time zero, i. e. the initial fractions of susceptibles and infectives are identical in all three clusters. Contacts between clusters occur according to the network matrix

$$\gamma^N = \begin{pmatrix} 0.80 & 0.10 & 0.10 \\ 0.10 & 0.85 & 0.05 \\ 0.10 & 0.05 & 0.85 \end{pmatrix},$$

where $\gamma^N = \gamma^S = \gamma^I$, which means that susceptible, infective and removed individuals show equal contact behaviour. This matrix is considered known, i. e. it is not statistically estimated in the following simulation study. The contact rates α_j are assumed to depend on the corresponding cluster j, while the average infectious period β^{-1} is assumed identical for the three groups. The model parameter hence equals $\boldsymbol{\theta} = (\alpha_1, \alpha_2, \alpha_3, \beta)'$. It is chosen to be $\boldsymbol{\theta} = (0.6, 0.5, 0.4, 0.2)'$.

- *Dataset 3 (multitype SIR model with $n = 10$ clusters):* The assumptions here are similar to dataset 2 but adopted to ten clusters. In particular, one has population sizes $N_j = 1{,}000$ for $j \in \{1, \ldots, 10\}$, state variable $\boldsymbol{x} = (s_1, \ldots, s_{10}, i_1, \ldots, i_{10})'$ with initial value

$$\begin{aligned}\boldsymbol{x}_0 = (&0.97, 0.93, 0.97, 0.93, 0.97, 0.93, 0.97, 0.93, 0.97, 0.93, \\ &0.03, 0.07, 0.03, 0.07, 0.03, 0.07, 0.03, 0.07, 0.03, 0.07)'\end{aligned}$$

at time zero, contact matrix

$$\gamma^N = \gamma^S = \gamma^I = \begin{pmatrix} 0.80 & 0.05 & 0.05 & 0.03 & 0.03 & 0.02 & 0.01 & 0.01 & 0 & 0 \\ 0.05 & 0.85 & 0.03 & 0.01 & 0.01 & 0.01 & 0.01 & 0.01 & 0.01 & 0.01 \\ 0.05 & 0.03 & 0.85 & 0.03 & 0.02 & 0.02 & 0 & 0 & 0 & 0 \\ 0.03 & 0.01 & 0.03 & 0.90 & 0.01 & 0.01 & 0.01 & 0 & 0 & 0 \\ 0.03 & 0.01 & 0.02 & 0.01 & 0.85 & 0.03 & 0.02 & 0.02 & 0.01 & 0 \\ 0.02 & 0.01 & 0.02 & 0.01 & 0.03 & 0.75 & 0.06 & 0.05 & 0.05 & 0 \\ 0.01 & 0.01 & 0 & 0.01 & 0.02 & 0.06 & 0.75 & 0.08 & 0.01 & 0.05 \\ 0.01 & 0.01 & 0 & 0 & 0.02 & 0.05 & 0.08 & 0.80 & 0 & 0.03 \\ 0 & 0.01 & 0 & 0 & 0.01 & 0.05 & 0.01 & 0 & 0.80 & 0.12 \\ 0 & 0.01 & 0 & 0 & 0 & 0 & 0.05 & 0.03 & 0.12 & 0.79 \end{pmatrix}$$

and model parameter $\boldsymbol{\theta} = (\alpha_1, \ldots, \alpha_{10}, \beta)'$ with

$$\boldsymbol{\theta} = (0.6, 0.6, 0.55, 0.55, 0.5, 0.5, 0.45, 0.45, 0.4, 0.4, 0.2)'.$$

8.1.2 Parameter Estimation

Chapter 7 presented methods for the Bayesian estimation of the parameters of diffusion processes by means of data augmentation. Special emphasis was put on the innovation scheme, which was presented and further developed in Sect. 7.4.4. This scheme is now applied to the just specified synthetic datasets in order to estimate the parameters α_j, $j \in \{1, \dots, n\}$, and β. The notation is adopted from Chap. 7.

For all datasets, α_j and β are assumed to be a priori exponentially distributed with expected values 0.5. In the MCMC algorithm,

$$\log \alpha_j^* \sim \mathcal{N}\big(\log \alpha_j, 0.0009\big) \quad \text{and} \quad \log \beta^* \sim \mathcal{N}\big(\log \beta, 0.0009\big)$$

for $j = 1, \dots, n$, where α_j and β represent the current values.

Consider dataset 1 first. To start with, both the fraction s of susceptibles and the fraction i of infectives are assumed to be given at the observation times. Figures 8.2 and 8.3 on p. 285 display trace plots, posterior density estimates and autocorrelation plots for α and β produced by the innovation scheme. In particular, the parameter update is performed according to Algorithm 7.6 on p. 253, and the diffusion path is updated with a modified bridge proposal as described in Sect. 7.1. The simulated Markov chains have length 10^5 but have been thinned by factor 50. The innovation scheme imputes data such that there are $m \in \{7, 14\}$ intermediate subintervals in between every two observation times.

In practice, the fraction of susceptible individuals is typically unknown. Hence, the above estimation procedure is carried out again with i observed and s considered latent. The modified diffusion bridge update in the presence of latent components is described in Sect. 7.2. Figures 8.4 and 8.5 on p. 286 show the obtained estimation results. This time, because of a large burn-in, the simulated Markov chains have length 10^6 and are thinned by factor 500. Table 8.1 lists the posterior means and 95 %-hpd intervals corresponding to the MCMC outputs from Figs. 8.2–8.5.

For datasets 2 and 3, parameter estimation is performed in an analogous manner as for dataset 1. The outcomes are summarised in Tables 8.2 and 8.3. All results in these tables are based on simulated Markov chains of length 10^5. Due to space restrictions, trace plots, posterior density estimates and autocorrelaton plots are exemplarily displayed for the model with $n = 10$ clusters and observed fraction s in Fig. 8.6.

Overall, the simulation study revealed that satisfactory estimation results for the contact rates α_j and the reciprocal infectious period β are obtained when both the fraction s of susceptibles and the fraction i of infective individuals are observed. All derived hpd intervals contain the true parameters, and in case of multiple clusters, the order of the estimated contact rates $\hat{\alpha}_j$ resembles the order of the true values. In practice however, the component s is latent, which makes parameter estimation more difficult. For the standard SIR model, estimation of both α and β is still possible. In case of the multitype SIR model, however, the contact rates α_j can obviously not be distinguished. Instead, one obtains similar confidence intervals

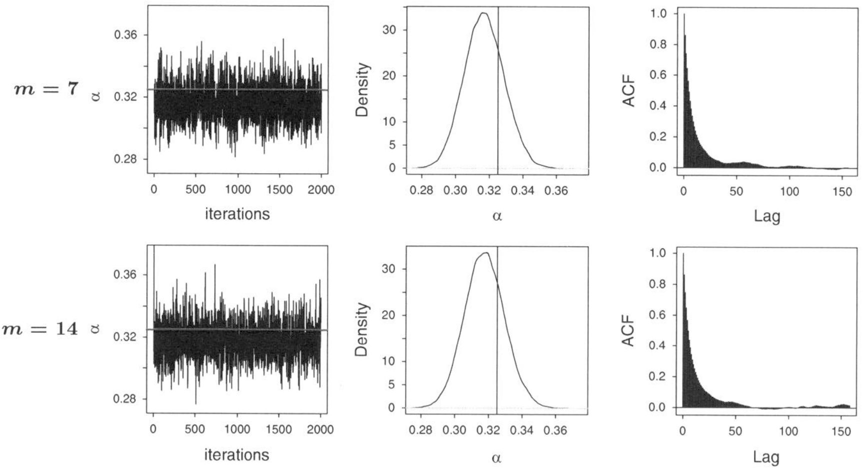

Fig. 8.2 Bayesian estimation of parameters of the standard SIR model when applied to dataset 1 with both s and i being observed. Details of the estimation procedure are described in the main text. The MCMC scheme introduces $m \in \{7, 14\}$ subintervals in between every two observations. This figure shows the trace plots of α (*left column*) with corresponding posterior density estimates (*middle column*) and autocorrelation plots (*right column*). The Markov chains have length 10^5 but have been thinned by factor 50 in the trace plots. *Red horizontal lines* in the trace plots and *black vertical lines* in the density plots indicate the true parameter values. Estimation of posterior densities and autocorrelation takes into account the full Markov chain, i.e. without thinning, after having discarded a 10 % burn-in phase

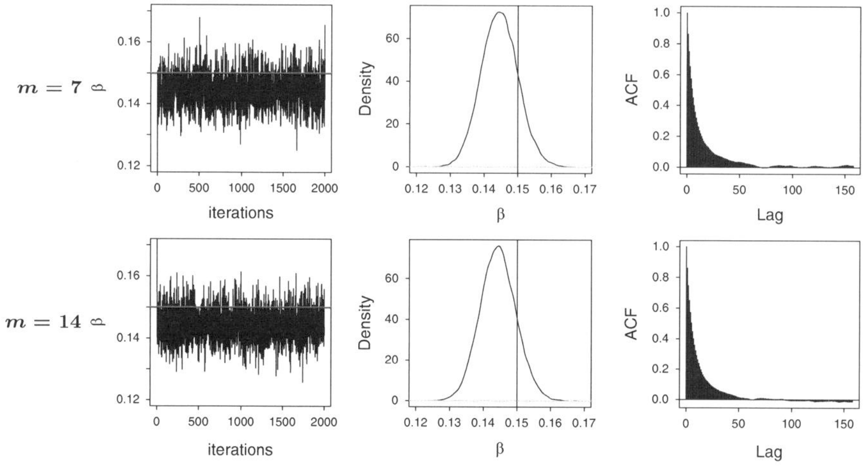

Fig. 8.3 Estimation results as described in Fig. 8.2, this time for the parameter β

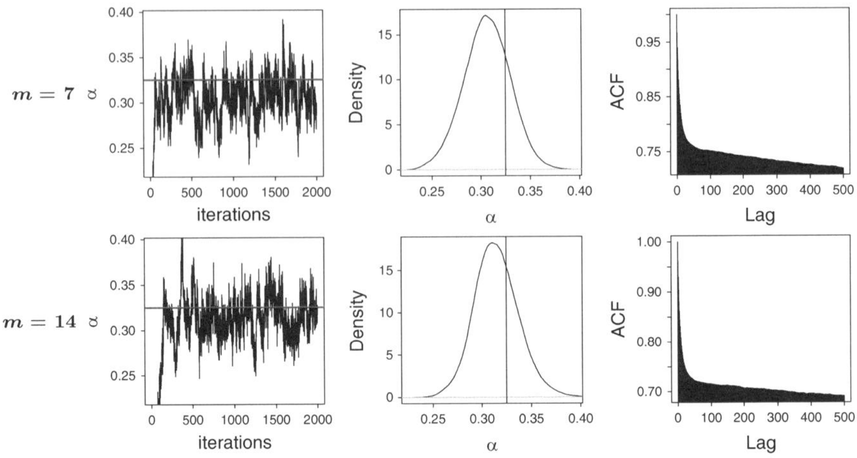

Fig. 8.4 Estimation results as described in Fig. 8.2, this time with the component s being latent

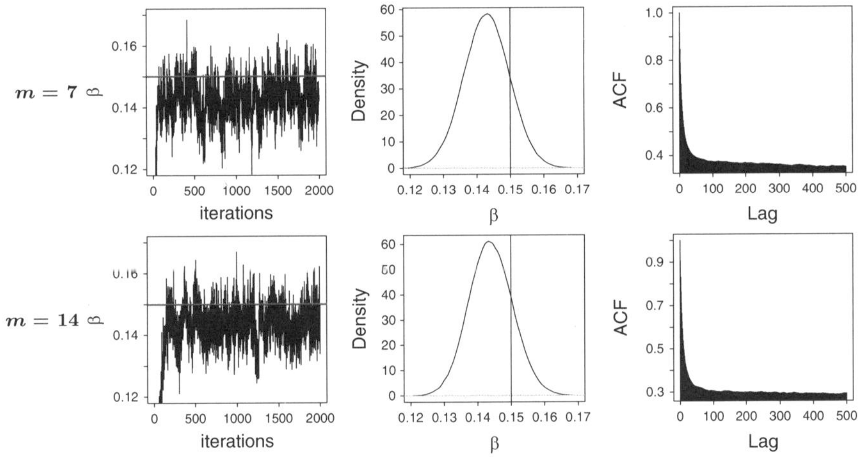

Fig. 8.5 Estimation results as described in Fig. 8.4, this time for the parameter β

for all $j = 1, \ldots, n$. These intervals cover a range which is approximately the average of all true α_j values. The parameter β, on the other hand, is satisfyingly estimated even for multiple clusters and s being latent. This raises hope that the infectious period can also be approximated precisely in the real data example in Sect. 8.3. In order to improve estimation of the contact rates, further information on the susceptible population is sought and might enter later work.

Table 8.1 Estimation results for dataset 1 as in Figs. 8.2–8.5. This table displays the posterior means and posterior 95 %-hpd intervals after a 10 % burn-in phase. The latter are computed according to Chen and Shao (1999). The true values of the parameters are displayed in the second column

		s observed		s latent	
Parameter	True value	$m = 7$	$m = 14$	$m = 7$	$m = 14$
α	0.325	0.317	0.318	0.306	0.315
		(0.29, 0.34)	(0.29, 0.34)	(0.26, 0.35)	(0.27, 0.36)
β	0.15	0.145	0.144	0.143	0.144
		(0.13, 0.16)	(0.13, 0.16)	(0.13, 0.16)	(0.13, 0.16)

Table 8.2 Estimation results for dataset 2 as described in the main text. This table displays the posterior means and posterior 95 %-hpd intervals after a 10 % burn-in phase. The true values of the parameters are displayed in the second column

		s observed		s latent	
Parameter	True value	$m = 7$	$m = 14$	$m = 7$	$m = 14$
α_1	0.6	0.57	0.58	0.43	0.45
		(0.52, 0.62)	(0.53, 0.63)	(0.39, 0.48)	(0.40, 0.49)
α_2	0.5	0.47	0.48	0.45	0.46
		(0.43, 0.51)	(0.44, 0.52)	(0.41, 0.49)	(0.42, 0.50)
α_3	0.4	0.40	0.41	0.44	0.45
		(0.37, 0.44)	(0.37, 0.45)	(0.40, 0.48)	(0.41, 0.49)
β	0.2	0.20	0.20	0.20	0.20
		(0.19, 0.21)	(0.19, 0.21)	(0.19, 0.21)	(0.19, 0.21)

8.2 Example: Influenza in a Boarding School

In 1978, the British Medical Journal (BMJ News and Notes 1978) published a report on an influenza outbreak in a boys' boarding school in Britain, which occurred in January and February 1978. The first case of influenza was introduced by a boy from Hong Kong who returned to school from holidays. Out of the 763 boys visiting the boarding school, 512 were infected within 14 days, while the approximately 130 teachers, house matrons and other adults remained unaffected with only one exception. The boys were immediately confined to bed as soon as they showed any symptoms of illness. As they furthermore lived in a closed community, where the susceptible population did obviously not include the adults, this influenza outbreak provides an ideal data situation. It has hence also attracted the attention of other authors: For example, Murray (2002) and Keeling and Rohani (2008) approximate the contact rate and infectious period in a deterministic SIR model by least squares estimation. Chen and Bokka (2005) utilise the resulting values from that book for the simulation of stochastic SIR epidemics. In this section, the influenza outbreak is modelled by the standard SIR diffusion process, and the model parameters are estimated by application of the innovation scheme. For comparison purposes, least squares estimation for the deterministic model is carried out as well.

Table 8.3 Estimation results for dataset 3 as described in the main text. This table displays the posterior means and posterior 95 %-hpd intervals after a 10 % burn-in phase. The true values of the parameters are displayed in the second column

		s observed		s latent	
Parameter	True value	$m = 7$	$m = 14$	$m = 7$	$m = 14$
α_1	0.60	0.60	0.60	0.43	0.44
		(0.55, 0.65)	(0.54, 0.65)	(0.39, 0.47)	(0.40, 0.48)
α_2	0.60	0.64	0.63	0.39	0.41
		(0.59, 0.69)	(0.59, 0.68)	(0.36, 0.42)	(0.37, 0.45)
α_3	0.55	0.54	0.54	0.42	0.48
		(0.50, 0.59)	(0.49, 0.58)	(0.39, 0.46)	(0.44, 0.52)
α_4	0.55	0.56	0.54	0.39	0.40
		(0.51, 0.60)	(0.50, 0.58)	(0.36, 0.42)	(0.37, 0.44)
α_5	0.50	0.54	0.53	0.45	0.44
		(0.50, 0.58)	(0.49, 0.57)	(0.41, 0.49)	(0.40, 0.48)
α_6	0.50	0.49	0.48	0.42	0.41
		(0.44, 0.54)	(0.44, 0.53)	(0.38, 0.47)	(0.37, 0.45)
α_7	0.45	0.45	0.45	0.45	0.40
		(0.41, 0.50)	(0.41, 0.49)	(0.41, 0.49)	(0.36, 0.45)
α_8	0.45	0.42	0.42	0.43	0.42
		(0.38, 0.46)	(0.38, 0.46)	(0.39, 0.47)	(0.38, 0.46)
α_9	0.40	0.39	0.39	0.45	0.49
		(0.35, 0.43)	(0.35, 0.43)	(0.41, 0.50)	(0.45, 0.53)
α_{10}	0.40	0.39	0.39	0.43	0.38
		(0.35, 0.43)	(0.35, 0.43)	(0.39, 0.48)	(0.33, 0.42)
β	0.20	0.20	0.19	0.19	0.19
		(0.19, 0.20)	(0.19, 0.20)	(0.19, 0.20)	(0.18, 0.19)

8.2.1 Data

The original paper (BMJ News and Notes 1978) graphically displays over a period of 2 weeks the daily number of pupils confined to bed. The exact counts are not available, but Table 8.4 shows numbers which are reconstructed from the graph. The observed fractions of infected boys are plotted in Fig. 8.7 on p. 292.

8.2.2 Parameter Estimation

In the following, the contact rate α and the inverse infectious period β are estimated by application of the standard SIR model to the above dataset. The fraction of susceptibles is considered latent.

Least Squares Estimation

As mentioned above, Murray (2002) applies the deterministic model (5.21) from p. 110 to the boarding school data and infers on the model parameters by least squares estimation. Translated to the parameterisation of this chapter, he obtains the estimates $\hat{\alpha} = 1.66$ and $\hat{\beta} = 0.44$. For comparison purposes, least squares estimation is also carried out here, yielding $\hat{\alpha} = 1.67$ and $\hat{\beta} = 0.45$. The small deviations are only natural as the original data is given graphically and the read out numbers will most probably differ by small amounts. The estimated basic reproductive ratio $\hat{\mathcal{R}}_0 = \hat{\alpha}/\hat{\beta}$ for the data from Table 8.4 equals 3.71, which explains the observed major outbreak.

The above estimates have been obtained by application of the Nelder-Mead algorithm (Nelder and Mead 1965). To that end, the trajectories of the deterministic process described by (5.21) have been calculated with the standard Euler scheme for ODEs with step length 0.02 and initial value $(s_0, i_0)' = (762/763, 1/763)'$. The estimated curve for the fraction of infectives is shown in Fig. 8.7, the corresponding mean sum of squared residuals equals $5 \cdot 10^{-4}$. The optimisation procedure yields 95 %-confidence intervals $[0.065, 42.798]$ for α and $[0.0002, 799.45]$ for β. For their derivation, the inverse Fisher information of $\log \alpha$ and $\log \beta$ (approximated by the Hessian of the function to minimise) has been evaluated at the point estimates, and the resulting confidence intervals have been back-transformed to the original scale. The term *Fisher information* is used here due to the analogy of least squares and maximum likelihood estimation with i.i.d. Gaussian errors, although no probability distribution has been specified above.

Bayesian Estimation

A more realistic model for the influenza outbreak than the just considered deterministic process is the diffusion approximation given by the SDE (5.19) on p. 109 since this model accounts for random fluctuations. As in the simulation study in Sect. 8.1, the innovation scheme is applied in order to estimate the parameters α and β. Again, these parameters are assumed to be a priori exponentially distributed with expected values 0.5. New proposals α^* and β^* are drawn in the MCMC scheme according to

$$\log \alpha^* \sim \mathcal{N}\big(\log \alpha, 0.0009\big) \quad \text{and} \quad \log \beta^* \sim \mathcal{N}\big(\log \beta, 0.0009\big)$$

with α and β denoting the current values. Figures 8.8 and 8.9 show resulting trace plots, posterior density estimates and autocorrelation plots for α and β. The simulated Markov chains have length 10^6 but have been thinned by factor 500. In order to decrease inter-observation time intervals, the innovation scheme imputes data such that there are m intermediate subintervals in between every two observation time points. Figures 8.8 and 8.9 show estimation results for $m = 2$ and $m = 20$. An increase of m should reduce a potential estimation bias. Since the outcomes for $m = 2$ and $m = 20$ are very similar, this variable is considered large enough.

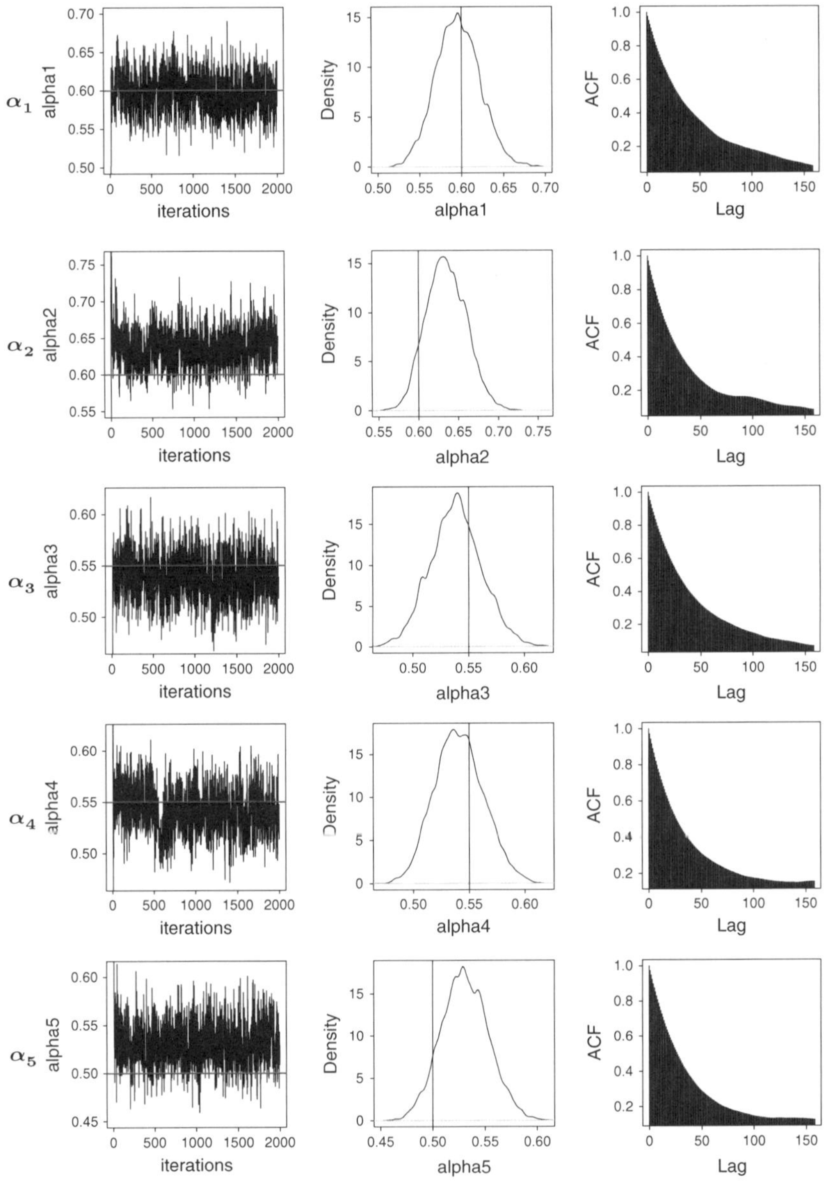

Fig. 8.6 Bayesian estimation of parameters of the multitype SIR model with $n = 10$ clusters when applied to dataset 3 with both s and i being observed. Details of the estimation procedure are described in the main text. The MCMC scheme introduces $m = 14$ subintervals in between every two observations. This figure shows the trace plots of all parameters (*left column*) with corresponding posterior density estimates (*middle column*) and autocorrelation plots (*right column*). The Markov chains have length 10^5 but have been thinned by factor 50 in the trace plots. *Red horizontal lines* in the trace plots and *black vertical lines* in the density plots indicate the true parameter values. Estimation of posterior densities and autocorrelation takes into account the full Markov chain, i.e. without thinning, after having discarded a 10 % burn-in phase

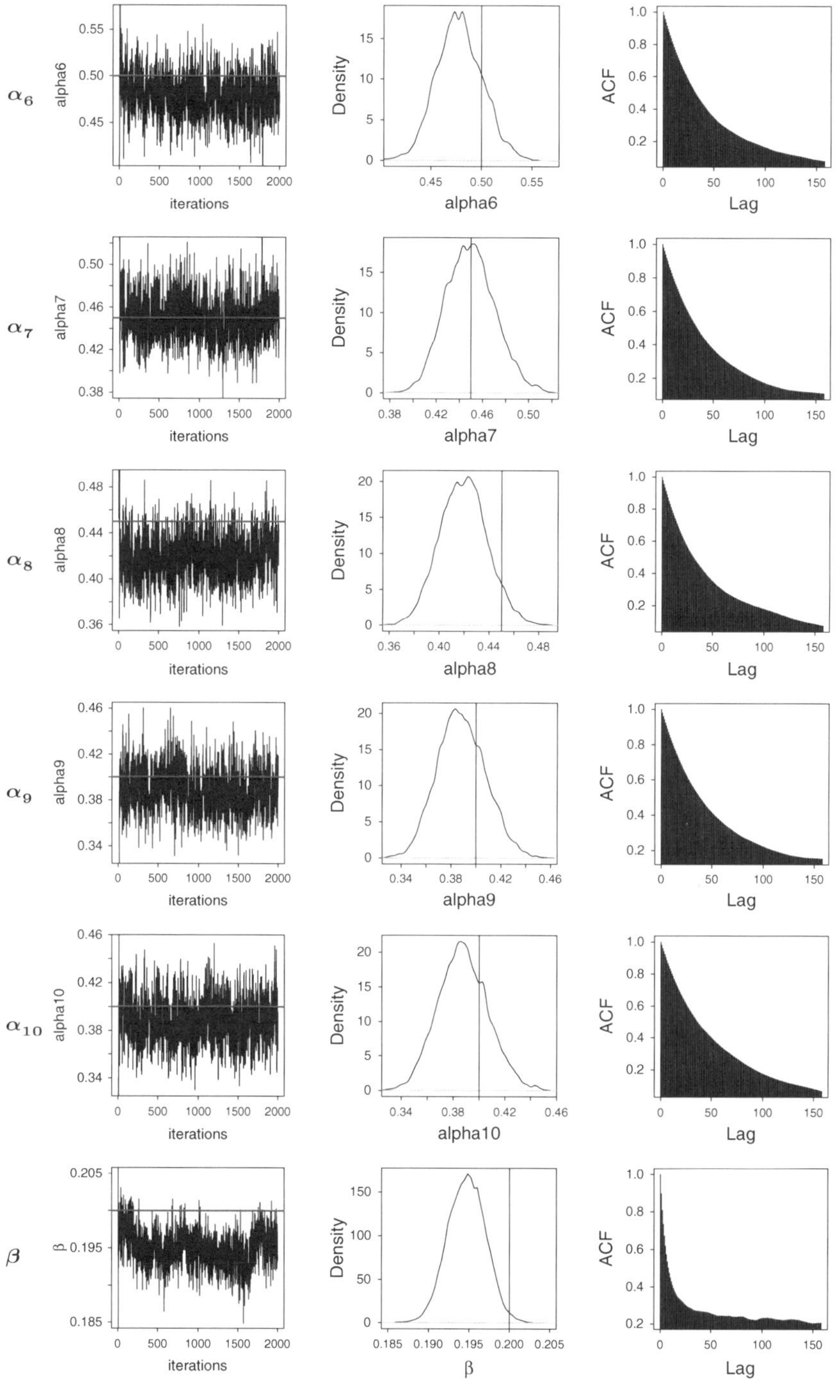

Fig. 8.6 (continued)

Table 8.4 Daily number of boys confined to bed, reconstructed from the graphic displayed in the original publication (BMJ News and Notes 1978). The total number of boys visiting the school was $N = 763$. The fractions of infective boys are plotted in Fig. 8.7

Date	Number of boys confined to bed
21 January	1
22 January	3
23 January	6
24 January	25
25 January	73
26 January	221
27 January	294
28 January	257
29 January	236
30 January	189
31 January	125
1 February	67
2 February	26
3 February	10
4 February	3

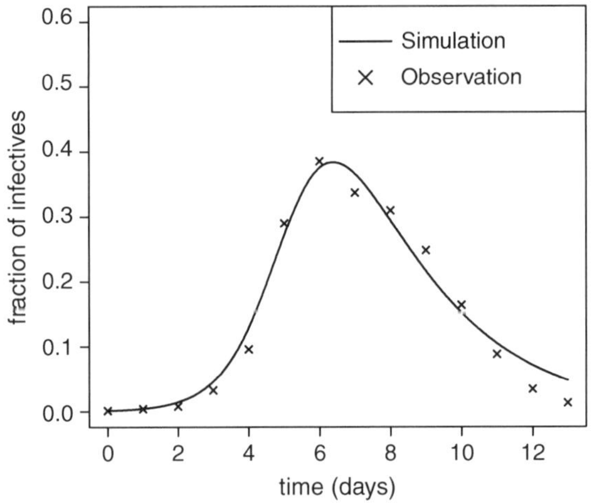

Fig. 8.7 The *crosses* show the observed fractions of infected boys in the boarding school as given in Table 8.4, where day 0 corresponds to 21 January. The *solid line* is the fitted deterministic course, based on the least squares estimates $\hat{\alpha} = 1.67$ and $\hat{\beta} = 0.45$. This curve has been calculated with the standard Euler scheme for ODEs with step length 0.02 and initial value $(s_0, i_0)' = (762/763, 1/763)'$. The resulting mean sum of squared residuals equals $5 \cdot 10^{-4}$

Table 8.5 lists the posterior means and 95 %-hpd intervals as obtained from the innovation scheme with $m = 2$ and $m = 20$. While the point estimates of the Bayesian and the least squares estimation approach are comparable, the Bayesian confidence intervals are much smaller and seem more reasonable than the one obtained through the inverse Fisher information. Thus, concerning the estimation of variation in the two considered approaches, the application of the stochastic model seems to be the more reliable approach.

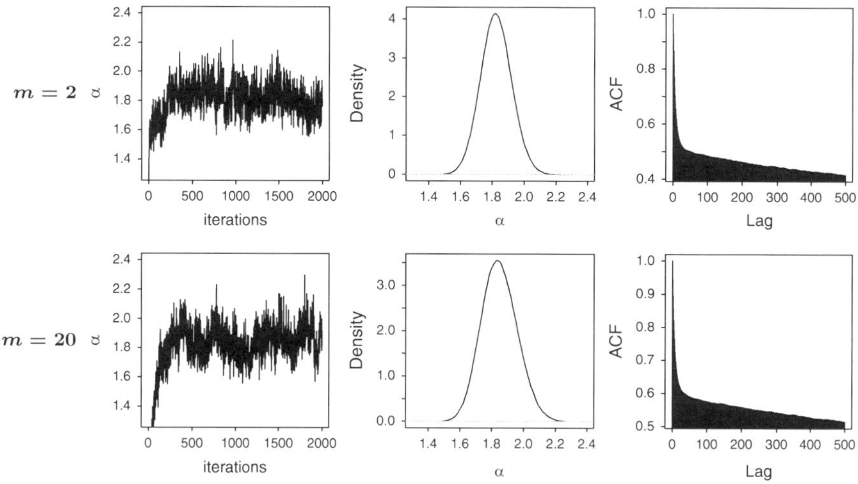

Fig. 8.8 Bayesian estimation of parameters of the standard SIR model as described in Sect. 8.2.2, applied to the boarding school data. The MCMC scheme conditions on the observed data from Table 8.4 and introduces $m \in \{2, 20\}$ subintervals in between every two observations. This figure shows the trace plots of α (*left column*) and the corresponding estimated posterior densities (*middle column*) and autocorrelation plots (*right column*). The Markov chains have length 10^6 but have been thinned by factor 500 in the trace plots. Estimation of posterior densities and autocorrelation takes into account the full Markov chain, i.e. without thinning, after having discarded a 10 % burn-in phase

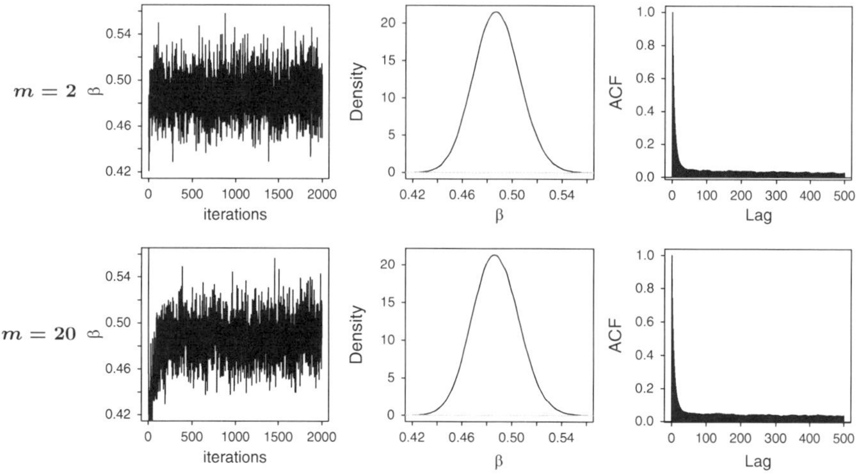

Fig. 8.9 Estimation results as described in Fig. 8.8, this time for the parameter β

Table 8.5 Estimation results as in Figs. 8.7–8.9. The upper part displays the posterior means and posterior 95 %-hpd intervals of α and β from the MCMC estimation with $m \in \{2, 20\}$ imputed inter-observation intervals after a 10 % burn-in phase. The hpd intervals are computed according to Chen and Shao (1999). For comparison purposes, the bottom part shows the least squares estimates of α and β with 95 %-confidence intervals as obtained from the inverse Fisher information of $\log \alpha$ and $\log \beta$ evaluated at the point estimates

Innovation scheme	$m = 2$	α: 1.82, (1.63,2.02)
		β: 0.49, (0.45,0.52)
	$m = 20$	α: 1.85, (1.63,2.07)
		β: 0.49, (0.45,0.52)
Least squares		α: 1.67, (0.06,42.80)
		β: 0.45, (0.0002,799.45)

8.3 Example: Influenza in Germany

As an example for the application of the multitype SIR model, this section investigates the spatial spread of influenza in Germany. To that end, administrative divisions of Germany are chosen to be represented by clusters. Contacts between geographical regions are approximated through data on daily commuter traffic from the German Federal Agency for Work. The contact rates and infectious periods for each cluster are statistically estimated based on available disease counts that were transferred to the Robert Koch Institute Berlin due to the German infection protection act.

The investigations in this section are intended to be a preliminary analysis for future research. To start with, the statistical inference focuses on the geographical area of Bavaria as one out of 16 states in Germany. Possible extensions are pointed out throughout the entire section and in the conclusion in Sect. 8.4.

The choice of geographical regions, the setup of the contact matrix and the data on cases of influenza are described in Sect. 8.3.1. Statistical inference on the model parameters is carried out in Sect. 8.3.2.

8.3.1 Data

This section describes the spatial structure, network matrix and data which are used for the statistical analysis in Sect. 8.3.2.

Geographical Regions

Germany is divided into the following administrative regions: At highest level, there are 16 states (*Bundesländer*). These are further partitioned into overall 40 counties (*Regierungsbezirke*). At an even finer level, there are 439 rural and urban districts

Table 8.6 IDs, names and population sizes of the seven counties in Bavaria and of all remaining counties in Germany. A map of these regions is shown in Fig. 8.10

ID	Name	Population size
91	Oberbayern	4,138,402
92	Niederbayern	1,185,467
93	Oberpfalz	1,085,609
94	Oberfranken	1,113,788
95	Mittelfranken	1,698,343
96	Unterfranken	1,340,912
97	Schwaben	1,767,193
–	Other counties	68,036,193

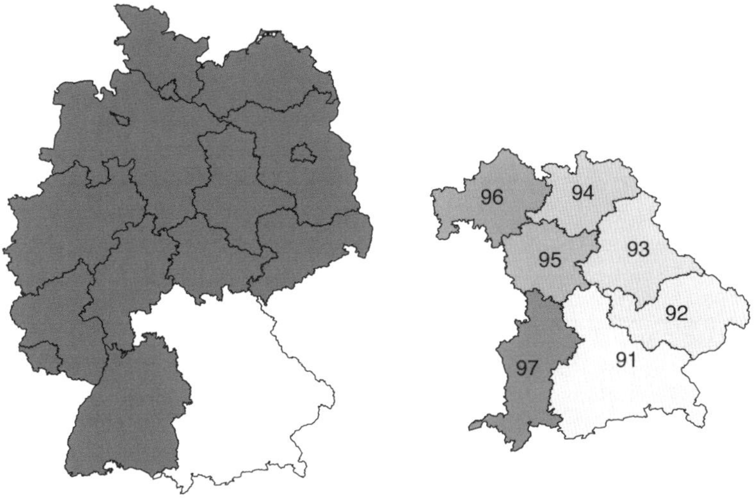

Fig. 8.10 Map of the eight regions for which disease counts are analysed in Sect. 8.3.2: Considered are the seven counties in Bavaria *(right graphic)*. Each region is labelled with an ID, and the corresponding names are listed in Table 8.6. The eighth region is the union of all remaining states in Germany *(left graphic)*

(*Landkreise, Stadtkreise*). In the datasets in this section, the island of Rügen is generally excluded, leading to only 438 districts. Due to reforms concerning the administrative organisation, the actual counties and districts of Germany are different today. The commuter data and disease counts described below, however, are available for the above mentioned regions.

As a proof of concept, the statistical analysis in Sect. 8.3.2 focuses on the seven counties of Bavaria. These are listed in Table 8.6 together with their population sizes. As an additional region, all remaining counties of Germany are summarised in one compartment, yielding an overall number of $n = 8$ geographical areas. These are illustrated in Fig. 8.10.

Connectivity Matrix

The connectivity matrix $\gamma^N = (\gamma^N_{jk})_{j,k=1,\ldots,n}$ reflects the traffic across the borders of the n geographical regions: γ^N_{jk} stands for the average percentage of

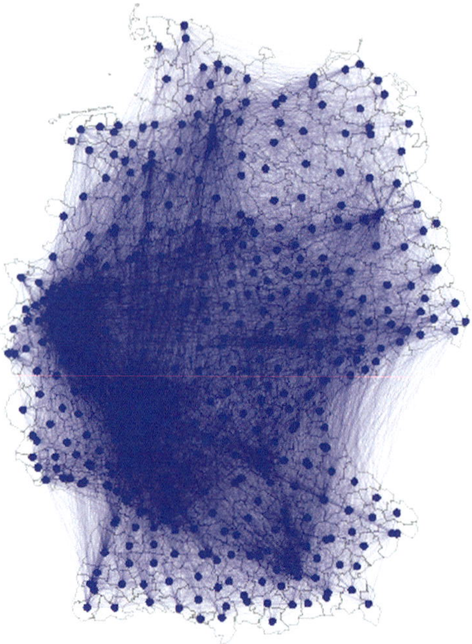

Fig. 8.11 Daily commuter traffic between the rural and urban districts of Germany. The thickness of each line represents the strength of migration between two regions

the population of region j travelling to region k per day. Row sums are equal to one such that the entries on the main diagonal represent the rates with which individuals stay in their home region. In order to account for different contact behaviour of susceptible, infected and removed individuals, the multitype SIR model further involves the contact matrices γ^S and γ^I. This refinement is neglected here such that $\gamma^N = \gamma^S = \gamma^I$.

The analysis in Sect. 8.3.2 requires such a network matrix representing the contacts between individuals living in the seven counties of Bavaria and in the remaining parts of Germany. Certainly, there is no exact data about daily migration between the different regions available. However, the daily flow of commuters seems to be a sensible indicator for such a network. This approach is especially meaningful on the district level because it is typically the urban districts which attract many commuters from surrounding suburbs, and it is also these urban regions which usually provide social facilities such as educational institutions, extended medical infrastructure, shopping areas and cultural events for people living in near rural areas.

In the following, we hence investigate data on commuter traffic which was purchased from the German Federal Agency for Work. This dataset takes into account the districts of residence and the locations of the employing companies as of 30 June 2006 for all employees who are subject to compulsory social insurance. The dataset includes 26,207,317 persons, that is 31.8 % of the total German population. Out of these, 9,896,745 people (37.8 %) are commuters, i. e. they work in a district other than the one they live in. The resulting network on the district level is shown in Fig. 8.11. The contact matrix for the counties of Bavaria can easily be derived from it by aggregation. Its entries are displayed in Table 8.7.

Table 8.7 Entries of the connectivity matrix for the seven counties in Bavaria and the union of all other counties. The places of residence are listed rowwise, the locations of the employing companies columnwise. The entries of each row sum up to one

	Oberbayern	Niederbayern	Oberpfalz	Oberfranken	Mittelfranken	Unterfranken	Schwaben	Other counties
Oberbayern	0.880	0.019	0.005	0.001	0.008	0.002	0.026	0.059
Niederbayern	0.239	0.628	0.108	$6 \cdot 10^{-5}$	0.006	$6 \cdot 10^{-5}$	0.003	0.016
Oberpfalz	0.066	0.054	0.693	0.045	0.114	0.002	0.002	0.023
Oberfranken	0.021	$2 \cdot 10^{-4}$	0.031	0.673	0.214	0.024	$5 \cdot 10^{-4}$	0.035
Mittelfranken	0.041	0.001	0.018	0.026	0.841	0.011	0.009	0.053
Unterfranken	0.017	$5 \cdot 10^{-5}$	0.001	0.034	0.025	0.716	0.001	0.206
Schwaben	0.170	0.001	0.001	$9 \cdot 10^{-5}$	0.008	0.001	0.620	0.199
Other	0.009	$3 \cdot 10^{-4}$	0.001	0.003	0.003	0.004	0.004	0.976

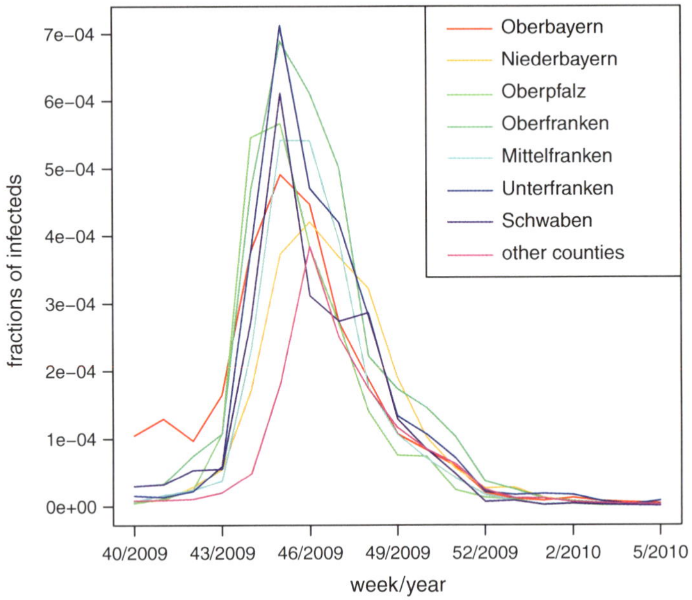

Fig. 8.12 Weekly fractions of influenza A cases for week 40/2009 until week 5/2010 obtained from SurvStat

Since the employment rates vary within Germany, the commuter dataset is probably less representative for some regions than for others. Moreover, the commuter data naturally involves only certain age groups. These imprecisions are considered negligible here but may be refined in future work.

Disease Counts

Data about occurrences of influenza in Germany is taken from the Robert Koch Institute: SurvStat, http://www3.rki.de/SurvStat, as of 29 July 2010. This database contains weekly case counts on the district, county and state level since 2001. However, contact and recovery rates typically vary between seasons (Dushoff et al. 2004) such that it is not always meaningful to base parameter estimation on a collection of data from different seasons. The statistical analysis in Sect. 8.3.2 considers the counts from week 40 in the year 2009 until week 5 in 2010. This influenza season is not only one of the latest available data; it has also started uncommonly early in the year and attracted particular attention because of the circulation of the 'swine flu' virus. The utilised dataset contains weekly counts for the eight specified regions over the considered period of 19 weeks. Only cases categorised as influenza A are considered since it was the influenza A virus that was most responsible for the national influenza epidemic in that season. The resulting fractions of infected persons are plotted in Fig. 8.12.

The above data suffers from high underreporting. The dataset is hence not immediately appropriate for the estimation of contact and recovery rates. It seems, however, interesting to study the outcomes of a statistical analysis to the above data. Such inference is accomplished in Sect. 8.3.2. Another point of interest concerns the changes in the parameter estimates when modifying the underlying data. Just as an example, assume that 10 % of the German population was affected by influenza during the season 2009/10. That would correspond to approximately 8.2 million infected people, but the SurvStat database contains only about 150,000 cases between week 17/2009 and 16/2010. In order to correct for this, the original dataset is multiplied by factor 55, and the statistical inference in Sect. 8.3.2 is repeated for the modified dataset.

In practice, the multiplication with a uniform factor for each region is presumably not appropriate as the levels of underreporting typically depend on region-specific reporting behaviour and also on the severity of the epidemic in the respective area. Advanced corrections may be investigated in future work.

Another difficulty, apart from the uncertainty in the numbers of infected individuals, concerns the number of susceptible persons. These numbers are generally unknown. In case of influenza, an infected individual acquires immunity to the strain he was affected by and can hence not become susceptible during the same wave of influenza again. However, there are steadily new antigen mutants of the influenza virus coming up (Stephenson and Nicholson 2001), which is why at the beginning of the next epidemic the individual is typically susceptible again. A person might, however, also have acquired partial immunity though. For simplicity, it is assumed in the investigations in Sect. 8.3.2 that there are no removed individuals at the beginning of the epidemic. The initial fractions of susceptibles in each region can hence be calculated from the fractions of infected persons. Again, more refined assumptions may be applied in the future.

8.3.2 Parameter Estimation

Inference on the parameters of the SIR diffusion model is now carried out as described in the simulation study in Sect. 8.1. The estimation is based on the original and modified datasets described in Sect. 8.3.1. The fractions of susceptibles are treated as latent variables.

Table 8.8 displays resulting posterior means and 95 %-hpd intervals for $\alpha_1, \ldots, \alpha_8$ and β in the multitype SIR model. The underlying simulated Markov chains have length 10^5, and the innovation scheme introduces $m = 7$ intermediate subintervals in between every two observations.

As expected, the results show substantial differences in parameter estimates between the original and the modified dataset. This emphasises the need for more precise data on influenza occurrences. While the estimated values for $\alpha_1, \ldots, \alpha_8$ are all at about the same range for the modified dataset, there is large variation in the estimated contact rates for the original dataset. Concerning estimation of β,

Table 8.8 Estimation results for the parameters of the multitype SIR model applied to the influenza datasets described in Sect. 8.3.1. The second column contains estimates for the original dataset, the third column shows those for the modified dataset, that is the original dataset multiplied by factor 55. The table displays the posterior means and posterior 95 %-hpd intervals after a 10 % burn-in phase. The underlying simulated Markov chains have length 10^5, and the innovation scheme introduced $m = 7$ subintervals in between every two observation time points

Parameter	Original dataset	Modified dataset
α_1	5.198	0.225
	(3.784, 5.621)	(0.221, 0.228)
α_2	0.735	0.236
	(0.675, 0.768)	(0.224, 0.244)
α_3	0.699	0.194
	(0.450, 0.776)	(0.188, 0.196)
α_4	0.180	0.201
	(0.150, 0.198)	(0.195, 0.207)
α_5	4.036	0.198
	(2.035, 4.808)	(0.193, 0.201)
α_6	0.031	0.249
	(0.028, 0.041)	(0.244, 0.254)
α_7	0.144	0.261
	(0.132, 0.156)	(0.257, 0.262)
α_8	12.025	0.358
	(11.769, 12.500)	(0.356, 0.359)
β	8.230	0.343
	(6.214, 8.741)	(0.343, 0.344)

there is again a large difference between the two datasets. As the time unit is 1 day, the value $0.34^{-1} \approx 2.9$ for the modified data seems much more plausible as an approximation of the length of the infectious period than $8.23^{-1} \approx 0.12$, which results from the original data.

For comparison purposes, in a further experiment, the above applied influenza dataset is aggregated over the eight distinct regions, yielding a one-dimensional time series for the entire area of Germany. The standard SIR model is applied to this dataset, again both in its original and a modified form, in order to evaluate the differences in the resulting parameter estimates. The according posterior means and hpd intervals are given in Table 8.9. There are again $m = 7$ subintervals introduced by the innovation scheme, and the simulated Markov chains have length $5 \cdot 10^5$.

Unsurprisingly, the estimates obtained from the multitype SIR model and the estimates from the standard model do not match, neither concerning the cluster-specific contact rates nor the global infectious period. The standard model cannot imitate the outcomes of the multitype model. This motivates the application of the more refined modelling approach.

Table 8.9 Estimation results for the parameters of the standard SIR model applied to the influenza dataset aggregated over all regions. The second column contains estimates for the original aggregated dataset, the third column shows those for the modified dataset, that is the original dataset multiplied by factor 55. The table displays the posterior means and posterior 95 %-hpd intervals after a 10 % burn-in phase. The underlying simulated Markov chains have length $5 \cdot 10^5$, and the innovation scheme introduced $m = 7$ subintervals in between every two observation time points

Parameter	Original dataset	Modified dataset
α	1.106	0.109
	(1.095, 1.125)	(0.109, 0.110)
β	1.100	0.072
	(1.092, 1.107)	(0.071, 0.072)

8.4 Conclusion and Outlook

This chapter investigated the statistical estimation of parameters in epidemic diffusion models. For carrying out these estimations, the newly developed techniques from Chap. 7 were required.

In this chapter, parameters were both estimated in a simulation study with synthetic datasets and in two applications with real data. The simulation study served as a benchmark for the quality of parameter estimates. It turned out that the contact and recovery rates in the standard and multitype SIR models can be estimated precisely as long as information on the fraction of susceptible individuals is provided. Otherwise, the estimates of the contact rates have to be considered with care, but estimation of the average infectious period seemed reliable.

In real data situations, one faces multiple difficulties, some of them have already been pointed out in the course of this chapter. These concern mainly the data on infectious cases and knowledge about the susceptible population. The solution of this problem requires the collaboration of data-collecting institutions and statisticians. An interesting approach has recently been proposed by Ginsberg et al. (2008) who utilise influenza-related queries to online search engines instead of relying on notified visits to the doctor.

Geographic modelling of epidemic outbreaks requires the specification of the spatial mixing of individuals. Possible advancements in the design of the connectivity matrix might be achieved by combinations of different data sources. In the literature, for example, there are several considerations of transportation networks: Baroyan and Rvachev (1967) and Baroyan et al. (1977) analysed the Russian train network for modelling the spread of influenza, and Rvachev and Longini (1985) extended this work to worldwide considerations. More recently, Grais et al. (2003), Brownstein et al. (2006) and Colizza et al. (2006a,b) worked out the impact of air travel and other modes of transportation on the spread of diseases today. Crépey and Barthélemy (2007) analysed influenza pandemics in the United States and France with respect to transmission channels via air and train traffic. A different approach was implemented by Brockmann et al. (2006) who drew conclusions about the travelling behaviour of humans within the United States from the dispersal of dollar notes, tracked through the website http://www.wheresgeorge.com. For a recent monograph on geographic models for the spread of diseases, see Sattenspiel (2009).

Future investigations of the influenza data in Sect. 8.3 will certainly involve the incorporation of further external information in the parameter estimation. For example, the contact rates are possibly correlated with the population densities of the respective regions, so these densities could be used as a priori knowledge. Alternatively, in order to reduce the number of unknown variables, administrative regions could be categorised as rural or urban with identical contact rates within each category. Concerning the contact matrix, the distinction between matrices γ^S and γ^I for susceptibles and infectives are particularly meaningful for travel routes of relatively large distances, i. e. contacts between non-adjacent regions. Further extensions such as the consideration of age groups have been mentioned throughout the chapter and in Sect. 2.2.3.

The ultimate objective of research on the spread of infectious diseases is typically the development of efficient intervention policies; see for example the discussion by Medlock and Galvani (2009) on control strategies like optimal vaccine distributions or the review article by Cauchemez et al. (2009) on the various aspects of school closure as part of an intervention plan. In case of a spatial multitype model such as the one considered in this chapter, additional options arise which correspond to modifications of the connectivity matrix. A change of connectivity can for example be accomplished by restriction of travel connections such as airport closures (e. g. Hufnagel et al. 2004).

Analysing epidemics using statistical inference techniques has shown the potential to provide more accurate estimates than available before. Several directions for future work have been pointed out.

References

Baroyan O, Rvachev L (1967) Deterministic epidemic models for a territory with a transport network (in Russian). Kibernetika 3:67–74

Baroyan O, Rvachev L, Ivannikov Y (1977) Modelling and forecasting of influenza epidemics for territory of the USSR (in Russian). Gamelaya Institute of Epidemiology and Microbiology, Moscow

BMJ News and Notes (1978) Influenza in a boarding school. Br Med J 1:587

Brockmann D, Hufnagel L, Geisel T (2006) The scaling laws of human travel. Nature 439:462–465

Brownstein J, Wolfe C, Mandl K (2006) Empirical evidence for the effect of airline travel on inter-regional influenza spread in the United States. PLoS Med 3:1826–1835

Cauchemez S, Ferguson N, Wachtel C, Tegnell A, Saour G, Duncan B, Nicoll A (2009) Closure of schools during an influenza pandemic. Lancet Infect Dis 9:473–81

Chen WY, Bokka S (2005) Stochastic modeling of nonlinear epidemiology. J Theor Biol 234: 455–470

Chen MH, Shao QM (1999) Monte Carlo estimation of Bayesian credible and hpd intervals. J Comput Graph Stat 8:69–92

Colizza V, Barrat A, Barthélemy M, Vespignani A (2006a) The modeling of global epidemics: stochastic dynamics and predictability. Bull Math Biol 68:1893–1921

Colizza V, Barrat A, Barthélemy M, Vespignani A (2006b) The role of the airline transportation network in the prediction and predictability of global epidemics. Proc Natl Acad Sci USA 103:2015–2020

Crépey P, Barthélemy M (2007) Detecting robust patterns in the spread of epidemics: a case study of influenza in the United States and France. Am J Epidemiol 166:1244–1251

Dushoff J, Plotkin J, Levin S, Earn D (2004) Dynamical resonance can account for seasonality of influenza epidemics. Proc Natl Acad Sci U S A 101:16915–16916

Ginsberg J, Mohebbi M, Patel R, Brammer L, Smolinski M, Brilliant L (2008) Detecting influenza epidemics using search engine query data. Nature 457:1012–1014

Grais R, Ellis J, Glass G (2003) Assessing the impact of airline travel on the geographic spread of pandemic influenza. Eur J Epidemiol 18:1065–1072

Hufnagel L, Brockmann D, Geisel T (2004) Forecast and control of epidemics in a globalized world. Proc Natl Acad Sci U S A 101:15124–15129

Keeling M, Rohani P (2008) Modeling infectious disease in humans and animals. Princeton University Press, Princeton

Medlock J, Galvani A (2009) Optimizing influenza vaccine distribution. Sci Express 10.1126/science.1175570:1–9

Murray J (2002) Mathematical biology: I. an introduction, 3rd edn. Interdisciplinary applied mathematics. Springer, Berlin/Heidelberg

Nelder J, Mead R (1965) A simplex method for function minimization. Comput J 7:308–313

Russell C, Jones T, Barr I, Cox N, Garten R, Gregory V, Gust I, Hamson A (2008) The global circulation of seasonal influenza A (H3N2) viruses. Science 320:340–346

Rvachev L, Longini I (1985) A mathematical model for the global spread of influenza. Math Biosci 75:3–22

Sattenspiel L (2009) The geographic spread of infectious diseases: models and applications. Princeton University Press, Princeton

Stephenson I, Nicholson K (2001) Influenza: vaccination and treatment. Eur Respir J 17: 1281–1293

Chapter 9
Application II: Analysis of Molecular Binding

The genetic material of humans and mammals is mainly contained in their cell nuclei, where most genome regulatory processes like DNA replication or transcription take place. These processes are controlled by complex protein networks. Hence, the comprehension of procedures like protein binding interactions in the nucleus are of large interest, and their investigation is the subject of active research. See for example Gorski and Misteli (2005) for an explanation of the importance of understanding this field.

Many findings about the behaviour of chromatin-binding proteins are based on in vitro experiments, i.e. on studies which are performed in an artificial environment outside a living organism. In vivo experiments, on the other hand, are carried out in a living cell and differ from in vitro settings with respect to, for example, binding sites and environmental conditions. It is desirable, though more challenging, to analyse data from in vivo experiments (Phair et al. 2004a; Mueller et al. 2008).

A suitable tool for the analysis of in vivo molecular binding is fluorescence microscopy. In this method, the protein of interest is labelled with a *green fluorescent protein (GFP)*. The spatio-temporal distribution of GFP-tagged molecules can then be observed in the living cell.

This chapter analyses the cell cycle dependent kinetics of the particular protein Dnmt1. Data is extracted by application of fluorescence microscopy. Kinetic compartment models for the dynamics of the protein are established and translated into stochastic and deterministic processes. Parameters of interest can then be estimated by application of appropriate estimation techniques to the model and the data.

In particular, the contents of this chapter are as follows: Sect. 9.1 presents the research questions of this chapter and tools for data acquisition. In Sect. 9.2, primary characteristics of the data are analysed which form the basis for the subsequent model construction. Based on biochemical principles, appropriate kinetic models are developed and further extended in Sects. 9.3–9.5. In particular, all kinetic models are initially designed as compartment models and then further approximated by stochastic and deterministic differential equations. The stochastic approximation

C. Fuchs, *Inference for Diffusion Processes: With Applications in Life Sciences*,
DOI 10.1007/978-3-642-25969-2_9,

is particularly important as it accounts for the apparently present randomness in the observed process. Simulation studies demonstrate the performance of suitable parameter estimation techniques and model choice criteria. Before applying these inference methods to real datasets, Sect. 9.6 discusses and further develops the preliminary preparation of the raw measurements from fluorescence microscopy experiments. Finally, Sect. 9.7 investigates the research problems of this chapter by means of the methodology from the previous sections applied to a variety of real datasets. Section 9.9 concludes and outlines future projects.

So far, diffusion approximations or comparable stochastic models have not been used in the literature for the analysis of observations from fluorescence microscopy experiments. Instead, deterministic differential equations are employed, and model parameters are approximated by least squares estimation. This approach, however, does not account for stochasticity and furthermore violates some of the basic model assumptions as outlined in Sect. 9.3.4. For comparison purposes, the latter procedure is contained in this book as well. A number of formulas and properties of the deterministic model are derived here. The emphasis of this work, however, lies on the application of Bayesian estimation techniques as developed in Chap. 7.

The focus of this chapter is on the presentation of mathematical models and the application of statistical estimation techniques to the collected data. Basic biological background information is given to an extent that suits the motivation and basic comprehension of the application. For details on biological aspects, the reader is referred to Schneider (2009) and Schneider et al. (2012).

9.1 Problem Statement

An important cellular process is *DNA methylation*, which is a DNA modification with diverse biological objectives. Proper cell function is only possible if the DNA methylation pattern is maintained over many cell cycles. Otherwise, the formation of tumor cells is one potential consequence. It has been shown that the protein *DNA (cytosine-5)-methyltransferase 1*, in short *Dnmt1*, plays a central role in the maintenance of DNA methylation patterns (see Kuch et al. 2008, and the references therein). Despite its importance, the dynamics of Dnmt1 is still unclear. In this chapter, we investigate the kinetic behaviour of Dnmt1 in living mice.

The following paragraphs describe the data acquisition process and the research questions that will be investigated in this chapter.

9.1.1 *Data Acquisition by Fluorescence Recovery After Photobleaching*

A popular technique for the analysis of the dynamics of molecules is *fluorescence recovery after photobleaching (FRAP)* (e.g. Sprague and McNally 2005), which

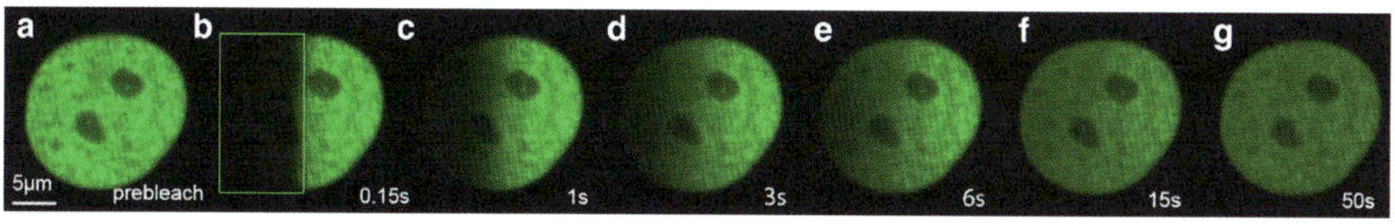

Fig. 9.1 Series of images obtained in a fluorescence recovery after photobleaching (FRAP) experiment: Initially, all chromatin proteins of interest in the cell nucleus are fluorescently labelled (*image A*). Then, one part of the nucleus is irreversibly bleached by a short laser pulse such that fluorescent emission of the proteins in that section becomes extinct (*B*). During a subsequent recovery phase, the fluorescent and non-fluorescent proteins diffuse (*C–F*) until they are uniformly mixed in the nucleus (*G*) (Modified from Schneider 2009)

is illustrated in Fig. 9.1. In this experiment, all chromatin proteins of interest in the cell nucleus are initially fluorescently labelled (image A). Then, one part of the nucleus is irreversibly bleached by a short laser pulse such that fluorescent emission of the proteins in that section becomes extinct (B). During a subsequent recovery phase, the fluorescent and non-fluorescent proteins diffuse (C–F) until they are uniformly mixed in the nucleus (G). The course of this diffusion and the duration until complete recovery allow conclusions about the mobility of the protein of interest.

All data was acquired in the context of a diploma thesis (Schneider 2009). Materials and methods concerning the preparation of cell cultures, the acquisition of images and subsequent image analysis are described in that work. General overviews can also be found in Phair et al. (2004a) and McNally (2008).

9.1.2 Research Questions

The following questions arise in the context of analysing the dynamic properties of Dnmt1 and will be statistically investigated in this chapter.

Estimation of Kinetic Parameters

Dnmt1 diffuses randomly through the cell nucleus until it binds to chromatin at a likewise random time point. The protein remains at this binding site for a stochastic time period until it unbinds and continues to diffuse. This procedure recurs throughout the whole experiment.

In order to be able to characterise the dynamics of Dnmt1, a fundamental issue is to determine the impacts of diffusion and binding on the recovery curves. Furthermore, it is important to have an estimate of the affinity of Dnmt1 to enter the bound state and of the average residence time that the protein remains at the binding site.

To that end, a preliminary analysis is performed in Sect. 9.2 to clarify the role of diffusion and binding. Kinetic models are formulated in Sects. 9.3–9.5. These incorporate the unknown measures as model parameters whose statistical estimation is the purpose of Sect. 9.7.

Number of Mobility Classes

There is possibly more than one type of binding partner for Dnmt1, i.e. the protein may sometimes associate to a partner of one type and sometimes to a partner of another type. These partners may differ with respect to the affinity of Dnmt1 to enter the bound state and the mean residence times in this state. All binding partners with identical or similar kinetic properties are gathered in one *mobility class*. This term seems more appropriate than *classes of binding sites* (e.g. Phair et al. 2004b) because different sites with identical kinetic properties cannot be distinguished using FRAP data (Schneider 2009). The number of mobility classes could hence be smaller than the number of different binding partners.

The number of mobility classes for Dnmt1 is of great interest. Related to that is the question about associated binding affinities and mean residence times for each class as well as the average fraction of free molecules and bound molecules of each type.

To that end, the kinetic model for one mobility class is extended to several mobility classes in Sect. 9.5. The identification of numbers of mobility classes from the FRAP curves is approached by model choice criteria in Sect. 9.7.

Cell Cycle Dependence

A eukariotic cell passes through a number of phases between every two cell divisions. These are part of the *cell cycle*, which is composed of a first gap phase (*G1 phase*) in which the cell grows, a synthesis phase (*S phase*) in which the DNA is duplicated, a second gap phase (*G2 phase*) where the cell grows further, and finally the mitosis phase (*M phase*) in which the cell divides. The S phase can further be partitioned into an *early S phase*, a *mid S phase* and a *late S phase*. The G1, S and G2 phases are again summarised as the *interphase*. Figure 9.2 depicts images of a cell nucleus during a part of the cell cycle.

This chapter is concerned with the cell cycle dependent kinetics of Dnmt1. In particular, FRAP data is collected during G1, early S and late S phases. The time series are displayed in Fig. 9.3. This chapter investigates whether Dnmt1 shows different binding behaviour depending on the phase, both with respect to binding affinities and mean residence times and regarding the number of mobility classes.

To that end, models are estimated for time series from distinct phases in Sect. 9.7, and the results are analysed with respect to cell cycle dependent statistical differences.

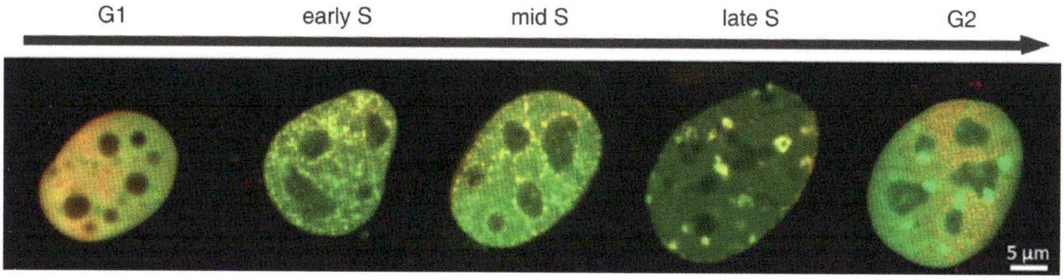

Fig. 9.2 Cell cycle dependent distribution of GFP-tagged Dnmt1 proteins (*green*) and replication sites (*red*) in a nucleus (From Schneider 2009)

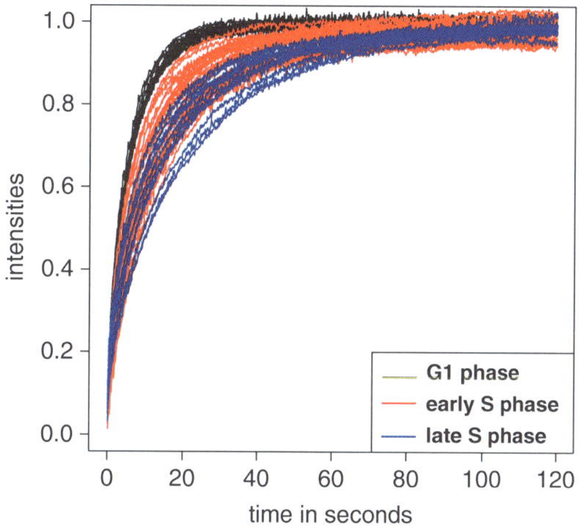

Fig. 9.3 Fluorescence intensities of GFP-tagged Dnmt1 measured in the bleached section of the nucleus during G1, early S and late S phases. The data is processed according to the triple normalisation described in Sect. 9.6.1 on pp. 348

9.2 Preliminary Analysis

The design of an appropriate kinetic model crucially depends on two factors: the impact of binding and the impact of diffusion on fluorescence recovery. These issues have to be clarified experimentally before formulating a mathematical model and statistically inferring on its parameters. This will be investigated in the following.

9.2.1 *Impact of Binding*

In order to determine whether binding interactions affect the fluorescence recovery dynamics of the protein of interest, Sprague and McNally (2005) suggest to compare several FRAP curves of this GFP-tagged protein with those of unconjugated, non-binding GFP. If recovery of the considered protein is substantially slower than recovery of GFP alone, binding events obviously influence the dynamics.

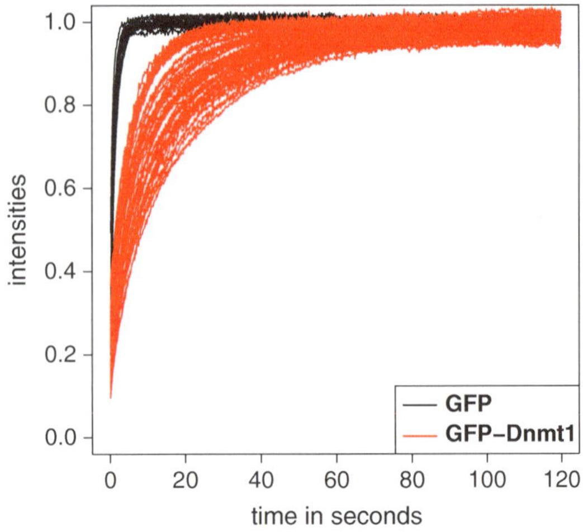

Fig. 9.4 Fluorescence intensities of GFP (*black*) and GFP-tagged Dnmt1 (*red*) measured in the bleached section of the nucleus during different phases of the cell cycle. The data is processed according to the triple normalisation described in Sect. 9.6.1 on pp. 348

Figure 9.4 displays FRAP curves of unconjugated GFP and of GFP-labelled Dnmt1. The difference in the speed of recovery is apparent. Hence, the kinetic models developed in this chapter take binding transactions into account.

9.2.2 *Impact of Diffusion*

There are two basic scenarios that one usually proceeds from: *diffusion-coupled* or *diffusion-uncoupled FRAP* (Sprague and McNally 2005). In a diffusion-coupled situation, the molecules diffuse across the nucleus with a rate that is of the same order as the rate with which binding occurs. In case of diffusion-uncoupled recovery, diffusion happens much faster than binding. Utilisation of the wrong pattern may entail misleading interpretation of the results.

Intuitively, one may assume that slow recovery indicates slow binding compared with the speed of diffusion, resulting in the diffusion-uncoupled case. However, as Sprague and McNally (2005), Beaudouin et al. (2006) and Lambert (2009) point out, a diffusion-uncoupled scenario is not necessarily implied by a long duration of the recovery phase. Instead, fluorescence recovery should be observed in different zones of the bleached section (Phair et al. 2004a) and for varying bleach spot sizes (Sprague and McNally 2005). Diffusion-uncoupled FRAP can be assumed if recovery is independent of the location of the zone and size of the spot.

For the application in this chapter, such control experiments have been carried out but initially gave no definite answer (Schneider 2009); there are indications for both scenarios. In the following sections, the mathematical models are based on diffusion-uncoupled recovery dynamics. Such investigations are of course also of interest for many proteins other than Dnmt1.

To be on the safe side, however, the data should also be analysed under the assumption of diffusion-coupled recovery. Respective kinetic models have mainly been developed in the literature for circular and line bleaching (e.g. Mueller et al. 2008), but also for more general bleaching geometries (Carrero et al. 2004). The data used in this chapter has been obtained by *half-nucleus FRAP* experiments, i.e. by bleaching (approximately) one half of the nucleus rather than a circle or strip. An according compartmental model is outlined in Sect. 9.8. Statistical analysis of this model is the subject of ongoing work (Schneider et al. 2012).

9.3 General Model

In this section, a general kinetic model for the dynamics of a protein in a cell nucleus is derived under the assumptions discussed in Sect. 9.2. The same compartmental model is utilised by, for example, Phair et al. (2004a,b) and Beaudouin et al. (2006), who translate it into a set of ordinary differential equations. However, more realistic models are achieved by the introduction of randomness. The importance of incorporating stochasticity into models for natural phenomena in life sciences has been emphasised throughout this book and in particular in Chap. 4. In the context of the application in this chapter, the presence of uncertainty is obvious from the time series displayed in Fig. 9.3 on p. 309 as the recovery curves clearly deviate from each other even within the same cell cycle.

For that reason, after having defined the compartmental model in Sect. 9.3.1, it is approximated by a diffusion process in Sect. 9.3.2 which mirrors the stochastic nature of the recovery dynamics. So far, diffusion approximations have not been applied to FRAP kinetics before. For the sake of comparability with the analyses of other authors, the deterministic analogue is given in Sect. 9.3.3. In Sect. 9.3.4, the virtues of the stochastic model are demonstrated in a simulation study.

9.3.1 Compartmental Description

The following model describes the behaviour of a protein of interest in a cell nucleus which has partly been bleached by a laser pulse. For shorter notation, this protein is simply referred to as *molecule*, ignoring all other types of molecules in the nucleus that are not directly expressed in the model. *Fluorescent molecules* are either fluorescent themselves or fluorescently labelled.

The molecule of interest has three properties:

1. It is *bleached* or *unbleached*.
2. It is located in the *bleached section* or in the *unbleached section*.
3. It is *free* or *bound*.

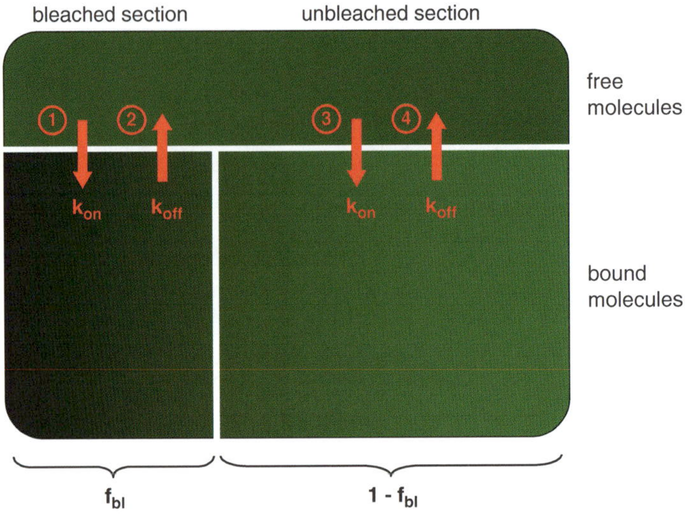

Fig. 9.5 Compartmental representation of the general kinetic model: The unbleached molecules in the nucleus are divided into three groups, namely into molecules that are free, molecules that are bound in the bleached section and molecules that are bound in the unbleached section. Due to the assumption of diffusion-uncoupled recovery (cf. Sect. 9.2.2), the location of a free molecule is not explicitly modelled. Four non-trivial transitions are possible: *(1)* A free molecule binds in the bleached section with rate k_{on}. *(2)* A bound molecule in the bleached section unbinds with rate k_{off}. *(3)* A free molecule binds in the unbleached section with rate k_{on}. *(4)* A bound molecule in the unbleached section unbinds with rate k_{off}. f_{bl} and $1-f_{\mathrm{bl}}$ express the fractions of molecules in the bleached and unbleached sections, respectively

Property (1) is an unchangeable attribute, i.e. a bleached or unbleached molecule remains bleached or unbleached, respectively, throughout the entire observation period. Bleached and unbleached molecules are assumed to behave identically, and therefore it suffices to focus on one type only. The following considerations model the dynamics of the unbleached molecules as these are visible through their fluorescence.

Properties (2) and (3) are changeable attributes, i.e. a molecule can change its location among the bleached and the unbleached section, and it can change its state among the free and the bound status: The cell nucleus is partitioned into a bleached and an unbleached area, determined by the bleaching laser pulse. Each molecule is located in either of these. When a molecule is free, it can diffuse freely within the nucleus. While it is bound, its location is fixed. Binding sites are assumed to be at fixed locations. Due to the diffusion-uncoupled scenario assumed in Sect. 9.2.2, diffusion of the free molecules happens so rapidly that their concentration is identical in the bleached and in the unbleached section. Hence, it is not necessary to model the location of a free molecule.

The above considerations motivate a kinetic model whose variables and transitions are described in the following. The model is illustrated in Fig. 9.5.

Variables

The unbleached molecules are divided into three disjoint groups, whose sizes are represented through

U^{free} : the number of unbleached free molecules,
$U^{\mathrm{bound}}_{\mathrm{bl}}$: the number of unbleached bound molecules in the bleached section,
$U^{\mathrm{bound}}_{\mathrm{unbl}}$: the number of unbleached bound molecules in the unbleached section.

These three variables sum up to the constant system size parameter

N_U : the number of unbleached molecules.

Hence, it is sufficient to model the time-evolution of two out of the three above quantities; the third variable is then easily obtained as the difference to N_U. The proportion of bleached molecules is expressed by

f_{bl} : the fraction of bleached molecules with respect to all molecules.

The number of bleached molecules in the nucleus equals the number of molecules in the bleached section at the time of bleaching. The number of molecules in the bleached section is assumed approximately constant over time. Hence, f_{bl} is also the fraction of (bleached or unbleached) molecules in the bleached section with respect to all molecules in the nucleus. Moreover, $f_{\mathrm{bl}}U^{\mathrm{free}}$ is the number of unbleached free molecules in the bleached section at any positive time, and $(1 - f_{\mathrm{bl}})U^{\mathrm{free}}$ is the number of unbleached free molecules in the unbleached section. The number of unbleached molecules is $N_U = (1 - f_{\mathrm{bl}})N$, where N is the total number of molecules in the nucleus.

Note that the structure of cell nuclei is such that the spatial distribution of the molecules is non-uniform. Thus, the parameter f_{bl} does not exactly express the fraction of the bleached area as measured in square micrometers.

Transitions and Parameters

We consider a kinetic model with the following non-trivial transitions (cf. Fig. 9.5):

1. An unbleached free molecule binds in the bleached section.
2. An unbleached bound molecule in the bleached section unbinds.
3. An unbleached free molecule binds in the unbleached section.
4. An unbleached bound molecule in the unbleached section unbinds.

Binding of a particular molecule occurs with *association rate* $k_{\mathrm{on}} \in \mathbb{R}_+$, unbinding with *dissociation rate* $k_{\mathrm{off}} \in \mathbb{R}_+$, irrespectively of the state of the other molecules. In particular, it is assumed that there are always sufficiently many binding sites available such that the occurrence of the first and third transition is independent of their number. To be more precise, the association rate is the product of an actual binding rate and the concentration of available binding sites. This product is assumed constant as the molecules are supposed to be in equilibrium (cf. e.g. Sprague and McNally 2005).

The expected time until a free molecule enters the bound state and the mean residence time of a molecule at a binding site are computed as $1/k_{\text{on}}$ and $1/k_{\text{off}}$, respectively. The objective is to statistically estimate the parameters k_{on} and k_{off}. The fraction f_{bl} is determined by image analysis via the loss of total fluorescence after bleaching.

Representation as Markov Jump Process

As pointed out above, the dispersion of unbleached molecules in the cell nucleus is completely described by two out of the three numbers U^{free}, $U_{\text{bl}}^{\text{bound}}$ and $U_{\text{unbl}}^{\text{bound}}$. In the following, we model a homogeneous Markov process with state $(U^{\text{free}}, U_{\text{bl}}^{\text{bound}})'$ and state space

$$\mathcal{D} = \{(U^{\text{free}}, U_{\text{bl}}^{\text{bound}})' \in [0, N_U]^2 \cap \mathbb{N}_0^2 \,|\, U^{\text{free}} + U_{\text{bl}}^{\text{bound}} \leq N_U\}.$$

Under the assumption that at most one event can occur within a small time interval of length Δt, this process is subject to transitions

1. $(U^{\text{free}}, U_{\text{bl}}^{\text{bound}})' \to (U^{\text{free}} - 1, U_{\text{bl}}^{\text{bound}} + 1)'$ w. prob. $k_{\text{on}} f_{\text{bl}} U^{\text{free}} \Delta t + o(\Delta t)$,
2. $(U^{\text{free}}, U_{\text{bl}}^{\text{bound}})' \to (U^{\text{free}} + 1, U_{\text{bl}}^{\text{bound}} - 1)'$ w. prob. $k_{\text{off}} U_{\text{bl}}^{\text{bound}} \Delta t + o(\Delta t)$,
3. $(U^{\text{free}}, U_{\text{bl}}^{\text{bound}})' \to (U^{\text{free}} - 1, U_{\text{bl}}^{\text{bound}})'$ w. prob. $k_{\text{on}} (1 - f_{\text{bl}}) U^{\text{free}} \Delta t + o(\Delta t)$,
4. $(U^{\text{free}}, U_{\text{bl}}^{\text{bound}})' \to (U^{\text{free}} + 1, U_{\text{bl}}^{\text{bound}})'$ w. prob. $k_{\text{off}} U_{\text{unbl}}^{\text{bound}} \Delta t + o(\Delta t)$,

where $o(\Delta t)/\Delta t \to 0$ as $\Delta t \to 0$. If none of these events occurs within time Δt, the process remains in state $(U^{\text{free}}, U_{\text{bl}}^{\text{bound}})'$.

9.3.2 *Diffusion Approximation*

So far, the considered dynamics in the cell nucleus has been modelled as a pure Markov jump process with discrete state space. Chapter 4, however, comprehensively motivated to alternatively use diffusion processes in case of large numbers of particles as given here. This facilitates the interpretation, simulation and statistical inference for the kinetic model. In the following, a diffusion approximation for the jump process with the above transitions is derived.

The first step is to convert the extensive state variables U^{free}, $U_{\text{bl}}^{\text{bound}}$ and $U_{\text{unbl}}^{\text{bound}}$ into intensive variables $u^{\text{free}} = U^{\text{free}}/N_U$, $u_{\text{bl}}^{\text{bound}} = U_{\text{bl}}^{\text{bound}}/N_U$ and $u_{\text{unbl}}^{\text{bound}} = U_{\text{unbl}}^{\text{bound}}/N_U$. These sum up to one. The new state space

$$\mathcal{C} = \{(u^{\text{free}}, u_{\text{bl}}^{\text{bound}})' \in [0, 1]^2 \cap \mathbb{R}_0^2 \,|\, u^{\text{free}} + u_{\text{bl}}^{\text{bound}} \leq 1\} \tag{9.1}$$

is considered continuous.

Section 4.3 introduced various techniques for the derivation of diffusion approximations. Under regularity conditions, which are met here, all methods yields the same result. In the following, we apply the Langevin approach from Sect. 4.3.3. According to this, the diffusion process with state variable $(u^{\text{free}}, u_{\text{bl}}^{\text{bound}})'$ solves the stochastic differential equation (SDE)

$$\begin{pmatrix} \mathrm{d}u^{\text{free}} \\ \mathrm{d}u_{\text{bl}}^{\text{bound}} \end{pmatrix} = \boldsymbol{\mu}\big(u^{\text{free}}, u_{\text{bl}}^{\text{bound}}\big)\mathrm{d}t + N_U^{-\frac{1}{2}}\boldsymbol{\sigma}\big(u^{\text{free}}, u_{\text{bl}}^{\text{bound}}\big)\mathrm{d}\boldsymbol{B}_t, \tag{9.2}$$

subject to an initial condition $(u_0^{\text{free}}, u_{\text{bl},0}^{\text{bound}})' \in \mathcal{C}$ at time $t_0 > 0$. In this equation, $\boldsymbol{B} = (\boldsymbol{B}_t)_{t \geq t_0}$ is two-dimensional Brownian motion representing fluctuations in binding and unbinding. In the Langevin approach, the drift vector $\boldsymbol{\mu}$ is obtained as

$$\begin{aligned} \boldsymbol{\mu}\big(u^{\text{free}}, u_{\text{bl}}^{\text{bound}}\big) &= k_{\text{on}} f_{\text{bl}} u^{\text{free}} \begin{pmatrix} -1 \\ 1 \end{pmatrix} + k_{\text{off}}\, u_{\text{bl}}^{\text{bound}} \begin{pmatrix} 1 \\ -1 \end{pmatrix} \\ &\quad + k_{\text{on}}(1 - f_{\text{bl}}) u^{\text{free}} \begin{pmatrix} -1 \\ 0 \end{pmatrix} + k_{\text{off}}\, u_{\text{unbl}}^{\text{bound}} \begin{pmatrix} 1 \\ 0 \end{pmatrix} \\ &= \begin{pmatrix} -(k_{\text{on}} + k_{\text{off}}) u^{\text{free}} + k_{\text{off}} \\ k_{\text{on}} f_{\text{bl}} u^{\text{free}} - k_{\text{off}}\, u_{\text{bl}}^{\text{bound}} \end{pmatrix}, \end{aligned}$$

where $u_{\text{bl}}^{\text{bound}} + u_{\text{unbl}}^{\text{bound}}$ has been replaced by $1 - u^{\text{free}}$. The diffusion coefficient $\boldsymbol{\sigma}$ is a square root of the diffusion matrix $\boldsymbol{\Sigma}$, i.e. $\boldsymbol{\Sigma} = \boldsymbol{\sigma}\boldsymbol{\sigma}'$, where

$$\begin{aligned} \boldsymbol{\Sigma}\big(u^{\text{free}}, u_{\text{bl}}^{\text{bound}}\big) &= k_{\text{on}} f_{\text{bl}} u^{\text{free}} \begin{pmatrix} 1 & -1 \\ -1 & 1 \end{pmatrix} + k_{\text{off}}\, u_{\text{bl}}^{\text{bound}} \begin{pmatrix} 1 & -1 \\ -1 & 1 \end{pmatrix} \\ &\quad + k_{\text{on}}(1 - f_{\text{bl}}) u^{\text{free}} \begin{pmatrix} 1 & 0 \\ 0 & 0 \end{pmatrix} + k_{\text{off}}\, u_{\text{unbl}}^{\text{bound}} \begin{pmatrix} 1 & 0 \\ 0 & 0 \end{pmatrix} \\ &= \begin{pmatrix} (k_{\text{on}} - k_{\text{off}}) u^{\text{free}} + k_{\text{off}} & -k_{\text{on}} f_{\text{bl}} u^{\text{free}} - k_{\text{off}}\, u_{\text{bl}}^{\text{bound}} \\ -k_{\text{on}} f_{\text{bl}} u^{\text{free}} - k_{\text{off}}\, u_{\text{bl}}^{\text{bound}} & k_{\text{on}} f_{\text{bl}} u^{\text{free}} + k_{\text{off}}\, u_{\text{bl}}^{\text{bound}} \end{pmatrix}. \end{aligned}$$

The square root $\boldsymbol{\sigma}$ of $\boldsymbol{\Sigma}$ is not unique. One possible candidate is

$$\begin{aligned} &\boldsymbol{\sigma}\big(u^{\text{free}}, u_{\text{bl}}^{\text{bound}}\big) \\ &= \begin{pmatrix} \sqrt{k_{\text{on}}(1 - f_{\text{bl}}) u^{\text{free}} + k_{\text{off}}(1 - u^{\text{free}} - u_{\text{bl}}^{\text{bound}})} & -\sqrt{k_{\text{on}} f_{\text{bl}} u^{\text{free}} + k_{\text{off}} u_{\text{bl}}^{\text{bound}}} \\ 0 & \sqrt{k_{\text{on}} f_{\text{bl}} u^{\text{free}} + k_{\text{off}} u_{\text{bl}}^{\text{bound}}} \end{pmatrix}. \end{aligned}$$

The particular choice of $\boldsymbol{\sigma}$ has no impact on the distribution of the diffusion process, cf. Sect. 3.2.5.

Observed Variable

The typical observation in a FRAP experiment is the mean grey value in the bleached section, measured over time. The value zero corresponds to the bleached section being completely dark, and the value one corresponds to it being completely lucid. Light colour is caused by the fluorescent, i.e. by the unbleached molecules. Hence, the observed value can be modelled as

$$\frac{\text{number of unbleached molecules in the bleached section}}{\text{total number of molecules in the bleached section}},$$

that is

$$\frac{f_{\text{bl}}U^{\text{free}} + U_{\text{bl}}^{\text{bound}}}{f_{\text{bl}}N} = \frac{f_{\text{bl}}u^{\text{free}} + u_{\text{bl}}^{\text{bound}}}{f_{\text{bl}}N/N_U} = \frac{f_{\text{bl}}u^{\text{free}} + u_{\text{bl}}^{\text{bound}}}{f_{\text{bl}}}(1 - f_{\text{bl}}).$$

This value would be equal to one if all molecules in the bleached section were unbleached. In practice, this will not be the case: At the time of bleaching, the number of unbleached molecules in the bleached section is zero. In the following course, the bleached and unbleached molecules will diffuse and eventually reach a state where the concentrations of bleached and unbleached molecules are identical in the bleached and the unbleached section, namely

$$\frac{\text{number of unbleached molecules in the nucleus}}{\text{total number of molecules in the nucleus}} = 1 - f_{\text{bl}}.$$

As a consequence, the final level of the observed value depends on the fraction f_{bl} of the bleached section. This value typically varies in each experiment and hence complicates the comparison of distinct experimental outcomes. For this reason, the measured mean grey value is divided by the normalising constant $1 - f_{\text{bl}}$ such that it will finally level off at value one, irrespectively of f_{bl}. Overall, one arrives at the normalised observation

$$q = \frac{f_{\text{bl}}u^{\text{free}} + u_{\text{bl}}^{\text{bound}}}{f_{\text{bl}}}. \tag{9.3}$$

Note that the variable q is subject to stochastic disturbances, i.e. it will finally fluctuate around the value one, and this level is not an upper bound. Theoretically, one rather has $0 \leq q \leq r$ for some $r \leq (1 - f_{\text{bl}})^{-1}$.

Transformation of Diffusion Approximation

With q being the only observed variable, both components of the process $(u^{\text{free}}, u_{\text{bl}}^{\text{bound}})'$ are latent. For statistical inference on the parameters k_{on} and k_{off}, it would be possible to, for example, estimate u^{free} and then to calculate a quasi-observed value of $u_{\text{bl}}^{\text{bound}}$ conditioned on the observed value of q and the estimated value of u^{free} through Eq. (9.3). A more convenient approach, however, is to take q as one out of two state variables. Hence, we in the following consider a diffusion process with state $(q, u^{\mathrm{free}})'$ with observed component q, latent component u^{free} and state space

$$\widetilde{\mathcal{C}} = \{(q, u^{\mathrm{free}})' \,|\, (u^{\mathrm{free}}, f_{\mathrm{bl}}(q - u^{\mathrm{free}}))' \in \mathcal{C}\}$$

with $\mathcal{C}$ defined as in (9.1). An SDE for this process can be obtained with Itô's formula from Sect. 3.2.10. Calculations have been moved to Sect. C.1.1 in the Appendix. The result is

$$\begin{pmatrix} \mathrm{d}q \\ \mathrm{d}u^{\mathrm{free}} \end{pmatrix} = \begin{pmatrix} k_{\mathrm{off}}(1-q) \\ -(k_{\mathrm{on}} + k_{\mathrm{off}})u^{\mathrm{free}} + k_{\mathrm{off}} \end{pmatrix} \mathrm{d}t + \frac{1}{\sqrt{N_U}} \begin{pmatrix} \tilde{\sigma}_{11} & \tilde{\sigma}_{12} \\ \tilde{\sigma}_{21} & \tilde{\sigma}_{22} \end{pmatrix} \mathrm{d}\boldsymbol{B}_t \tag{9.4}$$

with

$$\begin{aligned} \tilde{\sigma}_{11} = \tilde{\sigma}_{21} &= \sqrt{k_{\mathrm{off}}\,(1 - f_{\mathrm{bl}}q) + (k_{\mathrm{on}} - k_{\mathrm{off}})\,(1 - f_{\mathrm{bl}})u^{\mathrm{free}}} \\ \tilde{\sigma}_{12} &= \left(f_{\mathrm{bl}}^{-1} - 1\right)\sqrt{k_{\mathrm{off}} f_{\mathrm{bl}} q + (k_{\mathrm{on}} - k_{\mathrm{off}})\, f_{\mathrm{bl}} u^{\mathrm{free}}} \\ \tilde{\sigma}_{22} &= \qquad\quad -\sqrt{k_{\mathrm{off}} f_{\mathrm{bl}} q + (k_{\mathrm{on}} - k_{\mathrm{off}})\, f_{\mathrm{bl}} u^{\mathrm{free}}} \end{aligned}$$

and initial condition $(q_0, u_0^{\mathrm{free}})' \in \tilde{\mathcal{C}}$ at time t_0. The diffusion matrix for $(q, u^{\mathrm{free}})'$ reads

$$\frac{1}{N_U}\begin{pmatrix} k_{\mathrm{off}}\,(f_{\mathrm{bl}}^{-1} - 2)q + (k_{\mathrm{on}} - k_{\mathrm{off}})(f_{\mathrm{bl}}^{-1} - 1)u^{\mathrm{free}} + k_{\mathrm{off}} & k_{\mathrm{off}}\,(1-q) \\ k_{\mathrm{off}}\,(1-q) & (k_{\mathrm{on}} - k_{\mathrm{off}})u^{\mathrm{free}} + k_{\mathrm{off}} \end{pmatrix}.$$

This diffusion approximation can now be employed in order to statistically infer on k_{on} and k_{off} by application of the Bayesian estimation techniques described in Chap. 7. Before analysing experimental FRAP data in Sect. 9.7, the performance of the procedure and its benefits compared to a deterministic approach are demonstrated in a simulation study in Sect. 9.3.4.

Initial Conditions

Some remarks on the initial conditions for the process $(q, u^{\mathrm{free}})'$ are expedient. That is, on the one hand the number of unbleached molecules in the bleached compartment is assumed to be zero at time $t = 0$, i.e. at the time of bleaching. In particular, the number of unbleached *free* molecules in the bleached section is zero at time $t = 0$. On the other hand, the number of unbleached free molecules in the bleached part of the nucleus is modelled as $f_{\mathrm{bl}}U^{\mathrm{free}} > 0$ at any positive time.

That means, one has $u^{\text{free}}(0) \neq \lim_{t\to 0+} u^{\text{free}}(t)$, i.e. u^{free} is not right-continuous in $t = 0$, and hence $q(t)$ is not right-continuous in $t = 0$ either.

Diffusion processes are processes with almost surely continuous sample paths, and the deterministic differential equations in the next section refer to processes with even continuous sample paths. It is hence reasonable to formulate the initial conditions for all differential equations in this chapter for an initial time $t_0 > 0$. This poses no restriction on the applicability of the models as the first postbleach FRAP image is acquired at a positive time point, anyway.

9.3.3 Deterministic Approximation

For the purpose of comparing the performances of the diffusion approximation approach considered in this book and the deterministic approach generally employed in the literature, the deterministic counterpart of the above model is given here as well. That is, taking the limit $N_U \to \infty$ in the stochastic differential equation (SDE) (9.4), one obtains the ordinary differential equation (ODE)

$$\begin{pmatrix} \mathrm{d}q \\ \mathrm{d}u^{\text{free}} \end{pmatrix} = \begin{pmatrix} k_{\text{off}}(1-q) \\ -(k_{\text{on}}+k_{\text{off}})u^{\text{free}} + k_{\text{off}} \end{pmatrix} \mathrm{d}t \tag{9.5}$$

as a deterministic description of the FRAP dynamics. The starting values are again $(q_0, u_0^{\text{free}})' \in \mathcal{C}$. This model represents the macroscopic behaviour of the recovery process but, however, does not incorporate stochastic fluctuations. For this reason, the diffusion approximation model is clearly to be preferred.

Interestingly, the two-dimensional ODE (9.5) consists of two independent one-dimensional ODEs

$$\mathrm{d}q = k_{\text{off}}(1-q)\mathrm{d}t \tag{9.6}$$

and

$$\mathrm{d}u^{\text{free}} = \big(-(k_{\text{on}}+k_{\text{off}})u^{\text{free}} + k_{\text{off}}\big)\mathrm{d}t. \tag{9.7}$$

Since u^{free} is unobserved, Eq. (9.7) cannot directly be employed for estimation purposes. Instead, the FRAP curves are fitted to simulations from Eq. (9.6). The parameter k_{on} does not appear in this equation. That means, it cannot be estimated from recovery curves in the deterministic approach.

As a side note, Eqs. (9.6) and (9.7) possess the explicit solutions

$$q(t) = 1 + (q_0 - 1)\exp\big(-k_{\text{off}}(t-t_0)\big) \tag{9.8}$$

and

$$u^{\text{free}}(t) = \frac{k_{\text{off}}}{k_{\text{on}}+k_{\text{off}}} + \left(u_0^{\text{free}} - \frac{k_{\text{off}}}{k_{\text{on}}+k_{\text{off}}}\right)\exp\big(-(k_{\text{on}}+k_{\text{off}})(t-t_0)\big), \tag{9.9}$$

where $t \geq t_0 > 0$. As a consequence, for fitting the deterministic model to the observed data, there is no need to employ computationally demanding schemes for numerically solving the above ODE (9.6).

Equation (9.9) immediately allows an approximation of the deterministic fractions f^{free} and $f^{\text{bound}} = 1 - f^{\text{free}}$ of free and bound molecules. That is, f^{free} is the limit of $u^{\text{free}}(t)$ as $t \to \infty$, hence

$$f^{\text{free}} = \frac{k_{\text{off}}}{k_{\text{on}} + k_{\text{off}}} \qquad \text{and} \qquad f^{\text{bound}} = \frac{k_{\text{on}}}{k_{\text{on}} + k_{\text{off}}} \,.$$

If the nucleus is in chemical equilibrium at the time of bleaching, then u^{free} is constant, and hence $u^{\text{free}}(t) = k_{\text{off}}/(k_{\text{on}} + k_{\text{off}})$ for all $t \geq t_0$.

9.3.4 Simulation Study

Before applying estimation procedures to real datasets in Sect. 9.7, this section first investigates the performance of the statistical methods in a simulation study. This allows a direct comparison of parameter estimates with the true values used for the generation of synthetic data.

We use two datasets for the process (9.4) which have been simulated with initial value $(q_0, u_0^{\text{free}})' = (0.07, 0.05)'$ and parameters $(k_{\text{on}}, k_{\text{off}})' = (3.8, 0.2)'$ and $(k_{\text{on}}, k_{\text{off}})' = (0.3, 0.2)'$, respectively. The sample paths of q and u^{free} are displayed in Fig. 9.6a, d. The same plots display empirical pointwise 95 %-confidence bands for the trajectories of the diffusion process which have each been obtained from 100 simulated sample paths. Observations in the synthetic datasets are assumed to be given on the time interval $[0.15, 90]$ with an equidistant time step equal to 0.15, i.e. there are 600 observations including the initial value of the process. This roughly corresponds to the situation given in the real datasets in Sect. 9.7. The fraction of the bleached area is chosen to be $f_{\text{bl}} = 0.4$, and the number of molecules is set equal to $10{,}000$. As the sample path for u^{free} approximately remains at the same level in the first dataset, i.e. in Fig. 9.6a, this dataset resembles the real data situation most.

Bayesian Estimation

Chapter 7 introduced Bayesian methods for statistical inference on diffusion processes by means of data augmentation. In particular, the *innovation scheme* was presented and further developed in Sect. 7.4.4. This scheme is now applied to the two synthetic datasets in order to estimate the parameters k_{on} and k_{off}. The notation is adopted from Chap. 7.

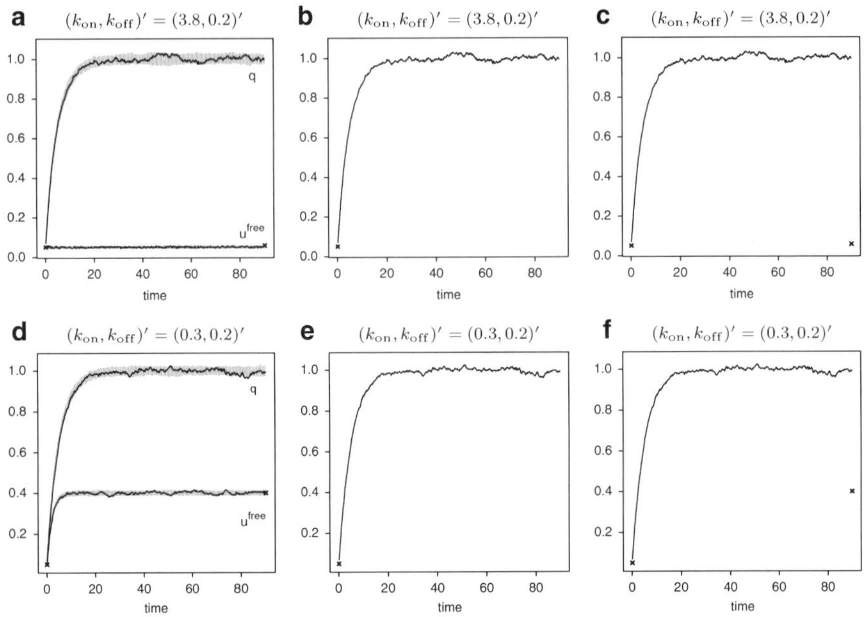

Fig. 9.6 Synthetic datasets used in the simulation study in this section. Simulations have been obtained by application of the Euler scheme from Sect. 3.3.2 with time step 0.025 and initial value $(q_0, u_0^{\text{free}})' = (0.07, 0.05)'$ at time $t_0 = 0.15$. Observations are assumed to be available at equidistant time steps of length 0.15 such that there are 600 observations on the time interval $[0.15, 90]$. The fraction of the bleached area equals $f_{\text{bl}} = 0.4$, and the number of molecules is $N = 10{,}000$. (**a**) Sample paths for q and u^{free} (*black lines*), simulated for $k_{\text{on}} = 3.8$ and $k_{\text{off}} = 0.2$. The *grey areas* represent empirical pointwise 95 %-confidence bands for the trajectories. These have been obtained from another 100 realisations of the diffusion process. (**b**) The same data as in (**a**), but as the component u^{free} is unobserved in practice, this dataset does only contain the discretely sampled path for q and the initial value of u^{free}. (**c**) The same data as in (**b**) but with additional information about the final value of u^{free}. (**d**) Sample paths for q and u^{free} (*black lines*), simulated for $k_{\text{on}} = 0.3$ and $k_{\text{off}} = 0.2$. The *grey areas* display confidence bands as in (**a**). (**e**) The same data as in (**d**), but this dataset does only contain the discretely sampled path for q and the initial value of u^{free}. (**f**) The same data as in (**e**) but with additional information about the final value of u^{free}

A priori, k_{on} and k_{off} are assumed to be exponentially distributed with expected values $\mathbb{E}(k_{\text{on}}) = 4$ and $\mathbb{E}(k_{\text{off}}) = 0.1$ in the first dataset and $\mathbb{E}(k_{\text{on}}) = 0.2$ and $\mathbb{E}(k_{\text{off}}) = 0.1$ in the second dataset. In the MCMC algorithm, new proposals k_{on}^* and k_{off}^* are drawn according to

$$\log k_{\text{on}}^* \sim \mathcal{N}\big(\log k_{\text{on}}, 0.0009\big) \quad \text{and} \quad \log k_{\text{off}}^* \sim \mathcal{N}\big(\log k_{\text{off}}, 0.0009\big),$$

where k_{on} and k_{off} represent the current values. Figure 9.7 displays trace plots for k_{on} and k_{off} produced by the innovation scheme. In particular, the parameter update is performed according to Algorithm 7.6 on p. 212, and the partially latent

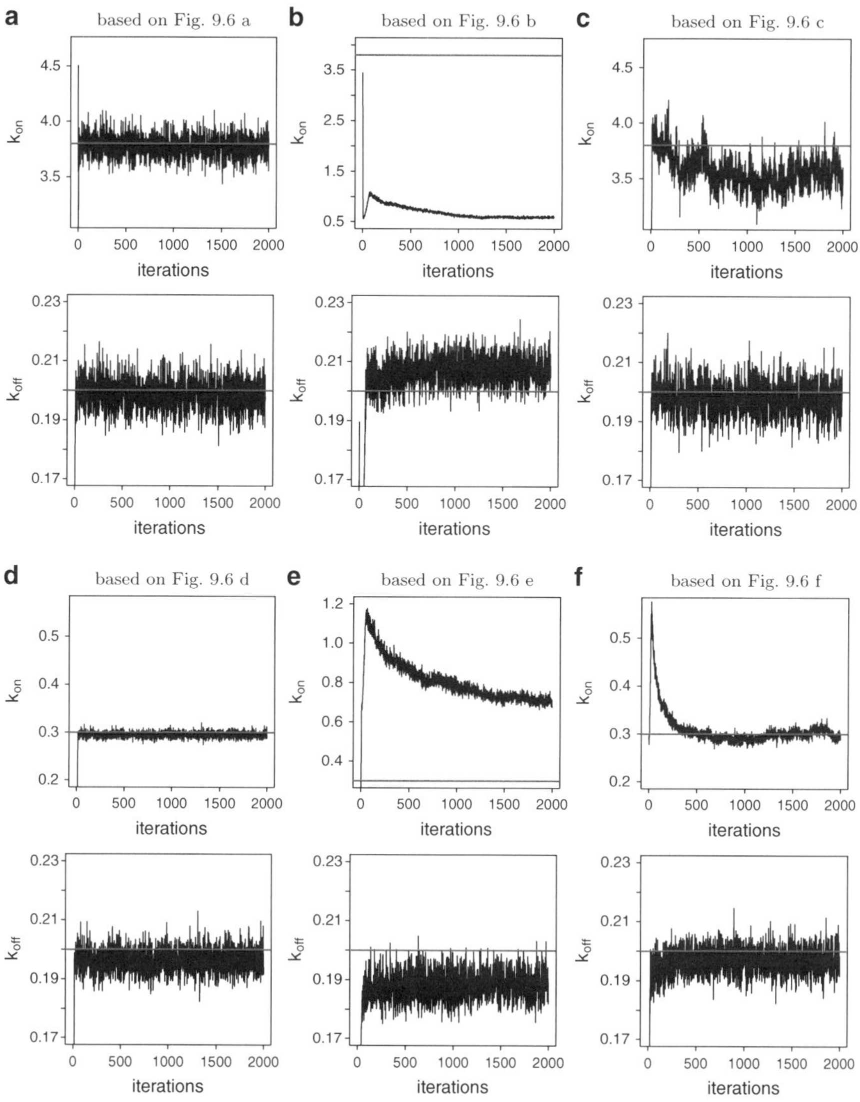

Fig. 9.7 Estimation of parameters of the diffusion process (9.4) by application of the innovation scheme based on the synthetic datasets displayed in Fig. 9.6. The MCMC algorithm introduces $m = 5$ subintervals in between every two observations. This figure shows the trace plots of k_{on} and k_{off}. The Markov chains in (**a**)–(**e**) have length 10^5 but have been thinned by factor 50; because of a large burn-in, the chain in (**f**) has length 10^6 and is thinned by factor 500. The true value of k_{on} equals 3.8 in Fig. 9.6a–c and 0.3 in Fig. 9.6d–f. The true value of k_{off} is 0.2. These are indicated by the *red horizontal lines*

Table 9.1 Estimation results as in Fig. 9.7. This table displays the posterior means and posterior 95 %-hpd intervals after a 10 % burn-in phase. The latter are computed according to Chen and Shao (1999). The true values of the parameters are displayed in the first column

True values	Estimates from Fig. 9.7a/d	Estimates from Fig. 9.7b/e	Estimates from Fig. 9.7c/f
$k_{\text{on}} = 3.8$	k_{on}: 3.78, (3.59,3.97)	k_{on}: 0.66, (0.56,0.85)	k_{on} : 3.55, (3.30,3.80)
$k_{\text{off}} = 0.2$	k_{off}: 0.20, (0.19,0.21)	k_{off} : 0.21, (0.20,0.22)	k_{off} : 0.20, (0.19,0.21)
$k_{\text{on}} = 0.3$	k_{on} : 0.29,(0.28,0.31)	k_{on} : 0.79, (0.68,0.93)	k_{on} :0.30, (0.28,0.32)
$k_{\text{off}} = 0.2$	k_{off}: 0.20, (0.19,0.20)	k_{off} : 0.19, (0.18, 0.20)	k_{off} : 0.20, (0.19,0.20)

diffusion path is updated with a modified bridge proposal as described in Sect. 7.2. The simulated Markov chains in Fig. 9.7a–e have length 10^5 but have been thinned by factor 50; because of a large burn-in, the chain in Fig. 9.7f has length 10^6 and is thinned by factor 500. For all estimations, data has been imputed such that there are $m = 5$ intermediate subintervals in between every two observation times.

When applied to the complete datasets from Fig. 9.6a, d, the innovation scheme estimates k_{on} and k_{off} very precisely as shown in Fig. 9.7a, d. In practice, however, the component u^{free} is unobserved and the datasets are as in Fig. 9.6b, e. When no information on u^{free} is given apart from its initial value, the innovation scheme can still roughly estimate k_{off} but experiences severe difficulties in the estimation of k_{on}. This is demonstrated in Fig. 9.7b, e. If, however, the endpoint of u^{free} is added to the set of observations, as it is done in Fig. 9.6c, f, it is again possible to obtain satisfactory estimation results for both k_{on} and k_{off}, see Fig. 9.7c, f. In practice, one expects u^{free} to be approximately constant as the cell nucleus is supposed to be in chemical equilibrium; approximations for the fraction u^{free} are therefore possible also in real applications as explained in Sect. 9.7. Table 9.1 displays the posterior means and 95 %-hpd intervals for k_{on} and k_{off} corresponding to the trace plots in Fig. 9.7.

Parameter estimation by application of the innovation scheme requires knowledge of the initial value of the latent component u^{free} and of the number of molecules N. Wrong assumptions about these two measures bias the estimation results as demonstrated in Table 9.2: First, the innovation scheme is applied based on the data from Fig. 9.6a but presuming $N = 5{,}000$ and $N = 20{,}000$ instead of the true value $N = 10{,}000$. Second, estimates are obtained based on the data from Fig. 9.6c with the starting value and endpoint of u^{free} set equal to 0.025 instead of approximately 0.05. Both modifications especially affect the estimates of k_{on}. It is hence important to carefully choose the value of N and u_0^{free} as also discussed in Sect. 9.7.

Least Squares Estimation

An alternative approach to the Bayesian estimation procedures in combination with a stochastic diffusion model is least squares estimation based on the deterministic model from Sect. 9.3.3. The latter approach is prevalent in the literature on the

Table 9.2 Estimation results under modified assumptions. In the left table, estimates for k_{on} and k_{off} are obtained based on the data from Fig. 9.6a but presuming $N = 5{,}000$ and $N = 20{,}000$ instead of the true value $N = 10{,}000$. In the right table, estimation is carried out based on the data from Fig. 9.6c with the starting value and endpoint of u^{free} set equal to 0.025 instead of approximately 0.05. The tables display the posterior means and posterior 95 %-hpd intervals after a 10 % burn-in phase of the MCMC algorithm with 10^5 iterations. The true values of the parameters are $k_{\mathrm{on}} = 3.8$ and $k_{\mathrm{off}} = 0.2$

Modification	Estimates for dataset from Fig. 9.6a
$N = 5{,}000$	k_{on} : 2.83, (2.64,3.02) k_{off} : 0.15, (0.14,0.16)
$N = 20{,}000$	k_{on} : 4.37, (4.22,4.52) k_{off} : 0.23, (0.22,0.24)

Modification	Estimates for dataset from Fig. 9.6c
Level of u^{free} set equal to 0.025	k_{on} : 7.29, (6.53,8.12) k_{off} : 0.20, (0.19,0.21)

analysis of molecular binding. For comparison purposes, it is also considered here. It should however be emphasised that the least squares approach violates two model assumptions which result directly from the original compartmental formulation. These concern the correspondence between least squares estimation and the assumption of independent and identically distributed deviations from the deterministic course. Neither independence nor homoscedasticity is given in the original model.

Let $x(t)$ denote the observed fluorescence intensity at time t and $q_{k_{\mathrm{off}}}(t)$ its simulated counterpart based on the parameter k_{off}. As pointed out before, the value of k_{on} has no impact on the deterministic course of q. In order to estimate the parameter k_{off}, the function $q_{k_{\mathrm{off}}}(t)$ is computed from Eq. (9.8) for different values of k_{off} with $q_0 = 0.07$ at time $t_0 = 0.15$. A least squares estimate for k_{off} is obtained as

$$\hat{k}_{\mathrm{off}} = \underset{k_{\mathrm{off}} \in \mathbb{R}_+}{\mathrm{argmin}}\ \mathrm{mSSR}, \qquad \text{where} \quad \mathrm{mSSR} = \frac{1}{n+1} \sum_{i=0}^{n} \left(q_{k_{\mathrm{off}}}(t_i) - x(t_i) \right)^2 \tag{9.10}$$

is the mean sum of squared residuals, and $t_0, \ldots, t_n$ are the observation times. Such an estimate $\hat{k}_{\mathrm{off}}$ can be determined by application of an optimisation method such as the Nelder-Mead algorithm (Nelder and Mead 1965), which is also chosen here. Applied to the datasets from Fig. 9.6a–c, the procedure yields $\hat{k}_{\mathrm{off}} = 0.20072$. For the data from Fig. 9.6d–f, the least squares estimate equals $\hat{k}_{\mathrm{off}} = 0.19443$. In both datasets, the true value is $k_{\mathrm{off}} = 0.2$. The resulting mean sums of squared residuals are 0.00012 in both cases. Figure 9.8a displays the agreement between the first synthetic curve q and its estimated counterpart.

An alternative to keeping the starting value $q_0 = 0.07$ fixed is to estimate this parameter as well. The Nelder-Mead algorithm can then be applied to find a tuple $(\hat{q}_0, \hat{k}_{\mathrm{off}})$ such that the mean sum of squared residuals is minimised. For the datasets from Fig. 9.6a–c, this procedure yields $\hat{q}_0 = 0.07542$ and $\hat{k}_{\mathrm{off}} = 0.19953$. For the data from Fig. 9.6d–f, the least squares estimates are $\hat{q}_0 = 0.06285$ and $\hat{k}_{\mathrm{off}} = 0.19593$. In both cases, the mean sum of squared residuals is decreased by only less than 10^{-6} compared to the model with fixed q_0.

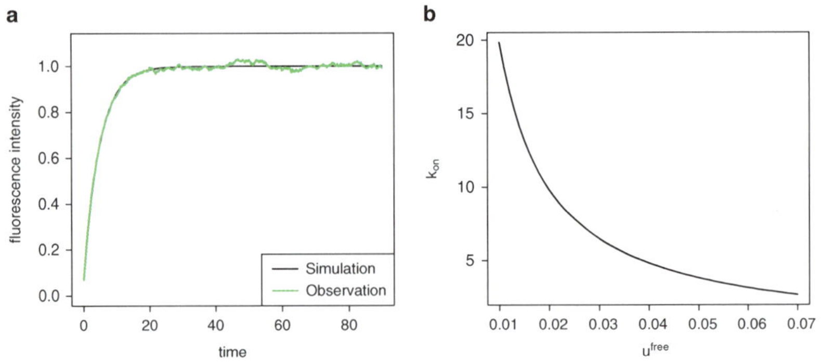

Fig. 9.8 (**a**) Deterministic course of q for $k_{\mathrm{off}} = 0.20072$ (*black*) compared to the (synthetic) observation from the dataset in Fig. 9.6a (*green*). The mean sum of squared residuals for this fit equals 0.00012. (**b**) Estimation of k_{on} according to Eq. (9.11) for $k_{\mathrm{off}} = 0.20072$ and different values of u^{free}

As mentioned above, Eq. (9.7) cannot directly be used for parameter estimation as u^{free} is unobserved. However, the nucleus is assumed to be in chemical equilibrium such that the fraction of free molecules (and also the fraction of unbleached free molecules) is approximately constant. In the deterministic model, this refers to $\mathrm{d}u^{\mathrm{free}}/\mathrm{d}t = 0$. Solving this equation yields

$$k_{\mathrm{on}} = k_{\mathrm{off}} \left(\frac{1}{u^{\mathrm{free}}} - 1 \right). \tag{9.11}$$

Hence, an approximation of k_{on} is possible if estimates are available for k_{off} and u^{free}. For the least squares estimate $\hat{k}_{\mathrm{off}} = 0.20072$ and the true value $u_0^{\mathrm{free}} = 0.05$, one obtains indeed $\hat{k}_{\mathrm{on}} = 3.81368$, which is close to the true value $k_{\mathrm{on}} = 3.8$. However, the value of $0 \leq u_0^{\mathrm{free}} \leq q_0$ is unknown in practice. Figure 9.8b shows estimates for k_{on} according to (9.11) for $k_{\mathrm{off}} = 0.20072$ and different values of u^{free}.

Conclusion

To summarise, both the Bayesian and the least squares estimation approaches are capable to correctly estimate the parameter k_{off} from the recovery curves in a FRAP experiment. Estimation of k_{on} is possible if information about u^{free} is available. Least squares estimation employs a deterministic model which does not account for random fluctuations of the recovery curves and erroneously assumes independence and homoscedasticity of the deviations between the observations and the determinstic course. Point estimates are, however, comparable to the Bayesian posterior means.

Before applying the estimation procedures to real datasets in Sect. 9.7, some further improvements of the kinetic model are considered in the next two sections.

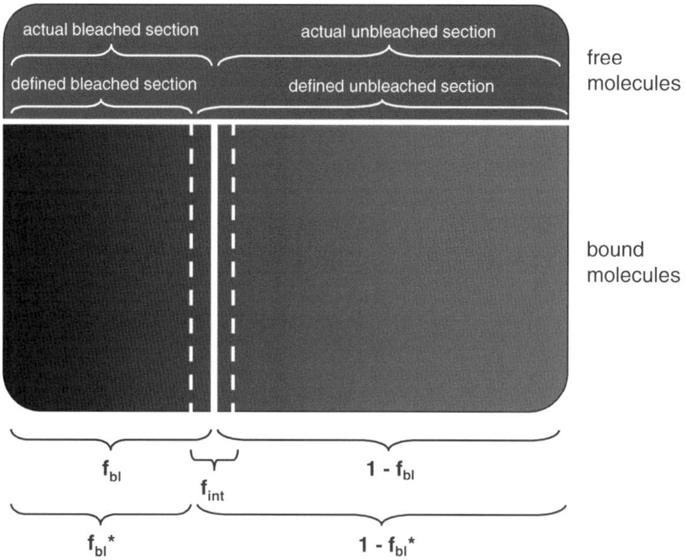

Fig. 9.9 Due to rapidly diffusing molecules, the border of the bleached section is presumably determined with error such that the actual bleached section is actually larger than the defined bleached section. The actual fraction f_{bl} of bleached molecules is assumed to be correctly identified. The size of the indefinite intermediate area is denoted by f_{int}. The fraction of molecules in the defined bleached section results as $f_{\mathrm{bl}}^* = f_{\mathrm{bl}} - 0.5\, f_{\mathrm{int}}$

9.4 Refinement of the General Model

While the parameters k_{on} and k_{off} shall be estimated statistically, the fraction f_{bl} of bleached molecules is determined experimentally: It is measured via the loss of fluorescence in the whole nucleus. We assume the value f_{bl} to be identified correctly, i.e. f_{bl} is indeed the fraction of bleached molecules with respect to all molecules. However, there is good reason for believing that the contour of the bleached section of the nucleus is determined with error: The free molecules diffuse very rapidly such that unbleached molecules presumably invade the bleached section in the short but nonzero time interval between the bleaching pulse and the first postbleach image. As a consequence, there arises an intermediate area in the nucleus which neither clearly belongs to the bleached section nor to the unbleached section. For this reason, we in the following distinguish between the *actual* and the *defined* bleached section of the nucleus. This is illustrated in Fig. 9.9.

The kinetic model can be adapted according to these considerations by introduction of a parameter

$$f_{\mathrm{bl}}^* = f_{\mathrm{bl}} - \frac{1}{2} f_{\mathrm{int}} ,$$

where f_{int} with $0 < f_{\mathrm{int}} \ll 2 f_{\mathrm{bl}}$ is a small positive constant called the *intermediate fraction*. This parameter stands for the magnitude of the intermediate area and may be determined experimentally or estimated statistically. f^*_{bl} represents the number of molecules in the defined bleached section with respect to all molecules in the cell nucleus. The number of unbleached free molecules in the defined bleached section is $f^*_{\mathrm{bl}} U^{\mathrm{free}}$.

The observed value q, given in Eq. (9.3) on p. 316, refers to spatially averaged grey values in the *defined* bleached section, that is

$$\frac{\text{number of unbleached molecules in the defined bleached section}}{\text{total number of molecules in the defined bleached section}}. \tag{9.12}$$

9.4.1 Compartmental Description

In order to be able to adequately model the fraction (9.12), (re-)introduce the following variables:

U^{free} : no. of unbleached free molecules,

$U^{\mathrm{bound}}_{\mathrm{bl}}$: no. of unbleached bound molecules in the *actual* bleached section,

$U^{\mathrm{bound}}_{\mathrm{bl}*}$: no. of unbleached bound molecules in the *defined* bleached section,

$U^{\mathrm{bound}}_{\mathrm{unbl}}$: no. of unbleached bound molecules in the *actual* unbleached section,

$U^{\mathrm{bound}}_{\mathrm{unbl}*}$: no. of unbleached bound molecules in the *defined* unbleached section.

One has $U^{\mathrm{free}} + U^{\mathrm{bound}}_{\mathrm{bl}} + U^{\mathrm{bound}}_{\mathrm{unbl}} = U^{\mathrm{free}} + U^{\mathrm{bound}}_{\mathrm{bl}*} + U^{\mathrm{bound}}_{\mathrm{unbl}*} = N_U = (1 - f_{\mathrm{bl}})N$, where N is the total number of molecules in the nucleus. It is hence reasonable to normalise the above numbers by dividing through N_U to obtain u^{free}, $u^{\mathrm{bound}}_{\mathrm{bl}}$, $u^{\mathrm{bound}}_{\mathrm{bl}*}$, $u^{\mathrm{bound}}_{\mathrm{unbl}}$ and $u^{\mathrm{bound}}_{\mathrm{unbl}*}$ with $u^{\mathrm{free}} + u^{\mathrm{bound}}_{\mathrm{bl}} + u^{\mathrm{bound}}_{\mathrm{unbl}} = u^{\mathrm{free}} + u^{\mathrm{bound}}_{\mathrm{bl}*} + u^{\mathrm{bound}}_{\mathrm{unbl}*} = 1$.

9.4.2 Diffusion Approximation

The observed mean grey value can now be expressed as

$$\frac{f^*_{\mathrm{bl}} U^{\mathrm{free}} + U^{\mathrm{bound}}_{\mathrm{bl}*}}{f^*_{\mathrm{bl}} N} = \frac{f^*_{\mathrm{bl}} u^{\mathrm{free}} + u^{\mathrm{bound}}_{\mathrm{bl}*}}{f^*_{\mathrm{bl}} N / N_U} = \frac{f^*_{\mathrm{bl}} u^{\mathrm{free}} + u^{\mathrm{bound}}_{\mathrm{bl}*}}{f^*_{\mathrm{bl}}} (1 - f_{\mathrm{bl}}).$$

This value has been normalised by the experimenter by dividing through $(1 - f_{\mathrm{bl}})$. Hence, the new target variable is represented by

$$q^* = \frac{f^*_{\mathrm{bl}} u^{\mathrm{free}} + u^{\mathrm{bound}}_{\mathrm{bl}*}}{f^*_{\mathrm{bl}}}. \tag{9.13}$$

Similarly to the variable q, the observed values of q^* will eventually level off at value one, and the theoretical range of the target variable remains $0 \leq q^* \leq r$ for some $r \leq (1 - f_{\text{bl}})^{-1}$.

The modified target variable q^* requires modelling the process $(u^{\text{free}}, u^{\text{bound}}_{\text{bl}*})'$ instead of $(u^{\text{free}}, u^{\text{bound}}_{\text{bl}})'$ with unaltered state space $\mathcal{C}$. A respective SDE can be obtained from (9.2) on p. 315 by simply replacing $u^{\text{bound}}_{\text{bl}}$ and f_{bl} by $u^{\text{bound}}_{\text{bl}*}$ and f^*_{bl}, respectively, in the drift and diffusion coefficients.

Analogously, a diffusion approximation can be set up from previous calculations for $(q^*, u^{\text{free}})'$ with state space

$$\widetilde{\mathcal{C}}^* = \{(q^*, u^{\text{free}})' \mid (u^{\text{free}}, f^*_{\text{bl}}(q^* - u^{\text{free}}))' \in \mathcal{C}\},$$

where $\mathcal{C}$ has been defined in (9.1) on p. 314. Now, if f_{int} and hence f^*_{bl} is to be estimated statistically, the state space of the process is not anymore independent of the unknown parameter. Independence has been one of the requirements for the estimation techniques in Chap. 7. In the present case, however, the dependence of $\widetilde{\mathcal{C}}^*$ on f_{int} does not impose any practical restrictions: Because of $0 \leq u^{\text{bound}}_{\text{bl}*} \leq 1 - u^{\text{free}}$, one has

$$u^{\text{free}} \leq q^* \leq \frac{(f^*_{\text{bl}} - 1)u^{\text{free}} + 1}{f^*_{\text{bl}}}.$$

This theoretical upper bound for q^* is, for realistic values of u^{free} and f^*_{bl}, much larger than its practical upper bound, which lies somewhat above one. The admissible upper value for q^* can hence confidently be replaced by the smaller $((f_{\text{bl}} - 1)u^{\text{free}} + 1)/f_{\text{bl}}$, which is independent of the parameters to estimate.

An SDE for $(q^*, u^{\text{free}})'$ then follows from Eq. (9.4) on p. 317 by replacing q and f_{bl} by q^* and f^*_{bl}, respectively. In particular, one obtains

$$\begin{pmatrix} dq^* \\ du^{\text{free}} \end{pmatrix} = \begin{pmatrix} k_{\text{off}}(1 - q^*) \\ -(k_{\text{on}} + k_{\text{off}})u^{\text{free}} + k_{\text{off}} \end{pmatrix} dt + \frac{1}{\sqrt{N_U}} \begin{pmatrix} \tilde{\sigma}^*_{11} & \tilde{\sigma}^*_{12} \\ \tilde{\sigma}^*_{21} & \tilde{\sigma}^*_{22} \end{pmatrix} d\boldsymbol{B}_t \tag{9.14}$$

with an initial condition $(q^*_0, u^{\text{free}}_0)' \in \widetilde{\mathcal{C}}^*$ at time $t_0 > 0$ and

$$\begin{aligned}
\tilde{\sigma}^*_{11} = \tilde{\sigma}^*_{21} &= \sqrt{k_{\text{off}}\,(1 - f^*_{\text{bl}}q^*) + (k_{\text{on}} - k_{\text{off}})\,(1 - f^*_{\text{bl}})u^{\text{free}}} \\
\tilde{\sigma}^*_{12} &= \left(\frac{1}{f^*_{\text{bl}}} - 1\right) \sqrt{k_{\text{off}}\, f^*_{\text{bl}}q^* + (k_{\text{on}} - k_{\text{off}})\, f^*_{\text{bl}}u^{\text{free}}} \\
\tilde{\sigma}^*_{22} &= \qquad - \sqrt{k_{\text{off}}\, f^*_{\text{bl}}q^* + (k_{\text{on}} - k_{\text{off}})\, f^*_{\text{bl}}u^{\text{free}}}\,.
\end{aligned}$$

The diffusion matrix for $(q^*, u^{\text{free}})'$ reads

$$\frac{1}{N_U}\begin{pmatrix} k_{\text{off}}\big((f^*_{\text{bl}})^{-1} - 2\big)q^* + (k_{\text{on}} - k_{\text{off}})\big((f^*_{\text{bl}})^{-1} - 1\big)u^{\text{free}} + k_{\text{off}} & k_{\text{off}}\,(1 - q^*) \\ k_{\text{off}}\,(1 - q^*) & (k_{\text{on}} - k_{\text{off}})u^{\text{free}} + k_{\text{off}} \end{pmatrix}.$$

9.4.3 Deterministic Approximation

Like in Sect. 9.3.3, the stochastic description in terms of a diffusion approximation immediately allows to read out a deterministic model as its limit. Here, we obtain a one-dimensional ODE with explicit solution

$$q^*(t) = 1 + (q_0^* - 1)\exp\big(-k_{\text{off}}\,(t - t_0)\big). \tag{9.15}$$

Once more, this equation does not contain the association rate k_{on}, and it does neither incorporate the intermediate fraction f_{int}. These parameters can therefore not be estimated by fitting the FRAP data to (9.15).

9.4.4 Simulation Study

The following considerations investigate the statistical estimation of the parameter f_{int}. As pointed out in the previous section, this is not possible by application of the deterministic approximation as the underlying model for the recovery curve. Hence, the diffusion model coupled with the innovation scheme is utilised.

A synthetic dataset is generated by application of the Euler scheme with all settings as described in Sect. 9.3.4. The parameters chosen for this simulation are $k_{\text{on}} = 3.8$, $k_{\text{off}} = 0.2$ and $f_{\text{int}} = 0.05$, and the starting value of the diffusion process equals once more $(q_0^*, u_0^{\text{free}})' = (0.07, 0.05)'$. The innovation scheme is applied to the data with the same preferences as in Sect. 9.3.4. A priori, the new parameter f_{int} is assumed to be exponentially distributed with expectation $\mathbb{E}(f_{\text{int}}) = 0.05$. It is updated according to

$$\log f_{\text{int}}^* \sim \mathcal{N}\big(\log f_{\text{int}}, 0.0001\big).$$

As before, estimation is carried out for different subsets of the simulated data which are shown in the top row of Fig. 9.10. The remaining graphics in this figure display trace plots for the parameters k_{on}, k_{off} and f_{int}.

Briefly summarised, estimation of f_{int} turns out to be difficult even if the component u^{free} is considered observed as in the dataset in Fig. 9.10a. On the other hand, the introduction of the additional parameter f_{int} does not seriously obstruct estimation of k_{on} and k_{off} in comparison to the results shown in Fig. 9.7. This is also demonstrated by the experiments in Figs. 9.11 and 9.12: Here, the general model (9.4), i.e. the model without the correction parameter f_{int}, is related to the dataset from Fig. 9.10a with $f_{\text{int}} = 0.05$, and the other way round, the refined model (9.14) is related to the dataset from Fig. 9.6a not incorporating the parameter f_{int}. In both cases, estimation of k_{on} and k_{off} works well, and the value of f_{int} in Fig. 9.12 is correctly estimated to approach zero.

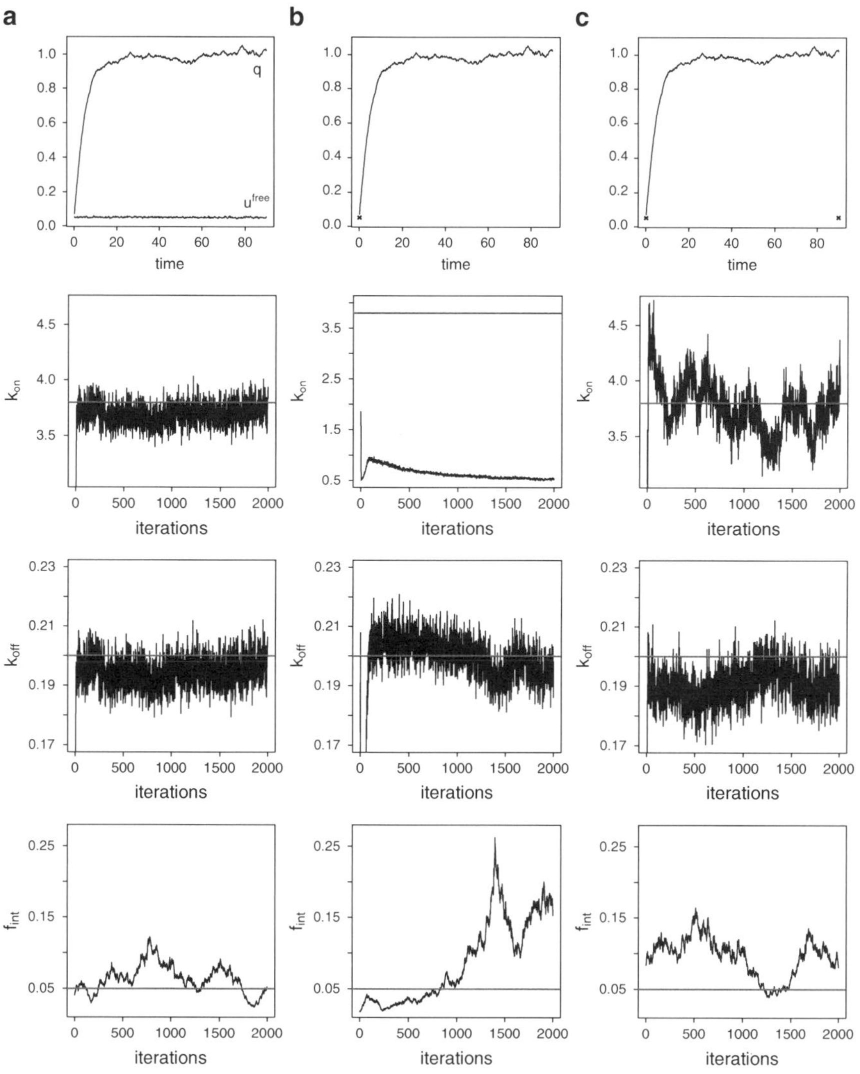

Fig. 9.10 *Top row*: Synthetic datasets for the diffusion model (9.14), obtained by application of the Euler scheme with time step 0.025 and initial value $(q_0^*, u_0^{\text{free}})' = (0.07, 0.05)'$ at time $t_0 = 0.15$. Observations are assumed to be available at equidistant time steps of length 0.15 such that there are 600 observations on the time interval $[0.15, 90]$. The fraction of the bleached area equals $f_{\text{bl}} = 0.4$, and the number of molecules is $N = 10{,}000$. (**a**) Sample paths for q^* and u^{free}, simulated for $k_{\text{on}} = 3.8$, $k_{\text{off}} = 0.2$ and $f_{\text{int}} = 0.05$. (**b**) The same data as in (**a**), but as the component u^{free} is unobserved in practice, this dataset does only contain the discretely sampled path for q^* and the initial value of u^{free}. (**c**) The same data as in (**b**) but with additional information about the final value of u^{free}. *Remaining rows*: Estimation of parameters of the diffusion process (9.14) by application of the innovation scheme based on the synthetic datasets displayed in the *top row*. The MCMC algorithm introduces $m = 5$ subintervals in between every two observations. This figure shows the trace plots of k_{on}, k_{off} and f_{int}. The Markov chains have length 10^5 but have been thinned by factor 50. The true values are indicated by the *red horizontal lines*

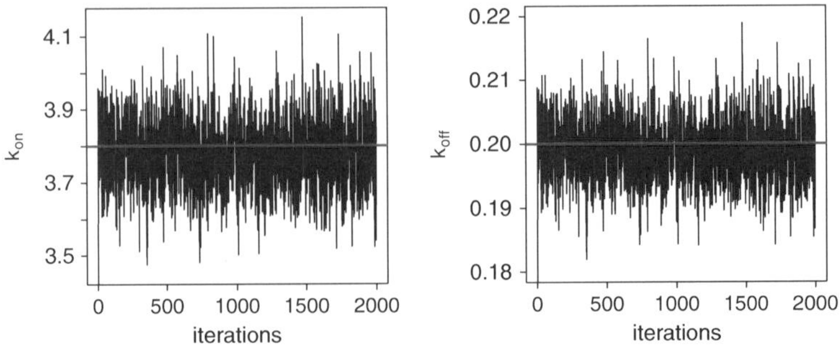

Fig. 9.11 Estimation results for k_{on} and k_{off}, where the general model (9.4), i.e. the model without the correction parameter f_{int}, is related to the dataset from Fig. 9.11a with $f_{\mathrm{int}} = 0.05$. Estimates are obtained by application of the innovation scheme. The MCMC algorithm introduces $m = 5$ subintervals in between every two observations. The Markov chains have length 10^5 but have been thinned by factor 50. The true values equal $k_{\mathrm{on}} = 3.8$ and $k_{\mathrm{off}} = 0.2$ and are indicated by the *red horizontal lines*

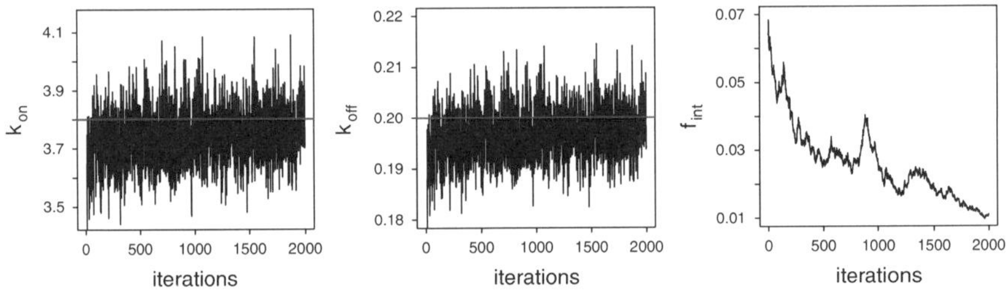

Fig. 9.12 Estimation results for k_{on}, k_{off} and f_{int}, where the refined model (9.14), i.e. the model with correction parameter f_{int}, is related to the dataset from Fig. 9.6a without f_{int}. Estimates are obtained by application of the innovation scheme. The MCMC algorithm introduces $m = 5$ subintervals in between every two observations. The Markov chains have length 10^5 but have been thinned by factor 50. The true values equal $k_{\mathrm{on}} = 3.8$, $k_{\mathrm{off}} = 0.2$ and $f_{\mathrm{int}} = 0$ and are indicated by the *red horizontal lines*

Precise estimation of f_{int} is subject to current research but not further considered in this book. In the application in Sect. 9.7, approximations for f_{int} are obtained by image analysis. In the statistical estimation of f_{int}, these can be employed as a priori knowledge. If the correction by f_{int} is neglected, the parameter is simply set equal to zero.

9.5 Extension of the General Model to Multiple Mobility Classes

One of the research questions listed in Sect. 9.1.2 was the investigation of the cell cycle dependent number of mobility classes of binding partners for Dnmt1 (cf. p. 308). If there is more than one mobility class, the protein binds and unbinds to different classes of binding partners with different association and dissociation rates. The kinetic models in Sects. 9.3 and 9.4 allow for one mobility class only. They are hence extended to multiple classes in this section. The same compartmental extension has been carried out by Phair et al. (2004a,b), who arrive at a system of ordinary differential equations describing the dynamics within the cell nucleus.

9.5.1 Compartmental Description

Suppose there are $M \in \mathbb{N}$ classes of kinetically different binding partners for the molecule of interest. Label these classes with numbers $i \in \{1, \ldots, M\}$ and refer to a molecule that is bound to a partner from class i as *bound of type* i, *type* i*-bound* or similarly. For $i = 1, \ldots, M$, define the following variables:

U^{free} : no. of unbleached free molecules,

$U_{\text{bl}*}^{\text{bound},i}$: no. of unbleached type i-bound molecules in the bleached section,

$U_{\text{unbl}*}^{\text{bound},i}$: no. of unbleached type i-bound molecules in the unbleached section.

For shorter notation, the terms *bleached section* and *unbleached section* now refer to the *defined* areas detected by image analysis (cf. the distinction between defined and actual areas in Sect. 9.4). The number of all unbleached molecules equals

$$U^{\text{free}} + \sum_{i=1}^{M} \left(U_{\text{bl}*}^{\text{bound},i} + U_{\text{unbl}*}^{\text{bound},i} \right) = N_U = (1 - f_{\text{bl}})N$$

with N again being the number of all bleached and unbleached molecules in the nucleus. Let

$$\left(U^{\text{free}}, U_{\text{bl}*}^{\text{bound},1}, \ldots, U_{\text{bl}*}^{\text{bound},M}, U_{\text{unbl}*}^{\text{bound},1}, \ldots, U_{\text{unbl}*}^{\text{bound},M} \right)'$$

be the state of a time-homogeneous Markov process with discrete state space. As all components add up to N_U, one of them could actually be left out. However, the following notation is more comprehensive with a state vector as defined above.

The following transitions are possible for $i = 1, \ldots, M$. For $M = 2$, these are illustrated in Fig. 9.13.

i_1. An unbleached free molecule binds of type i in the bleached section with rate $k_{\text{on},i}$.

i_2. An unbleached type i-bound molecule in the bleached section unbinds with rate $k_{\text{off},i}$.

i_3. An unbleached free molecule binds of type i in the unbleached section with rate $k_{\text{on},i}$.

i_4. An unbleached type i-bound molecule in the unbleached section unbinds with rate $k_{\text{off},i}$.

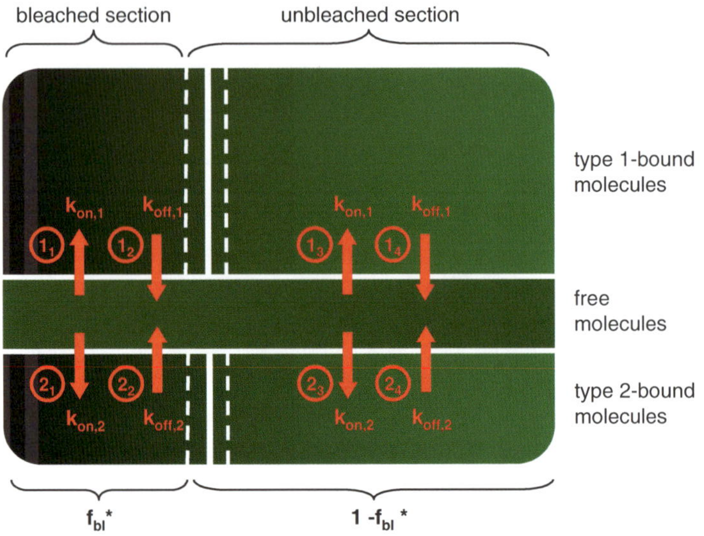

Fig. 9.13 Compartmental representation of the kinetic model with $M = 2$ mobility classes: The unbleached molecules in the nucleus are divided into five groups, namely into molecules that are free, molecules that are type 1-bound in the bleached section, molecules that are type 1-bound in the unbleached section, molecules that are type 2-bound in the bleached section, and molecules that are type 2-bound in the unbleached section. Due to the assumption of diffusion-uncoupled recovery (cf. Sect. 9.2.2), the location of a free molecule is not explicitly modelled. Eight non-trivial transitions are possible: (1_1) A free molecule binds of type 1 in the bleached section with rate $k_{\mathrm{on},1}$. (1_2) A type 1-bound molecule in the bleached section unbinds with rate $k_{\mathrm{off},1}$. (1_3) A free molecule binds of type 1 in the unbleached section with rate $k_{\mathrm{on},1}$. (1_4) A type 1-bound molecule in the unbleached section unbinds with rate $k_{\mathrm{off},1}$. (2_1) A free molecule binds of type 2 in the bleached section with rate $k_{\mathrm{on},2}$. (2_2) A type 2-bound molecule in the bleached section unbinds with rate $k_{\mathrm{off},2}$. (2_3) A free molecule binds of type 2 in the unbleached section with rate $k_{\mathrm{on},2}$. (2_4) A type 2-bound molecule in the unbleached section unbinds with rate $k_{\mathrm{off},2}$. f_{bl}^* and $1 - f_{\mathrm{bl}}^*$ express the fractions of molecules in the defined bleached and unbleached sections, respectively

The parameters $k_{\mathrm{on},i} \in \mathbb{R}_+$ and $k_{\mathrm{off},i} \in \mathbb{R}_+$ denote the association and dissociation rates corresponding to the ith mobility class. The transitions correspond to the following changes of the state variable:

$$\boldsymbol{\Delta}_{i_1} = \begin{pmatrix} -1 \\ \boldsymbol{e}_i \\ \boldsymbol{0} \end{pmatrix} \quad \text{for transition } i_1, \text{ which occurs with rate} \quad k_{\mathrm{on},i} f_{\mathrm{bl}}^* U^{\mathrm{free}},$$

$$\boldsymbol{\Delta}_{i_2} = \begin{pmatrix} 1 \\ -\boldsymbol{e}_i \\ \boldsymbol{0} \end{pmatrix} \quad \text{for transition } i_2, \text{ which occurs with rate} \quad k_{\mathrm{off},i} U_{\mathrm{bl}*}^{\mathrm{bound},i},$$

$$\boldsymbol{\Delta}_{i_3} = \begin{pmatrix} -1 \\ \mathbf{0} \\ \boldsymbol{e}_i \end{pmatrix} \quad \text{for transition } i_3, \text{ which occurs with rate} \quad k_{\text{on},i}(1 - f_{\text{bl}}^*)U^{\text{free}},$$

$$\boldsymbol{\Delta}_{i_4} = \begin{pmatrix} 1 \\ \mathbf{0} \\ -\boldsymbol{e}_i \end{pmatrix} \quad \text{for transition } i_4, \text{ which occurs with rate} \quad k_{\text{off},i}U_{\text{unbl}*}^{\text{bound},i},$$

where $\boldsymbol{e}_i = (0, \ldots, 0, 1, 0, \ldots, 0)' \in \mathbb{R}^M$ denotes the ith unit vector and $\mathbf{0} \in \mathbb{R}^M$ the null vector. This expresses the compartmental kinetic model in terms of a pure Markov jump process.

9.5.2 *Diffusion Approximation*

As motivated before, a desirable representation of the above model dynamics is by means of a diffusion approximation. An according process shall be specified in this section.

To that end, divide the numbers U^{free}, $U_{\text{bl}*}^{\text{bound},i}$ and $U_{\text{unbl}*}^{\text{bound},i}$ by N_U to obtain the fractions u^{free}, $u_{\text{bl}*}^{\text{bound},i}$ and $u_{\text{unbl}*}^{\text{bound},i}$ for all i. These fractions sum up to one. The observed variable in the FRAP experiment is the mean grey value in the bleached compartment, that is

$$q^* = u^{\text{free}} + \frac{1}{f_{\text{bl}}^*} \sum_{i=1}^{M} u_{\text{bl}*}^{\text{bound},i}. \tag{9.16}$$

This variable should be one of the components of the diffusion process. The proceeding is therefore as follows: First, one derives a diffusion approximation for the $(2M + 1)$-dimensional state variable

$$\boldsymbol{u} = \left(u^{\text{free}}, u_{\text{bl}*}^{\text{bound},1}, \ldots, u_{\text{bl}*}^{\text{bound},M}, u_{\text{unbl}*}^{\text{bound},1}, \ldots, u_{\text{unbl}*}^{\text{bound},M}\right)'$$

with state space

$$\mathcal{C}_M = \left\{ \boldsymbol{u} \in [0, 1]^{2M+1} \,\middle|\, u^{\text{free}} + \sum_{i=1}^{M} \left(u_{\text{bl}*}^{\text{bound},i} + u_{\text{unbl}*}^{\text{bound},i}\right) = 1 \right\}. \tag{9.17}$$

The resulting diffusion process is then transformed to a process with $2M$-dimensional state variable

$$\left(q^*, u^{\text{free}}, u_{\text{bl}*}^{\text{bound},1}, \ldots, u_{\text{bl}*}^{\text{bound},M-1}, u_{\text{unbl}*}^{\text{bound},1}, \ldots, u_{\text{unbl}*}^{\text{bound},M-1}\right)' \tag{9.18}$$

with an appropriate state space resulting from (9.16) and (9.17). Due to space restrictions, intermediate steps have been moved to Sect. C.1.2 in the Appendix. The resulting drift vector and diffusion matrix for (9.18) are

$$\begin{pmatrix} \mu^{\mathrm{q}} \\ \mu^{\mathrm{f}} \\ \boldsymbol{\mu}^{\mathrm{b}} \\ \boldsymbol{\mu}^{\mathrm{u}} \end{pmatrix} \quad \text{and} \quad \frac{1}{N_U} \begin{pmatrix} \Sigma^{\mathrm{qq}} & \Sigma^{\mathrm{qf}} & \boldsymbol{\Sigma}^{\mathrm{qb}} & \boldsymbol{\Sigma}^{\mathrm{qu}} \\ \Sigma^{\mathrm{fq}} & \Sigma^{\mathrm{ff}} & \boldsymbol{\Sigma}^{\mathrm{fb}} & \boldsymbol{\Sigma}^{\mathrm{fu}} \\ \boldsymbol{\Sigma}^{\mathrm{bq}} & \boldsymbol{\Sigma}^{\mathrm{bf}} & \boldsymbol{\Sigma}^{\mathrm{bb}} & \boldsymbol{\Sigma}^{\mathrm{bu}} \\ \boldsymbol{\Sigma}^{\mathrm{uq}} & \boldsymbol{\Sigma}^{\mathrm{uf}} & \boldsymbol{\Sigma}^{\mathrm{ub}} & \boldsymbol{\Sigma}^{\mathrm{uu}} \end{pmatrix}.$$

The components of the drift vector are

$\mu^{\mathrm{q}} \in \mathbb{R}$ with

$$\mu^{\mathrm{q}} = k_{\mathrm{off},M}\big(1 - q^*\big) + \sum_{i=1}^{M-1} \big(k_{\mathrm{off},i} - k_{\mathrm{off},M}\big) \left(u^{\mathrm{bound},i} - \frac{u_{\mathrm{bl}*}^{\mathrm{bound},i}}{f_{\mathrm{bl}}^*} \right)$$

$\mu^{\mathrm{f}} \in \mathbb{R}$ with

$$\mu^{\mathrm{f}} = -\left(\left(\sum_{i=1}^{M} k_{\mathrm{on},i} \right) + k_{\mathrm{off},M} \right) u^{\mathrm{free}} + k_{\mathrm{off},M} + \sum_{i=1}^{M-1} \big(k_{\mathrm{off},i} - k_{\mathrm{off},M}\big) u^{\mathrm{bound},i}$$

$$\boldsymbol{\mu}^{\mathrm{b}} = (\mu_i^{\mathrm{b}}) \in \mathbb{R}^{M-1} \ \text{ with } \ \mu_i^{\mathrm{b}} = k_{\mathrm{on},i} f_{\mathrm{bl}}^* u^{\mathrm{free}} - k_{\mathrm{off},i} u_{\mathrm{bl}*}^{\mathrm{bound},i}$$

$$\boldsymbol{\mu}^{\mathrm{u}} = (\mu_i^{\mathrm{u}}) \in \mathbb{R}^{M-1} \ \text{ with } \ \mu_i^{\mathrm{u}} = k_{\mathrm{on},i}(1 - f_{\mathrm{bl}}^*) u^{\mathrm{free}} - k_{\mathrm{off},i} u_{\mathrm{unbl}*}^{\mathrm{bound},i},$$

where $i = 1, \ldots, M-1$ and $u^{\mathrm{bound},i} = u_{\mathrm{bl}*}^{\mathrm{bound},i} + u_{\mathrm{unbl}*}^{\mathrm{bound},i}$. The main diagonal components of the diffusion matrix are

$\Sigma^{\mathrm{qq}} \in \mathbb{R}$ with

$$\Sigma^{\mathrm{qq}} = k_{\mathrm{off},M} \left(\frac{1}{f_{\mathrm{bl}}^*} - 2 \right) q^* + k_{\mathrm{off},M} + \left(\frac{1}{f_{\mathrm{bl}}^*} - 1 \right) \left(\left(\sum_{i=1}^{M} k_{\mathrm{on},i} \right) - k_{\mathrm{off},M} \right) u^{\mathrm{free}}$$
$$+ \sum_{i=1}^{M-1} \big(k_{\mathrm{off},i} - k_{\mathrm{off},M}\big) \left(u_{\mathrm{unbl}*}^{\mathrm{bound},i} + \left(\frac{1}{f_{\mathrm{bl}}^*} - 1 \right)^2 u_{\mathrm{bl}*}^{\mathrm{bound},i} \right)$$

$\Sigma^{\mathrm{ff}} \in \mathbb{R}$ with

$$\Sigma^{\mathrm{ff}} = \left(\left(\sum_{i=1}^{M} k_{\mathrm{on},i} \right) - k_{\mathrm{off},M} \right) u^{\mathrm{free}} + k_{\mathrm{off},M} + \sum_{i=1}^{M-1} \big(k_{\mathrm{off},i} - k_{\mathrm{off},M}\big) u^{\mathrm{bound},i}$$

$$\boldsymbol{\Sigma}^{\text{bb}} = (\Sigma_{ij}^{\text{bb}}) \in \mathbb{R}^{(M-1)\times(M-1)} \text{ with } \quad \Sigma_{ii}^{\text{bb}} = k_{\text{on},i} f_{\text{bl}}^* u^{\text{free}} + k_{\text{off},i} u_{\text{bl}*}^{\text{bound},i}$$

$$\text{and} \quad \Sigma_{ij}^{\text{bb}} = 0 \qquad \text{for } i \neq j$$

$$\boldsymbol{\Sigma}^{\text{uu}} = (\Sigma_{ij}^{\text{uu}}) \in \mathbb{R}^{(M-1)\times(M-1)} \text{ with } \quad \Sigma_{ii}^{\text{uu}} = k_{\text{on},i}(1-f_{\text{bl}}^*) u^{\text{free}} + k_{\text{off},i} u_{\text{unbl}*}^{\text{bound},i}$$

$$\text{and} \quad \Sigma_{ij}^{\text{uu}} = 0 \qquad \text{for } i \neq j,$$

where $i, j = 1, \ldots, M-1$. The remaining components of the diffusion matrix are

$$\Sigma^{\text{qf}} = \Sigma^{\text{fq}} \in \mathbb{R} \text{ with}$$

$$\Sigma^{\text{qf}} = k_{\text{off},M}\big(1-q^*\big) + \sum_{i=1}^{M-1}\big(k_{\text{off},i} - k_{\text{off},M}\big)\left(u_{\text{unbl}*}^{\text{bound},i} + \left(\frac{1}{f_{\text{bl}}^*} - 1\right) u_{\text{bl}*}^{\text{bound},i}\right)$$

$$\boldsymbol{\Sigma}^{\text{qb}} = (\boldsymbol{\Sigma}^{\text{bq}})' \in \mathbb{R}^{M-1} \text{ with } \quad \Sigma_i^{\text{bq}} = \left(\frac{1}{f_{\text{bl}}^*} - 1\right)\left(k_{\text{on},i} f_{\text{bl}}^* u^{\text{free}} + k_{\text{off},i} u_{\text{bl}*}^{\text{bound},i}\right)$$

$$\boldsymbol{\Sigma}^{\text{qu}} = (\boldsymbol{\Sigma}^{\text{uq}})' \in \mathbb{R}^{M-1} \text{ with } \quad \Sigma_i^{\text{uq}} = -k_{\text{on},i}\big(1 - f_{\text{bl}}^*\big) u^{\text{free}} - k_{\text{off},i} u_{\text{unbl}*}^{\text{bound},i}$$

$$\boldsymbol{\Sigma}^{\text{fb}} = (\boldsymbol{\Sigma}^{\text{bf}})' \in \mathbb{R}^{M-1} \text{ with } \quad \Sigma_i^{\text{bf}} = -\Sigma_{ii}^{\text{bb}}$$

$$\boldsymbol{\Sigma}^{\text{fu}} = (\boldsymbol{\Sigma}^{\text{uf}})' \in \mathbb{R}^{M-1} \text{ with } \quad \Sigma_i^{\text{uf}} = -\Sigma_{ii}^{\text{uu}}$$

$$\boldsymbol{\Sigma}^{\text{bu}} = (\boldsymbol{\Sigma}^{\text{ub}})' \in \mathbb{R}^{(M-1)\times(M-1)} \text{ with } \quad \boldsymbol{\Sigma}^{\text{ub}} = \mathbf{0}.$$

For $M = 1$, these formulas simplify to those derived in Sect. 9.4.2.

9.5.3 Deterministic Approximation

The drift function of the above diffusion approximation represents a deterministic description of the model dynamics involving M mobility classes. One obtains a set of $2M$ ODEs which are linear in each component, but other than in the case of one mobility class, these functions are not mutually independent when $M \geq 2$.

Nevertheless, some simple modifications allow exact simulation of the fluorescence intensity: In a deterministic setting, one can assume the nucleus being in equilibrium and the fractions of unbleached type i-bound molecules constant such that $u^{\text{bound},i} = f_i(1 - u^{\text{free}})$ for appropriate constants $f_1, \ldots, f_M$ with $f_1 + \ldots + f_M = 1$.

Plugging this in into the ODE for u^{free}, the function $u^{\text{free}}(t)$ becomes independent of the remaining components of the state variable. For given f_i,

$i \in \{1, \ldots, M\}$, and an appropriate initial condition, a realisation of u^{free} can then be obtained by calculating

$$u^{\text{free}}(t) = \left(u_0^{\text{free}} - \frac{B}{A}\right) \exp(-A(t - t_0)) + \frac{B}{A},$$

where

$$A = k_{\text{off},M} + \sum_{i=1}^{M-1} f_i \big(k_{\text{off},i} - k_{\text{off},M}\big) + \sum_{i=1}^{M} k_{\text{on},i}$$

$$B = k_{\text{off},M} + \sum_{i=1}^{M-1} f_i \big(k_{\text{off},i} - k_{\text{off},M}\big).$$

Once more, assume that $u^{\text{free}}(t) = u_0^{\text{free}} = B/A$ for all $t \geq t_0$. Then the ODE for $u_{\text{bl}*}^{\text{bound},i}$ is explicitly solved by

$$u_{\text{bl}*}^{\text{bound},i}(t) = \left(u_{\text{bl}*,0}^{\text{bound},i} - \frac{k_{\text{on},i}}{k_{\text{off},i}} f_{\text{bl}}^* u_0^{\text{free}}\right) \exp\big(-k_{\text{off},i}(t - t_0)\big) + \frac{k_{\text{on},i}}{k_{\text{off},i}} f_{\text{bl}}^* u_0^{\text{free}},$$

where $u_{\text{bl}*,0}^{\text{bound},i}$ denote suitable starting values. This equation is also true for $i = M$. Finally, with (9.16), one obtains

$$q^*(t) = \left(1 + \sum_{i=1}^{M} \frac{k_{\text{on},i}}{k_{\text{off},i}}\right) u_0^{\text{free}} + \sum_{i=1}^{M} \left(\frac{u_{\text{bl}*,0}^{\text{bound},i}}{f_{\text{bl}}^*} - \frac{k_{\text{on},i}}{k_{\text{off},i}} u_0^{\text{free}}\right) \exp\big(-k_{\text{off},i}(t - t_0)\big).$$

This curve can be fitted to the observed data. For $M = 1$, it reduces to (9.15). Note that the above formulas dispose of the state variables $u_{\text{unbl}*}^{\text{bound},i}$ but introduce the additional parameters f_i, $i \in \{1, \ldots, M\}$. Algorithm C.1 on p. 415 in the Appendix demonstrates how $q^*(t)$ can be calculated when only $k_{\text{on},1}, k_{\text{off},1}, \ldots, k_{\text{off},M}$, $u_{\text{bl}*,0}^{\text{bound},1}, f_1, \ldots, f_{M-1}$ and the initial value q_0^* are known and the nucleus is in chemical equilibrium.

An alternative, though computationally more costly, proceeding to the just described exact simulation of q^* is of course to solve the set of $2M$ ODEs numerically. This is especially applicable if the nucleus is not in equilibrium.

9.5.4 Simulation Study

Another simulation study is carried out in this section in order to evaluate the Bayesian and least squares estimation procedures on the kinetic model with multiple

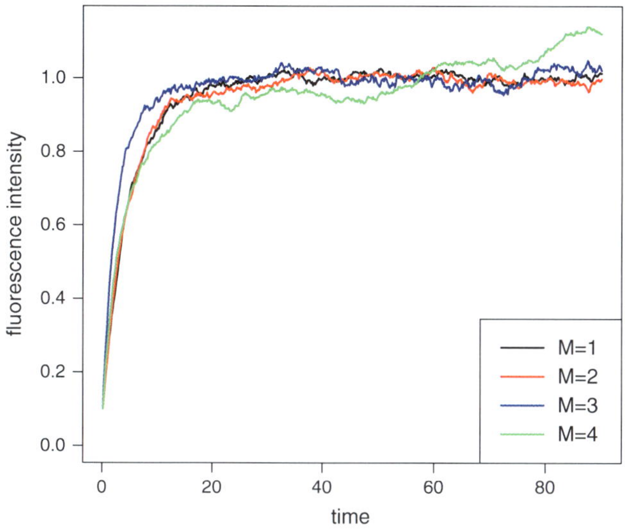

Fig. 9.14 Synthetic datasets used in the simulation study in this section. Simulations have been obtained by application of the Euler scheme from Sect. 3.3.2 with time step 0.025 to the diffusion model with $M \in \{1, \ldots, 4\}$ mobility classes. This figure displays the sample paths for the fluorescence intensity q^*. The model parameters and initial values at time t_0=0.15 are $(k_{\text{on}}, k_{\text{off}})'$=(3.8, 0.2)$'$ and $(q_0^*, u_0^{\text{free}})'$=(0.1, 0.05)$'$ for M=1 mobility class, $(k_{\text{on},1}, k_{\text{on},2}, k_{\text{off},1}, k_{\text{off},2})'$=(2.5, 1.3, 0.25, 0.15)$'$ and $(q_0^*, u_0^{\text{free}}, u_{\text{bl}*,0}^{\text{bound},1}, u_{\text{unbl}*,0}^{\text{bound},1})'$=(0.1, 0.05, 0.01, 0.45)$'$ for the model with $M = 2$ mobility classes, $(k_{\text{on},1}, k_{\text{on},2}, k_{\text{on},3}, k_{\text{off},1}, k_{\text{off},2}, k_{\text{off},3})'$=(3, 2.5, 1.3, 0.5, 0.4, 0.2)$'$ and $(q_0^*, u_0^{\text{free}}, u_{\text{bl}*,0}^{\text{bound},1}, u_{\text{bl}*,0}^{\text{bound},2}, u_{\text{unbl}*,0}^{\text{bound},1}, u_{\text{unbl}*,0}^{\text{bound},2})' = (0.1, 0.05, 0.005, 0.005, 0.2, 0.45)'$ for $M = 3$, and $(k_{\text{on},1}, k_{\text{on},2}, k_{\text{on},3}, k_{\text{on},4}, k_{\text{off},1}, k_{\text{off},2}, k_{\text{off},3}, k_{\text{off},4})'$ =(1.9, 1.0, 1.8, 0.2, 0.5, 0.2, 0.1, 0.01)$'$ and $(q_0^*, u_0^{\text{free}}, u_{\text{bl}*,0}^{\text{bound},1}, u_{\text{bl}*,0}^{\text{bound},2}, u_{\text{bl}*,0}^{\text{bound},3}, u_{\text{unbl}*,0}^{\text{bound},1}, u_{\text{unbl}*,0}^{\text{bound},2}, u_{\text{unbl}*,0}^{\text{bound},3})'$ set equal to $(0.1, 0.05, 0.004, 0.004, 0.004, 0.3, 0.2, 0.4)'$ for $M = 4$. Observations are assumed to be available at equidistant time steps of length 0.15 such that there are 600 observations on the time interval [0.15, 90]. The fraction of the bleached area equals $f_{\text{bl}} = 0.4$, the intermediate fraction is set to $f_{\text{int}} = 0$, and the number of molecules is $N = 10{,}000$. Note that we do *not* assume the nucleus to be in chemical equilibrium here

mobility classes. To that end, another four synthetic datasets are generated by the diffusion model with $M \in \{1, \ldots, 4\}$ mobility classes. The sample paths of q^* are displayed in Fig. 9.14. Details about the simulation, initial values and true parameter values are given in the caption of this figure.

In order to ensure identifiability of the distinct mobility classes, the additional restriction $k_{\text{off},1} > k_{\text{off},2} > \ldots > k_{\text{off},M}$ is introduced to the model. It is assumed that the values of $k_{\text{off},1}, \ldots, k_{\text{off},M}$ are mutually different because otherwise the model might be reduced to one with fewer mobility classes.

Table 9.3 Prior expectations for $k_{\text{off},1}, \ldots, k_{\text{off},M}$ in the model with M mobility classes assumed for the Bayesian inference as described in the main text

Parameter	Prior expectation			
	$M = 1$	$M = 2$	$M = 3$	$M = 4$
$k_{\text{off},1}$	0.2	0.3	0.3	0.30
$k_{\text{off},2}$	–	0.1	0.3	0.20
$k_{\text{off},3}$	–	–	0.1	0.10
$k_{\text{off},4}$	–	–	–	0.05

Bayesian Estimation

Like in the previous simulation studies, the innovation scheme from Sect. 7.4.4 is applied to the synthetic datasets in order to estimate the parameters $k_{\text{on},1}, \ldots, k_{\text{on},M}$ and $k_{\text{off},1}, \ldots, k_{\text{off},M}$. A priori, $k_{\text{on},i}$ and $k_{\text{off},i}$ are gamma distributed with expected values $\mathbb{E}(k_{\text{on},i}) = 2$ for all i and $\mathbb{E}(k_{\text{off},i})$ as specified in Table 9.3. The prior variances are chosen to be one tenth of the respective prior expectation. The $k_{\text{off},i}$ are furthermore subject to the above restriction concerning their order.

The MCMC algorithm draws new proposals $k^*_{\text{on},i}$ and $k^*_{\text{off},i}$ according to

$$\begin{aligned}
\log k^*_{\text{on},i} &\sim \mathcal{N}\big(\log k_{\text{on},i}, 0.0009\big) && \text{for } i = 1, \ldots, M \\
\log k^*_{\text{off},1} &\sim \mathcal{N}\big(\log k_{\text{off},1}, 0.0009\big) && \\
\text{logit}\big(k^*_{\text{off},i}\big) &\sim \mathcal{N}\big(\text{logit}\big(\min\{k_{\text{off},i}, k^*_{\text{off},i-1}\}\big), 0.0009\big) && \text{for } i = 2, \ldots, M,
\end{aligned}$$

where $k_{\text{on},i}$ and $k_{\text{off},i}$ represent the current values. The logit function has been defined in Eq. (7.19) on p. 188. It is chosen here with boundaries $u = 0$ and $v = k^*_{\text{off},i-1}$ such that the proposed $k^*_{\text{off},i}$ values automatically fulfil the assumption on their order.

For all estimations carried out in this section, the innovation scheme imputes data such that there are $m = 5$ subintervals in between every two observations, and it simulates Markov chains of length 10^5. Due to space restrictions, the resulting trace plots, empirical posterior densities and autocorrelation plots are not shown here. Posterior means and 95 %-hpd intervals are presented in Tables 9.4 and 9.5: Table 9.4 shows the results for the case where every model, i.e. each of the models with $M \in \{1, \ldots, 4\}$ classes, is applied to the dataset which has been generated by this model. Here, all components of the diffusion process are considered observed at the specified time points. The figures in Table 9.5, on the other hand, result from estimations where only the fluorescence intensity was considered observed and all other components were latent. It is hence possible here to apply each model to each dataset. Information on the end point of u^{free} was provided similarly to the proceeding in the simulation study in Sect. 9.3.4. The logarithm of the marginal likelihood, $\log \pi\big(\boldsymbol{Y}^{\text{obs}} | \mathcal{M}_M\big)$, in Table 9.5 is required for Bayesian model selection and explained in Eq. (9.19) on p. 344.

Table 9.4 Results of the Bayesian estimation procedure as described in the main text. More specifically, the parameters of the kinetic model with $M \in \{1, \ldots, 4\}$ mobility classes are estimated by application of the innovation scheme to the dataset which has been generated with the same number of classes. All components of the diffusion process are considered observed. The MCMC algorithm simulates Markov chains of length 10^5. The rightmost column displays the posterior means and 95 %-hpd intervals of the parameters after a 10 % burn-in phase

Model	Parameter	True value	Estimate
$M = 1$	$k_{\text{on},1}$	3.80	3.91 (3.72, 4.11)
	$k_{\text{off},1}$	0.20	0.21 (0.20, 0.22)
$M = 2$	$k_{\text{on},1}$	2.50	2.54 (2.40, 2.67)
	$k_{\text{on},2}$	1.30	1.32 (1.26, 1.37)
	$k_{\text{off},1}$	0.25	0.26 (0.24, 0.27)
	$k_{\text{off},2}$	0.15	0.15 (0.15, 0.16)
$M = 3$	$k_{\text{on},1}$	3.00	2.81 (2.64, 2.98)
	$k_{\text{on},2}$	2.50	2.41 (2.28, 2.55)
	$k_{\text{on},3}$	1.30	1.52 (1.46, 1.58)
	$k_{\text{off},1}$	0.50	0.47 (0.44, 0.50)
	$k_{\text{off},2}$	0.40	0.38 (0.36, 0.41)
	$k_{\text{off},3}$	0.20	0.23 (0.22, 0.24)
$M = 4$	$k_{\text{on},1}$	1.90	1.77 (1.72, 1.83)
	$k_{\text{on},2}$	1.00	1.48 (1.42, 1.56)
	$k_{\text{on},3}$	1.80	4.75 (4.60, 4.86)
	$k_{\text{on},4}$	0.20	2.49 (2.43, 2.57)
	$k_{\text{off},1}$	0.50	0.45 (0.44, 0.47)
	$k_{\text{off},2}$	0.20	0.29 (0.27, 0.30)
	$k_{\text{off},3}$	0.10	0.27 (0.26, 0.27)
	$k_{\text{off},4}$	0.01	0.27 (0.26, 0.27)

It turns out that satisfyingly precise estimation of the model parameters is possible for $M = 1$ and $M = 2$ when all components of the diffusion process are observed at discrete time points. For $M = 3$, estimates are more biased, and for $M = 4$, inference proves to be problematic. For that reason, the model and dataset with four mobility classes are omitted in the more challenging framework of Table 9.5. In that table, estimation of the model parameters in the correct datasets is still satisfactory although less information is provided than for the estimates in Table 9.4. Interestingly, the posterior mean of $k_{\text{on},1}$ in the model with $M = 1$ mobility class approximately equals the sum of the true $k_{\text{on},i}$ values for all datasets. Moreover, the mean of all $k_{\text{off},i}$ point estimates approximates the mean of all true $k_{\text{off},i}$ values, no matter which model is applied to which dataset. When applying the model with $M = 3$ classes to the datasets with $M = 1$ and $M = 2$ classes, one obtains adjoining or even overlapping hpd intervals for $k_{\text{on},1}$ and $k_{\text{on},2}$ and for $k_{\text{off},1}$ and $k_{\text{off},2}$. This suggests that both datasets do not require the model with three classes. Different model choice criteria are also considered later in this section.

Table 9.5 Results of the Bayesian estimation procedure as in Table 9.4, but this time with only the fluorescence intensity considered observed and all other components of the diffusion process regarded to be latent. It is hence possible here to apply each model to each dataset. The MCMC algorithm simulates Markov chains of length 10^5. This table displays the posterior means and 95 %-hpd intervals for each parameter after a 10 % burn-in phase. It furthermore shows the logarithm of the marginal likelihood, $\log \pi(\boldsymbol{Y}^{\text{obs}}|\mathcal{M}_M)$. This quantity is required for Bayesian model selection and explained in Eq. (9.19) on p. 344

Model	Parameter, Marg. log-likelihood	Dataset $M=1$	$M=2$	$M=3$
True values	$k_{\text{on},1}$	3.800	2.500	3.000
	$k_{\text{on},2}$	–	1.300	2.500
	$k_{\text{on},3}$	–	–	1.300
	$k_{\text{off},1}$	0.200	0.250	0.500
	$k_{\text{off},2}$	–	0.150	0.400
	$k_{\text{off},3}$	–	–	0.200
$M=1$	$k_{\text{on},1}$	4.00 (3.60, 4.43)	3.86 (3.34, 4.36)	6.72 (5.99, 7.59)
	$k_{\text{off},1}$	0.20 (0.19, 0.21)	0.20 (0.19, 0.21)	0.35 (0.33, 0.37)
	$\log \pi(\boldsymbol{Y}^{\text{obs}}|\mathcal{M}_1)$	14847.22	14835.48	14026.66
$M=2$	$k_{\text{on},1}$	2.29 (2.03, 2.50)	2.52 (2.22, 2.87)	3.63 (3.30, 4.04)
	$k_{\text{on},2}$	1.93 (1.74, 2.11)	1.65 (1.47, 1.81)	1.84 (1.66, 2.00)
	$k_{\text{off},1}$	0.23 (0.21, 0.24)	0.26 (0.23, 0.29)	0.43 (0.39, 0.46)
	$k_{\text{off},2}$	0.17 (0.15, 0.18)	0.15 (0.13, 0.16)	0.23 (0.21, 0.25)
	$\log \pi(\boldsymbol{Y}^{\text{obs}}|\mathcal{M}_2)$	14841.64	14780.84	13994.29
$M=3$	$k_{\text{on},1}$	2.24 (1.92, 2.53)	2.29 (2.03, 2.54)	2.80 (2.38, 3.35)
	$k_{\text{on},2}$	1.87 (1.63, 2.01)	1.92 (1.78, 2.05)	2.30 (2.04, 2.51)
	$k_{\text{on},3}$	1.36 (1.23, 1.51)	1.42 (1.27, 1.55)	1.98 (1.78, 2.18)
	$k_{\text{off},1}$	0.28 (0.25, 0.32)	0.27 (0.23, 0.31)	0.43 (0.36, 0.54)
	$k_{\text{off},2}$	0.24 (0.22, 0.25)	0.23 (0.21, 0.25)	0.35 (0.32, 0.38)
	$k_{\text{off},3}$	0.13 (0.12, 0.15)	0.13 (0.12, 0.14)	0.23 (0.21, 0.25)
	$\log \pi(\boldsymbol{Y}^{\text{obs}}|\mathcal{M}_3)$	14719.6	14738.16	14035.09

Table 9.6 Estimation results determined by least squares estimation applied to the deterministic model from Sect. 9.5.3 in combination with the synthetic data from Fig. 9.14. This table displays estimates for $k_{\text{off},1}, \ldots, k_{\text{off},M}$ and the corresponding mean sum of squared residuals (mSSR) as introduced in Eq. (9.10) on p. 323. For the models with $M \geq 2$ mobility classes, the output of the Nelder-Mead algorithm depends on the initial guesses for the unknown variables. Hence, several thousand initial guesses are randomly drawn and passed to the optimisation procedure. From the resulting return values, that estimate is chosen which produces the minimum mSSR. The results in this table have been obtained for $q_0^* = 0.1$ kept fixed. Table 9.7 contains estimates where q_0^* is a free parameter

	Parameter,	Dataset			
Model	mSSR	$M = 1$	$M = 2$	$M = 3$	$M = 4$
True values	$k_{\text{off},1}$	0.200	0.250	0.500	0.500
	$k_{\text{off},2}$	–	0.150	0.400	0.200
	$k_{\text{off},3}$	–	–	0.200	0.100
	$k_{\text{off},4}$	–	–	–	0.010
$M = 1$	$k_{\text{off},1}$	0.198	0.200	0.338	0.179
	mSSR	$1.05 \cdot 10^{-4}$	$2.26 \cdot 10^{-4}$	$4.01 \cdot 10^{-4}$	$3.17 \cdot 10^{-3}$
$M = 2$	$k_{\text{off},1}$	0.286	0.244	0.521	0.343
	$k_{\text{off},2}$	0.158	0.108	0.169	$6.71 \cdot 10^{-5}$
	mSSR	$9.33 \cdot 10^{-5}$	$1.71 \cdot 10^{-4}$	$3.02 \cdot 10^{-4}$	$7.08 \cdot 10^{-4}$
$M = 3$	$k_{\text{off},1}$	0.795	0.245	0.523	0.312
	$k_{\text{off},2}$	0.334	0.238	0.521	0.266
	$k_{\text{off},3}$	0.162	0.098	0.169	0.002
	mSSR	$9.23 \cdot 10^{-5}$	$1.70 \cdot 10^{-4}$	$3.02 \cdot 10^{-4}$	$6.60 \cdot 10^{-4}$
$M = 4$	$k_{\text{off},1}$	0.297	0.262	0.5878	0.622
	$k_{\text{off},2}$	0.288	0.246	0.4834	0.271
	$k_{\text{off},3}$	0.176	0.239	0.4832	0.260
	$k_{\text{off},4}$	0.156	0.117	0.1615	0.024
	mSSR	$9.33 \cdot 10^{-5}$	$1.71 \cdot 10^{-4}$	$3.03 \cdot 10^{-4}$	$1.07 \cdot 10^{-3}$

Least Squares Estimation

The parameters of the model with multiple mobility classes are now also approximated by least squares estimation. As for the general kinetic model on pp. 322, the Nelder-Mead algorithm is applied in order to find a combination of parameter values which minimises the mean sum of squared residuals (mSSR) in Eq. (9.10). For the model with M mobility classes, the parameters are $k_{\text{on},1}, \ldots, k_{\text{on},M}$ and $k_{\text{off},1}, \ldots, k_{\text{off},M}$.

As distinguished from the case of one mobility class, the output of the Nelder-Mead method depends on the initial guesses of all unknown variables when $M \geq 2$. Hence, several thousand initial guesses are randomly drawn and passed to the Nelder-Mead algorithm. Then, from the resulting return values, that estimate is chosen which produces the minimum mSSR.

Table 9.6 displays such estimation results, where the model with $M \in \{1, \ldots, 4\}$ mobility classes is applied to each of the datasets generated for $M \in \{1, \ldots, 4\}$

Table 9.7 Estimation results as in Table 9.6 but with q_0^* being a free parameter

Model	Parameter, mSSR	Dataset $M = 1$	$M = 2$	$M = 3$	$M = 4$
True values	q_0^*	0.100	0.100	0.100	0.100
	$k_{\text{off},1}$	0.200	0.250	0.500	0.500
	$k_{\text{off},2}$	–	0.150	0.400	0.200
	$k_{\text{off},3}$	–	–	0.200	0.100
	$k_{\text{off},4}$	–	–	–	0.010
$M = 1$	q_0^*	0.111	0.114	0.153	0.233
	$k_{\text{off},1}$	0.195	0.197	0.316	0.147
	mSSR	$1.03 \cdot 10^{-4}$	$2.23 \cdot 10^{-4}$	$3.77 \cdot 10^{-4}$	$2.91 \cdot 10^{-3}$
$M = 2$	q_0^*	0.092	0.085	0.094	0.261
	$k_{\text{off},1}$	0.312	0.262	0.534	0.256
	$k_{\text{off},2}$	0.161	0.117	0.170	0.006
	mSSR	$9.28 \cdot 10^{-5}$	$1.69 \cdot 10^{-4}$	$3.02 \cdot 10^{-4}$	$1.04 \cdot 10^{-3}$
$M = 3$	q_0^*	0.092	0.083	0.094	0.193
	$k_{\text{off},1}$	0.314	0.272	0.576	0.407
	$k_{\text{off},2}$	0.164	0.241	0.504	0.194
	$k_{\text{off},3}$	0.160	0.106	0.166	$3.17 \cdot 10^{-4}$
	mSSR	$9.31 \cdot 10^{-5}$	$1.68 \cdot 10^{-4}$	$3.02 \cdot 10^{-4}$	$6.92 \cdot 10^{-4}$
$M = 4$	q_0^*	0.099	0.085	0.093	0.179
	$k_{\text{off},1}$	0.906	0.279	0.605	0.632
	$k_{\text{off},2}$	0.312	0.255	0.573	0.276
	$k_{\text{off},3}$	0.197	0.218	0.506	0.212
	$k_{\text{off},4}$	0.134	0.101	0.180	0.019
	mSSR	$9.27 \cdot 10^{-5}$	$1.69 \cdot 10^{-4}$	$3.02 \cdot 10^{-4}$	$1.02 \cdot 10^{-3}$

classes, and the initial value $q_0^* = 0.1$ is kept fixed. Table 9.7 contains according estimates when q_0^* is determined by the Nelder-Mead procedure. Combinations of parameter values which produce similarly small values of mSSR as the optimal estimate show that there is relatively small variation in the $k_{\text{off},i}$ values but large variability in the $k_{\text{on},i}$ values. Hence, Tables 9.6 and 9.7 do not list approximations of $k_{\text{on},i}$. Figure C.2 on p. 416 in the Appendix presents the fittings of the deterministic curves to the observed data according to the estimates in Table 9.6.

Two issues becomes apparent when considering the results in Tables 9.6 and 9.7: The first is that, for any dataset, a model with M' mobility classes should theoretically produce a smaller mSSR than a model with $M < M'$ classes because the former is a generalisation of the latter. For the same reason, the model with q_0^* being a free variable should yield a smaller mSSR than the same model with fixed q_0^* when applied to the same dataset. However, this is not always the case in Tables 9.6 and 9.7, especially not for the model with four mobility classes. This indicates that the optimal estimates have not always been found for the models with larger numbers of mobility classes.

The second issue is that some parameter estimates contain almost identical $k_{\text{off},i}$ values, see for instance the estimate $(\hat{k}_{\text{off},1},\ldots,\hat{k}_{\text{off},4})'=(0.5878, 0.4834, 0.4832, 0.1615)'$ in Table 9.6 for the model with four mobility classes applied to the dataset with three classes. In this example, one may ask whether the parameters $k_{\text{off},2}$ and $k_{\text{off},3}$ should be summarised as one parameter, yielding the (true) model with three mobility classes. An obvious approach to answer this question is to investigate whether the confidence intervals of the estimates of the single components overlap. However, in the considered context, the mSSR is an extremely irregular function of the unknown variables such that a first investigation by means of the inverse Fisher information evaluated at the parameter estimates does not come to a practical conclusion. This issue is hence left for future work. Model choice is carried out as described in the following.

Model Choice

So far, parameter estimates and resulting fits of the model to the data have been considered for different numbers of mobility classes. Better agreement between observed and predicted values is achieved when using models with larger numbers of mobility classes (unless an appropriate estimate has not been found, as it is obviously the case for some of the least squares estimates, see the above comments). This is because the models are nested, i.e. the model with M mobility classes is a special case of any model with $M' > M$ mobility classes. Furthermore, an additional approximation of the initial value q_0^* yields improved fits because once again this setting is a generalisation of the model with a fixed starting value. However, the introduction of extra mobility classes or other variables involves an increase of model complexity. Parameter estimation becomes computationally more demanding in that case.

In what follows, well-established resources to balance between the accuracy of the fit and the complexity of the model are applied to the estimation results. In particular, *Bayes factors* (Jeffreys 1961) are utilised for selection of an appropriate diffusion model, and *Akaike's information criterion (AIC)* (Akaike 1973) and the *Bayesian information criterion (BIC)* (Schwarz 1978) are employed for choosing a deterministic model.

Bayes Factors

Let $\mathcal{M}_k$ and $\mathcal{M}_l$ denote two models which come into question to have generated a set $\boldsymbol{Y}^{\text{obs}}$ of observations. The *Bayes factor in favour of* $\mathcal{M}_k$ is defined as the ratio of marginal likelihoods

$$\mathcal{B}_{kl} = \frac{\pi\big(\boldsymbol{Y}^{\text{obs}}\,\big|\,\mathcal{M}_k\big)}{\pi\big(\boldsymbol{Y}^{\text{obs}}\,\big|\,\mathcal{M}_l\big)},$$

that is the posterior odds $\pi(\mathcal{M}_k|\boldsymbol{Y}^{\text{obs}})/\pi(\mathcal{M}_l|\boldsymbol{Y}^{\text{obs}})$ in case of identical a priori beliefs $p(\mathcal{M}_k) = p(\mathcal{M}_l)$. This ratio reflects the evidence in the data in favour of the model $\mathcal{M}_k$ as opposed to $\mathcal{M}_l$. An indication for $\mathcal{M}_k$ is given when $\mathcal{B}_{kl} > 1$. See, for example, Kass and Raftery (1995) for detailed interpretation schemes for the value of $\mathcal{B}_{kl}$.

Unfortunately, the marginal likelihood $\pi(\boldsymbol{Y}^{\text{obs}}|\mathcal{M})$ of the observed data $\boldsymbol{Y}^{\text{obs}}$ given an underlying model $\mathcal{M}$ is not always available. Hence, Chib (1995) investigates its approximation from MCMC output, also in the presence of imputed data. The following considerations adopt these ideas; a similar approach has also been chosen by Elerian et al. (2001).

Let $\boldsymbol{\theta}$ denote the vector of parameters in the model $\mathcal{M}$. One has

$$\pi\big(\boldsymbol{Y}^{\text{obs}}\,\big|\,\mathcal{M}\big) = \frac{\pi\big(\boldsymbol{Y}^{\text{obs}}\,\big|\,\boldsymbol{\theta},\mathcal{M}\big)\pi\big(\boldsymbol{\theta}\,\big|\,\mathcal{M}\big)}{\pi\big(\boldsymbol{\theta}\,\big|\,\boldsymbol{Y}^{\text{obs}},\mathcal{M}\big)}, \tag{9.19}$$

which holds for all values of $\boldsymbol{\theta}$ (Chib 1995). This ratio is best approximated at a high density value of $\boldsymbol{\theta}$. Hence, choose an appropriate value $\boldsymbol{\theta}^*$ such as the mode from the empirical posterior density of $\boldsymbol{\theta}$ and evaluate the right hand side of (9.19) at $\boldsymbol{\theta}^*$. To that end, $\pi(\boldsymbol{\theta}^*|\boldsymbol{Y}^{\text{obs}},\mathcal{M})$ can be obtained through kernel density estimation from the MCMC output. The prior density $\pi(\boldsymbol{\theta}^*|\mathcal{M})$ has been chosen by the experimenter in the MCMC procedure. Eventually, the likelihood can be approximated as

$$\pi\big(\boldsymbol{Y}^{\text{obs}}\,\big|\,\boldsymbol{\theta}^*,\mathcal{M}\big) \approx \frac{1}{K}\sum_{k=1}^{K}\pi\big(\boldsymbol{Y}^{\text{obs}},\boldsymbol{Y}^{\text{imp}(k)}\,\big|\,\boldsymbol{\theta}^*,\mathcal{M}\big) \tag{9.20}$$

for some large $K \in \mathbb{N}$, where $\boldsymbol{Y}^{\text{imp}(1)},\ldots,\boldsymbol{Y}^{\text{imp}(K)}$ is imputed data from the MCMC procedure. On the right hand side of this equation, the time grid of observed and imputed data is dense enough such that an Euler approximation of the true density is appropriate.

In the application in this chapter, different models refer to different numbers of mobility classes and hence to different dimensions of the diffusion process. Independently of the model, the only observed component of the process is the fluorescence intensity, i.e. the number of latent components and hence the amount of auxiliary data increases with each additional mobility class. In order to consider comparable quantities of imputed data on the right hand side of Eq. (9.20), the marginal likelihood $\pi(q^{*\text{obs}},q^{*\text{imp}(k)}|\boldsymbol{\theta}^*,\mathcal{M})$ of the observed and imputed values for q^* is employed in the calculations below instead of $\pi(\boldsymbol{Y}^{\text{obs}},\boldsymbol{Y}^{\text{imp}(k)}\,|\,\boldsymbol{\theta}^*,\mathcal{M})$. This is straightforward as the latter is a Gaussian density.

Polson and Roberts (1994) point out that, in case of two diffusion models $\mathcal{M}_k$ and $\mathcal{M}_l$ with different diffusion matrices, the Bayes factor $\mathcal{B}_{kl}$ degenerates when an infinite amount of data is imputed. This difficulty has the same source as the convergence problems described in Sect. 7.3. In the context of the relatively small amounts of imputed data in this chapter, however, this issue seems to be of no practical concern.

Table 9.5 on p. 340 contains approximations of $\log \pi(\boldsymbol{Y}^{\text{obs}}|\mathcal{M}_M)$ for $M \in \{1,2,3\}$, where $\mathcal{M}_M$ denotes the model with M mobility classes. These approximations are based on the parameter estimates from that table and the imputed data which was simulated in the course of the according estimation procedures.

Consider the logarithm of the Bayes factor, $\log \mathcal{B}_{kl} = \log \pi(\boldsymbol{Y}^{\text{obs}}|\mathcal{M}_k) - \log \pi(\boldsymbol{Y}^{\text{obs}}|\mathcal{M}_l)$ in favour of the model with k classes. For the dataset generated by the model with one mobility class, one has $\log \mathcal{B}_{12} = 5.58$, $\log \mathcal{B}_{13} = 127.62$ and $\log \mathcal{B}_{23} = 122.04$. That means that for this dataset the Bayes factor correctly favours the model with one mobility class. According to Kass and Raftery (1995), these values show very strong evidence in favour of $\mathcal{M}_1$ against $\mathcal{M}_2$ and $\mathcal{M}_3$, and also very strong evidence in favour of $\mathcal{M}_2$ against $\mathcal{M}_3$. For the dataset generated by the model with two mobility classes, however, the Bayes factors show the same ranking of models, i.e. the true model is not chosen here. For the data simulated with three classes, the favoured model is again the true model, i.e. the model with $M = 3$.

AIC and BIC

Let $q^*_{\hat{\boldsymbol{\rho}}}(t)$ denote the fluorescence intensity in the bleached region at time t as predicted by the deterministic model with parameter $\hat{\boldsymbol{\rho}}$, and let $x(t)$ denote the value at time $t \in \{t_0, \ldots, t_n\}$ that has actually been observed. The vector $\boldsymbol{\rho}$ does not only contain the original model parameters $k_{\text{on},1}, \ldots, k_{\text{on},M}$ and $k_{\text{off},1}, \ldots, k_{\text{off},M}$ but also all other unknowns $u_{\text{bl}*,0}^{\text{bound},1}, \ldots, u_{\text{bl}*,0}^{\text{bound},M-1}, f_1, \ldots, f_{M-1}$ and possibly q^*_0. Assume that $x(t) = q^*_{\hat{\boldsymbol{\rho}}}(t) + \varepsilon(t)$ for mutually independent $\varepsilon(t) \sim \mathcal{N}(0, \sigma^2)$ with unknown variance $\sigma^2 > 0$. Then, omitting additive constants, the AIC and BIC read

$$\text{AIC} = (n+1)\log\left(\frac{1}{n+1}\sum_{i=0}^{n}\left(q^*_{\hat{\boldsymbol{\rho}}}(t_i) - x(t_i)\right)^2\right) + 2\dim(\boldsymbol{\rho}) \tag{9.21}$$

and

$$\text{BIC} = (n+1)\log\left(\frac{1}{n+1}\sum_{i=0}^{n}\left(q^*_{\hat{\boldsymbol{\rho}}}(t_i) - x(t_i)\right)^2\right) + \log(n+1)\dim(\boldsymbol{\rho}) \tag{9.22}$$

(e.g. Fahrmeir et al. 2009). These indices evaluate the accuracy of the fit (measured by a small first summand) against the complexity of the model (measured by a large second summand). The latter is more pronounced in the BIC. At the end, one chooses the model with smallest AIC or BIC.

Tables 9.8 and 9.9 list the AIC and BIC for the estimation results from Tables 9.6 and 9.7, where $n + 1 = 600$. The comparison is of course redundant for those cases where no better agreement is found for a more complex model than for a simpler model. The minimum AIC or BIC in each column is printed in bold, showing

Table 9.8 AIC as defined in Eq. (9.21) for the estimation results from Tables 9.6 and 9.7. The minimum AIC in each column is printed in bold, marking the model that is chosen by the AIC

			Dataset			
M	q_0^*	$\dim(\boldsymbol{\rho})$	$M = 1$	$M = 2$	$M = 3$	$M = 4$
1	fixed	1	−5,495	−5,035	−4,691	−3,450
	free	2	−5,504	−5,041	−4,726	−3,500
2	fixed	6	**−5,558**	−5,192	**−4,851**	−4,340
	free	7	−5,556	**−5,197**	−4,849	−4,107
3	fixed	10	−5,556	−5,188	−4,843	**−4,374**
	free	11	−5,548	−5,193	−4,841	−4,344
4	fixed	14	−5,542	−5,176	−4,833	−4,077
	free	15	−5,540	−5,181	−4,833	−4,102

Table 9.9 BIC as defined in Eq. (9.22) for the estimation results from Tables 9.6 and 9.7. The minimum BIC in each column is printed in bold, marking the model that is chosen by the BIC

			Dataset			
M	q_0^*	$\dim(\boldsymbol{\rho})$	$M = 1$	$M = 2$	$M = 3$	$M = 4$
1	fixed	1	−5,491	−5,031	−4,687	−3,446
	free	2	−5,496	−5,032	−4,717	−3,491
2	fixed	6	**−5,531**	−5,166	**−4,825**	−4,313
	free	7	−5,525	**−5,167**	−4,818	−4,076
3	fixed	10	−5,512	−5,144	−4,799	**−4,330**
	free	11	−5,499	−5,145	−4,793	−4,295
4	fixed	14	−5,480	−5,115	−4,772	−4,016
	free	15	−5,474	−5,115	−4,767	−4,036

that for all datasets the AIC and BIC consistently select the same model, but not necessarily the one that was used for the generation of the respective dataset. It should, however, be emphasised that the model choice is sophisticated by the fact that optimal estimates for the model with four mobility classes have obviously not been identified.

Conclusion

To summarise, this simulation study showed that estimation of kinetic parameters in a FRAP experiment is possible even when a complex model with multiple mobility classes is assumed. Bayesian estimates proved to be much more promising than least squares estimates. In particular, it was not possible to determine reliable approximations of the $k_{\mathrm{on},i}$ values by least squares estimation. Moreover, the MCMC procedure applied in this section provided appropriate confidence intervals for all parameters. This was not feasible for the least squares estimates by standard procedures due to the extremely wiggly character of the target function. When combining a model and a dataset which involve different numbers of mobility classes, the average of the estimated posterior means still correctly approximated the average of the true values of the $k_{\text{off},i}$.

Model choice was carried out by application of Bayes factors, the BIC and the AIC. These rules did not always select the correct model, indicating that differences between models with different numbers of mobility classes are not substantial as long as reasonable parameters are chosen. As another criterion, one should investigate whether there are similar estimates for different $k_{\text{on},i}$ or $k_{\text{off},i}$ parameters and one could hence reduce the model by one class. In case of Bayesian estimation, overlapping confidence intervals were obtained where the model involved more classes than the dataset.

Overall, the kinetic model and estimation techniques are qualified for the statistical analysis of experimental FRAP data in Sect. 9.7. Before starting such investigation, the following section explains the preprocessing of the measurements.

9.6 Data Preparation

The previously described kinetic models start from the assumption of an idealised data situation in a sense that is particularised in what follows. In practice, this presumption is typically not met. Therefore, the raw measurements are to be normalised in an appropriate way before parameter estimation techniques are applied to the data.

This section explains three different normalisation procedures: *single normalisation* and *double normalisation* as described by Phair et al. (2004a), and *triple normalisation* as developed in Schneider et al. (2012). The single and double normalisations are specialisations of the triple normalisation; hence, we start with the presentation of the latter in Sect. 9.6.1 and then proceed with the double and single normalisations in Sects. 9.6.2 and 9.6.3, respectively. Comparisons between the three approaches are drawn in the course of this section. In the application in Sect. 9.7, all datasets are triple normalised. The impact of the triple normalisation on statistical inference as opposed to double normalisation is briefly evaluated in Sect. 9.7.3.

Throughout this section, let I_t^{T}, I_t^{B}, I_t^{U} and I_t^{bg} denote the intensities measured at time t in the total nucleus, in the bleached section, in the unbleached section and in a background area. See Fig. 9.15a for an illustration. Define $t = 0$ as the instant when the nucleus is exposed to the bleaching pulse. Consequently, negative values of t represent the time before bleaching, and positive values of t stand for the time after bleaching. In the idealised mathematical description, the bleaching by laser exposure is considered to be completed within a time interval of length zero. In practice, bleaching lasts for a short but positive time span, but this difference does not restrict the validity of the model. Figure 9.15b displays a dataset of unnormalised intensities measured in the four considered regions.

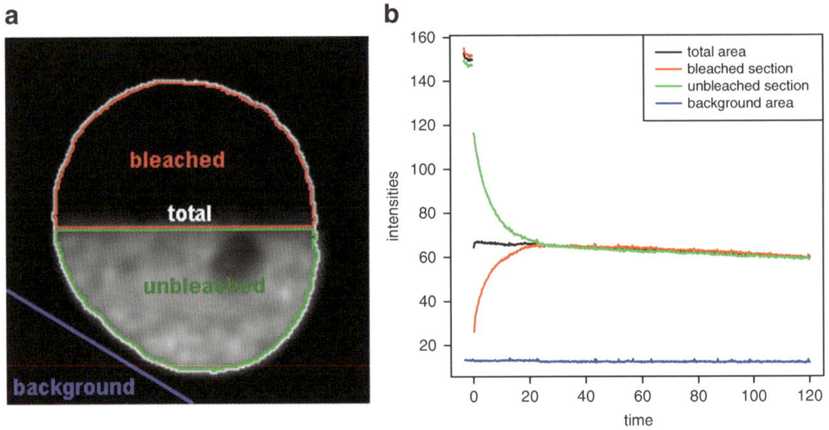

Fig. 9.15 (**a**) Illustration of the total, bleached, unbleached and background regions where the intensities I_t^{T}, I_t^{B}, I_t^{U} and I_t^{bg} are measured, respectively (Modified from Schneider 2009). (**b**) Dataset of unnormalised intensities in these four regions

9.6.1 Triple Normalisation

Briefly summarised, the triple normalisation procedure consists of five steps which correct for

(T.i) the background intensity,
(T.ii) the gain or loss of fluorescence due to natural processes and bleaching by acquisition,
(T.iii) the fact that not all proteins in the bleached section are bleached by the laser pulse,
(T.iv) the heterogeneity of structure and binding site distribution within the nucleus,
(T.v) the loss of fluorescence due to bleaching.

The above corrections are subsequently performed in the given order. They are motivated and specified in what follows. To that end, it suffices to consider the normalisation of the intensities in the total area and in the bleached section. The intensities in the unbleached section are normalised analogously to those in the bleached section, but for the sake of brevity, this is not shown here.

(T.i) Even in the absence of fluorescent proteins, the cell would not have zero intensity. This is due to read out noise of the camera and autofluorescence of the sample. The model, however, assumes that the mean grey value is zero when there are no fluorescent proteins in a considered region. Hence, subtract the background value from the measured intensities for all t:

$$I_t^{\mathrm{B}'} = I_t^{\mathrm{B}} - I_t^{\mathrm{bg}} \qquad \text{and} \qquad I_t^{\mathrm{T}'} = I_t^{\mathrm{T}} - I_t^{\mathrm{bg}}.$$

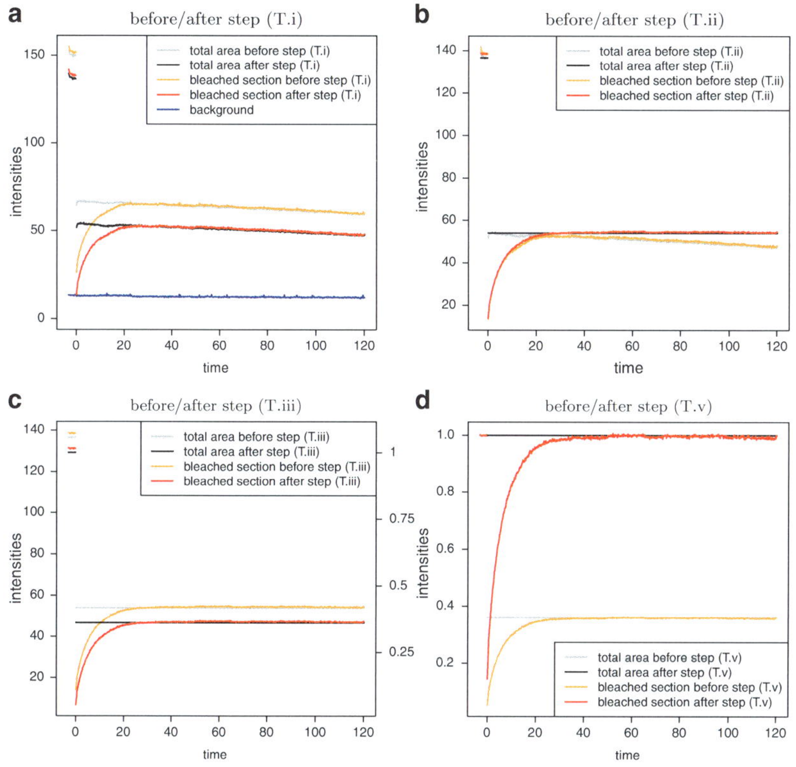

Fig. 9.16 Illustration of triple normalisation: (**a**) Intensities in the total area and bleached section before and after application of normalisation step (T.i). (**b**) Intensities before and after application of step (T.ii). (**c**) Intensities before and after application of step (T.iii). Furthermore, replacing the vertical axis on the left by the vertical axis on the right approximately corresponds to step (T.iv). (**d**) Intensities before and after application of step (T.v)

Here and in the following, each dash denotes one normalisation step that has been applied to the original variable. Figure 9.16a displays a background-subtracted dataset together with the original raw data.

(T.ii) While time elapses, there is variability in the total fluorescence due to flux of fluorescent particles into or out of the analysed cellular compartment and because of bleaching by acquisition. The model, in contrast, assumes a constant total amount of fluorescence apart from the loss due to the bleaching pulse at time zero. Therefore, all intensities are multiplied with an appropriate factor such that for $t < 0$ the total fluorescence equals some prebleach reference value $I^{\mathrm{T}'}_{\mathrm{pre}}$, and for $t > 0$ the total fluorescence equals a postbleach reference value $I^{\mathrm{T}'}_{\mathrm{post}}$:

$$I_t^{\mathrm{B}''} = \begin{cases} I_t^{\mathrm{B}'} \cdot \dfrac{I_{\mathrm{pre}}^{\mathrm{T}'}}{I_t^{\mathrm{T}'}} & \text{for } t < 0 \\[2ex] I_t^{\mathrm{B}'} \cdot \dfrac{I_{\mathrm{post}}^{\mathrm{T}'}}{I_t^{\mathrm{T}'}} & \text{for } t > 0 \end{cases} \qquad \text{and} \qquad I_t^{\mathrm{T}''} = \begin{cases} I_{\mathrm{pre}}^{\mathrm{T}'} & \text{for } t < 0 \\ I_{\mathrm{post}}^{\mathrm{T}'} & \text{for } t > 0. \end{cases}$$

Figure 9.16b illustrates the effect of this normalisation step.

The prebleach and postbleach reference values $I_{\mathrm{pre}}^{\mathrm{T}'}$ and $I_{\mathrm{post}}^{\mathrm{T}'}$ have to be chosen with care: During prebleach acquisition, an initial drop of intensity is typically observed until a steady state is reached. This pattern is due to the transition of a small fraction of GFP molecules to a non-fluorescent state (*triplet state*, Garcia-Parajo et al. 2000) and visible in Fig. 9.16a. Hence, the initial data points are discarded. To account for noise dependent intensity fluctuations, the prebleach reference value is then chosen as the mean of the last few prebleach values. In the application in this chapter, $I_{\mathrm{pre}}^{\mathrm{T}'}$ is defined as the mean of the last five background-subtracted total intensities $I_t^{\mathrm{T}'}$ before bleaching.

Directly after bleaching, one usually observes a short increase of the total intensity; compare with Fig. 9.16a. This phenomenon is due to a small fraction of molecules that have been reversibly bleached by the laser rather than irreversibly. Hence, the postbleach reference value should be chosen around the maximum intensity within a short period after bleaching. In this chapter, $I_{\mathrm{post}}^{\mathrm{T}'}$ is set equal to the mean of the background-subtracted total intensities in the 10th to 20th postbleach images.

Remark 9.1. Yet another modification is as follows: The just described initial increase of the total intensity is caused by reversibly bleached molecules that continue to fluoresce after a short interruption caused by the bleaching pulse. Directly after bleaching, these molecules are all located in the bleached section of the nucleus. Hence, the intensity curve for this section should be corrected more extensively than the total intensity curve. This holds for the time interval starting at the time τ_{0+} of the first postbleach image until a time point τ_{post} where the total intensity reaches its maximum. An appropriate rescaling is

$$I_t^{\mathrm{B}''} = I_t^{\mathrm{B}'} \cdot \frac{I_{\mathrm{post}}^{\mathrm{T}'}}{I_t^{\mathrm{T}'}} \cdot \frac{\nu(t)}{f_{\mathrm{bl}}} \qquad \text{for } t \in [\tau_{0+}, \tau_{\mathrm{post}}],$$

where $\nu : [\tau_{0+}, \tau_{\mathrm{post}}] \to [f_{\mathrm{bl}}, 1]$ is a strictly decreasing function fulfilling $\nu(\tau_{0+}) = 1$ and $\nu(\tau_{\mathrm{post}}) = f_{\mathrm{bl}}$, for example

$$\nu(t) = f_{\mathrm{bl}} \exp\left(-\log(f_{\mathrm{bl}}) \cdot \left(\frac{t - \tau_{\mathrm{post}}}{\tau_{0+} - \tau_{\mathrm{post}}} \right)^a \right)$$

for suitable $a \in \mathbb{N}$. This accounts for the progressive mixing of bleached and unbleached molecules and ensures that the intensity curve $I_t^{\mathrm{B}''}$ remains continuous at $t = \tau_{\mathrm{post}}$. In the present stage of the normalisation procedure, the value f_{bl} is yet unknown as it will be determined in Eq. (9.23) below. One might hence use an approximation of f_{bl} here as for example obtained by double normalisation or triple normalisation without the just discussed refinement. The normalisation variant described in this remark is not utilised in this book.

(T.iii) A small fraction of proteins remains unbleached though being located in the bleached section at the time of bleaching. Consequently, the variable $u_{\mathrm{bl}}^{\mathrm{bound}}$ is not zero at $t = 0^+$, the time directly after bleaching. Moreover, the value of $u_{\mathrm{bl}}^{\mathrm{bound}}$ at this time point differs in each experiment. In order to correct for this, subtract $I_{0+}^{\mathrm{B}''}$ from all intensities. The value $I_{0+}^{\mathrm{B}''}$ is, however, unknown due to the rapid invasion of unbleached free proteins into the bleached section. Hence, let $\hat{b}_0$ be an estimate of $I_{0+}^{\mathrm{B}''}$. For the datasets considered in this chapter, $\hat{b}_0$ is measured in an appropriate subregion of the bleached area, distral to the bleach border, in the first postbleach picture. The result is

$$I_t^{\mathrm{B}'''} = I_t^{\mathrm{B}''} - \hat{b}_0 \qquad \text{and} \qquad I_t^{\mathrm{T}'''} = I_t^{\mathrm{T}''} - \hat{b}_0$$

for all t. Figure 9.16c shows the changes in the data caused by this normalisation step. It is important that it is carried out after the correction for the loss of fluorescence in step (T.ii) because this loss also affects the proteins that escaped the laser. In particular, the estimate $\hat{b}_0$ has to be obtained from the data which is already corrected according to (T.i) and (T.ii).

(T.iv) Due to structural heterogeneity in the cell nucleus, caused for example by localised binding site clusters, the mean fluorescence in the bleached and unbleached sections may differ even before bleaching. As a consequence, their values also deviate from the intensity in the total area. The model, on the other hand, assumes homogeneity. Hence, modify all intensities such that their average levels before bleaching equal one, i.e.

$$I_t^{\mathrm{B}''''} = \frac{I_t^{\mathrm{B}'''}}{I_{\mathrm{pre}}^{\mathrm{B}'''}} \qquad \text{and} \qquad I_t^{\mathrm{T}''''} = \frac{I_t^{\mathrm{T}'''}}{I_{\mathrm{pre}}^{\mathrm{T}'''}}$$

for all t. As in step (T.ii), the reference values $I_{\mathrm{pre}}^{\mathrm{B}'''}$ and $I_{\mathrm{pre}}^{\mathrm{T}'''}$ are typically chosen to be the mean of the last few prebleach values $I_t^{\mathrm{B}'''}$ and $I_t^{\mathrm{T}'''}$, respectively. In Fig. 9.16c, this step approximately corresponds to replacing the vertical axis on the left by the vertical axis on the right, where $I_{\mathrm{pre}}^{\mathrm{T}'''}$ on the left corresponds to the value one on the right.

(T.v) The bleaching pulse abruptly decreases the fluorescence of the nucleus, but the model assumes the total intensity being one throughout the experiment.

Hence, normalise as follows:

$$I_t^{\mathrm{B}'''''} = \frac{I_t^{\mathrm{B}''''}}{I_t^{\mathrm{T}''''}} \qquad \text{and} \qquad I_t^{\mathrm{T}'''''} = 1$$

for all t. This step is illustrated in Fig. 9.16d.

Altogether, one has for $t > 0$

$$I_t^{\mathrm{B}'''''} = \frac{(I_t^{\mathrm{B}} - I_t^{\mathrm{bg}}) \cdot \dfrac{I_{\mathrm{post}}^{\mathrm{T}} - I_{\mathrm{post}}^{\mathrm{bg}}}{I_t^{\mathrm{T}} - I_t^{\mathrm{bg}}} - \hat{b}_0}{I_{\mathrm{pre}}^{\mathrm{B}''} - \hat{b}_0} \cdot \frac{I_{\mathrm{pre}}^{\mathrm{T}} - I_{\mathrm{pre}}^{\mathrm{bg}} - \hat{b}_0}{I_{\mathrm{post}}^{\mathrm{T}} - I_{\mathrm{post}}^{\mathrm{bg}} - \hat{b}_0} \, .$$

The above considerations also make clear how to determine the size f_{bl} of the bleached section: The average postbleach level of total fluorescence after step (T.iv) equals $I_{\mathrm{post}}^{\mathrm{T}''''}$. The only adjustment remaining to be done at that point is to correct for the intentional loss of fluorescence caused by the bleaching laser pulse. Consequently, $1 - f_{\mathrm{bl}}$ corresponds to this level, i.e. one sets

$$1 - f_{\mathrm{bl}} = I_{\mathrm{post}}^{\mathrm{T}''''} = \frac{I_{\mathrm{post}}^{\mathrm{T}'''}}{I_{\mathrm{pre}}^{\mathrm{T}'''}} \, . \tag{9.23}$$

In the subsequent application in Sect. 9.7, estimation techniques are applied to triple normalised datasets, and the value f_{bl} is determined according to (9.23).

9.6.2 *Double Normalisation*

A simplification of the triple normalisation in the previous section is the double normalisation as described for example by Phair et al. (2004a); see also McNally (2008). This procedure is described here for the sake of completeness and because a comparison of estimation results based on triple normalised and double normalised data is carried out in Sect. 9.7.3. In short, double normalisation corrects for

(D.i) the background intensity as in step (T.i),
(D.ii) the heterogeneity of structure and binding site distribution within the nucleus as in step (T.iv),
(D.iii) the loss of fluorescence due to bleaching and the gain or loss due to natural processes as in steps (T.ii) and (T.v).

The order of these items slightly differs from that in the triple normalisation, but this does not change the outcome as steps (D.ii) and (D.iii) both consist of multiplicative operations. All corrections are motivated as in the triple normalisation. Hence, the following description of the double normalisation merely displays the respective formulas. For more details, turn back to Sect. 9.6.1.

(D.i) Subtract the background intensity from all measured values for all t:

$$I_t^{\mathrm{B}'} = I_t^{\mathrm{B}} - I_t^{\mathrm{bg}} \qquad \text{and} \qquad I_t^{\mathrm{T}'} = I_t^{\mathrm{T}} - I_t^{\mathrm{bg}}.$$

(D.ii) Modify all intensities such that their average levels before bleaching equal one, i.e.

$$I_t^{\mathrm{B}''} = \frac{I_t^{\mathrm{B}'}}{I_{\mathrm{pre}}^{\mathrm{B}'}} \qquad \text{and} \qquad I_t^{\mathrm{T}''} = \frac{I_t^{\mathrm{T}'}}{I_{\mathrm{pre}}^{\mathrm{T}'}}.$$

As before, the prebleach reference values $I_{\mathrm{pre}}^{\mathrm{B}'}$ and $I_{\mathrm{pre}}^{\mathrm{T}'}$ are typically chosen as the mean of the last few prebleach intensities $I_t^{\mathrm{B}'}$ and $I_t^{\mathrm{T}'}$, respectively.

(D.iii) Scale all intensities such that the total fluorescence is equal to one for *all* t, i.e.

$$I_t^{\mathrm{B}'''} = \frac{I_t^{\mathrm{B}''}}{I_t^{\mathrm{T}''}} \qquad \text{and} \qquad I_t^{\mathrm{T}'''} = 1.$$

Altogether, one has

$$I_t^{\mathrm{B}'''} = \frac{(I_t^{\mathrm{B}} - I_t^{\mathrm{bg}})(I_{\mathrm{pre}}^{\mathrm{T}} - I_{\mathrm{pre}}^{\mathrm{bg}})}{(I_{\mathrm{pre}}^{\mathrm{B}} - I_{\mathrm{pre}}^{\mathrm{bg}})(I_t^{\mathrm{T}} - I_t^{\mathrm{bg}})}.$$

Similarly as for the triple normalised data, the above details indicate how to determine the size f_{bl} of the bleached section from the measured intensities: Let $I_{\mathrm{post}}^{\mathrm{T}''}$ denote an appropriate reference value for the postbleach level of total fluorescence. This may for example be the mean of $I_t^{\mathrm{T}''}$ as determined from the 10th to 20th postbleach images. Then, this reference level corresponds to $1-f_{\mathrm{bl}}$, i.e. one sets

$$1 - f_{\mathrm{bl}} = I_{\mathrm{post}}^{\mathrm{T}''} = \frac{I_{\mathrm{post}}^{\mathrm{T}'}}{I_{\mathrm{pre}}^{\mathrm{T}'}}.$$

Figure 9.17 displays a time series of the intensity measured in the bleached region, modified according to the double and triple normalisation and the single normalisation explained in the next section. The curves for the double and triple normalisation especially differ during the recovery phase until $t \approx 20$. The first postbleach intensities are 0.258 after double normalisation and 0.051 after triple normalisation. The estimates for f_{bl} are 0.605 and 0.663, respectively. Corresponding values for other datasets are listed in Table C.1 in the Appendix.

9.6.3 Single Normalisation

This section explains the single normalisation according to Phair et al. (2004a), which corrects for

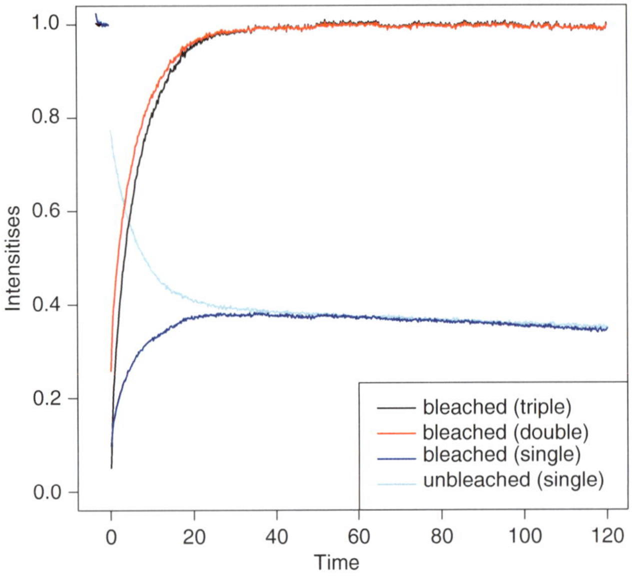

Fig. 9.17 Comparison of single, double and triple normalised data: The plot displays intensities measured in the bleached region, modified according to the single (*dark blue*), double (*red*) and triple (*black*) normalisation. The single normalised curve does not level off around one but equalises with the single normalised intensity in the unbleached region (*light blue*). The double and triple normalised curves especially differ during the recovery phase until $t \approx 20$. The first postbleach intensities are 0.258 after double normalisation and 0.051 after triple normalisation. The estimates for f_{bl} are 0.605 and 0.663, respectively

(S.i) the background intensity as in step (T.i),
(S.ii) the heterogeneity of structure and binding site distribution within the nucleus as in step (T.iv).

In contrast to the double and triple normalisations, the single normalisation considers the intensities in the bleached and unbleached sections rather than the measurements in the bleached and total compartments. The modifications are described in what follows. Once more, the reader is referred to Sect. 9.6.1 for a detailed explanation.

(S.i) Subtract the background intensity from the all measured values for all t:

$$I_t^{\mathrm{B}'} = I_t^{\mathrm{B}} - I_t^{\mathrm{bg}} \qquad \text{and} \qquad I_t^{\mathrm{U}'} = I_t^{\mathrm{U}} - I_t^{\mathrm{bg}}.$$

(S.ii) Modify the intensities in the bleached and unbleached section such that the average levels before bleaching equal one, i.e.

$$I_t^{\mathrm{B}''} = \frac{I_t^{\mathrm{B}'}}{I_{\mathrm{pre}}^{\mathrm{B}'}} \qquad \text{and} \qquad I_t^{\mathrm{U}''} = \frac{I_t^{\mathrm{U}'}}{I_{\mathrm{pre}}^{\mathrm{U}'}} .$$

To that end, the reference values $I_{\text{pre}}^{\text{B}'}$ and $I_{\text{pre}}^{\text{U}'}$ are determined as the mean of the last few prebleach values of $I_t^{\text{B}'}$ and $I_t^{\text{U}'}$, respectively.

Altogether, one has

$$I_t^{\text{B}''} = \frac{I_t^{\text{B}} - I_t^{\text{bg}}}{I_{\text{pre}}^{\text{B}} - I_{\text{pre}}^{\text{bg}}} \qquad \text{and} \qquad I_t^{\text{U}''} = \frac{I_t^{\text{U}} - I_t^{\text{bg}}}{I_{\text{pre}}^{\text{U}} - I_{\text{pre}}^{\text{bg}}} .$$

Other than double and triple normalisation, single normalisation does not correct for the intentional and unintentional loss of fluorescence over time, i.e. it does not scale the data such that the average postbleach level in the total area equals one. Hence, the data still contains information about the fraction of bleached molecules. The kinetic models in this chapter assume that all intensities are scaled such that the value in the bleached compartment will eventually level off around one. Thus, the models are not directly applicable to single normalised datasets.

Figure 9.17 displays a single normalised time series of intensities measured in the bleached and unbleached section of a nucleus.

In the following application in Sect. 9.7, all datasets are triple normalised as this technique is the most accurate one. For comparison purposes, statistical inference is also carried out for double normalised data in Sect. 9.7.3.

9.7 Application

This section analyses the kinetic behaviour of the protein Dnmt1, which was introduced in Sect. 9.1, based on the observations from FRAP experiments (cf. Sect. 9.1.1).

In Sects. 9.3–9.5, dedicated stochastic and deterministic models were constructed, and the performances of suitable estimation techniques were evaluated for these models in several simulation studies. Section 9.6 investigated how to appropriately process raw measurements from FRAP experiments such that the considered inference methods can be applied to the resulting time series. With the tools from Sects. 9.3–9.6 at hand, the present section deals with the investigation of the research questions presented in Sect. 9.1.2. These concern the estimates of the model parameters and the number of mobility classes, both depending on the phase of the cell cycle.

Section 9.7.1 describes the datasets that are available for statistical inference. In Sects. 9.7.2 and 9.7.3, Bayesian and least squares estimation in carried out. Section 9.7.4 concludes.

9.7.1 Data

There is a number of measurements from FRAP experiments available for the statistical investigation of the research questions of this chapter. The data has been acquired in the context of a diploma thesis (Schneider 2009) as described in Sect. 9.1.1. A protocol of the experimental setup is provided in that thesis.

In order to enable a cell cycle dependent analysis of the dynamic behaviour of Dnmt1, data has been collected during different phases of the cycle. In particular, there are 10 time series from G1 phase, 26 series from early S phase and 11 from late S phase. Each time series contains the measured intensities q^* in the bleached section of the nucleus over time, but no information on any other components of the multi-dimensional processes is provided. Unless otherwise stated, the data is triple normalised as described in Sect. 9.6.1. Figure 9.3 on p. 309 displays the 47 normalised recovery curves.

The observation times for the datasets are given by the exposure time of the laser plus a delay time such that data is available at equidistant time intervals of length 0.154. The number of measurements in each time series is typically around 780; exact numbers are listed in Table C.1 on p. 418 in the Appendix. The same tables also display the starting values q_0^* of the recovery curves and experimentally determined values of the bleached fraction f_{bl} and the intermediate fraction f_{int}.

9.7.2 Bayesian Estimation

In the following, the kinetic models described in this chapter are estimated based on the 47 provided datasets. The proceeding is as in the previous simulation studies in this chapter, especially as in Sect. 9.5.4. Estimation based on the diffusion model requires an approximation of the number N of molecules in the nucleus. This number has not been determined experimentally, and hence a statistical approximation is proposed in what follows. Estimation results for the parameters are presented afterwards.

Numbers of Molecules

The strength of random fluctuations in the stochastic model is controlled by the factor N_U^{-1} in the diffusion matrix, where $N_U = (1 - f_{\text{bl}})N$ is the number of unbleached molecules in the nucleus: Larger numbers of molecules correspond to a smaller impact of stochasticity. It has been demonstrated in Sect. 9.3.4 that wrong assumptions about N_U may cause wrong estimates for k_{off} and k_{on}. One hence requires a careful approximation of N_U.

In the following, the number N_U is extracted from the measured fluorescence intensities as follows: On the one hand, in the simplest case of one mobility class, the entry of the diffusion matrix corresponding to the fluctuations of q^* equals

$$\frac{1}{N_U}\left(k_{\text{off}}\left(f_{\text{bl}}^{-1}-2\right)q^* + (k_{\text{on}} - k_{\text{off}})\left(f_{\text{bl}}^{-1}-1\right)u^{\text{free}} + k_{\text{off}}\right) \tag{9.24}$$

as derived in Sect. 9.4.2. On the other hand, for observations $q^*_{t_0}, \ldots, q^*_{t_n}$ at times $t_0, \ldots, t_n$, this part of the diffusion matrix can be approximated empirically by

$$\frac{1}{n}\sum_{k=1}^{n}\frac{\left(q^*_{t_k} - q^*_{t_{k-1}}\right)^2}{t_k - t_{k-1}}, \tag{9.25}$$

which is motivated by Eq. (3.17) on p. 40. An estimate for (9.24) can be obtained when focusing on those measurements of q^* after the intensity has reached a stable plateau. Then the variable q^* in (9.24) may be replaced by the mean of all values on this plateau. f^*_{bl} can be obtained from image analysis, and k_{on}, k_{off} and u^{free} can be estimated by application of the deterministic techniques as discussed in the previous sections. Calculation of (9.25) should base on the same set of observations as (9.24) does. Equating (9.24) and (9.25) then gives an estimate of $N_U = (1-f_{\text{bl}})N$.

Applied to the synthetic datasets from Sects. 9.3.4–9.5.4, the just described procedure approximates the number of molecules surprisingly well, yielding values that deviate from the true value $N = 10{,}000$ by less than 2 %.

Table C.1 in the Appendix contains approximations of N for the real datasets considered in this section. In practice, numbers of molecules typically lie between 10,000 and 100,000 per nucleus (Phair et al. 2004b). In Table C.1, notably smaller numbers appear for some time series. The variation in the approximations is already apparent from the recovery curves, see Fig. C.1 on p. 416 for an example. Small values of N are most probably caused by measurement noise that is not corrected for by the data normalisation presented in Sect. 9.6. Hence, these numbers do not really represent the amounts of molecules in the nucleus but rather a lower bound. They however reflect the strength of fluctuations in the respective time series, and hence these values are employed for the subsequent Bayesian inference procedures.

Results

The innovation scheme is applied to the FRAP data with all settings as specified in Sect. 9.5.4. In particular, for each time series the model with $M \in \{1, 2, 3\}$ mobility classes is estimated. Table 9.10 exemplarily shows the results for two selected datasets. More specifically, it presents the estimated posterior means and posterior 95 %-hpd intervals for $k_{\text{on},1}, \ldots, k_{\text{on},M}$ and $k_{\text{off},1}, \ldots, k_{\text{off},M}$. Furthermore, the table displays the logarithms of the marginal likelihoods for each model as introduced in Eq. (9.19) on p. 344. Carrying out model choice by application of

Table 9.10 Bayesian estimation results for two selected real datasets. The first two columns specify the phase of the cell cycle and an index labelling the time series. Columns four to six list the posterior means and 95 %-hpd intervals for the parameters defined in the third column. Moreover, they contain approximated logarithms of the marginal likelihoods $\log \pi(\boldsymbol{Y}^{\text{obs}}|\mathcal{M}_M)$ for each model $\mathcal{M}_M$, $M \in \{1,2,3\}$. These can be used for model choice by means of Bayes factors; the respective selected models are shown in the last column. For the upper dataset, the selected model $\mathcal{M}_3$ however estimates overlapping confidence intervals for $k_{\text{on},1}, k_{\text{on},2}$ and $k_{\text{off},1}, k_{\text{off},2}$. As discussed in the main text, one might hence exclude $\mathcal{M}_3$ and choose $\mathcal{M}_1$ instead

			Model			Chosen
Phase	Index	Parameter	$M = 1$	$M = 2$	$M = 3$	model
G1	1	$k_{\text{on},1}$	2.86 (2.51, 3.19)	0.90 (0.74, 1.06)	1.31 (1.06, 1.61)	
		$k_{\text{on},2}$	–	2.20 (1.99, 2.46)	1.03 (0.85, 1.25)	
		$k_{\text{on},3}$	–	–	2.25 (2.01, 2.49)	
		$k_{\text{off},1}$	0.20 (0.19, 0.21)	0.37 (0.32, 0.40)	0.18 (0.15, 0.20)	
		$k_{\text{off},2}$	–	0.13 (0.11, 0.15)	0.16 (0.14, 0.18)	
		$k_{\text{off},3}$	–	–	0.13 (0.12, 0.14)	
		$\log \pi(\boldsymbol{Y}^{\text{obs}}\|\mathcal{M}_M)$	5969.52	5646.06	7044.44	$\mathcal{M}_3$ ($\mathcal{M}_1$)
late S	3	$k_{\text{on},1}$	2.31 (1.93, 2.72)	1.45 (1.09, 1.82)	2.43 (2.13, 2.72)	
		$k_{\text{on},2}$	–	2.14 (1.74, 2.52)	1.55 (1.31, 1.87)	
		$k_{\text{on},3}$	–	–	1.73 (1.39, 2.04)	
		$k_{\text{off},1}$	0.09 (0.08, 0.10)	0.09 (0.08, 0.11)	0.17 (0.15, 0.20)	
		$k_{\text{off},2}$	–	0.07 (0.06, 0.08)	0.09 (0.08, 0.11)	
		$k_{\text{off},3}$	–	–	0.04 (0.03, 0.04)	
		$\log \pi(\boldsymbol{Y}^{\text{obs}}\|\mathcal{M}_M)$	15019.27	5219.35	4965.65	$\mathcal{M}_1$

Bayes factors, one will clearly favour the model $\mathcal{M}_3$ with three classes for the upper dataset and the model $\mathcal{M}_1$ for the lower dataset. However, considering the confidence intervals for the upper dataset and $M = 3$, one notices that these intervals overlap for both $k_{\text{on},1}, k_{\text{on},2}$ and for $k_{\text{off},1}, k_{\text{off},2}$. Hence, the estimated model may be reduced to two mobility classes and hence be excluded from the range of appropriate models. In that case, again utilising the Bayes factor, one would favour $\mathcal{M}_1$.

For a more concise representation, Fig. 9.18 plots the estimated 95 %-hpd intervals for a number of arbitrarily selected datasets. There are obviously several time series where the confidence intervals overlap for some parameters. Table 9.11 on p. 361 lists the approximated logarithm of the marginal likelihoods for each of the models and datasets. The table furthermore indicates for $M = 2$ and $M = 3$ the numbers of distinctly estimated association and dissociation parameters, derived from potential intersections of the confidence intervals displayed in Fig. 9.18. When taking this criterion into account, the model choice obtained through Bayes factors is influenced only in two cases.

The representation of confidence intervals in Fig. 9.18 allows a direct comparison of the locations of the intervals. A cell cycle dependent impact is especially obvious for the parameter k_{off} in the model $\mathcal{M}_1$, indicating that molecules remain in the bound state for a longer time period during G1 phase than in S phase.

9.7.3 Least Squares Estimation

In the following, the results of the least squares estimation are presented. The estimation procedure is as in the simulation study in Sect. 9.5.4.

Figure 9.19 displays the least squares estimates for $k_{\text{off},1}, \ldots, k_{\text{off},M}$, based on the triple normalised datasets with $f_{\text{int}} = 0$ and fixed starting value q_0^* as displayed in Table C.1 on p. 418 in the Appendix. These parameters are estimated for the kinetic model with $M \in \{1, \ldots, 4\}$ mobility classes. The BIC, which was introduced in Eq. (9.22) on p. 345, is used to select the model with the most appropriate number of classes; results are listed in Table 9.12. This model choice is also visible in Fig. 9.19, where the estimates for $k_{\text{off},i}$ in a selected model are marked with a cross, and with a circle otherwise. Figure C.3 in the Appendix presents the fittings of the estimated to the observed recovery curve for one particular dataset and $M \in \{1, \ldots, 4\}$.

Concerning the estimated values of $k_{\text{off},1}, \ldots, k_{\text{off},M}$, there is obviously a difference between the phases of the cell cycle when considering the model with $M = 1$ mobility class, especially between G1 phase and the two S phases. This difference becomes less apparent for $M = 2$ and seems to disappear for $M \in \{3, 4\}$. For none of the datasets, the BIC chooses the model with one mobility class. For G1 phase, it typically selects $M = 2$ or $M = 3$, and for early S phase and late S phase the BIC mostly distinguishes $M = 3$ classes.

In the previous sections, different variants of the above used datasets and estimation settings were discussed: First, the triple normalisation from Sect. 9.6.1 could be replaced by the double normalisation from Sect. 9.6.2. Second, the intermediate fraction f_{int} was introduced as a correction factor in Sect. 9.4. This variable could be set equal to experimentally obtained values as listed in Table C.1 in the Appendix. Third, the starting value q_0^* of the recovery curve could be fitted by least squares estimation instead of being kept fixed to the first observed value.

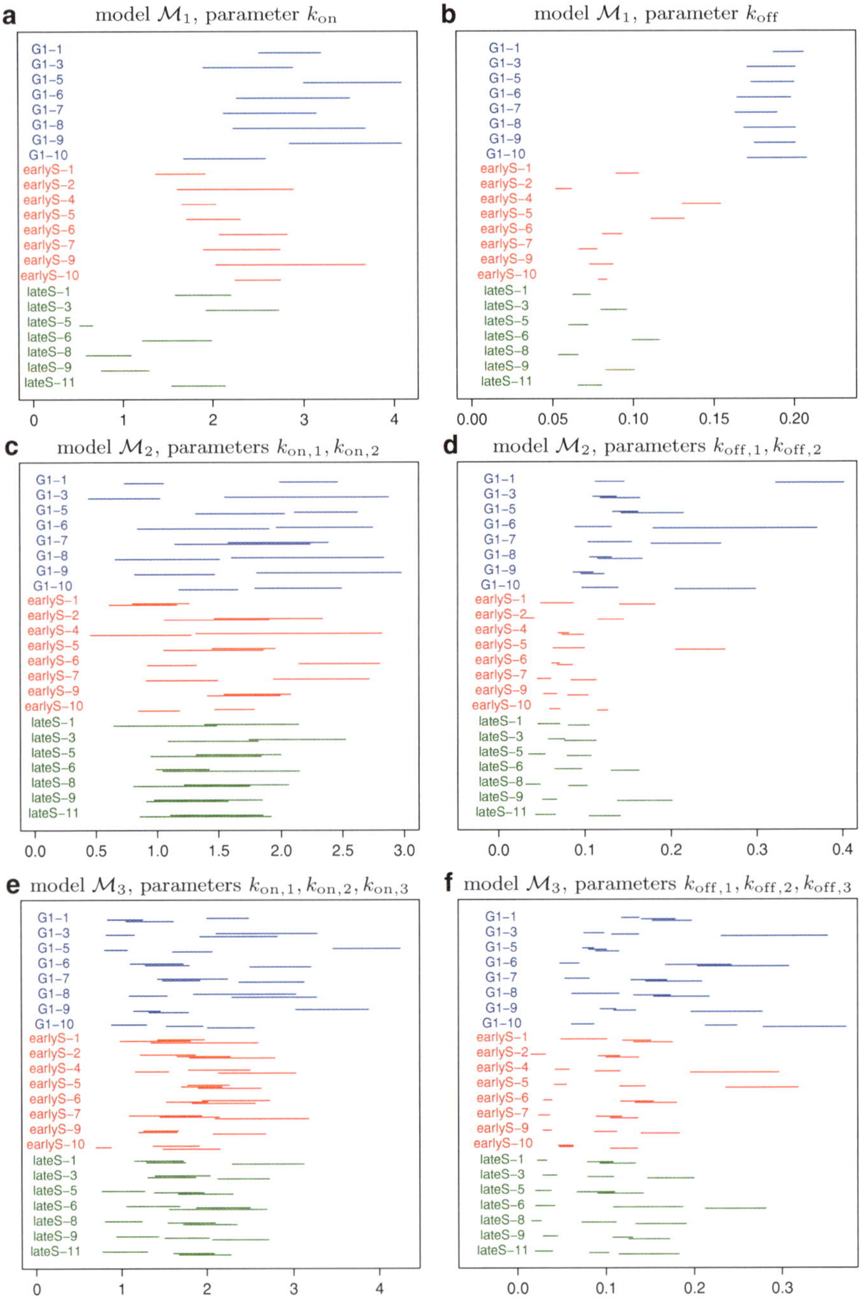

Fig. 9.18 Estimated 95 %-hpd intervals for the parameters $k_{\text{on},1}, \ldots, k_{\text{on},M}$ and $k_{\text{off},1}, \ldots, k_{\text{off},M}$ in the kinetic models with $M \in \{1, 2, 3\}$ mobility classes. The labels on the left indicate the phase of the cell cycle and an index marking the time series. (**a**) model $\mathcal{M}_1$, parameter k_{on} (**b**) model $\mathcal{M}_1$, parameter k_{off} (**c**) model $\mathcal{M}_2$, parameters $k_{\text{on},1}, k_{\text{on},2}$ (**d**) model $\mathcal{M}_2$, parameters $k_{\text{off},1}, k_{\text{off},2}$ (**e**) model $\mathcal{M}_3$, parameters $k_{\text{on},1}, k_{\text{on},2}, k_{\text{on},3}$ (**f**) model $\mathcal{M}_3$, parameters $k_{\text{off},1}, k_{\text{off},2}, k_{\text{off},3}$

Table 9.11 Bayesian estimation results for real datasets. The first two columns specify the phase of the cell cycle and an index labelling the time series. Columns three to five contain the approximated logarithms of the marginal likelihoods $\log \pi(\boldsymbol{Y}^{\text{obs}}|\mathcal{M}_M)$ for each model $\mathcal{M}_M$, $M \in \{1, 2, 3\}$. These can be used for model choice by means of Bayes factors; the respective selected models are shown in the last column. As discussed in the main text, however, some of the estimated confidence intervals overlap, and hence the according model may be reduced to one with fewer classes. Columns six to nine display the number of distinctly estimated association and dissociation parameters, derived from potential intersections of the confidence intervals displayed in Fig. 9.18. A model might be rejected when there are non-disjoint confidence intervals for both the association and dissociation parameters. Taking this criterion into account, the model choice obtained through Bayes factors is influenced only in two cases. The alternatively selected models are shown in parentheses in the last column

		$\log \pi(\boldsymbol{Y}^{\text{obs}}\|\mathcal{M}_M)$			Estimated class numbers				
					$\mathcal{M}_2$		$\mathcal{M}_3$		
Phase	Index	$M=1$	$M=2$	$M=3$	$k_{\text{on},i}$	$k_{\text{off},i}$	$k_{\text{on},i}$	$k_{\text{off},i}$	Model choice
G1	1	5969.5	5646.1	7044.4	2	2	2	2	$\mathcal{M}_3$ ($\mathcal{M}_1$)
G1	3	14699.1	15215.1	15050.2	2	1	2	3	$\mathcal{M}_2$
G1	5	15539.5	15954.3	17271.4	2	1	3	2	$\mathcal{M}_3$
G1	6	13050.5	13455.5	4455.2	2	2	2	2	$\mathcal{M}_2$
G1	7	14818.0	4828.1	5086.3	1	2	2	2	$\mathcal{M}_1$
G1	8	13892.1	14537.6	14415.8	2	1	2	2	$\mathcal{M}_2$
G1	9	15289.7	17004.1	16125.3	2	1	2	2	$\mathcal{M}_2$
G1	10	12140.4	12266.9	4067.0	2	2	3	3	$\mathcal{M}_2$
early S	1	16352.3	16598.5	15850.7	1	2	1	2	$\mathcal{M}_2$
early S	2	16372.4	15999.4	16271.4	1	2	1	2	$\mathcal{M}_1$
early S	4	14956.2	15968.9	4988.7	2	1	2	3	$\mathcal{M}_2$
early S	5	14895.9	14717.3	4797.9	1	2	1	3	$\mathcal{M}_1$
early S	6	16381.8	5580.6	5318.9	2	1	1	2	$\mathcal{M}_1$
early S	7	16050.2	5292.4	16114.2	2	2	2	2	$\mathcal{M}_3$ ($\mathcal{M}_1$)
early S	9	5091.5	5093.5	4962.5	1	2	2	3	$\mathcal{M}_2$
early S	10	13458.9	−658.3	1081.3	2	2	2	2	$\mathcal{M}_1$
late S	1	15689.3	15574.0	5486.4	1	2	2	2	$\mathcal{M}_1$
late S	3	15019.3	5219.4	4965.6	1	1	2	3	$\mathcal{M}_1$
late S	5	15553.0	5271.2	5203.9	1	2	2	2	$\mathcal{M}_1$
late S	6	15583.3	5122.6	5301.2	1	2	2	3	$\mathcal{M}_1$
late S	8	15723.1	5306.6	5241.6	1	2	2	3	$\mathcal{M}_1$
late S	9	14486.3	4764.0	4949.4	1	2	3	2	$\mathcal{M}_1$
late S	11	14987.6	4962.2	4966.6	1	2	2	3	$\mathcal{M}_1$

Figures C.4–C.6 on pp. 421 in the Appendix show the changed estimates for $k_{\text{off},1}, \ldots, k_{\text{off},M}$ due to these three modifications. Table C.2 displays the numbers of mobility classes chosen by the BIC with respect to these changes. It turns out that the largest impact on all outcomes originates in the third modification, where the starting value of the FRAP curve is estimated by least squares. Furthermore, for all modifications, deviations are more apparent for $M \in \{3, 4\}$ than for $M \in \{1, 2\}$,

Table 9.12 BIC as defined in Eq. (9.22) for the least squares estimates from Fig. 9.19. The first and second columns specify the phase of the cell cycle and a consecutive index for each dataset. The next four columns list the BIC as defined in Eq. (9.22) on p. 345 for the kinetic models with $M \in \{1, \ldots, 4\}$ mobility classes. The last column states the number of classes that is chosen by the BIC

		BIC				
Phase	Index	$M = 1$	$M = 2$	$M = 3$	$M = 4$	Chosen model
G1	1	−3,455	−4,225	−4,201	−4,136	2
G1	2	−7,798	−9,456	−9,603	−9,577	3
G1	3	−6,100	−7,965	−7,916	−7,846	2
G1	4	−3,512	−4,389	−4,369	−4,316	2
G1	5	−5,688	−8,340	−8,433	−8,287	3
G1	6	−6,299	−7,199	−7,216	−7,163	3
G1	7	−5,875	−7,965	−8,118	−7,959	3
G1	8	−6,092	−7,166	−7,302	−7,307	4
G1	9	−6,571	−8,304	−8,178	−8,009	2
G1	10	−6,096	−6,875	−6,848	−6,800	2
early S	1	−5,126	−8,504	−8,686	−8,602	3
early S	2	−4,743	−7,819	−7,812	−7,785	2
early S	3	−5,657	−7,392	−7,366	−7,307	2
early S	4	−5,480	−7,157	−7,434	−7,296	3
early S	5	−5,177	−7,747	−7,986	−7,935	3
early S	6	−4,822	−8,236	−8,478	−8,325	3
early S	7	−4,552	−8,174	−8,391	−8,287	3
early S	8	−5,086	−7,747	−7,756	−7,729	3
early S	9	−5,089	−7,760	−7,734	−7,689	2
early S	10	−4,934	−9,091	−9,430	−9,261	3
early S	11	−5,805	−8,378	−9,168	−9,037	3
early S	12	−5,037	−7,485	−7,623	−7,581	3
early S	13	−4,448	−7,461	−7,774	−7,747	3
early S	14	−5,028	−6,677	−6,789	−6,773	3
early S	15	−5,154	−7,605	−7,578	−7,470	2
early S	16	−4,565	−8,174	−8,313	−8,216	3
early S	17	−5,887	−8,551	−9,168	−8,864	3
early S	18	−5,741	−7,899	−8,575	−8,407	3
early S	19	−4,903	−6,990	−7,207	−6,905	3
early S	20	−4,589	−6,910	−7,084	−7,137	4
early S	21	−5,452	−6,984	−7,432	−7,289	3
early S	22	−5,637	−6,760	−6,907	−6,880	3
early S	23	−6,571	−7,942	−8,089	−8,009	3
early S	24	−5,446	−7,510	−7,852	−7,677	3
early S	25	−5,333	−7,350	−7,593	−7,538	3
early S	26	−4,817	−7,650	−7,738	−7,711	3
late S	1	−5,198	−8,155	−8,362	−8,261	3
late S	2	−5,112	−7,200	−7,420	−7,316	3
late S	3	−4,846	−7,127	−7,432	−7,275	3
late S	4	−5,127	−6,986	−7,292	−7,338	4
late S	5	−5,100	−7,691	−7,904	−7,856	3
late S	6	−5,072	−7,691	−8,128	−7,994	3

(continued)

Table 9.12 (continued)

Phase	Index	BIC $M = 1$	$M = 2$	$M = 3$	$M = 4$	Chosen model
late S	7	−5,005	−7,707	−7,949	−7,687	3
late S	8	−5,302	−7,829	−8,046	−7,969	3
late S	9	−4,807	−7,319	−7,468	−7,394	3
late S	10	−4,777	−7,200	−7,375	−7,338	3
late S	11	−4,718	−7,754	−7,884	−7,756	3

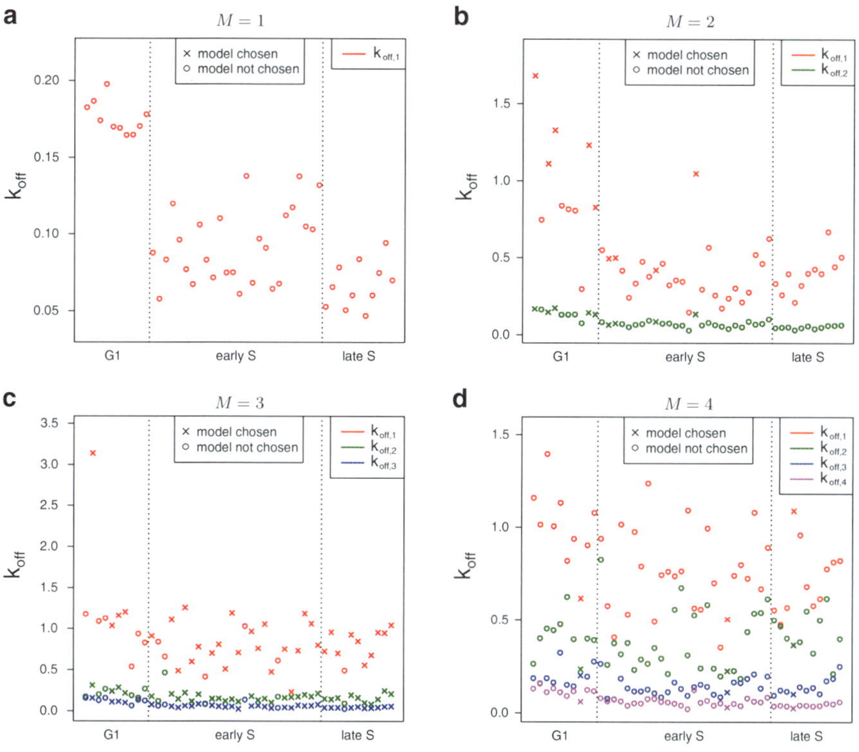

Fig. 9.19 Least squares estimates for $k_{\text{off},i}$, based on the triple normalised datasets with intermediate fraction $f_{\text{int}} = 0$ and fixed starting value q_0^* as displayed in Table C.1. (**a**)–(**d**) display the estimates for the parameters $k_{\text{off},1}, \ldots, k_{\text{off},M}$ in the deterministic kinetic model with $M = 1, \ldots, 4$ mobility classes. In each plot, the distinct time series are ordered according to their phase and index as in Table 9.12, and the respective results are presented from the left to the right. If a model is chosen by the BIC as listed in Table 9.12, the respective estimates are represented by a *cross*; otherwise, they are marked by a *circle*

but this may be due to a generally larger variability in the estimates for larger M. In practice, the third variation, i.e. estimation of q_0^*, is probably not eligible as it seems more important to find good agreement between the estimated and the observed curve for the initial phase rather than for the final phase of recovery.

9.7.4 *Conclusion*

The objective of this section was to investigate the research questions formulated in Sect. 9.1.2 on the cell cycle dependent binding behaviour of Dnmt1. To that end, the kinetic models and estimation techniques from Sects. 9.3–9.5 were applied to real datasets obtained from FRAP experiments.

Statistical inference on the model parameters was carried out both by Bayesian and least squares estimation. Resulting estimates are presented in the tables and figures in this section. The simulation studies in the previous sections demonstrated that precise estimation is possible especially for the dissociation rates $k_{\text{off},i}$, with more reliable outcomes being produced by the Bayesian techniques.

Cell cycle dependent differences in parameter estimates were especially observed between G1 phase and the two S phases: Both the Bayesian and the least squares procedure produce estimates for the dissociation rates which tend to be larger than those in S phase. This difference is obvious for $M = 1$ from the graphics in Figs. 9.18 and 9.19 on pp. 360 and 363.

Concerning the numbers of mobility classes in the three cell cycles, the considered model choice criteria yield contradictory statements: While the Bayes factor tends to choose more mobility classes in G1 phase than in early or late S phase, the BIC behaves the other way round. Model choice already proved to be problematic in the simulation studies carried out earlier in this chapter. Apart from that, a possible explanation for the indefinite outcomes is that one might have diffusion-coupled rather than diffusion-uncoupled FRAP for Dnmt1. This idea is pursued in ongoing work Schneider et al. (2012). The kinetic model is briefly considered in the following section.

9.8 Diffusion-Coupled FRAP

The role of diffusion of molecules in fluorescence recovery has been discussed in Sect. 9.2.2, where one has distinguished between diffusion-coupled and diffusion-uncoupled FRAP. Throughout this chapter, diffusion-uncoupled dynamics has been assumed for the construction of all kinetic models, because this scenario turned out to be eligible in control experiments. For the sake of completeness, however, one should also set up a model for diffusion-coupled recovery and estimate it for the datasets in this chapter. Respective kinetic models have been developed in the literature for circular and line bleaching (e.g. Mueller et al. 2008). This section proposes an according model for half-nucleus FRAP, as required in the context of this chapter. It is illustrated in Fig. 9.20. A similar approach has been taken by Carrero et al. (2004). An extension of our model to multiple mobility classes is shown in Sect. C.4.

In the diffusion-coupled model, one assumes that spatial diffusion of the GFP-tagged molecules across the cell nucleus happens at a rate that is of the same

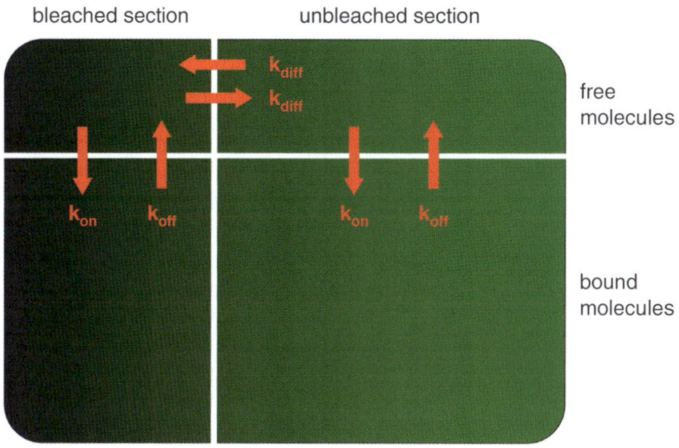

Fig. 9.20 Compartmental representation of the kinetic model for diffusion-coupled FRAP: The unbleached molecules in the nucleus are divided into four groups, namely into molecules that are free in the bleached section, molecules that are free in the unbleached section, molecules that are bound in the bleached section and molecules that are bound in the unbleached section. As opposed to the diffusion-uncoupled model in Fig. 9.5 on p. 312, the location of a free molecule is explicitly modelled. Free molecules can diffuse from the bleached section to the unbleached section and the other way round. Diffusion occurs with diffusion rate k_{diff}

order as the rates for binding and unbinding. In that case, other than in case of diffusion-uncoupled FRAP, it is required to explicitly model the location of a free molecule within the nucleus. An obvious approach is to divide the number U^{free} of unbleached free molecules into the number $U_{\text{bl}}^{\text{free}}$ of unbleached free molecules in the bleached section and the number $U_{\text{unbl}}^{\text{bound}}$ of unbleached free molecules in the unbleached section. The state of all unbleached molecules in the nucleus is then represented by $U_{\text{bl}}^{\text{free}}$, $U_{\text{unbl}}^{\text{free}}$, $U_{\text{bl}}^{\text{bound}}$ and $U_{\text{unbl}}^{\text{bound}}$. As these variables sum up to the number N_U of all unbleached molecules, complete information is provided when considering a Markov process with three-dimensional state vector $(U_{\text{bl}}^{\text{free}}, U_{\text{unbl}}^{\text{free}}, U_{\text{bl}}^{\text{bound}})'$ in the state space

$$\{(U_{\text{bl}}^{\text{free}}, U_{\text{unbl}}^{\text{free}}, U_{\text{bl}}^{\text{bound}})' \in [0, N_U]^3 \cap \mathbb{N}_0^3 \,|\, U_{\text{bl}}^{\text{free}} + U_{\text{unbl}}^{\text{free}} + U_{\text{bl}}^{\text{bound}} \leq N_U\}.$$

Binding and unbinding events are supposed to happen analogously to those in the general diffusion-uncoupled model in Sect. 9.3. Diffusion of free molecules between the bleached and unbleached section is assumed to occur with a *diffusion rate* k_{diff}. Every two molecules that are located at the same distance from the bleaching border are supposed to cross this border within a certain time interval with the same probability, no matter whether the direction of diffusion is from the bleached to the unbleached area or the other way round. If, however, the bleached fraction f_{bl} is not equal to one half, the sizes of the bleached and unbleached sections differ. Then, due to the geometry of the bleached area, several of the molecules in the larger section are located further away from the bleaching border than the molecules in the smaller area. In order to account for this disbalance, the diffusion events in the two possible directions are assumed to occur with the following rates:

$$(U_{\text{bl}}^{\text{free}}, U_{\text{unbl}}^{\text{free}}, U_{\text{bl}}^{\text{bound}})' \to (U_{\text{bl}}^{\text{free}}+1,\, U_{\text{unbl}}^{\text{free}}-1,\, U_{\text{bl}}^{\text{bound}})' \quad \text{with rate } k_{\text{diff}}\, f_{\text{bl}}\, U_{\text{unbl}}^{\text{free}},$$
$$(U_{\text{bl}}^{\text{free}}, U_{\text{unbl}}^{\text{free}}, U_{\text{bl}}^{\text{bound}})' \to (U_{\text{bl}}^{\text{free}}-1,\, U_{\text{unbl}}^{\text{free}}+1,\, U_{\text{bl}}^{\text{bound}})' \quad \text{with rate } k_{\text{diff}}\, (1-f_{\text{bl}})\, U_{\text{bl}}^{\text{free}}.$$

The value of k_{diff} depends on the geometry of the cell and is hence not immediately eligible for interpretation purposes. A diffusion approximation with the intensive state variable $(u_{\text{bl}}^{\text{free}}, u_{\text{unbl}}^{\text{free}}, u_{\text{bl}}^{\text{bound}})'$ results as the solution of an SDE with drift

$$\begin{pmatrix} -k_{\text{on}} u_{\text{bl}}^{\text{free}} + k_{\text{off}} u_{\text{bl}}^{\text{bound}} + k_{\text{diff}} \left(f_{\text{bl}} u_{\text{unbl}}^{\text{free}} - (1-f_{\text{bl}})\, u_{\text{bl}}^{\text{free}}\right) \\ -k_{\text{on}} u_{\text{unbl}}^{\text{free}} + k_{\text{off}} \left(1 - u_{\text{bl}}^{\text{free}} - u_{\text{unbl}}^{\text{free}} - u_{\text{bl}}^{\text{bound}}\right) - k_{\text{diff}} \left(f_{\text{bl}} u_{\text{unbl}}^{\text{free}} - (1-f_{\text{bl}})\, u_{\text{bl}}^{\text{free}}\right) \\ k_{\text{on}} u_{\text{bl}}^{\text{free}} - k_{\text{off}} u_{\text{bl}}^{\text{bound}} \end{pmatrix}$$

and diffusion matrix

$$\frac{1}{N_U} \begin{pmatrix} \Sigma_{11} & \Sigma_{12} & -\Sigma_{33} \\ \Sigma_{12} & \Sigma_{22} & 0 \\ -\Sigma_{33} & 0 & \Sigma_{33} \end{pmatrix},$$

where

$$\begin{aligned}
\Sigma_{11} &= k_{\text{on}} u_{\text{bl}}^{\text{free}} + k_{\text{off}} u_{\text{bl}}^{\text{bound}} + k_{\text{diff}} \left(f_{\text{bl}} u_{\text{unbl}}^{\text{free}} + (1-f_{\text{bl}})\, u_{\text{bl}}^{\text{free}}\right) \\
\Sigma_{22} &= k_{\text{on}} u_{\text{unbl}}^{\text{free}} + k_{\text{off}} \left(1 - u_{\text{bl}}^{\text{free}} - u_{\text{unbl}}^{\text{free}} - u_{\text{bl}}^{\text{bound}}\right) + k_{\text{diff}} \left(f_{\text{bl}} u_{\text{unbl}}^{\text{free}} + (1-f_{\text{bl}})\, u_{\text{bl}}^{\text{free}}\right) \\
\Sigma_{33} &= k_{\text{on}} u_{\text{bl}}^{\text{free}} + k_{\text{off}} u_{\text{bl}}^{\text{bound}} \\
\Sigma_{12} &= -k_{\text{diff}} \left(f_{\text{bl}} u_{\text{unbl}}^{\text{free}} + (1-f_{\text{bl}}) u_{\text{bl}}^{\text{free}}\right).
\end{aligned}$$

The initial value for this SDE is an element of the state space

$$\{(u_{\text{bl}}^{\text{free}}, u_{\text{unbl}}^{\text{free}}, u_{\text{bl}}^{\text{bound}})' \in [0,1]^3 \cap \mathbb{R}_0^3 \,|\, u_{\text{bl}}^{\text{free}} + u_{\text{unbl}}^{\text{free}} + u_{\text{bl}}^{\text{bound}} \le 1\}.$$

The observed variable is the fluorescence intensity

$$q = \frac{u_{\text{bl}}^{\text{free}} + u_{\text{bl}}^{\text{bound}}}{f_{\text{bl}}}.$$

In case of $k_{\text{diff}} \gg k_{\text{on}}, k_{\text{off}}$, i.e. for diffusion-uncoupled FRAP, the drift function can be approximated by

$$\begin{pmatrix} k_{\text{diff}}\left(f_{\text{bl}}u_{\text{unbl}}^{\text{free}}-(1-f_{\text{bl}})\,u_{\text{bl}}^{\text{free}}\right) \\ -k_{\text{diff}}\left(f_{\text{bl}}u_{\text{unbl}}^{\text{free}}-(1-f_{\text{bl}})\,u_{\text{bl}}^{\text{free}}\right) \\ 0 \end{pmatrix}$$

until the process reaches a state where the elements of this vector are small, that is $f_{\text{bl}}u_{\text{unbl}}^{\text{free}} \approx (1-f_{\text{bl}})u_{\text{bl}}^{\text{free}}$. This equality corresponds to $u_{\text{bl}}^{\text{free}} \approx f_{\text{bl}}u^{\text{free}}$ and $u_{\text{unbl}}^{\text{free}} \approx (1-f_{\text{bl}})u^{\text{free}}$, where $u^{\text{free}} = u_{\text{bl}}^{\text{free}} + u_{\text{unbl}}^{\text{free}}$. That is the basic assumption of the diffusion-uncoupled scenario.

The consideration of an intermediate fraction f_{int} is not meaningful in a diffusion-coupled model, because there is no assumption of a rapid invasion of unbleached molecules into the bleached area before acquisition of the first postbleach image. On the other hand, the diffusion-coupled model may be designed such that it can be applied to a more refined dataset containing the fluorescence recovery in several slices instead of just two regions of the nucleus.

9.9 Conclusion and Outlook

This chapter showed a second application of diffusion models in life sciences. It introduced a number of research questions concerning the binding behaviour of proteins within cell nuclei and described the FRAP technique as a convenient tool for data acquisition. Various stochastic and deterministic kinetic models for the dynamics of fluorescence recovery were derived. The application of diffusion models had not been considered in the FRAP literature before. The performances of Bayesian and least squares estimation techniques were analysed based on synthetic datasets in several simulation studies, and statistical model choice criteria were evaluated. An enhanced processing of raw FRAP measurements was proposed, and its impact on parameter estimation was investigated.

New insight could be gained especially concerning the cell cycle dependent average residence times of Dnmt1 remaining at binding sites, which were estimated as the inverse dissociation rates for each mobility class. Improved parameter estimates were achieved by utilisation of stochastic diffusion models in combination with Bayesian inference techniques. These were newly introduced to the field of FRAP analysis, where the application of deterministic models is prevalent.

Ongoing work Schneider et al. (2012) deals with the estimation of diffusion-coupled models for the Dnmt1 data. An according model for diffusion-coupled half-nucleus FRAP has been proposed in this book. In the diffusion-coupled scenario, it is also meaningful to divide the cell nucleus into several slices with different distances from the bleaching border and hence to apply the model to a richer dataset. This could yield more accurate parameter estimates.

Further analyses concern the biological interpretation of the estimation and model choice results. To that end, fluorescence recovery curves are investigated for mutants of Dnmt1, where certain binding interactions with DNA are possibly disturbed (Schneider 2009).

The present chapter was focused on the cell cycle dependent kinetic behaviour of the particular protein Dnmt1. Model derivations, statistical inference techniques and data processing have, however, been formulated in a universal context. The contents of this chapter hence provide a general framework for the kinetic analysis of a multitude of proteins of interest.

References

Akaike H (1973) Information theory and an extension of the maximum likelihood principle. In: Petrov B, Csaki F (eds) 2nd international symposium on information theory, Akademiai Kiado, pp 267–281. Reprinted, with an introduction by J. deLeeuw, in *Breakthroughs in Statistics, Volume I*, edited by Samuel Kotz and Norman L. Johnson. Springer, 1992, pp. 599-624

Beaudouin J, Mora-Bermúdez F, Klee T, Daigle N, Ellenberg J (2006) Dissecting the contribution of diffusion and interactions to the mobility of nuclear proteins. Biophys J 90:1878–1894

Carrero G, Crawford E, Hendzel MJ (2004) Characterizing fluorescence recovery curves for nuclear proteins undergoing binding events. Bull Math Biol 66:1515–1545

Chen MH, Shao QM (1999) Monte Carlo estimation of Bayesian credible and hpd intervals. J Comput Graph Stat 8:69–92

Chib S (1995) Marginal likelihood from the Gibbs output. J Am Stat Assoc 90:1313–1321

Elerian O, Chib S, Shephard N (2001) Likelihood inference for discretely observed nonlinear diffusions. Econometrica 69:959–993

Fahrmeir L, Kneib T, Lang S (2009) Regression. Modelle, Methoden und Anwendungen, 2nd edn. Springer, Heidelberg

Garcia-Parajo M, Segers-Nolten G, Veerman J, Greve J, van Hulst N (2000) Real-time light-driven dynamics of the fluorescence emission in single green fluorescent protein molecules. Proc Natl Acad Sci U S A 97:7237–7242

Gorski S, Misteli T (2005) Systems biology in the cell nucleus. J Cell Sci 118:4083–4092

Jeffreys H (1961) Theory of probability. Prentice Hall, Princeton

Kass R, Raftery A (1995) Bayes factors. J Am Stat Assoc 90:773–795

Kuch D, Schermelleh L, Manetto S, Leonhardt H, Carell T (2008) Synthesis of DNA dumb-bell based inhibitors for the human DNA methyltransferase Dnmt1. Angew Chem Int Ed 47:1515–1518

Lambert N (2009) Uncoupling diffusion and binding in FRAP experiments. Nat Methods 6:183

McNally J (2008) Quantitative FRAP in analysis of molecular binding dynamics in vivo. Method Cell Biol 85:329–351

Mueller F, Wach P, McNally J (2008) Evidence for a common mode of transcription factor interaction with chromatin as revealed by improved quantitative fluorescence recovery after photobleaching. Biophys J 94:3323–3339

Nelder J, Mead R (1965) A simplex method for function minimization. Comput J 7:308–313

Phair R, Gorski S, Misteli T (2004a) Measurement of dynamic protein binding to chromatin in vivo, using photobleaching microscopy. Method Enzymol 375:393–414

Phair R, Scaffidi P, Elbi C, Vecerová J, Dey A, Ozato K, Brown D, Hager G, Bustin M, Misteli T (2004b) Global nature of dynamic protein-chromatin interactions in vivo: three-dimensional genome scanning and dynamic interaction networks of chromatin proteins. Mol Cell Biol 24:6393–6402

Polson N, Roberts G (1994) Bayes factors for discrete observations from diffusion processes. Biometrika 81:11–26

Schneider K (2009) Analysis of cell cycle dependent kinetics of Dnmt1 by FRAP and kinetic modeling. Diploma Thesis, LMU Munich

Schneider K, Fuchs C, Dobay A, Rottach A, Qin W, Álvarez-Castro J, Nalaskowski M, Schmid V, Leonhardt H, Schermelleh L (2012) Dissection of cell cycle dependent dynamics of Dnmt1 by FRAP and diffusion-coupled modeling (in preparation)

Schwarz G (1978) Estimating the dimension of a model. Ann Stat 6:461–464

Sprague B, McNally J (2005) FRAP analysis of binding: proper and fitting. Trends Cell Biol 15:84–91

Chapter 10
Summary and Future Work

Stochastic modelling and statistical estimation are important tools for the understanding of complex processes in life sciences. This book motivated the use of diffusion processes for both purposes and contributed to their applicability in practice. The following section summarises the achievements of this book, oriented towards the initially set objectives which were formulated in Sect. 1.1. Section 10.2 points out directions for future work.

10.1 Summary

Starting from a specific real-world phenomenon, one often requires a mathematical Markov model which appropriately represents the time-continuous dynamics of the considered system. To that end, many authors either choose a computationally costly exact description in terms of stochastic jump processes or an over-simplistic state-continuous deterministic model. A convenient trade-off between these two extremes is provided by diffusion processes. These are both stochastic and state-continuous but mathematically more sophisticated.

In particular, there are no general guidelines for practitioners which describe the correct approximation of stochastic jump processes by diffusions. More specifically, existing approaches are partly formulated for one-dimensional diffusions, and they always assume systems whose dimension is sufficiently characterised by one single size parameter. Both of these properties do not match the requirements of, for example, the multitype susceptible–infected–removed (SIR) model considered in this work. Chapters 2 and 3 of the present book motivate the application of diffusion models in life sciences and provide a compact overview of their mathematical background. Chapter 4 elucidates the derivation of diffusion approximations and complements existing approaches by new formulations, multi-dimensional extensions and the generalisation to systems which involve multiple size parameters.

C. Fuchs, *Inference for Diffusion Processes: With Applications in Life Sciences*,
DOI 10.1007/978-3-642-25969-2_10,

Importantly for practitioners, this chapter for the first time presents a systematic procedure for the derivation of diffusion approximations in a universal framework. The methodology is further exemplified in Chap. 5.

With a diffusion model at hand, which is represented in parametric form as the solution of a stochastic differential equation, the next objective is to statistically estimate its parameters based on time-discrete observations of the process. Chapter 6 investigates and reviews established frequentist methodology on this subject in a multi-dimensional framework. It turns out that the application of such techniques is problematic, if not impossible, in typical data situations in life sciences, which can involve sparse and non-equidistant observations, measurements with error and unobserved components of multi-dimensional processes. An appropriate alternative to tackle this problem, however, is given by the application of a well-known Bayesian approach which introduces auxiliary data points as additional observations. These are estimated by application of Markov chain Monte Carlo (MCMC) techniques which alternately update the auxiliary data and the model parameter. Chapter 7 reviews this idea in detail. Unfortunately, the procedure suffers from convergence problems which originate in a deterministic relationship between the model parameters and the quadratic variation of a continuously observed diffusion path. A practical solution for this problem had not yet been proven for multi-dimensional diffusion processes. Chapter 7 formulates a neat modification of the Bayesian approach for conditioned diffusions on infinite-dimensional state spaces. This book provides the mathematical proof that the so-constructed MCMC scheme converges. Its performance is proven in several simulation studies.

In order to show the potential of modelling and estimation by means of diffusions, Chaps. 8 and 9 utilise the theoretical insights gained in the previous chapters for the statistical investigation of real problems from life sciences. Chapter 8 analyses the spread of influenza among humans, based on one dataset containing the numbers of occurrences of influenza in a British boarding school in 1978 over a period of two weeks, and a second dataset concerning one of the latest influenza epidemics in Germany in 2009/2010. Spatial modelling is carried out in the latter case by using an extension of the standard SIR model, developed in Chap. 5, which allows for host heterogeneity. In another application, Chap. 9 explores the kinetic properties of the protein Dnmt1 which is an important factor for DNA methylation. Appropriate data is acquired by use of fluorescence microscopy. The statistical investigations of Chap. 9 provide new insights into the understanding of the binding behaviour of Dnmt1.

10.2 Future Work

Based on the contributions of this book, diffusion processes can more easily and more efficiently be applied for modelling and estimation purposes in life sciences. Future perspectives of this work mainly concern the utilisation of the developed methods to further areas of applications, the variety of which is manifold.

Practitioners should feel encouraged to dare apply diffusion processes in their research areas: This book provides general guidelines for the setup of appropriate diffusion models, and it supplies adequate information for their statistical inference. Thanks to the achievements in Chap. 7, the considered Bayesian estimation approach is not limited by convergence problems anymore. For practical usability, the proposed scheme has been formulated in algorithmic form. All algorithms have been implemented in `R`.

Concerning the two fields of applications considered in this book, several possible extensions have already been pointed out in the respective chapters, see Sects. 2.2.3, 5.4, 8.4 and 9.9. Hence, only a few perspectives shall be commented on here.

The utilisation of diffusion approximations coupled with statistical inference techniques in the spatio-temporal modelling of the spread of infectious diseases is new. Hence, research is in the early stages, and multiple enhancements are conceivable. These could for example concern the choice of clusters and their connectivities, the specification of model parameters and the quality of the underlying data. Such advancements will help improving the comprehension and prediction of epidemic outbreaks. This book provides a first step in that direction.

In the second application, diffusion models and their statistical inference have also been newly introduced to the analysis of fluorescence microscopy data. One direction of future research is to investigate diffusion-coupled recovery, which was explained in Chap. 9. Furthermore, in close collaboration with scientists from molecular biology, comparisons between wild type proteins and appropriate mutants can be drawn by application of statistical methods.

Overall, the combined application of diffusion modelling and statistical inference promises to supply new insight in many exciting areas of life sciences in the future. This book has demonstrated the potential of this approach.

Appendix A
Benchmark Models

This chapter briefly introduces well-known diffusion processes that serve as benchmark models in this book. In particular, for each process, the characterising stochastic differential equation (SDE) and the transition density as defined in (3.14) on p. 39 are given. From that, it follows immediately how to simulate paths of the processes.

A.1 Geometric Brownian Motion

One-dimensional *geometric Brownian motion* $X = (X_t)_{t \geq 0}$ is defined through the SDE

$$\mathrm{d}X_t = \alpha X_t dt + \sigma X_t \mathrm{d}B_t, \quad X_0 = x_0, \tag{A.1}$$

with parameters $\alpha \in \mathbb{R}, \sigma \in \mathbb{R}_+$ and state space $\mathcal{X} = \mathbb{R}_+$ for $x_0 \in \mathbb{R}_+$. In financial mathematics, it generally serves as a model for asset prices with interest rate α and volatility σ and forms the basis of the famous Black-Scholes model (Black and Scholes 1973; Merton 1973).

SDE (A.1) has the explicit solution

$$X_t = x_0 \exp\left((\alpha - \frac{1}{2}\sigma^2)t + \sigma B_t \right)$$

for all $t \geq 0$. The transition density reads

$$p(s, x, t, y) = \frac{1}{\sqrt{2\pi(t-s)}\sigma y} \exp\left(-\frac{\left(\log y - \log x - (\alpha - \frac{1}{2}\sigma^2)(t-s)\right)^2}{2\sigma^2(t-s)} \right)$$

for $x, y \in \mathcal{X}$ and $t > s \geq 0$; that is the density of a log-normal distribution, i.e.

$$X_t\{X_s = x\} \sim \mathcal{LN}\left(\log x + \left(\alpha - \frac{1}{2}\sigma^2 \right)(t-s), \sigma^2(t-s) \right).$$

C. Fuchs, *Inference for Diffusion Processes: With Applications in Life Sciences*,
DOI 10.1007/978-3-642-25969-2,

The conditional expectation and variance of the state of the process are

$$\mathbb{E}(X_t|X_s=x)=x\mathrm{e}^{\alpha(t-s)} \text{ and } \mathrm{Var}(X_t|X_s=x)=x^2\mathrm{e}^{2\alpha(t-s)}\left(\mathrm{e}^{\sigma^2(t-s)}-1\right).$$

A.2 Ornstein-Uhlenbeck Process

The one-dimensional time-homogeneous *Ornstein-Uhlenbeck process* $X=(X_t)_{t\geq 0}$ with state space $\mathcal{X}$=$\mathbb{R}$ is described by the SDE

$$\mathrm{d}X_t=\alpha(\beta-X_t)dt+\sigma\mathrm{d}B_t \quad , X_0=x_0, \tag{A.2}$$

for parameters $\beta\in\mathbb{R}$ and $\alpha,\sigma\in\mathbb{R}_+$. It was first used by Uhlenbeck and Ornstein (1930) to describe the movement of a diffusing particle, where $\beta=0$, α is the friction coefficient divided by the mass of the particle, and σ stands for the strength of the fluctuations. Vasicek (1977) applied this model later to interest rates with long-run equilibrium value β, speed of adjustment α and volatility σ.

The solution of SDE (A.2) is

$$X_t=x_0\mathrm{e}^{-\alpha t}+\beta\left(1-\mathrm{e}^{-\alpha t}\right)+\sigma\int_0^t \mathrm{e}^{-\alpha(t-\tau)}\mathrm{d}B_\tau$$

for all $t\geq 0$. Due to the deterministic integrand, this is a Gaussian process with transition density

$$p(s,x,t,y)=\phi_{(\mu(t-s;x),\Sigma(t-s;x))}(y)$$

for all $x,y\in\mathcal{X}$ and $t>s\geq 0$; that is the normal density with mean

$$\mu(t-s;x)=\mathbb{E}(X_t|X_s=x)=x\mathrm{e}^{-\alpha(t-s)}+\beta\left(1-\mathrm{e}^{-\alpha(t-s)}\right)$$

and variance

$$\Sigma(t-s;x)=\mathrm{Var}(X_t|X_s=x)=\frac{\sigma^2}{2\alpha}\left(1-\mathrm{e}^{-2\alpha(t-s)}\right)$$

evaluated at y.

A.3 Cox-Ingersoll-Ross Process

The one-dimensional *Cox-Ingersoll-Ross (CIR) process*, also called *square-root process*, fulfils the SDE

$$\mathrm{d}X_t=\alpha(\beta-X_t)dt+\sigma\sqrt{X_t}\mathrm{d}B_t, \quad X_0=x_0,$$

with positive parameters α, β, σ, state space $\mathcal{X} = \mathbb{R}_0$ and $x_0 \in \mathbb{R}_+$. It was introduced by Cox et al. (1985) to model a randomly moving interest rate, where the model parameters are interpreted as in the Ornstein-Uhlenbeck model (Sect. A.2). It is reasonable to assume $2\alpha\beta > \sigma^2$ since then $\mathcal{X} = \mathbb{R}_+$.

Under this assumption the transition density of the process is

$$p(s, x, t, y) = c \left(\frac{v}{u}\right)^{\frac{\nu}{2}} \exp\left(-(u+v)\right) I_\nu\left(2\sqrt{uv}\right) \tag{A.3}$$

for $t > s \geq 0$, where

$$c = \frac{2\alpha}{\sigma^2(1 - \mathrm{e}^{-\alpha(t-s)})}, \quad u = cx\mathrm{e}^{-\alpha(t-s)}, \quad v = cy, \quad \nu = \frac{2\alpha\beta}{\sigma^2} - 1\,.$$

I_ν denotes the modified Bessel function of the first kind of order ν, i.e.

$$I_\nu(z) = \sum_{k=0}^{\infty} \left(\frac{z}{2}\right)^{2k+\nu} \frac{1}{k!\,\Gamma(k+\nu+1)}$$

for $z \in \mathbb{R}$, where Γ is the Gamma function.

Formula (A.3) implies that $Y_t = 2cX_t$ conditioned on $Y_s = 2cx$ has the non-central chi-square distribution with non-centrality parameter $2cx\exp(-\alpha(t-s))$ and $4\alpha\beta/\sigma^2$ degrees of freedom. The conditional expectation and variance of the state of the original process are

$$\mathbb{E}(X_t|X_s = x) = 2\,\frac{\alpha\beta}{c\sigma^2} + x\mathrm{e}^{-\alpha(t-s)}$$

and

$$\mathrm{Var}(X_t|X_s = x) = \frac{2}{c}\left(\frac{\alpha\beta}{c\sigma^2} + x\mathrm{e}^{-\alpha(t-s)}\right).$$

References

Black F, Scholes M (1973) The pricing of options and corporate liabilities. J Polit Econ 81:637–654

Cox J, Ingersoll J, Ross S (1985) A theory of the term structure of interest rates. Econometrica 53:385–408

Merton R (1973) Theory of rational option pricing. Bell J Econ Manag Sci 4:141–183

Uhlenbeck G, Ornstein L (1930) On the theory of the Brownian motion. Phys Rev 36:823–841

Vasicek O (1977) An equilibrium characterization of the term structure. J Financ Econ 5:177–188

Appendix B
Miscellaneous

This chapter contains several auxiliary definitions, proofs and calculations which are required in the course of this book. They are not contained in the main material because they do not stand at the core of this work.

B.1 Difference Operators

The following definitions and lemmas are used in the context of expanding the master equation and infinitesimal generator of a Markov jump process in Sects. 4.3.1, 4.3.2, 4.4.1 and 4.4.2. The notation, proofs and further results are entirely new but moved to the Appendix due to space restrictions.

Definition B.1. Let $f : \mathbb{R}^n \to \mathbb{R}$ be an infinitely often differentiable function which is smooth enough such that the order of differentiation with respect to different arguments does not matter. For fixed $\boldsymbol{\varepsilon} = (\varepsilon_1, \ldots, \varepsilon_n)' \in \mathbb{R}^n$, define the *difference operator* $D_{\boldsymbol{k}}^m$ *of order* m with $\boldsymbol{k} = (k_1, \ldots, k_n)' \in \mathbb{N}_0^n$ and $|\boldsymbol{k}| = \sum_{i=1}^n k_i = m$ recursively as follows:

$$D_{\boldsymbol{0}}^0 f(\boldsymbol{x}) = f(\boldsymbol{x}), \quad D_{\boldsymbol{e}_i}^1 f(\boldsymbol{x}) = f(\boldsymbol{x}+\boldsymbol{e}_i \diamond \boldsymbol{\varepsilon}) - f(\boldsymbol{x}), \quad D_{\boldsymbol{k}+\boldsymbol{e}_i}^{m+1} f(\boldsymbol{x}) = D_{\boldsymbol{e}_i}^1 D_{\boldsymbol{k}}^m f(\boldsymbol{x})$$

for $m \geq 0$, where $\boldsymbol{e}_i = (0, \ldots, 1, \ldots, 0)'$ denotes the ith unit vector of dimension n and $\boldsymbol{u} \diamond \boldsymbol{v} = (u_1 v_1, \ldots, u_n v_n)'$ for arbitrary $\boldsymbol{u}, \boldsymbol{v} \in \mathbb{R}^n$. If the fixed parameter $\boldsymbol{\varepsilon}$ is ambiguous, attach it as a second subscript to the operator, i.e. write $D_{\boldsymbol{k},\boldsymbol{\varepsilon}}^{|\boldsymbol{k}|}$.

The difference operator allows the notation of difference quotients in correspondence to according derivatives: As ε_i tends to zero for all $i = 1, \ldots, n$,

$$\frac{D_{\boldsymbol{k}}^m f(\boldsymbol{x})}{\varepsilon_1^{k_1} \cdots \varepsilon_n^{k_n}} \longrightarrow \frac{\partial^m f(\boldsymbol{x})}{\partial x_1^{k_1} \cdots \partial x_n^{k_n}} . \tag{B.1}$$

C. Fuchs, *Inference for Diffusion Processes: With Applications in Life Sciences*,
DOI 10.1007/978-3-642-25969-2, © Springer-Verlag Berlin Heidelberg 2013

The following lemmas concern explicit formulas for these difference operators.

Lemma B.2. *The difference operator can be expressed as*

$$D_{\boldsymbol{k}}^{m} f(\boldsymbol{x}) = \sum_{j_1=0}^{k_1} \cdots \sum_{j_n=0}^{k_n} (-1)^{m-\sum_{i=1}^{n} j_i} \binom{k_1}{j_1} \cdots \binom{k_n}{j_n} f(\boldsymbol{x} + \boldsymbol{j} \diamond \boldsymbol{\varepsilon}) \quad \text{(B.2)}$$

$$= \sum_{l=0}^{m} \sum_{\boldsymbol{j} \in \mathcal{K}_l} (-1)^{m-l} \binom{k_1}{j_1} \cdots \binom{k_n}{j_n} f(\boldsymbol{x} + \boldsymbol{j} \diamond \boldsymbol{\varepsilon}), \quad \text{(B.3)}$$

where

$$\mathcal{K}_l = \left\{ \boldsymbol{j} = (j_1, \ldots, j_n)' \in \mathbb{N}_0^n \;\middle|\; |\boldsymbol{j}| = l \text{ and } 0 \le j_i \le k_i \text{ for all } i = 1, \ldots, n \right\}$$

for $l = 0, \ldots, m$ *and fixed* $\boldsymbol{k}$.

Proof. Formula (B.3) follows directly from (B.2), whose validity in turn is proven by complete induction on $m = |\boldsymbol{k}|$: Eq. (B.2) trivially holds for $m = 0$ and $m = 1$. Assume that it is true for any fixed m. Then, by definition and induction hypothesis,

$$D_{\boldsymbol{k}+\boldsymbol{e}_i}^{m+1} f(\boldsymbol{x}) = D_{\boldsymbol{e}_i}^{1} D_{\boldsymbol{k}}^{m} f(\boldsymbol{x})$$

$$= \sum_{j_1=0}^{k_1} \cdots \sum_{j_i=0}^{k_i} \cdots \sum_{j_n=0}^{k_n} (-1)^{m-\sum_{h=1}^{n} j_h} \binom{k_1}{j_1} \cdots \binom{k_i}{j_i} \cdots \binom{k_n}{j_n} f(\boldsymbol{x} + (\boldsymbol{j} + \boldsymbol{e}_i) \diamond \boldsymbol{\varepsilon})$$

$$- \sum_{j_1=0}^{k_1} \cdots \sum_{j_i=0}^{k_i} \cdots \sum_{j_n=0}^{k_n} (-1)^{m-\sum_{h=1}^{n} j_h} \binom{k_1}{j_1} \cdots \binom{k_i}{j_i} \cdots \binom{k_n}{j_n} f(\boldsymbol{x} + \boldsymbol{j} \diamond \boldsymbol{\varepsilon})$$

for any $i \in \{1, \ldots, n\}$. With an index shift of j_i in the second line, this becomes

$$\sum_{j_1=0}^{k_1} \cdots \sum_{j_{i-1}=0}^{k_{i-1}} \sum_{j_{i+1}=0}^{k_{i+1}} \cdots \sum_{j_n=0}^{k_n} (-1)^{m+1-\sum_{h \neq i} j_h} \binom{k_1}{j_1} \cdots \binom{k_{i-1}}{j_{i-1}} \binom{k_{i+1}}{j_{i+1}} \cdots \binom{k_n}{j_n}$$

$$\times \left(f(x_1 + j_1\varepsilon_1, \ldots, x_i, \ldots, x_n + j_n\varepsilon_n) + \sum_{j_i=1}^{k_i} (-1)^{-j_i} \left[\binom{k_i}{j_i - 1} + \binom{k_i}{j_i} \right] f(\boldsymbol{x} + \boldsymbol{j} \diamond \boldsymbol{\varepsilon}) \right.$$

$$\left. + (-1)^{-(k_i+1)} f(x_1 + j_1\varepsilon_1, \ldots, x_i + (k_i + 1)\varepsilon_i, \ldots, x_n + j_n\varepsilon_n) \right),$$

which equals

$$\sum_{j_1=0}^{k_1} \cdots \sum_{j_i=0}^{k_1+1} \cdots \sum_{j_n=0}^{k_n} (-1)^{m+1-\sum_{h=1}^{n} j_h} \binom{k_1}{j_1} \cdots \binom{k_i + 1}{j_i} \cdots \binom{k_n}{j_n} f(\boldsymbol{x} + \boldsymbol{j} \diamond \boldsymbol{\varepsilon}).$$

This proves the proposition. □

Lemma B.3. *Each expression of the form $f(\boldsymbol{x} + \boldsymbol{k} \diamond \boldsymbol{\varepsilon}) - f(\boldsymbol{x})$ can be expanded as the sum of differences $D^{m'}_{\boldsymbol{k}'} f(\boldsymbol{x})$ with $|\boldsymbol{k}'| = m' \leq m = |\boldsymbol{k}|$. In other words, for each $\boldsymbol{k}$ there is a set $I_{\boldsymbol{k}} \subseteq \mathbb{N}_0^n$ with*

$$\boldsymbol{k}' \in I_{\boldsymbol{k}} \Rightarrow |\boldsymbol{k}'| \leq |\boldsymbol{k}|$$

such that

$$f(\boldsymbol{x} + \boldsymbol{k} \diamond \boldsymbol{\varepsilon}) - f(\boldsymbol{x}) = \sum_{\boldsymbol{k}' \in I_{\boldsymbol{k}}} D^{|\boldsymbol{k}'|}_{\boldsymbol{k}'} f(\boldsymbol{x}).$$

Proof. The proposition is again shown by complete induction on m. The statement is true for $m = 1$ as

$$f(\boldsymbol{x} + \boldsymbol{e}_i \diamond \boldsymbol{\varepsilon}) - f(\boldsymbol{x}) = D^1_{\boldsymbol{e}_i} f(\boldsymbol{x})$$

for any $i \in \{1, \ldots, n\}$. Presume that it holds for all $m' \leq m$. Let $\boldsymbol{k} = (k_1, \ldots, k_n)'$ arbitrary but fixed with $|\boldsymbol{k}| = m + 1$ and

$$\mathcal{K}_l = \left\{\boldsymbol{j} = (j_1, \ldots, j_n)' \in \mathbb{N}_0^n \,\middle|\, |\boldsymbol{j}| = l \ \wedge\ 0 \leq j_i \leq k_i \text{ for all } i = 1, \ldots, n\right\} \quad \text{(B.4)}$$

for $l = 0, \ldots, m + 1$. Then, with Lemma B.2,

$$\begin{aligned}
&f(\boldsymbol{x} + \boldsymbol{k} \diamond \boldsymbol{\varepsilon}) - f(\boldsymbol{x}) \\
&= \left(\sum_{l=0}^{m+1} \sum_{\boldsymbol{j} \in \mathcal{K}_l} (-1)^{m+1-l} \binom{k_1}{j_1} \cdots \binom{k_n}{j_n} f(\boldsymbol{x} + \boldsymbol{j} \diamond \boldsymbol{\varepsilon})\right) \\
&\quad - \left(\sum_{l=0}^{m} \sum_{\boldsymbol{j} \in \mathcal{K}_l} (-1)^{m+1-l} \binom{k_1}{j_1} \cdots \binom{k_n}{j_n} f(\boldsymbol{x} + \boldsymbol{j} \diamond \boldsymbol{\varepsilon})\right) - f(\boldsymbol{x}) \\
&= D^{m+1}_{\boldsymbol{k}} f(\boldsymbol{x}) - \left(\sum_{l=0}^{m} \sum_{\boldsymbol{j} \in \mathcal{K}_l} (-1)^{m+1-l} \binom{k_1}{j_1} \cdots \binom{k_n}{j_n} \Big[f(\boldsymbol{x} + \boldsymbol{j} \diamond \boldsymbol{\varepsilon}) - f(\boldsymbol{x})\Big]\right) \quad \text{(B.5)} \\
&\quad - \left(1 + \sum_{l=0}^{m} \sum_{\boldsymbol{j} \in \mathcal{K}_l} (-1)^{m+1-l} \binom{k_1}{j_1} \cdots \binom{k_n}{j_n}\right) f(\boldsymbol{x}). \quad \text{(B.6)}
\end{aligned}$$

Line (B.5) is the sum of differences of order less than or equal to $m + 1$ due to the induction hypothesis. Line (B.6) equals zero since

$$1 + \sum_{l=0}^{m} (-1)^{m+1-l} \sum_{\boldsymbol{j} \in \mathcal{K}_l} \binom{k_1}{j_1} \cdots \binom{k_n}{j_n} = \sum_{l=0}^{m+1} (-1)^{m+1-l} \binom{m+1}{l} = 0,$$

using the generalised Vandermonde's identity and the binomial formula. This concludes the proof. □

The proof of Lemma B.3 already indicates how to expand an expression $f(\boldsymbol{x} + \boldsymbol{k} \diamond \boldsymbol{\varepsilon}) - f(\boldsymbol{x})$ with $|\boldsymbol{k}| = m$ in order to obtain a representation as the sum of differences $D^{m'}_{\boldsymbol{k}'}$.

Algorithm B.1. *This algorithm converts an expression $f(\boldsymbol{x} + \boldsymbol{k} \diamond \varepsilon) - f(\boldsymbol{x})$ to a sum of differences that can each be expressed by the difference operator from Definition B.1. In the following, the variable C stands for the already converted part of this expression, and R denotes the remaining part which still has to be converted. At any time, the remainder R consists of terms $f(\boldsymbol{x} + \boldsymbol{j} \diamond \varepsilon)$ with $|\boldsymbol{j}| \leq |\boldsymbol{k}|$, and $C + R = f(\boldsymbol{x} + \boldsymbol{k} \diamond \varepsilon) - f(\boldsymbol{x})$.*

Initially, one has $C = 0$ and $R = f(\boldsymbol{x} + \boldsymbol{k} \diamond \varepsilon) - f(\boldsymbol{x})$. While $R \neq 0$, execute the following steps:

- *Select the term $\alpha f(\boldsymbol{x} + \boldsymbol{k}^* \diamond \varepsilon)$ of R which has highest order $|\boldsymbol{k}^*|$. If this choice is not unique, choose any term of highest order. Set*
  $$a \leftarrow f(\boldsymbol{x} + \boldsymbol{k}^* \diamond \varepsilon)$$
  and choose $\alpha \in \mathbb{Z} \setminus \{0\}$ according to the prefactor of a in R.
- *Set*
  $$b \leftarrow \sum_{l=0}^{m^*-1} \sum_{\boldsymbol{j} \in \mathcal{K}_l^*} (-1)^{m^*-l} \binom{k_1^*}{j_1} \cdots \binom{k_n^*}{j_n} f(\boldsymbol{x} + \boldsymbol{j} \diamond \varepsilon)$$
  with $m^* = |\boldsymbol{k}^*|$ and $\mathcal{K}_l^*$ defined as in (B.4) *with k_i replaced by k_i^*.*
- *Set*
  $$C \leftarrow C + \alpha(a + b) = C + \alpha D_{\boldsymbol{k}^*}^{m^*} f(\boldsymbol{x})$$
  and
  $$R \leftarrow R - \alpha(a + b).$$

When $R = 0$, the variable C has the desired form. This algorithm terminates in finite time since the summands of b are of order less than $|\boldsymbol{k}^*|$.

Lemma B.3 does not ensure uniqueness of the expansion of $f(\boldsymbol{x}+\boldsymbol{k}\diamond\varepsilon) - f(\boldsymbol{x})$. If there is more than one possible representation, all of them are equally correct. In the context of the expansions in Chap. 4, however, caution is advised: There, we are not taking full limits but include in the result all terms up to a certain order of a small parameter. To be more precise, terms of the form

$$D_{\boldsymbol{k}}^m f(\boldsymbol{x}) = \varepsilon_1^{k_1} \cdots \varepsilon_n^{k_n} \frac{D_{\boldsymbol{k}}^m f(\boldsymbol{x})}{\varepsilon_1^{k_1} \cdots \varepsilon_n^{k_n}} \approx \varepsilon_1^{k_1} \cdots \varepsilon_n^{k_n} \frac{\partial^m f(\boldsymbol{x})}{\partial x_1^{k_1} \cdots \partial x_n^{k_n}}$$

with $\varepsilon_1, \ldots, \varepsilon_n \in \{-\tilde{\varepsilon}, \tilde{\varepsilon}\}$ for some small positive $\tilde{\varepsilon}$ are considered. In order to take limits of the difference quotients consistently, all differences should be expanded over intervals $[a_1, b_1] \times \ldots \times [a_n, b_n]$ with identical sums of lengths $|b_1 - a_1| + \ldots + |b_n - a_n|$. According to Lemma B.2, $D_{\boldsymbol{k}}^m f(\boldsymbol{x})$ covers an interval with cumulated length $k_1|\varepsilon_1| + \ldots + k_n|\varepsilon_n| = m\tilde{\varepsilon}$, i.e. $D_{\boldsymbol{k}}^m f(\boldsymbol{x})$ and $D_{\boldsymbol{k}'}^{m'} f(\boldsymbol{x})$ do not fulfil the required uniformity for $m \neq m'$.

Example B.1. In Chap. 5, the above expansions of terms of the form $f(\boldsymbol{x} + \boldsymbol{k} \diamond \varepsilon) - f(\boldsymbol{x})$ are employed in order to derive diffusion approximations for Markov jump processes. There, we especially deal with models where $|\boldsymbol{k}| \leq 2$ for all differences. In such cases proceed as follows: As proposed in Definition B.1, attach the parameter $\boldsymbol{\varepsilon}$ as a second subscript to the difference operator. Where both difference operators $D^1_{\boldsymbol{k},\boldsymbol{\varepsilon}}$ and $D^2_{\boldsymbol{k}',\boldsymbol{\varepsilon}'}$ with $|\varepsilon_j| = |\varepsilon'_j|$ for all $j = 1, \ldots, n$ appear, the first order difference operator can again be expanded as

$$\begin{aligned} D^1_{\boldsymbol{e}_i,\boldsymbol{\varepsilon}} f(\boldsymbol{x}) &= \frac{1}{2}\Big[D^1_{\boldsymbol{e}_i,\boldsymbol{\varepsilon}} f(\boldsymbol{x}) + D^1_{\boldsymbol{e}_i,-\boldsymbol{\varepsilon}} f(\boldsymbol{x})\Big] + \frac{1}{2}\Big[D^1_{\boldsymbol{e}_i,\boldsymbol{\varepsilon}} f(\boldsymbol{x}) - D^1_{\boldsymbol{e}_i,-\boldsymbol{\varepsilon}} f(\boldsymbol{x})\Big] \\ &= \frac{1}{2} D^2_{2\boldsymbol{e}_i,\boldsymbol{\varepsilon}} f(\boldsymbol{x}-\boldsymbol{\varepsilon}) + \frac{1}{2} D^1_{\boldsymbol{e}_i,2\boldsymbol{\varepsilon}} f(\boldsymbol{x}-\boldsymbol{\varepsilon}) \end{aligned}$$

for arbitrary $i \in \{1, \ldots, n\}$. The last row consists of expansions over intervals with cumulated lengths identical to that of the one covered by $D^2_{\boldsymbol{k}',\boldsymbol{\varepsilon}'}$. The according approximation is

$$\begin{aligned} D^1_{\boldsymbol{e}_i,\boldsymbol{\varepsilon}} f(\boldsymbol{x}) &= \frac{\varepsilon_i^2}{2} \frac{D^2_{2\boldsymbol{e}_i,\boldsymbol{\varepsilon}} f(\boldsymbol{x}-\boldsymbol{\varepsilon})}{\varepsilon_i^2} + \frac{2\varepsilon_i}{2} \frac{D^1_{\boldsymbol{e}_i,2\boldsymbol{\varepsilon}} f(\boldsymbol{x}-\boldsymbol{\varepsilon})}{2\varepsilon_i} \\ &\approx \frac{\varepsilon_i^2}{2} \frac{\partial^2 f(\boldsymbol{x})}{\partial x_i^2} + \varepsilon_i \frac{\partial f(\boldsymbol{x})}{\partial x_i}. \end{aligned} \tag{B.7}$$

Proceed similarly even in the absence of second order differences (e.g. for the expansion of the master equation of the Poisson process, which cannot satisfyingly be approximated by a diffusion though).

Example B.2. Consider a jump process with state variable $\boldsymbol{x} = (x_1, \ldots, x_n)'$ and possible jumps

$$\Big\{\boldsymbol{\Delta} = (\Delta_1, \ldots, \Delta_n)' \,\Big|\, \Delta_i \in \{-\varepsilon, 0, \varepsilon\} \text{ for all } i = 1, \ldots, n \text{ and } \sum_{i=1}^{n} |\Delta_i| \in \{\varepsilon, 2\varepsilon\}\Big\}$$

for some fixed $\varepsilon > 0$. Section 4.3.1 describes how to approximate the so-called master equation of this jump process by a partial differential equation. Let $f : \mathbb{R}^n \to \mathbb{R}$ be a twice differentiable function. The occurring difference terms in the approximation procedure are the following:

- For $\boldsymbol{\Delta} = \varepsilon \boldsymbol{e}_i$:

$$f(x_i - \varepsilon) - f(x_i) = D^1_{\boldsymbol{e}_i,-\boldsymbol{\Delta}} f(\boldsymbol{x}) \overset{\text{(B.7)}}{\approx} \frac{\varepsilon^2}{2} \frac{\partial^2 f(\boldsymbol{x})}{\partial x_i^2} - \varepsilon \frac{\partial f(\boldsymbol{x})}{\partial x_i},$$

  where only the component x_i of interest is displayed as an argument of f.

- For $\boldsymbol{\Delta} = \varepsilon(\boldsymbol{e}_i + \boldsymbol{e}_j)$, where $i \neq j$:

$$\begin{aligned} & f(x_i - \varepsilon, x_j - \varepsilon) - f(x_i, x_j) \\ = & \left[f(x_i - \varepsilon, x_j - \varepsilon) - f(x_i, x_j - \varepsilon) - f(x_i - \varepsilon, x_j) + f(x_i, x_j)\right] \\ & \quad + \left[f(x_i, x_j - \varepsilon) - f(x_i, x_j)\right] + \left[f(x_i - \varepsilon, x_j) - f(x_i, x_j)\right] \\ = & \left(D^2_{\boldsymbol{e}_i+\boldsymbol{e}_j,-\boldsymbol{\Delta}} + D^1_{\boldsymbol{e}_j,-\boldsymbol{\Delta}} + D^1_{\boldsymbol{e}_i,-\boldsymbol{\Delta}}\right) f(\boldsymbol{x}) \\ \overset{(B.7)}{\approx} & \left(\varepsilon^2 \frac{\partial^2}{\partial x_i \partial x_j} + \frac{\varepsilon^2}{2} \frac{\partial^2}{\partial x_j^2} - \varepsilon \frac{\partial}{\partial x_j} + \frac{\varepsilon^2}{2} \frac{\partial^2}{\partial x_i^2} - \varepsilon \frac{\partial}{\partial x_i}\right) f(\boldsymbol{x}). \end{aligned}$$

- Similarly, for $\boldsymbol{\Delta} = \varepsilon(\boldsymbol{e}_i - \boldsymbol{e}_j)$, where $i \neq j$:

$$f(x_i-\varepsilon, x_j+\varepsilon) - f(x_i, x_j) \approx \left(-\varepsilon^2 \frac{\partial^2}{\partial x_i \partial x_j} + \frac{\varepsilon^2}{2} \frac{\partial^2}{\partial x_j^2} + \varepsilon \frac{\partial}{\partial x_j} + \frac{\varepsilon^2}{2} \frac{\partial^2}{\partial x_i^2} - \varepsilon \frac{\partial}{\partial x_i}\right) f(\boldsymbol{x}).$$

- For $\boldsymbol{\Delta} = -\varepsilon \boldsymbol{e}_i$ and $\boldsymbol{\Delta} = -\varepsilon(\boldsymbol{e}_i + \boldsymbol{e}_j)$, replace ε by $-\varepsilon$ above.

In Sect. 4.3.1, the role of f is taken by the product of the scaled transition rate $w = \varepsilon w_N$ and the transition density p of the process. The above difference terms are summed up, divided by ε and rearranged such that one arrives at the partial differential equation (4.20), that is

$$\frac{\partial p(t, \boldsymbol{x})}{\partial t} = -\sum_{i=1}^{n} \frac{\partial \left[\mu_i(\boldsymbol{x}, t) p(t, \boldsymbol{x})\right]}{\partial x_i} + \frac{\varepsilon}{2} \sum_{i,j=1}^{n} \frac{\partial^2 \left[\Sigma_{ij}(\boldsymbol{x}, t) p(t, \boldsymbol{x})\right]}{\partial x_i \partial x_j}.$$

The above jumps contribute to the unknown vector $\boldsymbol{\mu}(\boldsymbol{x}, t) = (\mu_i(\boldsymbol{x}, t))_{i=1,\ldots,n}$ and matrix $\boldsymbol{\Sigma}(\boldsymbol{x}, t) = (\Sigma_{ij}(\boldsymbol{x}, t))_{i,j=1,\ldots,n}$ as follows:

Jump	Add to $\boldsymbol{\mu}(\boldsymbol{x}, t)$	Add to $\boldsymbol{\Sigma}(\boldsymbol{x}, t)$
$\boldsymbol{\Delta}_1 = \varepsilon \boldsymbol{e}_i$	$\boldsymbol{e}_i w(t, \boldsymbol{x}, \boldsymbol{\Delta}_1)$	$\boldsymbol{e}_i \boldsymbol{e}_i' w(t, \boldsymbol{x}, \boldsymbol{\Delta}_1)$
$\boldsymbol{\Delta}_2 = -\varepsilon \boldsymbol{e}_i$	$-\boldsymbol{e}_i w(t, \boldsymbol{x}, \boldsymbol{\Delta}_2)$	$\boldsymbol{e}_i \boldsymbol{e}_i' w(t, \boldsymbol{x}, \boldsymbol{\Delta}_2)$
$\boldsymbol{\Delta}_3 = \varepsilon(\boldsymbol{e}_i + \boldsymbol{e}_j)$	$(\boldsymbol{e}_i + \boldsymbol{e}_j) w(t, \boldsymbol{x}, \boldsymbol{\Delta}_3)$	$(\boldsymbol{e}_i + \boldsymbol{e}_j)(\boldsymbol{e}_i + \boldsymbol{e}_j)' w(t, \boldsymbol{x}, \boldsymbol{\Delta}_3)$
$\boldsymbol{\Delta}_4 = -\varepsilon(\boldsymbol{e}_i + \boldsymbol{e}_j)$	$-(\boldsymbol{e}_i + \boldsymbol{e}_j) w(t, \boldsymbol{x}, \boldsymbol{\Delta}_4)$	$(\boldsymbol{e}_i + \boldsymbol{e}_j)(\boldsymbol{e}_i + \boldsymbol{e}_j)' w(t, \boldsymbol{x}, \boldsymbol{\Delta}_4)$
$\boldsymbol{\Delta}_5 = \varepsilon(\boldsymbol{e}_i - \boldsymbol{e}_j)$	$(\boldsymbol{e}_i - \boldsymbol{e}_j) w(t, \boldsymbol{x}, \boldsymbol{\Delta}_5)$	$(\boldsymbol{e}_i - \boldsymbol{e}_j)(\boldsymbol{e}_i - \boldsymbol{e}_j)' w(t, \boldsymbol{x}, \boldsymbol{\Delta}_5)$

This result coincides with the one that would have been obtained by application of the Langevin approach or Kramers-Moyal expansion, which are introduced in Sects. 4.3.3 and 4.3.4. It is also valid for the approximation of the infinitesimal generator, considered in Sect. 4.3.2.

B.2 Lipschitz Continuity for SIR Models

A most tractable way to prove the existence and uniqueness of a strong solution of a given SDE is to verify Lipschitz continuity of the drift $\boldsymbol{\mu}$ and diffusion coefficient $\boldsymbol{\sigma}$ (cf. Sect. 3.2.3). One thus has to show that there is a positive constant C such that for all t in the time set and all $\boldsymbol{x}, \boldsymbol{y}$ in the state space

$$\|\boldsymbol{\mu}(\boldsymbol{x},t) - \boldsymbol{\mu}(\boldsymbol{y},t)\| + \|\boldsymbol{\sigma}(\boldsymbol{x},t) - \boldsymbol{\sigma}(\boldsymbol{y},t)\| \leq C\|\boldsymbol{x} - \boldsymbol{y}\|, \tag{B.8}$$

where $\|\boldsymbol{A}\|^2 = \text{tr}(\boldsymbol{A}'\boldsymbol{A})$ denotes the Euclidean norm. The solution is non-explosive if

$$\|\boldsymbol{\mu}(\boldsymbol{x},t)\|^2 + \|\boldsymbol{\sigma}(\boldsymbol{x},t)\|^2 \leq D\big(1 + \|\boldsymbol{x}\|^2\big) \tag{B.9}$$

for all t and $\boldsymbol{x}$. These two properties are investigated in the following for the diffusion models derived in Chap. 5. The results are discussed in Sect. 5.3.

B.2.1 Standard SIR Model

Let $\boldsymbol{x}_1 = (s_1, i_1)'$ and $\boldsymbol{x}_2 = (s_2, i_2)'$ denote arbitrary elements of the state space of the standard SIR model. Formula (B.8) is true if and only if there are positive constants C_1 and C_2 such that for all t

$$\|\boldsymbol{\mu}(\boldsymbol{x}_1,t) - \boldsymbol{\mu}(\boldsymbol{x}_2,t)\|^2 \leq C_1\|\boldsymbol{x}_1 - \boldsymbol{x}_2\|^2 \quad \text{and} \quad \|\boldsymbol{\sigma}(\boldsymbol{x}_1,t) - \boldsymbol{\sigma}(\boldsymbol{x}_2,t)\|^2 \leq C_2\|\boldsymbol{x}_1 - \boldsymbol{x}_2\|^2.$$

The first inequality is

$$\begin{aligned} &2\alpha^2(s_1 i_1 - s_2 i_2)^2 - 2\alpha\beta(s_1 i_1 - s_2 i_2)(i_1 - i_2) + \beta^2(i_1 - i_2)^2 \\ &\quad \leq C_1\left((s_1 - s_2)^2 + (i_1 - i_2)^2\right). \end{aligned} \tag{B.10}$$

In what follows, the three summands on the left are considered separately. First, one has

$$\begin{aligned} (s_1 i_1 - s_2 i_2)^2 &= \big((s_1 - s_2)i_1 + s_2(i_1 - i_2)\big)^2 \\ &\leq (s_1 - s_2)^2 + (i_1 - i_2)^2 + 2i_1 s_2(s_1 - s_2)(i_1 - i_2). \end{aligned}$$

The product $(s_1 - s_2)(i_1 - i_2)$ is either negative and can be ignored, or it is positive and less than or equal to $\max\{(s_1 - s_2)^2, (i_1 - i_2)^2\}$. In any case, there is a constant $\kappa_1 > 0$ such that

$$2\alpha^2(s_1 i_1 - s_2 i_2)^2 \leq \kappa_1\Big((s_1 - s_2)^2 + (i_1 - i_2)^2\Big). \tag{B.11}$$

For the second summand on the left of (B.10), one has

$$-(s_1 i_1 - s_2 i_2)(i_1 - i_2) = -(s_1 - s_2)(i_1 - i_2)i_1 - s_2(i_1 - i_2)^2 \leq (s_2 - s_1)(i_1 - i_2)i_1.$$

The product is of importance only if $(s_2 - s_1)(i_1 - i_2)$ is positive. In that case, one has $(s_2 - s_1)(i_1 - i_2)i_1 \leq \max\{(s_1 - s_2)^2, (i_1 - i_2)^2\}$, and hence

$$-2\alpha\beta(s_1 i_1 - s_2 i_2)(i_1 - i_2) \leq \kappa_2\Big((s_1 - s_2)^2 + (i_1 - i_2)^2\Big)$$

for an appropriate $\kappa_2 > 0$. The third summand on the left of (B.10) can trivially be bounded by the term on the right. Altogether, the inequality (B.10) is satisfied, i.e. the drift vector $\boldsymbol{\mu}$ fulfils the required Lipschitz condition.

For Lipschitz continuity of the diffusion coefficient $\boldsymbol{\sigma}$, one needs to show

$$\frac{1}{N}\Big(2\alpha\big(\sqrt{s_1 i_1}-\sqrt{s_2 i_2}\big)^2+\beta\big(\sqrt{i_1}-\sqrt{i_2}\big)^2\Big) \leq C_2\Big((s_1-s_2)^2+(i_1-i_2)^2\Big). \quad \text{(B.12)}$$

It is, however, well known that the function $f(x) = \sqrt{x}$ is *not* Lipschitz continuous in $x = 0$. Hence, the inequality (B.12) cannot be true for any of the variables s_1, s_2, i_1, i_2 being equal to zero. If, on the other hand, one requires $s_1, s_2, i_1, i_2 > \varepsilon$ for some small but positive fixed ε, Lipschitz continuity is given as shown in the following. In that case, one has

$$\begin{aligned}\big(\sqrt{s_1 i_1} - \sqrt{s_2 i_2}\big)^2 &= \left(\frac{s_1 i_1 - s_2 i_2}{\sqrt{s_1 i_1} + \sqrt{s_2 i_2}}\right)^2 \\ &\leq \frac{(s_1 i_1 - s_2 i_2)^2}{4\varepsilon^2} \overset{\text{(B.11)}}{\leq} \kappa_3\Big((s_1 - s_2)^2 + (i_1 - i_2)^2\Big)\end{aligned}$$

for suitable $\kappa_3 > 0$. Similarly, there is some $\kappa_4 > 0$ such that

$$\big(\sqrt{i_1} - \sqrt{i_2}\big)^2 = \left(\frac{i_1 - i_2}{\sqrt{i_1} + \sqrt{i_2}}\right)^2 \leq \frac{(i_1 - i_2)^2}{4\varepsilon} \leq \kappa_4\Big((s_1 - s_2)^2 + (i_1 - i_2)^2\Big).$$

That proves (B.12).

Provided that a solution of an SDE exists, it does not explode when (B.9) is true. This condition is met for the considered SIR model as shown next: For $\boldsymbol{x} = (s, i)'$, the inequality (B.9) reads

$$2\alpha^2 s^2 i^2 - 2\alpha\beta s i^2 + \beta^2 i^2 + \frac{2\alpha s i + \beta i}{N} \leq D\left(1 + s^2 + i^2\right).$$

Because of $s, i \in [0, 1]$, the left hand side of this expression is bounded, and the inequality is trivially fulfilled.

B.2.2 Multitype SIR Model

Instead of formally proving conditions (B.8) and (B.9) also for the multitype SIR model, this section heuristically motivates why Lipschitz continuity must hold for this model on the restricted state space, and why the solution does not explode. That is because the components of the drift vector $\boldsymbol{\mu}$ and diffusion coefficient $\boldsymbol{\sigma}$ in the multitype SIR model have the same structure as those in the standard SIR model. In particular, all components of $\boldsymbol{\mu}$ and $\boldsymbol{\sigma}$ contain the fractions of susceptibles and infectives either linearly or as a product $s_j i_m$ with $j, m \in \{1, \ldots, n\}$. Hence, the validation of (B.8) and (B.9) works as in Sect. B.2.1 but is definitely more elaborate.

B.3 On the Choice of the Update Interval

This section deals with the appropriate choice of an update interval for a sample path in the context of the MCMC scheme introduced in Sect. 7.1. The choice of the update interval is discussed in Sect. 7.1.6.

Assume we have $S + 1$ observed or imputed consecutive data points $\boldsymbol{Y}_0, \boldsymbol{Y}_1, \ldots, \boldsymbol{Y}_S$. Setting the update interval equal to (a, b) implies proposing new values for $\{\boldsymbol{Y}_{a+1}, \ldots, \boldsymbol{Y}_{b-1}\}$. The interval (a, b) should be chosen in a way such that $a, b \in \{0, 1, \ldots, S\}$ and $b - a \geq 2$. Furthermore, the number of points in (a, b) shall be bounded by $R \leq S - 1$. Algorithm 7.1 on p. 191 presents a simple procedure to randomly draw such an interval (a, b). However, this strategy updates data points near the boundaries of the time interval less frequently than those in the centre. This fact is elucidated in the following.

Let $1 \leq k \leq S - 1$. The probability that k is included in the interior of (a, b) equals

$$\begin{aligned}
\mathbb{P}\big(k \in (a, b)\big) &= \sum_{i=0}^{k-1} \sum_{j=k+1}^{S} \mathbb{P}\big(a = i \wedge b = j\big) \\
&= \sum_{i=0}^{k-1} \sum_{j=k+1}^{S} \mathbb{P}\big(a = i\big)\, \mathbb{P}\big(b = j \,\big|\, a = i\big) \\
&= \sum_{i=0}^{k-1} \sum_{j=k+1}^{S} \frac{1}{S-1} \cdot \frac{\mathbb{1}\big(i + 2 \leq j \leq \min\{i + R + 1, S\}\big)}{\min\{i + R + 1, S\} - (i + 1)} \\
&= \frac{1}{S-1} \sum_{i=0}^{k-1} \sum_{j=k+1}^{\min\{i+R+1,S\}} \frac{1}{\min\{i + R + 1, S\} - (i + 1)},
\end{aligned} \tag{B.13}$$

where $\mathbb{1}$ denotes the indicator function, i.e. $\mathbb{1}(A)$ equals one if A is true and zero otherwise. The inner sum in (B.13) equals zero if $\min\{i + R + 1, S\} < k + 1$, that is $i < k - R$. Hence, one has

$$\sum_{j=k+1}^{\min\{i+R+1,S\}} \frac{1}{\min\{i + R + 1, S\} - i - 1} = \begin{cases} 0 & \text{if } i < k - R, \\ \dfrac{\min\{i + R + 1, S\} - k}{\min\{i + R + 1, S\} - i - 1} & \text{otherwise.} \end{cases}$$

Overall,

$$\mathbb{P}\big(k \in (a,b)\big) = \frac{1}{S-1} \sum_{i=\max\{0,k-R\}}^{k-1} \frac{\min\{i+R+1,S\}-k}{\min\{i+R+1,S\}-i-1} \,.$$

This probability is constant for $R \leq k \leq S-R$, that is where $\max\{0,k-R\} = k-R$ and $\min\{i+R+1,S\} = i+R+1$ for all $i = 0,\ldots,k-1$. For $k < R$ or $k > S-R$, however, the probability is generally lower, because the number of possible intervals (a,b) covering points near the boundaries of $(0,S)$ is less than the number of intervals covering points in the centre. To correct for this disparity, extend both boundaries of $(0,S)$ by $R-1$ and draw an interval (a^*,b^*) within $(1-R, S+R-1)$. If $a^*, b^* \notin \{0,\ldots,S\}$, adjust them respectively. The corrected procedure is carried out by Algorithm 7.2. The achievement is that $\mathbb{P}(k \in (a,b))$ is constant for $k = 1,\ldots,S-1$.

B.4 Posteriori Densities for the Ornstein-Uhlenbeck Process

This section provides the calculation of the exact and approximate full conditional densities for the model parameters of the one-dimensional Ornstein-Uhlenbeck process. These are utilised in Chap. 7 for the illustration of the MCMC scheme considered there.

The Ornstein-Uhlenbeck $X = (X_t)_{t\geq 0}$ process is the solution of the SDE

$$\mathrm{d}X_t = \alpha(\beta - X_t)\mathrm{d}t + \sigma \mathrm{d}B_t, \qquad X_0 = x_0,$$

for parameters $\beta \in \mathbb{R}$, $\alpha, \sigma \in \mathbb{R}_+$ and initial value $x_0 \in \mathbb{R}$. This is a Gaussian process; its explicit form and transition density are shown in Sect. A.2.

In the following we consider both proper and improper priors for the model parameters. When improper priors are involved, the joint posterior density of all parameters might be improper as well even if the full conditional densities are not. In that case the posterior distribution is not well-defined. We hence start with the joint posterior density and investigate for which priors its integral is finite.

Exact Posterior Density

Assume we have observations $Y_0,\ldots,Y_m$ of an Ornstein-Uhlenbeck process at times $t_0,\ldots,t_m$, where Y_0 is the predefined initial value. Then the exact joint posterior density of α, β and σ^2 is

$$\begin{aligned}
&\pi\big(\alpha,\beta,\sigma^2 \,\big|\, Y_0,\ldots,Y_m\big)\\
&\quad\propto \pi\big(Y_0,\ldots,Y_m \,\big|\, \alpha,\beta,\sigma^2\big)p(\alpha,\beta,\sigma^2)\\
&\quad\propto \left(\prod_{k=0}^{m-1}\pi\big(Y_{k+1}\,\big|\,Y_k,\alpha,\beta,\sigma^2\big)\right)p(\alpha,\beta,\sigma^2)\\
&\quad= \left(\prod_{k=0}^{m-1}\phi\big(Y_{k+1}\,\big|\,\mu(\Delta t_k,Y_k,\alpha,\beta),\Sigma(\Delta t_k,Y_k,\alpha,\sigma)\big)\right)p(\alpha,\beta,\sigma^2),
\end{aligned}$$

where ϕ denotes the Gaussian density and

$$\mu(\Delta t_k,Y_k,\alpha,\beta) = Y_k \mathrm{e}^{-\alpha\Delta t_k} + \beta\left(1-\mathrm{e}^{-\alpha\Delta t_k}\right)$$

and

$$\Sigma(\Delta t_k,Y_k,\alpha,\sigma) = \frac{\sigma^2}{2\alpha}\left(1-\mathrm{e}^{-2\alpha\Delta t_k}\right).$$

The joint posterior density $\pi(\alpha,\beta,\sigma^2|Y_0,\ldots,Y_m)$ of all parameters is hence proportional to

$$\frac{p(\alpha,\beta,\sigma^2)\exp\left(-\dfrac{\alpha}{\sigma^2}\displaystyle\sum_{k=0}^{m-1}\frac{\left(Y_{k+1}-Y_k\mathrm{e}^{-\alpha\Delta t_k}-\beta\left(1-\mathrm{e}^{-\alpha\Delta t_k}\right)\right)^2}{1-\mathrm{e}^{-2\alpha\Delta t_k}}\right)}{\left(\dfrac{\sigma^2}{\alpha}\right)^{m/2}\displaystyle\prod_{k=0}^{m-1}\sqrt{1-\mathrm{e}^{-2\alpha\Delta t_k}}}. \tag{B.14}$$

Suppose that a priori the model parameters are mutually independent. More specifically, let

$$\beta \sim \mathcal{N}(\beta_0,\rho_\beta^2) \qquad \text{and} \qquad \sigma^2 \sim \mathrm{IG}(\kappa_0,\nu_0) \tag{B.15}$$

for $\beta_0 \in \mathbb{R}$, $\rho_\beta \in \mathbb{R}_+ \cup \{+\infty\}$ and $(\kappa_0,\nu_0) \in \mathbb{R}_+^2 \cup \{(-1,0),(0,0)\}$. With these parameter ranges we explicitly include the improper priors $p(\beta) \propto 1$, $p(\sigma^2) \propto 1$ and $p(\sigma^2) \propto \sigma^{-2}$. The choice of $p(\alpha)$ will be considered later.

We want to investigate if the joint posteriori density (B.14) is proper for our choice of prior densities, i.e. whether the integral of (B.14) over all α, β and σ^2 is finite. Consider first the joint marginal posterior density of α and β,

$$\begin{aligned}
\pi\big(\alpha,\beta\,\big|\,Y_0,\ldots,Y_m\big) &= \int_0^\infty \pi\big(\alpha,\beta,\sigma^2\,\big|\,Y_0,\ldots,Y_m\big)\mathrm{d}\sigma^2\\
&\propto \frac{p(\alpha,\beta)\alpha^{\frac{m}{2}}}{\displaystyle\prod_{k=0}^{m-1}\sqrt{1-\mathrm{e}^{-2\alpha\Delta t_k}}}\int_0^\infty (\sigma^2)^{-\left(\frac{m}{2}+\kappa_0+1\right)}\exp\left(-\frac{K+\nu_0}{\sigma^2}\right)\mathrm{d}\sigma^2,
\end{aligned}$$

where

$$K = \alpha \sum_{k=0}^{m-1} \frac{\left(Y_{k+1} - Y_k \mathrm{e}^{-\alpha \Delta t_k} - \beta(1 - \mathrm{e}^{-\alpha \Delta t_k})\right)^2}{1 - \mathrm{e}^{-2\alpha \Delta t_k}} .$$

The integrand is the unnormalised density of an inverse gamma distribution with parameters $m/2 + \kappa_0$ and $K + \nu_0$. As m is usually greater than two, both parameters are positive. Hence,

$$\pi\big(\alpha, \beta \,\big|\, Y_0, \ldots, Y_m\big) \propto \frac{p(\alpha, \beta)\alpha^{\frac{m}{2}}}{\prod_{k=0}^{m-1} \sqrt{1 - \mathrm{e}^{-2\alpha \Delta t_k}}} \big(\nu_0 + K\big)^{-\left(\frac{m}{2} + \kappa_0\right)} .$$

Next, integrate out β. One has

$$\begin{aligned} \pi\big(\alpha \,\big|\, Y_0, \ldots, Y_m\big) &= \int_{-\infty}^{\infty} \pi\big(\alpha, \beta \,\big|\, Y_0, \ldots, Y_m\big) \mathrm{d}\beta \\ &\propto \frac{p(\alpha)\alpha^{\frac{m}{2}}}{\prod_{k=0}^{m-1} \sqrt{1 - \mathrm{e}^{-2\alpha \Delta t_k}}} \int_{-\infty}^{\infty} \exp\left(-\frac{(\beta - \beta_0)^2}{2\rho_\beta^2}\right) \big(\nu_0 + K\big)^{-\left(\frac{m}{2} + \kappa_0\right)} \mathrm{d}\beta . \end{aligned}$$

The first factor in the integrand is less than or equal to one for all choices of β_0 and ρ_β. It can hence be omitted when we are interested only in an upper bound for the posterior of α. The second factor can be rewritten as

$$\big(\nu_0 + K\big)^{-\left(\frac{m}{2} + \kappa_0\right)} = \left(\nu_0 + \alpha \sum_{k=0}^{m-1} \frac{(\beta - \beta_k)^2}{c_k}\right)^{-\left(\frac{m}{2} + \kappa_0\right)} ,$$

where

$$\beta_k = \frac{Y_{k+1} - Y_k \mathrm{e}^{-\alpha \Delta t_k}}{1 - \mathrm{e}^{-\alpha \Delta t_k}} \quad \text{and} \quad c_k = \frac{1 + \mathrm{e}^{-\alpha \Delta t_k}}{1 - \mathrm{e}^{-\alpha \Delta t_k}} .$$

Further rearranging yields

$$\begin{aligned} &\big(\nu_0 + K\big)^{-\left(\frac{m}{2} + \kappa_0\right)} \\ &= \left(1 + \frac{1}{m + 2\kappa_0 - 1} \cdot \frac{m + 2\kappa_0 - 1}{\frac{\nu_0}{a_1 \alpha} + \frac{a_3}{a_1} - \frac{a_2^2}{a_1^2}} \cdot \left(\beta - \frac{a_2}{a_1}\right)^2\right)^{-\frac{m + 2\kappa_0}{2}} \end{aligned} \tag{B.16}$$

with

$$a_1 = \sum_{k=0}^{m-1} \frac{1}{c_k}, \quad a_2 = \sum_{k=0}^{m-1} \frac{\beta_k}{c_k} \quad \text{and} \quad a_3 = \sum_{k=0}^{m-1} \frac{\beta_k^2}{c_k} .$$

This is the unnormalised density of the univariate t-distribution with mean a_2/a_1, scale parameter $\sqrt{m + 2\kappa_0 - 1}^{\,-1} \sqrt{\nu_0/a_1\alpha + a_3/a_1 - a_2^2/a_1^2}$ and $m + 2\kappa_0 - 1$ degrees of freedom. The scale parameter is well-defined as $a_1 a_3 - a_2^2 \geq 0$ due to the Cauchy-Schwarz inequality. The integral of (B.16) over all β is proportional to the scale parameter, i.e.

$$\pi\big(\alpha \big| Y_0, \ldots, Y_m\big) \le C\, p(\alpha) \left(\prod_{k=0}^{m-1} \sqrt{\frac{\alpha}{1 - \mathrm{e}^{-2\alpha \Delta t_k}}}\right) \sqrt{\frac{\nu_0}{a_1 \alpha} + \frac{a_3}{a_1} - \frac{a_2^2}{a_1^2}} \tag{B.17}$$

for some constant $C \in \mathbb{R}_+$. For $\alpha \in \mathbb{R}_+$ fixed, i.e. $p(\alpha) = \delta(\alpha - \alpha_0)$ being the Dirac delta function with positive α_0, the integral of this expression is finite, that means the joint posterior density of α, β and σ^2 is proper. Otherwise, a sufficient criterion to obtain a proper posterior is that (B.17) is normalisable.

In the simulation study in Sect. 7.1.7, the parameter $\alpha \in \mathbb{R}_+$ is considered fixed. In that case, one obtains a proper posterior if β and σ^2 are chosen according to (B.15). The latter explicitly includes improper priors.

Exact Full Conditional Densities

We now derive the full conditional densities for the three parameters of the Ornstein-Uhlenbeck process. The existence of a proper full conditional density does however not automatically imply a proper joint posterior distribution. That is why the following formulas should only be applied in an MCMC algorithm after one has confirmed that the chosen combination of prior distributions implies a proper posterior.

All full conditional densities are proportional to the joint posterior density and hence to the expression (B.14). They are obtained by dropping all multiplicative terms which are constant with respect to the considered parameter. Suppose that a priori the model parameters are mutually independent. Then the full conditional density for the parameter β is

$$\begin{aligned}
&\pi\big(\beta \,\big|\, \alpha, \sigma^2, Y_0, \ldots, Y_m\big) \\
&\propto p(\beta) \exp\left(-\frac{\alpha}{\sigma^2} \sum_{k=0}^{m-1} \frac{\big(Y_{k+1} - Y_k \mathrm{e}^{-\alpha \Delta t_k} - \beta\big(1 - \mathrm{e}^{-\alpha \Delta t_k}\big)\big)^2}{1 - \mathrm{e}^{-2\alpha \Delta t_k}}\right) \\
&\propto p(\beta) \exp\left(-\frac{\alpha}{\sigma^2}\left[\beta^2 \sum_{k=0}^{m-1} \frac{\big(1 - \mathrm{e}^{-\alpha \Delta t_k}\big)^2}{1 - \mathrm{e}^{-2\alpha \Delta t_k}}\right.\right. \\
&\qquad\qquad \left.\left. -2\beta \sum_{k=0}^{m-1} \frac{\big(Y_{k+1} - Y_k \mathrm{e}^{-\alpha \Delta t_k}\big)\big(1 - \mathrm{e}^{-\alpha \Delta t_k}\big)}{1 - \mathrm{e}^{-2\alpha \Delta t_k}}\right]\right) \\
&= p(\beta) \exp\left(-\frac{1}{2} \cdot \frac{2\alpha}{\sigma^2} \left(\sum_{k=0}^{m-1} \frac{1 - \mathrm{e}^{-\alpha \Delta t_k}}{1 + \mathrm{e}^{-\alpha \Delta t_k}}\right) \left[\beta^2 - 2\beta \frac{\sum_{k=0}^{m-1} \frac{Y_{k+1} - Y_k \mathrm{e}^{-\alpha \Delta t_k}}{1 + \mathrm{e}^{-\alpha \Delta t_k}}}{\sum_{k=0}^{m-1} \frac{1 - \mathrm{e}^{-\alpha \Delta t_k}}{1 + \mathrm{e}^{-\alpha \Delta t_k}}}\right]\right).
\end{aligned}$$

In case of a flat prior $p(\beta) \propto 1$ for $\beta \in \mathbb{R}$, this is an unnormalised Gaussian density,

$$\beta|\alpha, \sigma^2, Y_0, \ldots, Y_m \sim \mathcal{N}\left(\frac{\displaystyle\sum_{k=0}^{m-1} \frac{Y_{k+1} - Y_k \mathrm{e}^{-\alpha \Delta t_k}}{1 + \mathrm{e}^{-\alpha \Delta t_k}}}{\displaystyle\sum_{k=0}^{m-1} \frac{1 - \mathrm{e}^{-\alpha \Delta t_k}}{1 + \mathrm{e}^{-\alpha \Delta t_k}}}, \frac{\dfrac{\sigma^2}{2\alpha}}{\displaystyle\sum_{k=0}^{m-1} \frac{1 - \mathrm{e}^{-\alpha \Delta t_k}}{1 + \mathrm{e}^{-\alpha \Delta t_k}}} \right). \quad \text{(B.18)}$$

For $\beta \sim \mathcal{N}(\beta_0, \rho_\beta^2)$ with $\beta_0 \in \mathbb{R}$ and $\rho_\beta \in \mathbb{R}_+$, the full conditional density becomes

$$\begin{aligned}
&\pi\big(\beta \,\big|\, \alpha, \sigma^2, Y_0, \ldots, Y_m\big) \\
&\propto \exp\left(-\frac{1}{2} \left[\beta^2 \left(\frac{1}{\rho_\beta^2} + \frac{2\alpha}{\sigma^2} \sum_{k=0}^{m-1} \frac{1 - \mathrm{e}^{-\alpha \Delta t_k}}{1 + \mathrm{e}^{-\alpha \Delta t_k}} \right) \right.\right. \\
&\qquad \left.\left. -2\beta \left(\frac{\beta_0}{\rho_\beta^2} + \frac{2\alpha}{\sigma^2} \sum_{k=0}^{m-1} \frac{Y_{k+1} - Y_k \mathrm{e}^{-\alpha \Delta t_k}}{1 + \mathrm{e}^{-\alpha \Delta t_k}} \right) \right] \right) \\
&= \exp\left(-\frac{1}{2} \left(\frac{1}{\rho_\beta^2} + \frac{2\alpha}{\sigma^2} \sum_{k=0}^{m-1} \frac{1 - \mathrm{e}^{-\alpha \Delta t_k}}{1 + \mathrm{e}^{-\alpha \Delta t_k}} \right) \right. \\
&\qquad \left. \cdot \left[\beta^2 - 2\beta \frac{\dfrac{\beta_0}{\rho_\beta^2} + \dfrac{2\alpha}{\sigma^2} \displaystyle\sum_{k=0}^{m-1} \frac{Y_{k+1} - Y_k \mathrm{e}^{-\alpha \Delta t_k}}{1 + \mathrm{e}^{-\alpha \Delta t_k}}}{\dfrac{1}{\rho_\beta^2} + \dfrac{2\alpha}{\sigma^2} \displaystyle\sum_{k=0}^{m-1} \frac{1 - \mathrm{e}^{-\alpha \Delta t_k}}{1 + \mathrm{e}^{-\alpha \Delta t_k}}} \right] \right).
\end{aligned}$$

The resulting density is again Gaussian, in particular

$$\begin{aligned}
&\beta \,\big|\, \alpha, \sigma^2, Y_0, \ldots, Y_m \\
&\sim \mathcal{N}\left(\frac{\dfrac{\sigma^2 \beta_0}{2\alpha\rho_\beta^2} + \displaystyle\sum_{k=0}^{m-1} \frac{Y_{k+1} - Y_k \mathrm{e}^{-\alpha \Delta t_k}}{1 + \mathrm{e}^{-\alpha \Delta t_k}}}{\dfrac{\sigma^2}{2\alpha\rho_\beta^2} + \displaystyle\sum_{k=0}^{m-1} \frac{1 - \mathrm{e}^{-\alpha \Delta t_k}}{1 + \mathrm{e}^{-\alpha \Delta t_k}}}, \frac{\dfrac{\sigma^2}{2\alpha}}{\dfrac{\sigma^2}{2\alpha\rho_\beta^2} + \displaystyle\sum_{k=0}^{m-1} \frac{1 - \mathrm{e}^{-\alpha \Delta t_k}}{1 + \mathrm{e}^{-\alpha \Delta t_k}}} \right).
\end{aligned}$$

Note that for $\rho_\beta = \infty$ this expression equals Eq. (B.18). For the parameter σ^2, the full conditional density fulfils

$$\pi\left(\sigma^2 \,\middle|\, \alpha, \beta, Y_0, \ldots, Y_m\right) \propto p(\sigma^2)\,\sigma^{-m} \exp\left(-\frac{\alpha}{\sigma^2}\sum_{k=0}^{m-1}\frac{\left(Y_{k+1}-Y_k \mathrm{e}^{-\alpha \Delta t_k} - \beta\left(1-\mathrm{e}^{-\alpha \Delta t_k}\right)\right)^2}{1-\mathrm{e}^{-2\alpha \Delta t_k}}\right).$$

If one chooses a flat prior $p(\sigma^2) \propto 1$ for $\sigma^2 \in \mathbb{R}_+$, the above is an unnormalised inverse gamma density of

$$\sigma^2 \,\middle|\, \alpha, \beta, Y_0, \ldots, Y_m \sim \mathrm{IG}\left(\frac{m}{2}-1\,,\ \alpha\sum_{k=0}^{m-1}\frac{\left(Y_{k+1}-Y_k \mathrm{e}^{-\alpha \Delta t_k} - \beta\left(1-\mathrm{e}^{-\alpha \Delta t_k}\right)\right)^2}{1-\mathrm{e}^{-2\alpha \Delta t_k}}\right).$$

For $\sigma^2 \sim \mathrm{IG}(\kappa_0, \nu_0)$ for $\kappa_0, \nu_0 \in \mathbb{R}_+$, the full conditional density is proportional to

$$\left(\sigma^2\right)^{-(m/2+\kappa_0+1)} \exp\left(-\frac{1}{\sigma^2}\left(\nu_0 + \alpha\sum_{k=0}^{m-1}\frac{\left(Y_{k+1}-Y_k \mathrm{e}^{-\alpha \Delta t_k} - \beta\left(1-\mathrm{e}^{-\alpha \Delta t_k}\right)\right)^2}{1-\mathrm{e}^{-2\alpha \Delta t_k}}\right)\right),$$

that is

$$\sigma^2 \,\middle|\, \alpha, \beta, Y_0, \ldots, Y_m \sim \mathrm{IG}\left(\frac{m}{2}+\kappa_0, \nu_0 + \alpha\sum_{k=0}^{m-1}\frac{\left(Y_{k+1}-Y_k \mathrm{e}^{-\alpha \Delta t_k} - \beta\left(1-\mathrm{e}^{-\alpha \Delta t_k}\right)\right)^2}{1-\mathrm{e}^{-2\alpha \Delta t_k}}\right).$$

For $\kappa_0 = -1$ and $\nu_0 = 0$, one again arrives at the result derived for the flat prior. The full conditional density of α,

$$\pi\left(\alpha \,\middle|\, \beta, \sigma^2, Y_0, \ldots, Y_m\right) \propto \frac{p(\alpha)\alpha^{m/2} \exp\left(-\frac{\alpha}{\sigma^2}\sum_{k=0}^{m-1}\frac{\left(Y_{k+1}-Y_k \mathrm{e}^{-\alpha \Delta t_k} - \beta\left(1-\mathrm{e}^{-\alpha \Delta t_k}\right)\right)^2}{1-\mathrm{e}^{-2\alpha \Delta t_k}}\right)}{\prod_{k=0}^{m-1}\sqrt{1-\mathrm{e}^{-2\alpha \Delta t_k}}},$$

cannot be recognised to be of any standard distribution type.

Approximate Posterior Density

The exact transition density is usually unknown, but can for small Δt_k be approximated by application of the Euler scheme

$$Y_{k+1} \sim \mathcal{N}\left(Y_k + \alpha(\beta - Y_k)\Delta t_k, \sigma^2 \Delta t_k\right)$$

for $k = 0, \dots, m-1$. The approximate joint posterior density of α, β and σ^2 then is

$$\begin{aligned}
&\pi\big(\alpha, \beta, \sigma^2 \big| Y_0, \dots, Y_m\big) \\
&\quad \propto \left(\prod_{k=0}^{m-1} \phi\big(Y_{k+1} \,\big|\, Y_k + \alpha(\beta - Y_k)\Delta t_k, \sigma^2 \Delta t_k\big) \right) p(\alpha, \beta, \sigma^2) \\
&\quad \propto p(\alpha, \beta, \sigma^2)\sigma^{-m} \exp\left(-\frac{1}{2} \sum_{k=0}^{m-1} \frac{\big(Y_{k+1} - Y_k - \alpha(\beta - Y_k)\Delta t_k\big)^2}{\sigma^2 \Delta t_k} \right). \qquad \text{(B.19)}
\end{aligned}$$

The remarks on proper and improper posterior densities on p. 388 naturally also apply for the approximate densities. Thus, we first consider in which cases the approximate posterior is normalisable before deriving the approximate full conditional distributions.

Choose the prior densities as in (B.15). Integrating the joint posterior density over all σ^2 yields

$$\pi\big(\alpha, \beta \,\big|\, Y_0, \dots, Y_m\big) \propto p(\alpha, \beta) \int_0^\infty (\sigma^2)^{-(\frac{m}{2} + \kappa_0 + 1)} \exp\left(-\frac{1}{\sigma^2} \cdot (\nu_0 + K) \right) d\sigma^2,$$

where

$$K = \frac{1}{2} \sum_{k=0}^{m-1} \frac{\big(Y_{k+1} - Y_k - \alpha(\beta - Y_k)\Delta t_k\big)^2}{\Delta t_k}.$$

The integrand is an unnormalised inverse gamma density with parameters $m/2 + \kappa_0$ and $\nu_0 + K$. As in the consideration of the exact posterior density on p. 390, these hyperparameters are usually well-defined. Therefore,

$$\pi\big(\alpha, \beta \,\big|\, Y_0, \dots, Y_m\big) \propto p(\alpha, \beta)\big(\nu_0 + K\big)^{-\frac{m + 2\kappa_0}{2}}.$$

Now integrate this expression over all β, that is

$$\pi\big(\alpha \,\big|\, Y_0, \dots, Y_m\big) \propto p(\alpha) \int_{-\infty}^{\infty} \exp\left(-\frac{(\beta - \beta_0)^2}{2\rho_\beta^2} \right) \big(\nu_0 + K\big)^{-\frac{m + 2\kappa_0}{2}} d\beta.$$

The exponential function in the integrand is less than or equal to one for all values of β_0 and ρ_β. Suppress this factor to obtain an upper bound of the integral. Furthermore, rewrite

$$\big(\nu_0 + K\big)^{-\frac{m + 2\kappa_0}{2}} = \left(\nu_0 + \frac{\alpha^2}{2} \sum_{k=0}^{m-1} \Delta t_k (\beta - \beta_k)^2 \right)^{-\frac{m + 2\kappa_0}{2}}$$

with $\beta_k = Y_k + (Y_{k+1} - Y_k)/\alpha\Delta t_k$, and furthermore

$$(\nu_0 + K)^{-\frac{m+2\kappa_0}{2}} = \left(1 + \frac{1}{m+2\kappa_0-1} \cdot \frac{m+2\kappa_0-1}{\frac{2\nu_0}{b_1\alpha^2} + \frac{b_3}{b_1} - \frac{b_2^2}{b_1^2}} \cdot \left(\beta - \frac{b_2}{b_1}\right)^2\right)^{-\frac{m+2\kappa_0}{2}}$$

with

$$b_1 = \sum_{k=0}^{m-1} \Delta t_k\,, \quad b_2 = \sum_{k=0}^{m-1} \Delta t_k \beta_k \quad \text{and} \quad b_3 = \sum_{k=0}^{m-1} \Delta t_k \beta_k^2\,.$$

This is the unnormalised density of a univariate t-distribution with mean b_2/b_1, scale parameter $\sqrt{m+2\kappa_0-1}^{-1}\sqrt{2\nu_0/b_1\alpha^2 + b_3/b_1 - b_2^2/b_1^2}$ and $m+2\kappa_0-1$ degrees of freedom. Once again, one can easily verify with the Cauchy-Schwarz inequality that the scale parameter is well-defined. Thus there exists a constant $C \in \mathbb{R}_+$ such that

$$\pi\big(\alpha \,\big|\, Y_0, \ldots, Y_m\big) \le C\, p(\alpha) \sqrt{\frac{2\nu_0}{b_1\alpha^2} + \frac{b_3}{b_1} - \frac{b_2^2}{b_1^2}}\,.$$

If $p(\alpha)$ is chosen such that the integral of this expression over all α is finite, the joint posterior distribution of all model parameters is proper. In the simulation study in Sect. 7.1.7 this condition is fulfilled as α is considered fixed.

Approximate Full Conditional Densities

In case the prior densities are chosen such that the posterior distribution is proper, the full conditionals are proportional to (B.19). Let the prior densities of all parameters be independent. Then for the full conditional density of α one obtains

$$\begin{aligned}
&\pi\big(\alpha \,\big|\, \beta, \sigma^2, Y_0, \ldots, Y_m\big) \\
&\quad \propto p(\alpha) \exp\left(-\frac{1}{2}\left[\alpha^2 \sum_{k=0}^{m-1} \frac{\big(\beta - Y_k\big)^2 \Delta t_k}{\sigma^2} - 2\alpha \sum_{k=0}^{m-1} \frac{\big(Y_{k+1} - Y_k\big)\big(\beta - Y_k\big)}{\sigma^2}\right]\right) \\
&\quad = p(\alpha) \exp\left(-\frac{1}{2}\left(\sum_{k=0}^{m-1} \frac{\big(\beta - Y_k\big)^2 \Delta t_k}{\sigma^2}\right)\left[\alpha^2 - 2\alpha \frac{\sum_{k=0}^{m-1} \big(Y_{k+1} - Y_k\big)\big(\beta - Y_k\big)}{\sum_{k=0}^{m-1} \big(\beta - Y_k\big)^2 \Delta t_k}\right]\right).
\end{aligned}$$

If $p(\alpha) \propto 1$ for $\alpha \in \mathbb{R}_+$, this corresponds to the truncated Gaussian distribution

$$\alpha \,|\, \beta, \sigma^2, Y_0, \ldots, Y_m$$
$$\sim \mathcal{N}_{\text{trunc}}\left(\frac{\sum_{k=0}^{m-1} (Y_{k+1} - Y_k)(\beta - Y_k)}{\sum_{k=0}^{m-1} (\beta - Y_k)^2 \Delta t_k},\ \sigma^2 \left(\sum_{k=0}^{m-1} (\beta - Y_k)^2 \Delta t_k \right)^{-1} \right),$$

and for $\alpha \sim \mathcal{N}_{\text{trunc}}(\alpha_0, \rho_\alpha^2)$ one obtains

$$\alpha \,|\, \beta, \sigma^2, Y_0, \ldots, Y_m$$
$$\sim \mathcal{N}_{\text{trunc}}\left(\frac{\rho_\alpha^2 \sum_{k=0}^{m-1} (Y_{k+1} - Y_k)(\beta - Y_k) + \alpha_0 \sigma^2}{\rho_\alpha^2 \sum_{k=0}^{m-1} (\beta - Y_k)^2 \Delta t_k + \sigma^2},\ \frac{\sigma^2 \rho_\alpha^2}{\rho_\alpha^2 \sum_{k=0}^{m-1} (\beta - Y_k)^2 \Delta t_k + \sigma^2} \right).$$

For $\rho_\alpha = \infty$, this is the result for a flat prior. The full conditional density of β equals

$$\pi\big(\beta \,|\, \alpha, \sigma^2, Y_0, \ldots, Y_m\big)$$
$$\propto p(\beta) \exp\left(-\frac{1}{2} \sum_{k=0}^{m-1} \frac{\alpha^2 (\beta - Y_k)^2 \Delta t_k - 2(Y_{k+1} - Y_k)\alpha(\beta - Y_k)}{\sigma^2} \right)$$
$$\propto p(\beta) \exp\left(-\frac{1}{2\sigma^2} \left[\beta^2 \sum_{k=0}^{m-1} \alpha^2 \Delta t_k - 2\beta \sum_{k=0}^{m-1} \big(\alpha(Y_{k+1} - Y_k) + \alpha^2 Y_k \Delta t_k \big) \right] \right)$$
$$= p(\beta) \exp\left(-\frac{1}{2\sigma^2}\, \alpha^2 (t_m - t_0) \left[\beta^2 - 2\beta\, \frac{1}{t_m - t_0} \left(\frac{Y_m - Y_0}{\alpha} + \sum_{k=0}^{m-1} Y_k \Delta t_k \right) \right] \right).$$

With a flat prior $p(\beta) \propto 1$ for $\beta \in \mathbb{R}$, this leads to

$$\beta \,|\, \alpha, \sigma^2, Y_0, \ldots, Y_m \sim \mathcal{N}\left(\frac{\frac{Y_m - Y_0}{\alpha} + \sum_{k=0}^{m-1} Y_k \Delta t_k}{t_m - t_0},\ \frac{\sigma^2}{\alpha^2 (t_m - t_0)} \right),$$

for $\beta \sim \mathcal{N}(\beta_0, \rho_\beta^2)$ to

$$\beta \,|\, \alpha, \sigma^2, Y_0, \ldots, Y_m$$
$$\sim \mathcal{N}\left(\frac{\alpha^2 \rho_\beta^2 \left(\frac{Y_m - Y_0}{\alpha} + \sum_{k=0}^{m-1} Y_k \Delta t_k \right) + \sigma^2 \beta_0}{\alpha^2 \rho_\beta^2 (t_m - t_0) + \sigma^2},\ \frac{\sigma^2 \rho_\beta^2}{\alpha^2 \rho_\beta^2 (t_m - t_0) + \sigma^2} \right).$$

Again, this expression yields the same result as for the flat prior when setting $\rho_\beta = \infty$. Eventually, the full conditional density of σ^2 fulfils

$$\pi\big(\sigma^2 \,\big|\, \alpha, \beta, Y_0, \ldots, Y_m\big)$$
$$\propto p(\sigma^2)\,\sigma^{-m} \exp\left(-\frac{1}{2\sigma^2}\sum_{k=0}^{m-1}\frac{\big(Y_{k+1} - Y_k - \alpha(\beta - Y_k)\Delta t_k\big)^2}{\Delta t_k}\right).$$

If $p(\sigma^2) \propto 1$ for $\sigma^2 \in \mathbb{R}_+$, it follows immediately that

$$\sigma^2 \,\big|\, \alpha, \beta, Y_0, \ldots, Y_m \sim \mathrm{IG}\left(\frac{m}{2} - 1\,,\, \frac{1}{2}\sum_{k=0}^{m-1}\frac{\big(Y_{k+1} - Y_k - \alpha(\beta - Y_k)\Delta t_k\big)^2}{\Delta t_k}\right),$$

and in case of $\sigma^2 \sim \mathrm{IG}(\kappa_0, \nu_0)$ one obtains

$$\sigma^2 \,\big|\, \alpha, \beta, Y_0, \ldots, Y_m \sim \mathrm{IG}\left(\frac{m}{2} + \kappa_0,\ \nu_0 + \frac{1}{2}\sum_{k=0}^{m-1}\frac{\big(Y_{k+1} - Y_k - \alpha(\beta - Y_k)\Delta t_k\big)^2}{\Delta t_k}\right).$$

For $\kappa_0 = -1$ and $\nu_0 = 0$, this expression is the outcome for a flat prior for σ^2.

B.5 Inefficiency Factors

In order to graphically represent the serial correlation of consecutive draws of an imputed data point $\boldsymbol{Y}_k$ from MCMC schemes as in Chap. 7, Elerian et al. (2001) utilise the *inefficiency factor*

$$\iota(\boldsymbol{Y}_k) = 1 + 2\sum_{j=1}^{\infty}\rho_j(\boldsymbol{Y}_k)$$

of according posterior estimates, where $\rho_j(\boldsymbol{Y}_k)$ is the autocorrelation of $\boldsymbol{Y}_k$ at lag j. The inefficiency factor is the factor by which one has to multiply the length of an MCMC chain in order to achieve equivalent results as from i.i.d. draws. Elerian et al. estimate ι as

$$\hat{\iota}(\boldsymbol{Y}_k) = 1 + \frac{2n}{n-1}\sum_{j=1}^{\kappa} K\left(\frac{j}{\kappa}\right)\hat{\rho}_j(\boldsymbol{Y}_k),$$

where n is the length of the Markov chain, κ is an appropriate bandwidth until which the autocorrelation significantly contributes to the serial dependence, $\hat{\rho}_j$ is an estimate of ρ_j, and K is the Parzen kernel, that is (Parzen 1964)

$$K(u) = \begin{cases} 1 - 6u^2 + 6|u|^3 & \text{for } |u| \leq \frac{1}{2} \\ 2(1-|u|)^3 & \text{for } \frac{1}{2} < |u| \leq 1 \\ 0 & \text{otherwise.} \end{cases}$$

B.6 Path Proposals in the Latent Data Framework

In this section, appropriate proposal densities for diffusion paths are derived as required in Sect. 7.2.1. The notation is adopted from there. In short, the following considerations avail proposing a path segment $\{\boldsymbol{Y}_{a+1}, \ldots, \boldsymbol{Y}_{r-1}, \boldsymbol{L}_r, \boldsymbol{Y}_{r+1}, \ldots, \boldsymbol{Y}_{b-1}\}$, where the vector $\boldsymbol{Y}_r = (\boldsymbol{V}_r', \boldsymbol{L}_r')'$ consists of an observed part $\boldsymbol{V}_r \in \mathbb{R}^{d_1}$ and a latent part $\boldsymbol{L}_r \in \mathbb{R}^{d_2}$.

For shorter notation, abbreviate $\boldsymbol{\mu}_k = \boldsymbol{\mu}(\boldsymbol{Y}_k, \boldsymbol{\theta})$ and $\boldsymbol{\Sigma}_k = \boldsymbol{\Sigma}(\boldsymbol{Y}_k, \boldsymbol{\theta})$ for all k. Furthermore, decompose $\boldsymbol{\mu}$ and $\boldsymbol{\Sigma}$ into

$$\boldsymbol{\mu} = \begin{pmatrix} \boldsymbol{\mu}^v \\ \boldsymbol{\mu}^l \end{pmatrix} \qquad \text{and} \qquad \boldsymbol{\Sigma} = \begin{pmatrix} \boldsymbol{\Sigma}^{vv} & \boldsymbol{\Sigma}^{vl} \\ \boldsymbol{\Sigma}^{lv} & \boldsymbol{\Sigma}^{ll} \end{pmatrix}$$

such that $\boldsymbol{\mu}^v \in \mathbb{R}^{d_1}$, $\boldsymbol{\mu}^l \in \mathbb{R}^{d_2}$, $\boldsymbol{\Sigma}^{vv} \in \mathbb{R}^{d_1 \times d_1}$ and $\boldsymbol{\Sigma}^{ll} \in \mathbb{R}^{d_2 \times d_2}$.

Approximation of $\mathfrak{L}(\boldsymbol{L}_r | \boldsymbol{Y}_k, \boldsymbol{V}_r, \boldsymbol{Y}_b, \boldsymbol{\theta})$ for $k < r$

Let $a \leq k < r$. Similarly to the derivation of the modified bridge proposal on p. 181, one has

$$\begin{aligned} \pi(\boldsymbol{Y}_r | \boldsymbol{Y}_k, \boldsymbol{Y}_b, \boldsymbol{\theta}) &\propto \pi(\boldsymbol{Y}_b \,|\, \boldsymbol{Y}_r, \boldsymbol{\theta}) \pi(\boldsymbol{Y}_r | \boldsymbol{Y}_k, \boldsymbol{\theta}) \\ &\approx \phi(\boldsymbol{Y}_b \,|\, \boldsymbol{Y}_r + \boldsymbol{\mu}_r \Delta_{rb}, \boldsymbol{\Sigma}_r \Delta_{rb}) \cdot \phi(\boldsymbol{Y}_r \,|\, \boldsymbol{Y}_k + \boldsymbol{\mu}_k \Delta_{kr}, \boldsymbol{\Sigma}_k \Delta_{kr}), \end{aligned}$$

where $\Delta_{rb} = t_b - t_r$ and $\Delta_{kr} = t_r - t_k$. The Gaussian densities ϕ stem from the Euler approximation (7.3). Approximate $\boldsymbol{\mu}_r$ and $\boldsymbol{\Sigma}_r$ by $\boldsymbol{\mu}_k$ and $\boldsymbol{\Sigma}_k$. Then $\pi(\boldsymbol{Y}_r | \boldsymbol{Y}_k, \boldsymbol{Y}_b, \boldsymbol{\theta})$ is approximately proportional to

$$\begin{aligned} &\exp\Bigg(-\frac{1}{2}\Bigg[\Big(\boldsymbol{Y}_r - (\boldsymbol{Y}_b - \boldsymbol{\mu}_k \Delta_{rb})\Big)' \frac{\boldsymbol{\Sigma}_k^{-1}}{\Delta_{rb}} \Big(\boldsymbol{Y}_r - (\boldsymbol{Y}_b - \boldsymbol{\mu}_k \Delta_{rb})\Big) \\ &\qquad + \Big(\boldsymbol{Y}_r - (\boldsymbol{Y}_k + \boldsymbol{\mu}_k \Delta_{kr})\Big)' \frac{\boldsymbol{\Sigma}_k^{-1}}{\Delta_{kr}} \Big(\boldsymbol{Y}_r - (\boldsymbol{Y}_k + \boldsymbol{\mu}_k \Delta_{kr})\Big)\Bigg]\Bigg) \\ &\propto \exp\Bigg(-\frac{1}{2}\Bigg[\boldsymbol{Y}_r'\Big((\Delta_{rb}^{-1} + \Delta_{kr}^{-1})\boldsymbol{\Sigma}_k^{-1}\Big)\boldsymbol{Y}_r \end{aligned}$$

$$- 2\boldsymbol{Y}_r' \boldsymbol{\Sigma}_k^{-1} \left(\frac{\boldsymbol{Y}_b - \boldsymbol{\mu}_k \Delta_{rb}}{\Delta_{rb}} + \frac{\boldsymbol{Y}_k + \boldsymbol{\mu}_k \Delta_{kr}}{\Delta_{kr}} \right) \Bigg] \Bigg)$$

$$\propto \exp\left(-\frac{1}{2} \frac{\Delta_{kb}}{\Delta_{rb}\Delta_{kr}} \left[\boldsymbol{Y}_r' \boldsymbol{\Sigma}_k^{-1} \left(\boldsymbol{Y}_r - 2 \frac{\Delta_{kr}\boldsymbol{Y}_b + \Delta_{rb}\boldsymbol{Y}_k}{\Delta_{kb}} \right) \right] \right),$$

where $\Delta_{kb} = \Delta_{kr} + \Delta_{rb} = t_b - t_k$. This is the unnormalised density of the normal distribution

$$\boldsymbol{Y}_r \,|\, \boldsymbol{Y}_k, \boldsymbol{Y}_b, \boldsymbol{\theta} \sim \mathcal{N}\left(\boldsymbol{Y}_k + \frac{\boldsymbol{Y}_b - \boldsymbol{Y}_k}{\Delta_{kb}} \Delta_{kr} \,,\, \frac{\Delta_{rb}\Delta_{kr}}{\Delta_{kb}} \boldsymbol{\Sigma}_k \right),$$

i.e.

$$\begin{pmatrix} \boldsymbol{V}_r \\ \boldsymbol{L}_r \end{pmatrix} \Bigg| \boldsymbol{Y}_k, \boldsymbol{Y}_b, \boldsymbol{\theta} \sim \mathcal{N}\left(\begin{pmatrix} \boldsymbol{V}_k + \dfrac{\boldsymbol{V}_b - \boldsymbol{V}_k}{\Delta_{kb}} \Delta_{kr} \\ \boldsymbol{L}_k + \dfrac{\boldsymbol{L}_b - \boldsymbol{L}_k}{\Delta_{kb}} \Delta_{kr} \end{pmatrix} ,\, \frac{\Delta_{rb}\Delta_{kr}}{\Delta_{kb}} \begin{pmatrix} \boldsymbol{\Sigma}_k^{vv} & \boldsymbol{\Sigma}_k^{vl} \\ \boldsymbol{\Sigma}_k^{lv} & \boldsymbol{\Sigma}_k^{ll} \end{pmatrix} \right).$$

The conditional distribution of $\boldsymbol{L}_r$ given $\boldsymbol{V}_r$ (and $\boldsymbol{Y}_k, \boldsymbol{Y}_b, \boldsymbol{\theta}$) follows from this joint distribution by application of multivariate normal theory. That is

$$\boldsymbol{L}_r \,|\, \boldsymbol{Y}_k, \boldsymbol{V}_r, \boldsymbol{Y}_b, \boldsymbol{\theta} \sim \mathcal{N}(\boldsymbol{\eta}_k \,,\, \boldsymbol{\Lambda}_k)$$

with

$$\boldsymbol{\eta}_k = \boldsymbol{L}_k + \frac{\boldsymbol{L}_b - \boldsymbol{L}_k}{\Delta_{kb}} \Delta_{kr} + \boldsymbol{\Sigma}_k^{lv} \left(\boldsymbol{\Sigma}_k^{vv} \right)^{-1} \left(\boldsymbol{V}_r - \boldsymbol{V}_k - \frac{\boldsymbol{V}_b - \boldsymbol{V}_k}{\Delta_{kb}} \Delta_{kr} \right)$$

and

$$\boldsymbol{\Lambda}_k = \frac{\Delta_{rb}\Delta_{kr}}{\Delta_{kb}} \left(\boldsymbol{\Sigma}_k^{ll} - \boldsymbol{\Sigma}_k^{lv} \left(\boldsymbol{\Sigma}_k^{vv} \right)^{-1} \boldsymbol{\Sigma}_k^{vl} \right).$$

Approximation of $\mathfrak{L}(\boldsymbol{Y}_{k+1} | \boldsymbol{Y}_k, \boldsymbol{V}_r, \boldsymbol{\theta})$ for $k < r - 1$

Let $a \leq k < r - 1$. Application of the Euler scheme yields

$$\begin{aligned} \boldsymbol{Y}_{k+1} \,|\, \boldsymbol{Y}_k, \boldsymbol{\theta} &\sim \mathcal{N}\big(\boldsymbol{Y}_k + \boldsymbol{\mu}_k \Delta t_k, \boldsymbol{\Sigma}_k \Delta t_k\big) \\ \boldsymbol{V}_r \,|\, \boldsymbol{Y}_{k+1}, \boldsymbol{Y}_k, \boldsymbol{\theta} &\sim \mathcal{N}\big(\boldsymbol{V}_{k+1} + \boldsymbol{\mu}_{k+1}^v \Delta_{kr-}, \boldsymbol{\Sigma}_{k+1}^{vv} \Delta_{kr-}\big), \end{aligned}$$

where $\Delta t_k = t_{k+1} - t_k$ and $\Delta_{kr-} = t_r - t_{k+1}$. Approximate $\boldsymbol{\mu}_{k+1}$ and $\boldsymbol{\Sigma}_{k+1}$ by $\boldsymbol{\mu}_k$ and $\boldsymbol{\Sigma}_k$, respectively, such that

$$\boldsymbol{Y}_{k+1} \,|\, \boldsymbol{Y}_k, \boldsymbol{\theta} \sim \mathcal{N}\big(\boldsymbol{Y}_k + \boldsymbol{\mu}_k \Delta t_k, \boldsymbol{\Sigma}_k \Delta t_k\big)$$

$$\boldsymbol{V}_r \,|\, \boldsymbol{Y}_{k+1}, \boldsymbol{Y}_k, \boldsymbol{\theta} \sim \mathcal{N}\big(\boldsymbol{V}_{k+1} + \boldsymbol{\mu}_k^v \Delta_{kr-}, \boldsymbol{\Sigma}_k^{vv} \Delta_{kr-}\big).$$

The joint distribution of $\boldsymbol{Y}_{k+1}$ and $\boldsymbol{V}_r$ conditioned on $\boldsymbol{Y}_k$ and $\boldsymbol{\theta}$ is again Gaussian. The conditional expected value and variance of $(\boldsymbol{Y}_{k+1}', \boldsymbol{V}_r')'$ can be obtained as follows: The iterated expectation theorem yields

$$\begin{aligned}
\mathbb{E}\big(\boldsymbol{V}_r \,\big|\, \boldsymbol{Y}_k, \boldsymbol{\theta}\big) &= \mathbb{E}\big(\mathbb{E}\big(\boldsymbol{V}_r \,\big|\, \boldsymbol{Y}_{k+1}\big) \,\big|\, \boldsymbol{Y}_k, \boldsymbol{\theta}\big) \\
&= \mathbb{E}\big(\boldsymbol{V}_{k+1} + \boldsymbol{\mu}_k^v \Delta_{kr-} \,\big|\, \boldsymbol{Y}_k, \boldsymbol{\theta}\big) = \boldsymbol{V}_k + \boldsymbol{\mu}_k^v \Delta_{kr},
\end{aligned}$$

where $\Delta_{kr} = \Delta_{kr-} + \Delta t_k = t_r - t_k$. Furthermore, the variance decomposition formula (law of total variance) leads to

$$\begin{aligned}
&\mathrm{Var}\big((\boldsymbol{Y}_{k+1}', \boldsymbol{V}_r')' \,\big|\, \boldsymbol{Y}_k, \boldsymbol{\theta}\big) \\
&\quad = \mathrm{Var}\big(\mathbb{E}\big((\boldsymbol{Y}_{k+1}', \boldsymbol{V}_r')' \,\big|\, \boldsymbol{Y}_{k+1}\big) \,\big|\, \boldsymbol{Y}_k, \boldsymbol{\theta}\big) + \mathbb{E}\big(\mathrm{Var}\big((\boldsymbol{Y}_{k+1}', \boldsymbol{V}_r')' \,\big|\, \boldsymbol{Y}_{k+1}\big) \,\big|\, \boldsymbol{Y}_k, \boldsymbol{\theta}\big) \\
&\quad = \mathrm{Var}\left(\begin{pmatrix} \boldsymbol{Y}_{k+1} \\ \boldsymbol{V}_{k+1} + \boldsymbol{\mu}_k^v \Delta_{kr-} \end{pmatrix} \,\middle|\, \boldsymbol{Y}_k, \boldsymbol{\theta}\right) + \mathbb{E}\left(\begin{pmatrix} \mathbf{0} & \mathbf{0} \\ \mathbf{0} & \boldsymbol{\Sigma}_k^{vv} \Delta_{kr-} \end{pmatrix} \,\middle|\, \boldsymbol{Y}_k, \boldsymbol{\theta}\right) \\
&\quad = \begin{pmatrix} \boldsymbol{\Sigma}_k \Delta t_k & \boldsymbol{D}_k' \Delta t_k \\ \boldsymbol{D}_k \Delta t_k & \boldsymbol{\Sigma}_k^{vv} \Delta_{kr} \end{pmatrix},
\end{aligned}$$

where $\boldsymbol{D}_k = (\boldsymbol{\Sigma}_k^{vv}, \boldsymbol{\Sigma}_k^{vl})$. Altogether,

$$\begin{pmatrix} \boldsymbol{Y}_{k+1} \\ \boldsymbol{V}_r \end{pmatrix} \,\middle|\, \boldsymbol{Y}_k, \boldsymbol{\theta} \sim \mathcal{N}\left(\begin{pmatrix} \boldsymbol{Y}_k + \boldsymbol{\mu}_k \Delta t_k \\ \boldsymbol{V}_k + \boldsymbol{\mu}_k^v \Delta_{kr} \end{pmatrix}, \begin{pmatrix} \boldsymbol{\Sigma}_k \Delta t_k & \boldsymbol{D}_k' \Delta t_k \\ \boldsymbol{D}_k \Delta t_k & \boldsymbol{\Sigma}_k^{vv} \Delta_{kr} \end{pmatrix}\right). \tag{B.20}$$

This implies

$$\boldsymbol{Y}_{k+1} \,|\, \boldsymbol{Y}_k, \boldsymbol{V}_r, \boldsymbol{\theta} \sim \mathcal{N}(\boldsymbol{\rho}_k, \boldsymbol{\Gamma}_k)$$

with

$$\begin{aligned}
\boldsymbol{\rho}_k &= \boldsymbol{Y}_k + \boldsymbol{\mu}_k \Delta t_k + \frac{\Delta t_k}{\Delta_{kr}} \boldsymbol{D}_k' (\boldsymbol{\Sigma}_k^{vv})^{-1} \big(\boldsymbol{V}_r - \boldsymbol{V}_k - \boldsymbol{\mu}_k^v \Delta_{kr}\big) \\
&= \begin{pmatrix} \boldsymbol{V}_k + \dfrac{\boldsymbol{V}_r - \boldsymbol{V}_k}{\Delta_{kr}} \Delta t_k \\ \boldsymbol{L}_k + \boldsymbol{\mu}_k^l \Delta t_k + \boldsymbol{\Sigma}_k^{lv} (\boldsymbol{\Sigma}_k^{vv})^{-1} \left(\dfrac{\boldsymbol{V}_r - \boldsymbol{V}_k}{\Delta_{kr}} - \boldsymbol{\mu}_k^v\right) \Delta t_k \end{pmatrix}
\end{aligned}$$

and

$$\begin{aligned}
\boldsymbol{\Gamma}_k &= \left(\boldsymbol{\Sigma}_k - \frac{\Delta t_k}{\Delta_{kr}} \boldsymbol{D}_k' (\boldsymbol{\Sigma}_k^{vv})^{-1} \boldsymbol{D}_k\right) \Delta t_k \\
&= \begin{pmatrix} \boldsymbol{\Sigma}_k^{vv} \Delta_{kr-} & \boldsymbol{\Sigma}_k^{vl} \Delta_{kr-} \\ \boldsymbol{\Sigma}_k^{lv} \Delta_{kr-} & \boldsymbol{\Sigma}_k^{ll} \Delta_{kr} - \boldsymbol{\Sigma}_k^{lv} (\boldsymbol{\Sigma}_k^{vv})^{-1} \boldsymbol{\Sigma}_k^{vl} \Delta t_k \end{pmatrix} \frac{\Delta t_k}{\Delta_{kr}}.
\end{aligned}$$

Approximation of $\mathfrak{L}(\boldsymbol{Y}_{k+1}|\boldsymbol{Y}_k, \boldsymbol{V}_r, \boldsymbol{Y}_b, \boldsymbol{\theta})$ for $k < r-1$

Let $a \leq k < r-1$. Application of the Euler scheme yields approximately

$$\begin{aligned}
\boldsymbol{Y}_{k+1} \,|\, \boldsymbol{Y}_k, \boldsymbol{\theta} &\sim \mathcal{N}\big(\boldsymbol{Y}_k + \boldsymbol{\mu}_k \Delta t_k, \boldsymbol{\Sigma}_k \Delta t_k\big) \\
\boldsymbol{V}_r \,|\, \boldsymbol{Y}_{k+1}, \boldsymbol{Y}_k, \boldsymbol{\theta} &\sim \mathcal{N}\big(\boldsymbol{V}_{k+1} + \boldsymbol{\mu}_k^v \Delta_{kr-}, \boldsymbol{\Sigma}_k^{vv} \Delta_{kr-}\big) \\
\boldsymbol{Y}_b \,|\, \boldsymbol{Y}_{k+1}, \boldsymbol{Y}_k, \boldsymbol{\theta} &\sim \mathcal{N}\big(\boldsymbol{Y}_{k+1} + \boldsymbol{\mu}_k \Delta_{kb-}, \boldsymbol{\Sigma}_k \Delta_{kb-}\big),
\end{aligned}$$

where $\Delta t_k = t_{k+1} - t_k$, $\Delta_{kr-} = t_r - t_{k+1}$ and $\Delta_{kb-} = t_b - t_{k+1}$. As in the preceding derivations, $\boldsymbol{\mu}_{k+1}$ and $\boldsymbol{\Sigma}_{k+1}$ have been replaced by $\boldsymbol{\mu}_k$ and $\boldsymbol{\Sigma}_k$ here. Conditionally on $\boldsymbol{Y}_k$ and $\boldsymbol{\theta}$, the three random vectors $\boldsymbol{Y}_{k+1}$, $\boldsymbol{V}_r$ and $\boldsymbol{Y}_b$ are jointly Gaussian distributed. The joint distribution of $\boldsymbol{Y}_{k+1}$ and $\boldsymbol{V}_r$ is already known from (B.20). The remaining distribution parameters can be achieved as above by application of the iterated expectation theorem and the variance decomposition formula. In particular,

$$\begin{aligned}
\mathbb{E}\big(\boldsymbol{Y}_b \,|\, \boldsymbol{Y}_k, \boldsymbol{\theta}\big) &= \mathbb{E}\big(\mathbb{E}\big(\boldsymbol{Y}_b \,|\, \boldsymbol{Y}_{k+1}\big) \,|\, \boldsymbol{Y}_k, \boldsymbol{\theta}\big) \\
&= \mathbb{E}\big(\boldsymbol{Y}_{k+1} + \boldsymbol{\mu}_k \Delta_{kb-} \,|\, \boldsymbol{Y}_k, \boldsymbol{\theta}\big) = \boldsymbol{Y}_k + \boldsymbol{\mu}_k \Delta_{kb},
\end{aligned}$$

where $\Delta_{kb} = \Delta t_k + \Delta_{kb-} = t_b - t_k$, and

$$\begin{aligned}
&\mathrm{Var}\big((\boldsymbol{Y}_{k+1}', \boldsymbol{Y}_b')' \,|\, \boldsymbol{Y}_k, \boldsymbol{\theta}\big) \\
&\quad = \mathrm{Var}\big(\mathbb{E}\big((\boldsymbol{Y}_{k+1}', \boldsymbol{Y}_b')' \,|\, \boldsymbol{Y}_{k+1}\big) \,|\, \boldsymbol{Y}_k, \boldsymbol{\theta}\big) + \mathbb{E}\big(\mathrm{Var}\big((\boldsymbol{Y}_{k+1}', \boldsymbol{Y}_b')' \,|\, \boldsymbol{Y}_{k+1}\big) \,|\, \boldsymbol{Y}_k, \boldsymbol{\theta}\big) \\
&\quad = \mathrm{Var}\left(\begin{pmatrix} \boldsymbol{Y}_{k+1} \\ \boldsymbol{Y}_{k+1} + \boldsymbol{\mu}_k \Delta_{kb-} \end{pmatrix} \,\middle|\, \boldsymbol{Y}_k, \boldsymbol{\theta}\right) + \mathbb{E}\left(\begin{pmatrix} \boldsymbol{0} & \boldsymbol{0} \\ \boldsymbol{0} & \boldsymbol{\Sigma}_k \Delta_{kb-} \end{pmatrix} \,\middle|\, \boldsymbol{Y}_k, \boldsymbol{\theta}\right) \\
&\quad = \begin{pmatrix} \boldsymbol{\Sigma}_k \Delta t_k & \boldsymbol{\Sigma}_k \Delta t_k \\ \boldsymbol{\Sigma}_k \Delta t_k & \boldsymbol{\Sigma}_k \Delta_{kb} \end{pmatrix}.
\end{aligned}$$

In order to derive the conditional covariance of $\boldsymbol{V}_r$ and $\boldsymbol{Y}_b$, consider the approximate distributions

$$\begin{aligned}
\boldsymbol{Y}_r \,|\, \boldsymbol{Y}_k, \boldsymbol{\theta} &\sim \mathcal{N}\big(\boldsymbol{Y}_k + \boldsymbol{\mu}_k \Delta_{kr}, \boldsymbol{\Sigma}_k \Delta_{kr}\big) \\
\boldsymbol{Y}_b \,|\, \boldsymbol{Y}_r, \boldsymbol{Y}_k, \boldsymbol{\theta} &\sim \mathcal{N}\big(\boldsymbol{Y}_r + \boldsymbol{\mu}_k \Delta_{rb}, \boldsymbol{\Sigma}_k \Delta_{rb}\big)
\end{aligned}$$

with $\Delta_{rb} = t_b - t_r$. Then

$$\begin{aligned}
&\mathrm{Var}\big((\boldsymbol{Y}_r', \boldsymbol{Y}_b')' \,|\, \boldsymbol{Y}_k, \boldsymbol{\theta}\big) \\
&\quad = \mathrm{Var}\big(\mathbb{E}\big((\boldsymbol{Y}_r', \boldsymbol{Y}_b')' \,|\, \boldsymbol{Y}_r\big) \,|\, \boldsymbol{Y}_k, \boldsymbol{\theta}\big) + \mathbb{E}\big(\mathrm{Var}\big((\boldsymbol{Y}_r', \boldsymbol{Y}_b')' \,|\, \boldsymbol{Y}_r\big) \,|\, \boldsymbol{Y}_k, \boldsymbol{\theta}\big) \\
&\quad = \begin{pmatrix} \boldsymbol{\Sigma}_k \Delta_{kr} & \boldsymbol{\Sigma}_k \Delta_{kr} \\ \boldsymbol{\Sigma}_k \Delta_{kr} & \boldsymbol{\Sigma}_k \Delta_{kb} \end{pmatrix}.
\end{aligned}$$

In summary,

$$\begin{pmatrix} \boldsymbol{Y}_{k+1} \\ \boldsymbol{V}_r \\ \boldsymbol{Y}_b \end{pmatrix} \Bigg| \boldsymbol{Y}_k, \boldsymbol{\theta} \sim \mathcal{N}\left(\begin{pmatrix} \boldsymbol{Y}_k + \boldsymbol{\mu}_k \Delta t_k \\ \boldsymbol{V}_k + \boldsymbol{\mu}_k^v \Delta_{kr} \\ \boldsymbol{Y}_k + \boldsymbol{\mu}_k \Delta_{kb} \end{pmatrix}, \begin{pmatrix} \boldsymbol{\Sigma}_k \Delta t_k & \boldsymbol{D}_k' \Delta t_k & \boldsymbol{\Sigma}_k \Delta t_k \\ \boldsymbol{D}_k \Delta t_k & \boldsymbol{\Sigma}_k^{vv} \Delta_{kr} & \boldsymbol{D}_k \Delta_{kr} \\ \boldsymbol{\Sigma}_k \Delta t_k & \boldsymbol{D}_k' \Delta_{kr} & \boldsymbol{\Sigma}_k \Delta_{kb} \end{pmatrix} \right).$$

The resulting conditional distribution of $\boldsymbol{Y}_{k+1}$ reads

$$\boldsymbol{Y}_{k+1} \,|\, \boldsymbol{Y}_k, \boldsymbol{V}_r, \boldsymbol{Y}_b, \boldsymbol{\theta} \sim \mathcal{N}(\boldsymbol{\xi}_k, \boldsymbol{\Psi}_k)$$

with

$$\boldsymbol{\xi}_k = \boldsymbol{Y}_k + \boldsymbol{\mu}_k \Delta t_k + \left(\boldsymbol{D}_k' \Delta t_k, \boldsymbol{\Sigma}_k \Delta t_k \right) \begin{pmatrix} \boldsymbol{\Sigma}_k^{vv} \Delta_{kr} & \boldsymbol{D}_k \Delta_{kr} \\ \boldsymbol{D}_k' \Delta_{kr} & \boldsymbol{\Sigma}_k \Delta_{kb} \end{pmatrix}^{-1} \begin{pmatrix} \boldsymbol{V}_r - \boldsymbol{V}_k - \boldsymbol{\mu}_k^v \Delta_{kr} \\ \boldsymbol{Y}_b - \boldsymbol{Y}_k - \boldsymbol{\mu}_k \Delta_{kb} \end{pmatrix}$$

and

$$\boldsymbol{\Psi}_k = \boldsymbol{\Sigma}_k \Delta t_k - \left(\boldsymbol{D}_k' \Delta t_k, \boldsymbol{\Sigma}_k \Delta t_k \right) \begin{pmatrix} \boldsymbol{\Sigma}_k^{vv} \Delta_{kr} & \boldsymbol{D}_k \Delta_{kr} \\ \boldsymbol{D}_k' \Delta_{kr} & \boldsymbol{\Sigma}_k \Delta_{kb} \end{pmatrix}^{-1} \begin{pmatrix} \boldsymbol{D}_k \Delta t_k \\ \boldsymbol{\Sigma}_k \Delta t_k \end{pmatrix}.$$

B.7 Derivation of Radon-Nikodym Derivatives

This section provides the proof of Corollary 7.5 on p. 250. In particular, it derives explicit expressions for the Radon-Nikodym derivatives $d\tilde{\mathbb{P}}_{\boldsymbol{\theta}}/\mathrm{d}\mathbb{D}_{\boldsymbol{\mu},\boldsymbol{\theta}}$ and $\mathrm{d}\tilde{\mathbb{P}}_{\boldsymbol{\theta}}/\mathrm{d}\mathbb{D}_{\mathbf{0},\boldsymbol{\theta}}$, where the measures $\tilde{\mathbb{P}}_{\boldsymbol{\theta}}$, $\mathbb{D}_{\boldsymbol{\mu},\boldsymbol{\theta}}$ and $\mathbb{D}_{\mathbf{0},\boldsymbol{\theta}}$ are defined in Table 7.6 on p. 243. These derivatives are employed as parts of acceptance probabilities in Sect. 7.4.4. The notation is adopted from there.

Under the assumptions from p. 246, Theorems 5 and 6 in Delyon and Hu (2006) prove that $\tilde{\mathbb{P}}_{\boldsymbol{\theta}} \ll \mathbb{D}_{\boldsymbol{\mu},\boldsymbol{\theta}}$ and $\tilde{\mathbb{P}}_{\boldsymbol{\theta}} \ll \mathbb{D}_{\mathbf{0},\boldsymbol{\theta}}$, i.e. the requested derivatives exist. The assumptions are supposed to hold here as well. In particular, $\boldsymbol{\sigma}$ is assumed to be invertible and is hence a square matrix. Delyon and Hu also provide explicit formulas for $\mathrm{d}\tilde{\mathbb{P}}_{\boldsymbol{\theta}}/\mathrm{d}\mathbb{D}_{\boldsymbol{\mu},\boldsymbol{\theta}}(\boldsymbol{X}_{[0,T]})$ and $\mathrm{d}\tilde{\mathbb{P}}_{\boldsymbol{\theta}}/\mathrm{d}\mathbb{D}_{\mathbf{0},\boldsymbol{\theta}}(\boldsymbol{X}_{[0,T]})$, but these are up to proportionality constants which depend on the parameter $\boldsymbol{\theta}$, the initial value $\boldsymbol{X}_0 = \boldsymbol{x}_0$ and the final value $\boldsymbol{X}_T = \boldsymbol{x}$. These constants are required in the context of Sect. 7.4.4. Furthermore, the derivatives shall be applied to $\boldsymbol{X}_{(0,T-\varepsilon]}$ instead of $\boldsymbol{X}_{[0,T]}$, where $\varepsilon > 0$ is a small but fixed time step. This causes further changes in the resulting formulas.

In Chap. 7, solely time-homogeneous diffusions are considered, i.e. the drift function $\boldsymbol{\mu}$, diffusion coefficient $\boldsymbol{\sigma}$ and diffusion matrix $\boldsymbol{\Sigma}$ of the target diffusion do not depend on time t. The following results can however be obtained also for time-inhomogeneous diffusions without relevant additional overhead. Hence, in this section, the time variable is included in the notation $\boldsymbol{\mu}(\boldsymbol{X}_t, t)$, $\boldsymbol{\sigma}(\boldsymbol{X}_t, t)$ and $\boldsymbol{\Sigma}(\boldsymbol{X}_t, t)$. Dependence on the parameter $\boldsymbol{\theta}$, on the other hand, is suppressed because the parameter is considered fixed in the following derivations. The parameter is however easily re-incorporated in the notation as an argument of $\boldsymbol{\mu}$, $\boldsymbol{\sigma}$ and $\boldsymbol{\Sigma}$.

In the following, we will show that

$$\frac{\mathrm{d}\tilde{\mathbb{P}}_{\boldsymbol{\theta}}}{\mathrm{d}\mathbb{D}_{\boldsymbol{\mu},\boldsymbol{\theta}}}\big(\boldsymbol{X}_{(0,T-\varepsilon]}\big) = \exp\left(-\int_0^{T-\varepsilon} \frac{D_1(\boldsymbol{X}_t,t)+D_2(\boldsymbol{X}_t,t)+D_3(\boldsymbol{X}_t,t)}{2(T-t)}\right)\left(\frac{T}{\varepsilon}\right)^{-\frac{d(d-1)}{2}} \cdot \frac{\phi\big(\boldsymbol{x}\,\big|\,\boldsymbol{x}_0, T\boldsymbol{\Sigma}(\boldsymbol{x}_0,0)\big)}{f_{\boldsymbol{\theta}}(\boldsymbol{x})} \frac{|\boldsymbol{\Sigma}(\boldsymbol{x}_0,0)|^{\frac{1}{2}}}{|\boldsymbol{\Sigma}(\boldsymbol{X}_{T-\varepsilon},T-\varepsilon)|^{\frac{1}{2}}}$$

and

$$\frac{\mathrm{d}\tilde{\mathbb{P}}_{\boldsymbol{\theta}}}{\mathrm{d}\mathbb{D}_{\boldsymbol{0},\boldsymbol{\theta}}}\big(\boldsymbol{X}_{(0,T-\varepsilon]}\big)$$

$$= \exp\left(-\int_0^{T-\varepsilon} \frac{D_2(\boldsymbol{X}_t,t)+D_3(\boldsymbol{X}_t,t)}{2(T-t)}\right)\left(\frac{T}{\varepsilon}\right)^{-\frac{d(d-1)}{2}}$$
$$\cdot \exp\left(\int_0^{T-\varepsilon} \boldsymbol{\mu}'(\boldsymbol{X}_t,t)\boldsymbol{\Sigma}^{-1}(\boldsymbol{X}_t,t)\mathrm{d}\boldsymbol{X}_t - \frac{1}{2}\int_0^{T-\varepsilon} \boldsymbol{\mu}'(\boldsymbol{X}_t,t)\boldsymbol{\Sigma}^{-1}(\boldsymbol{X}_t,t)\boldsymbol{\mu}(\boldsymbol{X}_t,t)\mathrm{d}t\right)$$
$$\cdot \frac{\phi\big(\boldsymbol{x}\,\big|\,\boldsymbol{x}_0, T\boldsymbol{\Sigma}(\boldsymbol{x}_0,0)\big)}{f_{\boldsymbol{\theta}}(\boldsymbol{x})} \frac{|\boldsymbol{\Sigma}(\boldsymbol{x}_0,0)|^{\frac{1}{2}}}{|\boldsymbol{\Sigma}(\boldsymbol{X}_{T-\varepsilon},T-\varepsilon)|^{\frac{1}{2}}},$$

where

$$D_1(\boldsymbol{X}_t,t) = -2(\boldsymbol{x}-\boldsymbol{X}_t)'\boldsymbol{\Sigma}^{-1}(\boldsymbol{X}_t,t)\boldsymbol{\mu}(\boldsymbol{X}_t,t)\mathrm{d}t$$

$$D_2(\boldsymbol{X}_t,t) = (\boldsymbol{x}-\boldsymbol{X}_t)'\big(\mathrm{d}\boldsymbol{\Sigma}^{-1}(\boldsymbol{X}_t,t)\big)(\boldsymbol{x}-\boldsymbol{X}_t)$$

$$D_3(\boldsymbol{X}_t,t) = -\sum_{i=1}^{d}\sum_{j=1}^{d} \frac{(\boldsymbol{x}-\boldsymbol{X}_t)'\left(\frac{\partial \boldsymbol{\Sigma}^{-1}(\boldsymbol{X}_t,t)}{\partial x^{(j)}}\boldsymbol{e}_i + \frac{\partial \boldsymbol{\Sigma}^{-1}(\boldsymbol{X}_t,t)}{\partial x^{(i)}}\boldsymbol{e}_j\right)}{T-t}\mathrm{d}X_t^{(i)}\mathrm{d}X_t^{(j)}.$$

In these formulas, $f_{\boldsymbol{\theta}}(\boldsymbol{x})$ is the Lebesgue density of the final point $\boldsymbol{x}$ under the unconditioned law $\mathbb{P}_{\boldsymbol{\theta}}$ (defined in Table 7.6 on p. 243), $\phi(\boldsymbol{y}|\boldsymbol{\nu},\boldsymbol{\Lambda})$ is the multivariate Gaussian density evaluated at $\boldsymbol{y}$ with mean $\boldsymbol{\nu}$ and covariance matrix $\boldsymbol{\Lambda}$, $\boldsymbol{e}_i$ is the ith unit vector of dimension d, $|\boldsymbol{A}|$ is the determinant of a square matrix $\boldsymbol{A}$, $\mathrm{d}X_t^{(i)}$ is the ith component of $\mathrm{d}\boldsymbol{X}_t$, and $\partial/\partial x^{(i)}$ denotes differentiation with respect to the ith component of the state variable.

We start with the derivation of $\mathrm{d}\tilde{\mathbb{P}}_{\boldsymbol{\theta}}/\mathrm{d}\mathbb{D}_{\boldsymbol{\mu},\boldsymbol{\theta}}(\boldsymbol{X}_{(0,T-\varepsilon]}) = (\mathrm{d}\tilde{\mathbb{P}}_{\boldsymbol{\theta}}/\mathrm{d}\mathbb{P}_{\boldsymbol{\theta}})(\mathrm{d}\mathbb{P}_{\boldsymbol{\theta}}/\mathrm{d}\mathbb{D}_{\boldsymbol{\mu},\boldsymbol{\theta}})(\boldsymbol{X}_{(0,T-\varepsilon]})$. First, investigate the relationship between $\tilde{\mathbb{P}}_{\boldsymbol{\theta}}$ and $\mathbb{P}_{\boldsymbol{\theta}}$. As already shown in Eq. (7.62) on p. 248, one heuristically has

$$\frac{\mathrm{d}\tilde{\mathbb{P}}_{\boldsymbol{\theta}}}{\mathrm{d}\mathbb{P}_{\boldsymbol{\theta}}}\big(\boldsymbol{X}_{(0,T-\varepsilon]}\big) = \frac{f_{\boldsymbol{\theta}}(\boldsymbol{x}|\boldsymbol{X}_{T-\varepsilon})}{f_{\boldsymbol{\theta}}(\boldsymbol{x})}. \tag{B.21}$$

So continue with the derivation of $\mathrm{d}\mathbb{P}_{\boldsymbol{\theta}}/\mathrm{d}\mathbb{D}_{\boldsymbol{\mu},\boldsymbol{\theta}}$. Delyon and Hu (2006, Theorem 1) provide a generalisation of Girsanov's formula which holds under weaker conditions than those in Sect. 3.2.12 and which is applicable in the present case. With this theorem, we however obtain the same result as under blind application of (3.25) to $(\mathrm{d}\mathbb{P}_{\boldsymbol{\theta}}/\mathrm{d}\mathbb{W}_{\boldsymbol{\theta}})(\mathrm{d}\mathbb{W}_{\boldsymbol{\theta}}/\mathrm{d}\mathbb{D}_{\boldsymbol{\mu},\boldsymbol{\theta}})$, where $\mathbb{W}_{\boldsymbol{\theta}}$ is the driftless analogue of $\mathbb{P}_{\boldsymbol{\theta}}$ (cf. Table 7.6); that is

$$\begin{aligned}
&\log\left(\frac{\mathrm{d}\mathbb{P}_{\boldsymbol{\theta}}}{\mathrm{d}\mathbb{D}_{\boldsymbol{\mu},\boldsymbol{\theta}}}\right)\big(\boldsymbol{X}_{(0,T-\varepsilon]}\big)\\
&= -\int_0^{T-\varepsilon}\left(\frac{\boldsymbol{x}-\boldsymbol{X}_t}{T-t}\right)'\boldsymbol{\Sigma}^{-1}(\boldsymbol{X}_t,t)\mathrm{d}\boldsymbol{X}_t + \int_0^{T-\varepsilon}\left(\frac{\boldsymbol{x}-\boldsymbol{X}_t}{T-t}\right)'\boldsymbol{\Sigma}^{-1}(\boldsymbol{X}_t,t)\boldsymbol{\mu}(\boldsymbol{X}_t,t)\mathrm{d}t\\
&+\frac{1}{2}\int_0^{T-\varepsilon}\left(\frac{\boldsymbol{x}-\boldsymbol{X}_t}{T-t}\right)'\boldsymbol{\Sigma}^{-1}(\boldsymbol{X}_t,t)\left(\frac{\boldsymbol{x}-\boldsymbol{X}_t}{T-t}\right)\mathrm{d}t.
\end{aligned}\tag{B.22}$$

The integral in (B.22) is now rewritten as in the proof of Theorem 5 in Delyon and Hu (2006). This is as follows: Consider $\mathrm{d}g(t,\boldsymbol{X}_t)$, where

$$g(t,\boldsymbol{X}_t) = \frac{1}{T-t}(\boldsymbol{x}-\boldsymbol{X}_t)'\boldsymbol{\Sigma}^{-1}(\boldsymbol{X}_t,t)(\boldsymbol{x}-\boldsymbol{X}_t).$$

With the Itô formula from Sect. 3.2.10 we obtain

$$\begin{aligned}
\mathrm{d}g(t,\boldsymbol{X}_t) &= \frac{\partial g(t,\boldsymbol{X}_t)}{\partial t}\mathrm{d}t + \sum_{i=1}^{d}\frac{\partial g(t,\boldsymbol{X}_t)}{\partial x^{(i)}}\mathrm{d}X_t^{(i)}\\
&+\frac{1}{2}\sum_{i=1}^{d}\sum_{j=1}^{d}\frac{\partial^2 g(t,\boldsymbol{X}_t)}{\partial x^{(i)}\partial x^{(j)}}\mathrm{d}X_t^{(i)}\mathrm{d}X_t^{(j)}
\end{aligned}\tag{B.23}$$

with

$$\begin{aligned}
\frac{\partial g(t,\boldsymbol{X}_t)}{\partial t}\mathrm{d}t &= \frac{(\boldsymbol{x}-\boldsymbol{X}_t)'\boldsymbol{\Sigma}^{-1}(\boldsymbol{X}_t,t)(\boldsymbol{x}-\boldsymbol{X}_t)}{(T-t)^2}\mathrm{d}t\\
&\quad+\frac{(\boldsymbol{x}-\boldsymbol{X}_t)'\dfrac{\partial\boldsymbol{\Sigma}^{-1}(\boldsymbol{X}_t,t)}{\partial t}(\boldsymbol{x}-\boldsymbol{X}_t)}{T-t}\mathrm{d}t,
\end{aligned}\tag{B.24}$$

$$\frac{\partial g(t,\boldsymbol{X}_t)}{\partial x^{(i)}}\mathrm{d}X_t^{(i)} = -\frac{2(\boldsymbol{x}-\boldsymbol{X}_t)'\boldsymbol{\Sigma}^{-1}(\boldsymbol{X}_t,t)\boldsymbol{e}_i}{T-t}\mathrm{d}X_t^{(i)}$$

$$+\frac{(\boldsymbol{x}-\boldsymbol{X}_t)'\frac{\partial\boldsymbol{\Sigma}^{-1}(\boldsymbol{X}_t,t)}{\partial x^{(i)}}(\boldsymbol{x}-\boldsymbol{X}_t)}{T-t}\mathrm{d}X_t^{(i)}, \tag{B.25}$$

$$\frac{1}{2}\frac{\partial^2 g(t,\boldsymbol{X}_t)}{\partial x^{(i)}\partial x^{(j)}}\mathrm{d}X_t^{(i)}\mathrm{d}X_t^{(j)} = -\frac{(\boldsymbol{x}-\boldsymbol{X}_t)'\frac{\partial\boldsymbol{\Sigma}^{-1}(\boldsymbol{X}_t,t)}{\partial x^{(j)}}\boldsymbol{e}_i}{T-t}\mathrm{d}X_t^{(i)}\mathrm{d}X_t^{(j)}$$

$$-\frac{(\boldsymbol{x}-\boldsymbol{X}_t)'\frac{\partial\boldsymbol{\Sigma}^{-1}(\boldsymbol{X}_t,t)}{\partial x^{(i)}}\boldsymbol{e}_j}{T-t}\mathrm{d}X_t^{(i)}\mathrm{d}X_t^{(j)}$$

$$+\frac{(\boldsymbol{x}-\boldsymbol{X}_t)'\frac{\partial^2\boldsymbol{\Sigma}^{-1}(\boldsymbol{X}_t,t)}{\partial x^{(i)}\partial x^{(j)}}(\boldsymbol{x}-\boldsymbol{X}_t)}{2(T-t)}\mathrm{d}X_t^{(i)}\mathrm{d}X_t^{(j)} \tag{B.26}$$

$$+\frac{\boldsymbol{e}_i'\boldsymbol{\Sigma}^{-1}(\boldsymbol{X}_t,t)\boldsymbol{e}_j}{T-t}\mathrm{d}X_t^{(i)}\mathrm{d}X_t^{(j)}. \tag{B.27}$$

The following simplifications are possible: Summarise the expressions in lines (B.24)–(B.26) including the summation signs as in (B.23) as

$$\frac{(\boldsymbol{x}-\boldsymbol{X}_t)'\left(\mathrm{d}\boldsymbol{\Sigma}^{-1}(\boldsymbol{X}_t,t)\right)(\boldsymbol{x}-\boldsymbol{X}_t)}{T-t}$$

according to Itô's formula. Furthermore, apply the mean-square rules (3.24) from p. 44 to obtain

$$\begin{aligned}\mathrm{d}X_t^{(i)}\mathrm{d}X_t^{(j)} &= \left(\sum_k \sigma_{ik}(\boldsymbol{X}_t,t)\mathrm{d}B_t^{(k)}\right)\left(\sum_k \sigma_{jk}(\boldsymbol{X}_t,t)\mathrm{d}B_t^{(k)}\right)\\ &= \sum_k \sigma_{ik}(\boldsymbol{X}_t,t)\sigma_{jk}(\boldsymbol{X}_t,t)\mathrm{d}t = \Sigma_{ij}(\boldsymbol{X}_t,t)\mathrm{d}t,\end{aligned} \tag{B.28}$$

where $\mathrm{d}B_t^{(i)}$ is the ith component of $\mathrm{d}\boldsymbol{B}_t$, and σ_{ij} and Σ_{ij} denote the entries of $\boldsymbol{\sigma}$ and $\boldsymbol{\Sigma}$ in row i and column j. With this, line (B.27) simplifies to $\mathrm{d}t/(T-t)$. Overall,

$$d\frac{(\boldsymbol{x}-\boldsymbol{X}_t)'\boldsymbol{\Sigma}^{-1}(\boldsymbol{X}_t,t)(\boldsymbol{x}-\boldsymbol{X}_t)}{T-t}$$

$$=\frac{(\boldsymbol{x}-\boldsymbol{X}_t)'\boldsymbol{\Sigma}^{-1}(\boldsymbol{X}_t,t)(\boldsymbol{x}-\boldsymbol{X}_t)}{(T-t)^2}\mathrm{d}t+\frac{(\boldsymbol{x}-\boldsymbol{X}_t)'\left(\mathrm{d}\boldsymbol{\Sigma}^{-1}(\boldsymbol{X}_t,t)\right)(\boldsymbol{x}-\boldsymbol{X}_t)}{T-t}$$

$$
-\sum_{i=1}^{d} \frac{2(\boldsymbol{x}-\boldsymbol{X}_t)'\boldsymbol{\Sigma}^{-1}(\boldsymbol{X}_t,t)\boldsymbol{e}_i}{T-t}\,\mathrm{d}X_t^{(i)} + d^2\cdot\frac{\mathrm{d}t}{T-t}
$$

$$
-\sum_{i=1}^{d}\sum_{j=1}^{d} \frac{(\boldsymbol{x}-\boldsymbol{X}_t)'\left(\frac{\partial\boldsymbol{\Sigma}^{-1}(\boldsymbol{X}_t,t)}{\partial x^{(j)}}\boldsymbol{e}_i + \frac{\partial\boldsymbol{\Sigma}^{-1}(\boldsymbol{X}_t,t)}{\partial x^{(i)}}\boldsymbol{e}_j\right)}{T-t}\,\mathrm{d}X_t^{(i)}\mathrm{d}X_t^{(j)}.
$$

The first summand on the right hand side of this equation equals the integrand in (B.22). Hence use this expression to obtain

$$
\begin{aligned}
&\log\left(\frac{\mathrm{d}\mathbb{P}_{\boldsymbol{\theta}}}{\mathrm{d}\mathbb{D}_{\boldsymbol{\mu},\boldsymbol{\theta}}}\right)(\boldsymbol{X}_{(0,T-\varepsilon]}) \\
&= -\int_0^{T-\varepsilon}\left(\frac{\boldsymbol{x}-\boldsymbol{X}_t}{T-t}\right)'\boldsymbol{\Sigma}^{-1}(\boldsymbol{X}_t,t)\mathrm{d}\boldsymbol{X}_t + \int_0^{T-\varepsilon}\left(\frac{\boldsymbol{x}-\boldsymbol{X}_t}{T-t}\right)'\boldsymbol{\Sigma}^{-1}(\boldsymbol{X}_t,t)\boldsymbol{\mu}(\boldsymbol{X}_t,t)\mathrm{d}t \\
&\quad + \frac{1}{2}\int_0^{T-\varepsilon}\mathrm{d}\frac{(\boldsymbol{x}-\boldsymbol{X}_t)'\boldsymbol{\Sigma}^{-1}(\boldsymbol{X}_t,t)(\boldsymbol{x}-\boldsymbol{X}_t)}{T-t} \\
&\quad - \frac{1}{2}\int_0^{T-\varepsilon}\frac{(\boldsymbol{x}-\boldsymbol{X}_t)'\left(\mathrm{d}\boldsymbol{\Sigma}^{-1}(\boldsymbol{X}_t,t)\right)(\boldsymbol{x}-\boldsymbol{X}_t)}{T-t} \\
&\quad + \int_0^{T-\varepsilon}\frac{(\boldsymbol{x}-\boldsymbol{X}_t)'\boldsymbol{\Sigma}^{-1}(\boldsymbol{X}_t,t)}{T-t}\,\mathrm{d}\boldsymbol{X}_t - \frac{d^2}{2}\int_0^{T-\varepsilon}\frac{\mathrm{d}t}{T-t} \\
&\quad + \frac{1}{2}\sum_{i=1}^{d}\sum_{j=1}^{d}\int_0^{T-\varepsilon}\frac{(\boldsymbol{x}-\boldsymbol{X}_t)'\left(\frac{\partial\boldsymbol{\Sigma}^{-1}(\boldsymbol{X}_t,t)}{\partial x^{(j)}}\boldsymbol{e}_i + \frac{\partial\boldsymbol{\Sigma}^{-1}(\boldsymbol{X}_t,t)}{\partial x^{(i)}}\boldsymbol{e}_j\right)}{T-t}\,\mathrm{d}X_t^{(i)}\mathrm{d}X_t^{(j)}.
\end{aligned}
$$

This leads to

$$
\begin{aligned}
&\log\left(\frac{\mathrm{d}\mathbb{P}_{\boldsymbol{\theta}}}{\mathrm{d}\mathbb{D}_{\boldsymbol{\mu},\boldsymbol{\theta}}}\right)(\boldsymbol{X}_{(0,T-\varepsilon]}) \\
&= -\int_0^{T-\varepsilon}\left(\frac{D_1(\boldsymbol{X}_t,t)+D_2(\boldsymbol{X}_t,t)+D_3(\boldsymbol{X}_t,t)}{2(T-t)}\right) - \frac{d^2}{2}\log\left(\frac{T}{\varepsilon}\right) \\
&\quad + \frac{1}{2}\left(\frac{(\boldsymbol{x}-\boldsymbol{X}_{T-\varepsilon})'\boldsymbol{\Sigma}^{-1}(\boldsymbol{X}_{T-\varepsilon},T-\varepsilon)(\boldsymbol{x}-\boldsymbol{X}_{T-\varepsilon})}{\varepsilon}\right. \\
&\qquad \left. - \frac{(\boldsymbol{x}-\boldsymbol{x}_0)'\boldsymbol{\Sigma}^{-1}(\boldsymbol{x}_0,0)(\boldsymbol{x}-\boldsymbol{x}_0)}{T}\right)
\end{aligned} \tag{B.29}
$$

with D_1, D_2 and D_3 as defined on p. 403. The last line equals the logarithms of two unnormalised Gaussian densities. That implies

$$\frac{\mathrm{d}\mathbb{P}_{\boldsymbol{\theta}}}{\mathrm{d}\mathbb{D}_{\boldsymbol{\mu},\boldsymbol{\theta}}}\left(\boldsymbol{X}_{(0,T-\varepsilon]}\right)$$

$$= \exp\left(-\int_0^{T-\varepsilon} \frac{D_1(\boldsymbol{X}_t,t)+D_2(\boldsymbol{X}_t,t)+D_3(\boldsymbol{X}_t,t)}{2(T-t)}\right)\left(\frac{T}{\varepsilon}\right)^{-\frac{d^2}{2}}$$

$$\cdot \frac{\phi\big(\boldsymbol{x}\,\big|\,\boldsymbol{x}_0, T\boldsymbol{\Sigma}(\boldsymbol{x}_0,0)\big)}{\phi\big(\boldsymbol{x}\,\big|\,\boldsymbol{X}_{T-\varepsilon}, \varepsilon\boldsymbol{\Sigma}(\boldsymbol{X}_{T-\varepsilon},T-\varepsilon)\big)}\cdot\left(\frac{T}{\varepsilon}\right)^{\frac{d}{2}}\frac{|\boldsymbol{\Sigma}(\boldsymbol{x}_0,0)|^{\frac{1}{2}}}{|\boldsymbol{\Sigma}(\boldsymbol{X}_{T-\varepsilon},T-\varepsilon)|^{\frac{1}{2}}}.$$

As $\varepsilon > 0$ is typically small, one can assume $f_{\boldsymbol{\theta}}(\boldsymbol{x}|\boldsymbol{X}_{T-\varepsilon}) \approx \phi(\boldsymbol{x}|\boldsymbol{X}_{T-\varepsilon}, \varepsilon\boldsymbol{\Sigma}(\boldsymbol{X}_{T-\varepsilon}, T-\varepsilon))$. Then, with Eq. (B.21),

$$\frac{\mathrm{d}\tilde{\mathbb{P}}_{\boldsymbol{\theta}}}{\mathrm{d}\mathbb{D}_{\boldsymbol{\mu},\boldsymbol{\theta}}}\left(\boldsymbol{X}_{(0,T-\varepsilon]}\right)$$

$$= \exp\left(-\int_0^{T-\varepsilon} \frac{D_1(\boldsymbol{X}_t,t)+D_2(\boldsymbol{X}_t,t)+D_3(\boldsymbol{X}_t,t)}{2(T-t)}\right)\left(\frac{T}{\varepsilon}\right)^{-\frac{d}{2}(d-1)}$$

$$\cdot \frac{\phi\big(\boldsymbol{x}\,\big|\,\boldsymbol{x}_0, T\boldsymbol{\Sigma}(\boldsymbol{x}_0,0)\big)}{f_{\boldsymbol{\theta}}(\boldsymbol{x})}\frac{|\boldsymbol{\Sigma}(\boldsymbol{x}_0,0)|^{\frac{1}{2}}}{|\boldsymbol{\Sigma}(\boldsymbol{X}_{T-\varepsilon},T-\varepsilon)|^{\frac{1}{2}}}.$$

This is the first of the two formulas which were claimed at the beginning of this section. Utilise this result for the derivation of $(\mathrm{d}\tilde{\mathbb{P}}_{\boldsymbol{\theta}}/\mathrm{d}\mathbb{D}_{\mathbf{0},\boldsymbol{\theta}})(\boldsymbol{X}_{(0,T-\varepsilon]})$: The generalised Girsanov formula from Delyon and Hu (2006, Theorem 1) yields

$$\log\left(\frac{\mathrm{d}\mathbb{D}_{\boldsymbol{\mu},\boldsymbol{\theta}}}{\mathrm{d}\mathbb{D}_{\mathbf{0},\boldsymbol{\theta}}}\right)\left(\boldsymbol{X}_{(0,T-\varepsilon]}\right) = \int_0^{T-\varepsilon} \boldsymbol{\mu}'(\boldsymbol{X}_t,t)\boldsymbol{\Sigma}^{-1}(\boldsymbol{X}_t,t)\mathrm{d}\boldsymbol{X}_t$$

$$-\frac{1}{2}\int_0^{T-\varepsilon} \boldsymbol{\mu}'(\boldsymbol{X}_t,t)\boldsymbol{\Sigma}^{-1}(\boldsymbol{X}_t,t)\boldsymbol{\mu}(\boldsymbol{X}_t,t)\mathrm{d}t + \int_0^{T-\varepsilon}\frac{D_1(\boldsymbol{X}_t,t)}{2(T-t)}.$$

Hence

$$\frac{\mathrm{d}\tilde{\mathbb{P}}_{\boldsymbol{\theta}}}{\mathrm{d}\mathbb{D}_{\mathbf{0},\boldsymbol{\theta}}}\left(\boldsymbol{X}_{(0,T-\varepsilon]}\right) = \left(\frac{\mathrm{d}\tilde{\mathbb{P}}_{\boldsymbol{\theta}}}{\mathrm{d}\mathbb{D}_{\boldsymbol{\mu},\boldsymbol{\theta}}}\frac{\mathrm{d}\mathbb{D}_{\boldsymbol{\mu},\boldsymbol{\theta}}}{\mathrm{d}\mathbb{D}_{\mathbf{0},\boldsymbol{\theta}}}\right)\left(\boldsymbol{X}_{(0,T-\varepsilon]}\right)$$

$$= \exp\left(-\int_0^{T-\varepsilon}\frac{D_2(\boldsymbol{X}_t,t)+D_3(\boldsymbol{X}_t,t)}{2(T-t)}\right)\cdot\frac{\phi(\boldsymbol{x}|\,\boldsymbol{x}_0,T\boldsymbol{\Sigma}(\boldsymbol{x}_0,0))}{f_{\boldsymbol{\theta}}(\boldsymbol{x})}\frac{|\boldsymbol{\Sigma}(\boldsymbol{x}_0,0)|^{\frac{1}{2}}}{|\boldsymbol{\Sigma}(\boldsymbol{X}_{T-\varepsilon},T-\varepsilon)|^{\frac{1}{2}}}$$

$$\cdot \exp\left(\int_0^{T-\varepsilon} \boldsymbol{\mu}'(\boldsymbol{X}_t,t)\boldsymbol{\Sigma}^{-1}(\boldsymbol{X}_t,t)\mathrm{d}\boldsymbol{X}_t - \frac{1}{2}\int_0^{T-\varepsilon} \boldsymbol{\mu}'(\boldsymbol{X}_t,t)\boldsymbol{\Sigma}^{-1}(\boldsymbol{X}_t,t)\boldsymbol{\mu}(\boldsymbol{X}_t,t)\mathrm{d}t\right)\left(\frac{T}{\varepsilon}\right)^{-\frac{d}{2}(d-1)}.$$

That was to be shown.

Remark B.1. Delyon and Hu (2006) utilise (B.29) with $\varepsilon \to 0$ to perform a transition from expectations with respect to $\mathbb{P}_{\boldsymbol{\theta}}$ to expectations with respect to $\tilde{\mathbb{P}}_{\boldsymbol{\theta}}$. This way they arrive at

$$\frac{\mathrm{d}\tilde{\mathbb{P}}_{\boldsymbol{\theta}}}{\mathrm{d}\mathbb{D}_{\boldsymbol{\mu},\boldsymbol{\theta}}}(\boldsymbol{X}_{[0,T]}) = \frac{\exp\left(-\int_0^T \frac{D_1(\boldsymbol{X}_t,t)+D_2(\boldsymbol{X}_t,t)+D_3(\boldsymbol{X}_t,t)}{2(T-t)}\right)}{\mathbb{E}_{\mathbb{D}_{\boldsymbol{\mu},\boldsymbol{\theta}}}\left(\exp\left(-\int_0^T \frac{D_1(\boldsymbol{X}_t,t)+D_2(\boldsymbol{X}_t,t)+D_3(\boldsymbol{X}_t,t)}{2(T-t)}\right)\right)} \tag{B.30}$$

and

$$\frac{\mathrm{d}\tilde{\mathbb{P}}_{\boldsymbol{\theta}}}{\mathrm{d}\mathbb{D}_{\mathbf{0},\boldsymbol{\theta}}}(\boldsymbol{X}_{[0,T]}) = \frac{\exp\left(-\int_0^T \frac{D_2(\boldsymbol{X}_t,t)+D_3(\boldsymbol{X}_t,t)}{2(T-t)} + D^*(\boldsymbol{X}_{[0,T]})\right)}{\mathbb{E}_{\mathbb{D}_{\mathbf{0},\boldsymbol{\theta}}}\left(\exp\left(-\int_0^T \frac{D_2(\boldsymbol{X}_t,t)+D_3(\boldsymbol{X}_t,t)}{2(T-t)} + D^*(\boldsymbol{X}_{[0,T]})\right)\right)}, \tag{B.31}$$

where

$$D^*(\boldsymbol{X}_{[0,T]}) = \int_0^T \boldsymbol{\mu}'(\boldsymbol{X}_t,t)\boldsymbol{\Sigma}^{-1}(\boldsymbol{X}_t,t)\mathrm{d}\boldsymbol{X}_t - \frac{1}{2}\int_0^T \boldsymbol{\mu}'(\boldsymbol{X}_t,t)\boldsymbol{\Sigma}^{-1}(\boldsymbol{X}_t,t)\boldsymbol{\mu}(\boldsymbol{X}_t,t)\mathrm{d}t.$$

Be aware that the denominators of (B.30) and (B.31) depend on the parameter $\boldsymbol{\theta}$.

B.8 Derivation of Acceptance Probability

This section aims to derive the acceptance probability (7.78) from p. 256 as an implication of Eq. (7.76). The notation here is the adopted from those formulas.

Let $\boldsymbol{X}^{\mathrm{imp}*} \sim \mathbb{D}_{\boldsymbol{\mu},\boldsymbol{\theta}}$. Then $\boldsymbol{Z}^{\mathrm{imp}*} = g^{-1}(\boldsymbol{X}^{\mathrm{imp}*}, \boldsymbol{\theta})$ induces some law $\mathbb{H}_{\boldsymbol{\theta}}$. The proposal density for $\boldsymbol{Z}^{\mathrm{imp}*}$ is hence $q = \mathrm{d}\mathbb{H}_{\boldsymbol{\theta}}/\mathrm{d}\mathbb{L}$. Then

$$\frac{(\mathrm{d}\mathbb{Z}_{\boldsymbol{\theta}}/\mathrm{d}\mathbb{L})(\boldsymbol{Z}^{\mathrm{imp}*})}{q(\boldsymbol{Z}^{\mathrm{imp}*}|\boldsymbol{Z}^{\mathrm{imp}}, \boldsymbol{x}_0, \boldsymbol{x}, \boldsymbol{\theta})} = \left(\frac{\mathrm{d}\mathbb{Z}_{\boldsymbol{\theta}}}{\mathrm{d}\mathbb{W}}\frac{\mathrm{d}\mathbb{W}}{\mathrm{d}\mathbb{L}}\frac{\mathrm{d}\mathbb{L}}{\mathrm{d}\mathbb{H}_{\boldsymbol{\theta}}}\right)(\boldsymbol{Z}^{\mathrm{imp}*}) = \left(\frac{\mathrm{d}\mathbb{Z}_{\boldsymbol{\theta}}}{\mathrm{d}\mathbb{W}}\frac{\mathrm{d}\mathbb{W}}{\mathrm{d}\mathbb{H}_{\boldsymbol{\theta}}}\right)(\boldsymbol{Z}^{\mathrm{imp}*})$$

$$= \left(\frac{\mathrm{d}\tilde{\mathbb{P}}_{\boldsymbol{\theta}}}{\mathrm{d}\mathbb{D}_{\mathbf{0},\boldsymbol{\theta}}}\frac{\mathrm{d}\mathbb{D}_{\mathbf{0},\boldsymbol{\theta}}}{\mathrm{d}\mathbb{D}_{\boldsymbol{\mu},\boldsymbol{\theta}}}\right)(\boldsymbol{X}^{\mathrm{imp}*}) = \frac{\mathrm{d}\tilde{\mathbb{P}}_{\boldsymbol{\theta}}}{\mathrm{d}\mathbb{D}_{\boldsymbol{\mu},\boldsymbol{\theta}}}(\boldsymbol{X}^{\mathrm{imp}*}).$$

The change of measures follows with the same argument as in (7.59) and (7.60) on p. 247. Plug in the above equality into the acceptance probability (7.76) on p. 256 to obtain

$$\zeta(\boldsymbol{Z}^{\mathrm{imp}*}, \boldsymbol{Z}^{\mathrm{imp}}) = 1 \wedge \left(\frac{\mathrm{d}\tilde{\mathbb{P}}_{\boldsymbol{\theta}}}{\mathrm{d}\mathbb{D}_{\boldsymbol{\mu},\boldsymbol{\theta}}}(\boldsymbol{X}^{\mathrm{imp}*})\right) \Big/ \left(\frac{\mathrm{d}\tilde{\mathbb{P}}_{\boldsymbol{\theta}}}{\mathrm{d}\mathbb{D}_{\boldsymbol{\mu},\boldsymbol{\theta}}}(\boldsymbol{X}^{\mathrm{imp}})\right).$$

That agrees with (7.78).

References

Delyon B, Hu Y (2006) Simulation of conditioned diffusion and application to parameter estimation. Stoch Process Appl 116:1660–1675

Elerian O, Chib S, Shephard N (2001) Likelihood inference for discretely observed nonlinear diffusions. Econometrica 69:959–993

Parzen E (1964) On statistical spectral analysis. In: Bellman R (ed) Stochastic processes in mathematical physics and engineering: proceedings of symposia in applied mathematics, vol XVI. American Mathematical Society, Providence, pp 221–246

Appendix C
Supplementary Material for Application II

This chapter contains additional calculations, figures and tables for the analysis of molecular binding in Chap. 9.

C.1 Diffusion Approximations

In this section, two diffusion approximations are derived which are utilised as kinetic models in Chap. 9.

C.1.1 One Mobility Class

In Sect. 9.3.2, the SDE

$$\begin{pmatrix} \mathrm{d}u^{\mathrm{free}} \\ \mathrm{d}u^{\mathrm{bound}}_{\mathrm{bl}} \end{pmatrix} = \begin{pmatrix} \mu_1 \\ \mu_2 \end{pmatrix} \mathrm{d}t + \frac{1}{\sqrt{N_U}} \begin{pmatrix} \sigma_{11} & -\sigma_{22} \\ 0 & \sigma_{22} \end{pmatrix} \begin{pmatrix} \mathrm{d}B_1(t) \\ \mathrm{d}B_2(t) \end{pmatrix}$$

was derived, where

$$\begin{aligned} \mu_1 &= -(k_{\mathrm{on}} + k_{\mathrm{off}})u^{\mathrm{free}} + k_{\mathrm{off}} \\ \mu_2 &= k_{\mathrm{on}} f_{\mathrm{bl}} u^{\mathrm{free}} - k_{\mathrm{off}} u^{\mathrm{bound}}_{\mathrm{bl}} \\ \sigma_{11} &= \sqrt{k_{\mathrm{on}}(1-f_{\mathrm{bl}})u^{\mathrm{free}} + k_{\mathrm{off}}(1-u^{\mathrm{free}}-u^{\mathrm{bound}}_{\mathrm{bl}})} \\ \sigma_{22} &= \sqrt{k_{\mathrm{on}} f_{\mathrm{bl}} u^{\mathrm{free}} + k_{\mathrm{off}} u^{\mathrm{bound}}_{\mathrm{bl}}}\,. \end{aligned} \tag{C.1}$$

C. Fuchs, *Inference for Diffusion Processes: With Applications in Life Sciences*, DOI 10.1007/978-3-642-25969-2, © Springer-Verlag Berlin Heidelberg 2013

In the following, this SDE is transformed into one for the process $(q, u^{\text{free}})'$ by application of Itô's formula from Sect. 3.2.10, where

$$q = u^{\text{free}} + \frac{1}{f_{\text{bl}}} u_{\text{bl}}^{\text{bound}}.$$

Itô's formula yields

$$\begin{aligned} \mathrm{d}q &= \mathrm{d}u^{\text{free}} + \frac{1}{f_{\text{bl}}} \mathrm{d}u_{\text{bl}}^{\text{bound}} \\ &= \left(\mu_1 + \frac{1}{f_{\text{bl}}} \mu_2\right) \mathrm{d}t + \frac{1}{\sqrt{N_U}} \left(\sigma_{11} \mathrm{d}B_1(t) + \left(\frac{1}{f_{\text{bl}}} - 1\right) \sigma_{22} \mathrm{d}B_2(t)\right). \end{aligned}$$

Thus, one has

$$\begin{aligned} \begin{pmatrix} \mathrm{d}q \\ \mathrm{d}u^{\text{free}} \end{pmatrix} &= \begin{pmatrix} \tilde{\mu}_1 \\ \tilde{\mu}_2 \end{pmatrix} \mathrm{d}t + \frac{1}{\sqrt{N_U}} \begin{pmatrix} \tilde{\sigma}_{11} & \tilde{\sigma}_{12} \\ \tilde{\sigma}_{21} & \tilde{\sigma}_{22} \end{pmatrix} \begin{pmatrix} \mathrm{d}B_1(t) \\ \mathrm{d}B_2(t) \end{pmatrix} \\ &= \begin{pmatrix} \mu_1 + \mu_2 / f_{\text{bl}} \\ \mu_1 \end{pmatrix} \mathrm{d}t + \frac{1}{\sqrt{N_U}} \begin{pmatrix} \sigma_{11} & (f_{\text{bl}}^{-1} - 1)\sigma_{22} \\ \sigma_{11} & -\sigma_{22} \end{pmatrix} \begin{pmatrix} \mathrm{d}B_1(t) \\ \mathrm{d}B_2(t) \end{pmatrix}. \end{aligned}$$

Hence, the components of the drift vector and diffusion coefficient of the transformed process $(q, u^{\text{free}})'$ can easily be obtained from (C.1). In these formulas, however, the variable $u_{\text{bl}}^{\text{bound}}$ is to be replaced by $f_{\text{bl}}(q - u^{\text{free}})$. Then

$$\begin{aligned} \tilde{\mu}_1 &= k_{\text{off}} (1 - q) \\ \tilde{\mu}_2 &= -(k_{\text{on}} + k_{\text{off}}) u^{\text{free}} + k_{\text{off}} \\ \tilde{\sigma}_{11} = \tilde{\sigma}_{21} &= \sqrt{k_{\text{off}} (1 - f_{\text{bl}} q) + (k_{\text{on}} - k_{\text{off}}) (1 - f_{\text{bl}}) u^{\text{free}}} \\ \tilde{\sigma}_{12} &= \left(\frac{1}{f_{\text{bl}}} - 1\right) \sqrt{k_{\text{off}} f_{\text{bl}} q + (k_{\text{on}} - k_{\text{off}}) f_{\text{bl}} u^{\text{free}}} \\ \tilde{\sigma}_{22} &= - \sqrt{k_{\text{off}} f_{\text{bl}} q + (k_{\text{on}} - k_{\text{off}}) f_{\text{bl}} u^{\text{free}}}. \end{aligned}$$

The diffusion matrix of $(q, u^{\text{free}})'$ equals

$$\frac{1}{N_U} \begin{pmatrix} \tilde{\Sigma}_{11} & \tilde{\Sigma}_{12} \\ \tilde{\Sigma}_{21} & \tilde{\Sigma}_{22} \end{pmatrix},$$

where

$$\tilde{\Sigma}_{11} = k_{\text{off}} \left(\frac{1}{f_{\text{bl}}} - 2 \right) q + \left(k_{\text{on}} - k_{\text{off}} \right) \left(\frac{1}{f_{\text{bl}}} - 1 \right) u^{\text{free}} + k_{\text{off}}$$

$$\tilde{\Sigma}_{12} = \tilde{\Sigma}_{21} = k_{\text{off}} \left(1 - q \right)$$

$$\tilde{\Sigma}_{22} = \left(k_{\text{on}} - k_{\text{off}} \right) u^{\text{free}} + k_{\text{off}} .$$

C.1.2 Multiple Mobility Classes

In what follows, the Markov jump model from Sect. 9.5.1 is translated into a diffusion process. To that end, a diffusion approximation for a process with $(2M+1)$-dimensional state variable

$$\left(u^{\text{free}}, u_{\text{bl}*}^{\text{bound},1}, \ldots, u_{\text{bl}*}^{\text{bound},M}, u_{\text{unbl}*}^{\text{bound},1}, \ldots, u_{\text{unbl}*}^{\text{bound},M} \right)' . \tag{C.2}$$

is formulated. Afterwards, this process is transformed to a diffusion approximation for the $2M$-dimensional state variable

$$\left(q^*, u^{\text{free}}, u_{\text{bl}*}^{\text{bound},1}, \ldots, u_{\text{bl}*}^{\text{bound},M-1}, u_{\text{unbl}*}^{\text{bound},1}, \ldots, u_{\text{unbl}*}^{\text{bound},M-1} \right)' . \tag{C.3}$$

The notation in this section is adopted from Sect. 9.5.2.

With the Langevin approach from Sect. 4.3.3, one arrives at a diffusion process for (C.2) with drift vector and diffusion matrix

$$\begin{pmatrix} \tilde{\mu}^{\text{f}} \\ \tilde{\boldsymbol{\mu}}^{\text{b}} \\ \tilde{\boldsymbol{\mu}}^{\text{u}} \end{pmatrix} \quad \text{and} \quad \frac{1}{N_U} \begin{pmatrix} \tilde{\Sigma}^{\text{ff}} & \tilde{\boldsymbol{\Sigma}}^{\text{fb}} & \tilde{\boldsymbol{\Sigma}}^{\text{fu}} \\ \tilde{\boldsymbol{\Sigma}}^{\text{bf}} & \tilde{\boldsymbol{\Sigma}}^{\text{bb}} & \tilde{\boldsymbol{\Sigma}}^{\text{bu}} \\ \tilde{\boldsymbol{\Sigma}}^{\text{uf}} & \tilde{\boldsymbol{\Sigma}}^{\text{ub}} & \tilde{\boldsymbol{\Sigma}}^{\text{uu}} \end{pmatrix} .$$

More precisely, the components of the drift vector are

$$\tilde{\mu}^{\text{f}} \in \mathbb{R} \quad \text{with} \qquad \tilde{\mu}^{\text{f}} = -\left(\sum_{i=1}^{M} k_{\text{on},i} \right) u^{\text{free}} + \sum_{i=1}^{M} k_{\text{off},i} u^{\text{bound},i}$$

$$\tilde{\boldsymbol{\mu}}^{\text{b}} = (\tilde{\mu}_i^{\text{b}}) \in \mathbb{R}^M \text{ with} \qquad \tilde{\mu}_i^{\text{b}} = k_{\text{on},i} f_{\text{bl}}^* u^{\text{free}} - k_{\text{off},i} u_{\text{bl}*}^{\text{bound},i}$$

$$\tilde{\boldsymbol{\mu}}^{\text{u}} = (\tilde{\mu}_i^{\text{u}}) \in \mathbb{R}^M \text{ with} \qquad \tilde{\mu}_i^{\text{u}} = k_{\text{on},i}(1 - f_{\text{bl}}^*) u^{\text{free}} - k_{\text{off},i} u_{\text{unbl}*}^{\text{bound},i} ,$$

where $i = 1, \ldots, M$ and $u^{\text{bound},i} = u_{\text{bl}*}^{\text{bound},i} + u_{\text{unbl}*}^{\text{bound},i}$. The components of the diffusion matrix are

$$\tilde{\Sigma}^{\mathrm{ff}} \in \mathbb{R} \qquad \text{with} \qquad \tilde{\Sigma}^{\mathrm{ff}} = \left(\sum_{i=1}^{M} k_{\mathrm{on},i}\right) u^{\mathrm{free}} + \sum_{i=1}^{M} k_{\mathrm{off},i} u^{\mathrm{bound},i}$$

$$\tilde{\boldsymbol{\Sigma}}^{\mathrm{bb}} = (\tilde{\Sigma}_{ij}^{\mathrm{bb}}) \in \mathbb{R}^{M \times M} \text{ with } \qquad \tilde{\Sigma}_{ii}^{\mathrm{bb}} = k_{\mathrm{on},i} f_{\mathrm{bl}}^{*} u^{\mathrm{free}} + k_{\mathrm{off},i} u_{\mathrm{bl}*}^{\mathrm{bound},i}$$

$$\text{and} \qquad \tilde{\Sigma}_{ij}^{\mathrm{bb}} = 0 \qquad \text{for } i \neq j$$

$$\tilde{\boldsymbol{\Sigma}}^{\mathrm{uu}} = (\tilde{\Sigma}_{ij}^{\mathrm{uu}}) \in \mathbb{R}^{M \times M} \text{ with } \qquad \tilde{\Sigma}_{ii}^{\mathrm{uu}} = k_{\mathrm{on},i}(1 - f_{\mathrm{bl}}^{*}) u^{\mathrm{free}} + k_{\mathrm{off},i} u_{\mathrm{unbl}*}^{\mathrm{bound},i}$$

$$\text{and} \qquad \tilde{\Sigma}_{ij}^{\mathrm{uu}} = 0 \qquad \text{for } i \neq j,$$

where $i, j = 1, \ldots, M$. Furthermore,

$$\tilde{\boldsymbol{\Sigma}}^{\mathrm{fb}} = (\tilde{\boldsymbol{\Sigma}}^{\mathrm{bf}})' \in \mathbb{R}^{M} \qquad \text{with} \quad \tilde{\Sigma}_{i}^{\mathrm{bf}} = -\tilde{\Sigma}_{ii}^{\mathrm{bb}}$$

$$\tilde{\boldsymbol{\Sigma}}^{\mathrm{fu}} = (\tilde{\boldsymbol{\Sigma}}^{\mathrm{uf}})' \in \mathbb{R}^{M} \qquad \text{with} \quad \tilde{\Sigma}_{i}^{\mathrm{uf}} = -\tilde{\Sigma}_{ii}^{\mathrm{uu}}$$

$$\tilde{\boldsymbol{\Sigma}}^{\mathrm{bu}} = (\tilde{\boldsymbol{\Sigma}}^{\mathrm{ub}})' \in \mathbb{R}^{M \times M} \text{ with } \quad \tilde{\Sigma}_{ij}^{\mathrm{ub}} = 0 \qquad \text{for all } i, j.$$

When proceeding from the state variable (C.2)–(C.3), the components $u_{\mathrm{bl}*}^{\mathrm{bound},M}$ and $u_{\mathrm{unbl}*}^{\mathrm{bound},M}$ are to be replaced by

$$u_{\mathrm{bl}*}^{\mathrm{bound},M} = f_{\mathrm{bl}}^{*}\left(q^{*} - u^{\mathrm{free}}\right) - \sum_{i=1}^{M-1} u_{\mathrm{bl}*}^{\mathrm{bound},i}$$

and

$$u_{\mathrm{unbl}*}^{\mathrm{bound},M} = 1 - f_{\mathrm{bl}}^{*} q^{*} + (f_{\mathrm{bl}}^{*} - 1) u^{\mathrm{free}} - \sum_{i=1}^{M-1} u_{\mathrm{unbl}*}^{\mathrm{bound},i},$$

which implies

$$u^{\mathrm{bound},M} = 1 - u^{\mathrm{free}} - \sum_{i=1}^{M-1} u^{\mathrm{bound},i}.$$

The above drift vector and diffusion matrix for (C.2) can then be transformed to a drift vector and diffusion matrix

$$\begin{pmatrix} \mu^{\mathrm{q}} \\ \mu^{\mathrm{f}} \\ \boldsymbol{\mu}^{\mathrm{b}} \\ \boldsymbol{\mu}^{\mathrm{u}} \end{pmatrix} \quad \text{and} \quad \frac{1}{N_U} \begin{pmatrix} \Sigma^{\mathrm{qq}} & \Sigma^{\mathrm{qf}} & \boldsymbol{\Sigma}^{\mathrm{qb}} & \boldsymbol{\Sigma}^{\mathrm{qu}} \\ \Sigma^{\mathrm{fq}} & \Sigma^{\mathrm{ff}} & \boldsymbol{\Sigma}^{\mathrm{fb}} & \boldsymbol{\Sigma}^{\mathrm{fu}} \\ \boldsymbol{\Sigma}^{\mathrm{bq}} & \boldsymbol{\Sigma}^{\mathrm{bf}} & \boldsymbol{\Sigma}^{\mathrm{bb}} & \boldsymbol{\Sigma}^{\mathrm{bu}} \\ \boldsymbol{\Sigma}^{\mathrm{uq}} & \boldsymbol{\Sigma}^{\mathrm{uf}} & \boldsymbol{\Sigma}^{\mathrm{ub}} & \boldsymbol{\Sigma}^{\mathrm{uu}} \end{pmatrix}$$

for (C.3). The lengthy calculations are not shown here, but the results are given in Sect. 9.5.2.

C.2 Calculation of Deterministic Process

The following algorithm shows how the fluorescence intensity q^* can be calculated from knowledge of $k_{\text{on},1}$, $k_{\text{off},1}, \ldots, k_{\text{off},M}$, $u_{\text{bl}*,0}^{\text{bound},1}$ and $f_1, \ldots, f_{M-1}$. That means that there are $2M+1$ free parameters plus the initial value q_0^* which may either be kept fixed or estimated as well. The algorithm requires the nucleus to be in chemical equilibrium.

Algorithm C.1. *Assume that the variables $k_{\text{on},1}$, $k_{\text{off},1}, \ldots, k_{\text{off},M}$, $u_{\text{bl}*,0}^{\text{bound},1}$, $f_1, \ldots, f_{M-1}$ and the initial value q_0^* are known. The fluorescence curve $q^*(t)$ can then be determined for all $t \geq 0$ as follows:*

1. *Calculate $f_M = 1 - f_1 - \ldots - f_{M-1}$.*
2. *For $i = 2, \ldots, M$, derive*

$$u_{\text{bl}*,0}^{\text{bound},i} = u_{\text{bl}*,0}^{\text{bound},1} \frac{f_i}{f_1} .$$

3. *Set*

$$u_0^{\text{free}} = q_0^* - \frac{1}{f_{\text{bl}}} \sum_{i=1}^{M} u_{\text{bl}*,0}^{\text{bound},i} .$$

4. *Obtain the sum of all $k_{\text{on},i}$ through*

$$\sum_{i=1}^{M} k_{\text{on},i} = B \cdot \frac{1 - u_0^{\text{free}}}{u_0^{\text{free}}} ,$$

where

$$B = k_{\text{off},M} + \sum_{i=1}^{M-1} f_i \big(k_{\text{off},i} - k_{\text{off},M}\big).$$

5. *Derive the values $u_{\text{bl}*,\infty}^{\text{bound},i} = \lim_{t\to\infty} u_{\text{bl}*}^{\text{bound},i}(t)$ as*

$$u_{\text{bl}*,\infty}^{\text{bound},1} = \frac{k_{\text{on},1}}{k_{\text{off},1}} f_{\text{bl}}\, u_0^{\text{free}} \qquad \textit{and} \qquad u_{\text{bl}*,\infty}^{\text{bound},i} = u_{\text{bl}*,\infty}^{\text{bound},1} \cdot \frac{f_i}{f_1}$$

for $i = 2, \ldots, M$.

6. *Calculate*

$$k_{\text{on},i} = \frac{u_{\text{bl}*,\infty}^{\text{bound},i}\, k_{\text{off},i}}{f_{\text{bl}}\, u_0^{\text{free}}}$$

for $i = 2, \ldots, M$.

7. *Finally, determine the fluorescence intensity via*

$$q^*(t) = \left(1 + \sum_{i=1}^{M} \frac{k_{\text{on},i}}{k_{\text{off},i}}\right) u_0^{\text{free}} + \sum_{i=1}^{M} \left(\frac{u_{\text{bl}*,0}^{\text{bound},i}}{f_{\text{bl}}^*} - \frac{k_{\text{on},i}}{k_{\text{off},i}}\, u_0^{\text{free}}\right) \exp\big(-k_{\text{off},i}(t - t_0)\big).$$

C.3 Estimation Results

This section shows additional estimation results for the application in Sect. 9.7. These are integrated in the main text in that section.

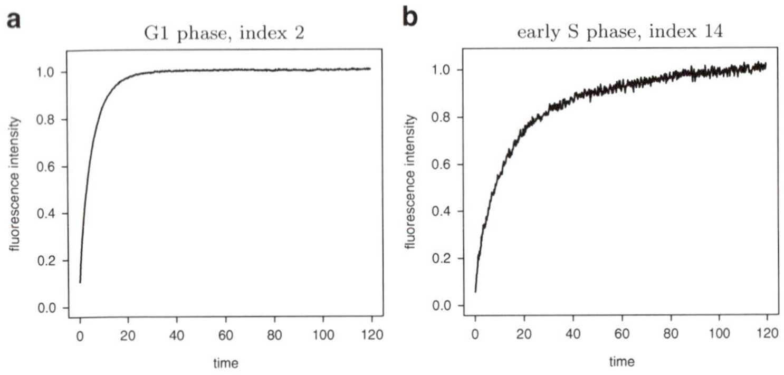

Fig. C.1 Two out of the 47 datasets used in the application in Sect. 9.7, namely the second dataset in G1 phase, (**a**), and the 14th dataset of early S phase, (**b**). The curves substantially differ with respect to their roughness. Consequently, they produce notably different estimates for the number N of molecules (cf. Sect. 9.7.2). These are 38,453 in (**a**) and 180 in (**b**) as also displayed in Table C.1

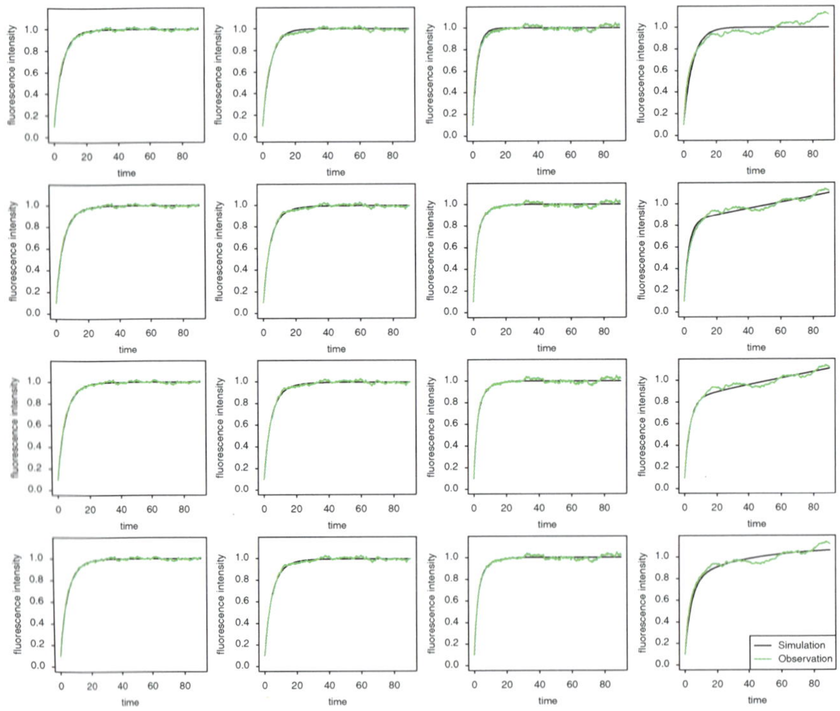

Fig. C.2 Fittings of the predicted deterministic curve (*black*) to the observed data (*green*) according to the estimates in Table 9.6 on p. 341

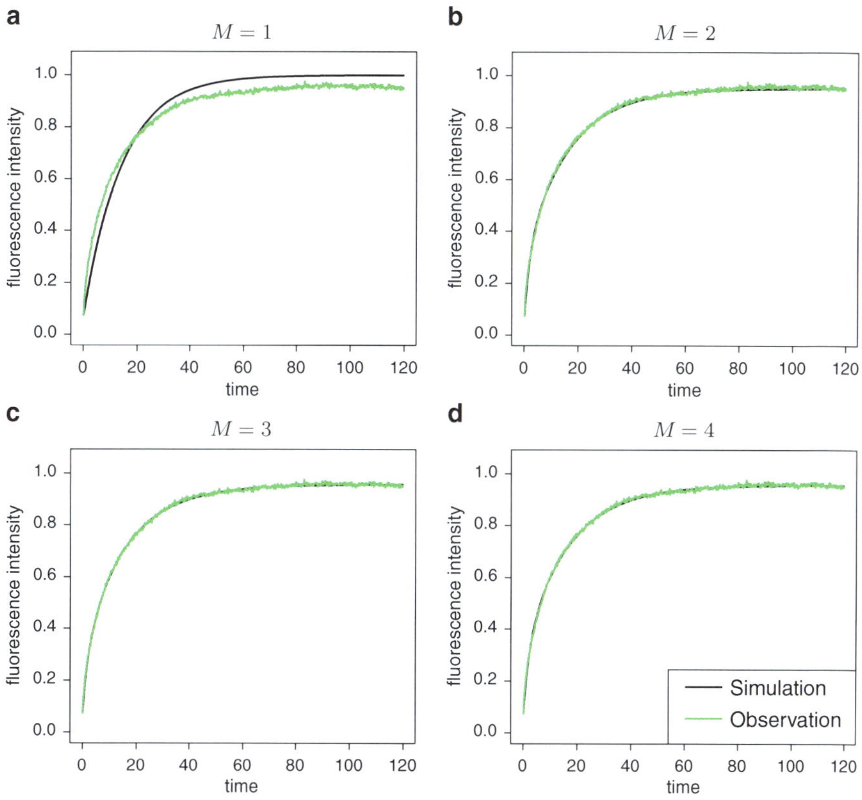

Fig. C.3 Fittings of the predicted deterministic curve (*black*) to the observed data (*green*) for the 11th dataset in late S phase. The data was triple normalised, the intermediate fraction f_{int} set to zero, and the starting value of the FRAP curve was kept fixed. Least squares estimation was carried out for the kinetic model with (**a**) $M = 1$, (**b**) $M = 2$, (**c**) $M = 3$ and (**d**) $M = 4$ mobility classes. The according mean sums of squared residuals (mSSR) from Eq. (9.10) on p. 323 are (**a**) 0.002, (**b**) $4.4 \cdot 10^{-5}$, (**c**) $3.6 \cdot 10^{-5}$ and (**d**) $4.1 \cdot 10^{-5}$. The BIC chooses $M = 3$

C.4 Diffusion-Coupled Model

In Sect. 9.8, a kinetic model for diffusion-coupled FRAP was derived. In that section, the dynamics is represented by a compartmental description, as a diffusion approximation and as a deterministic process. The model can be extended to the case of multiple mobility classes in the same manner as for diffusion-uncoupled recovery in Sect. 9.5.

Table C.1 Key figures for the real datasets in Chap. 9: The first and second columns specify the phase of the cell cycle in which the FRAP experiment has been carried out and a consecutive index for each dataset. The third column lists the number of observations in the respective time series. Columns four to seven show some quantities for the triple normalised data, columns eight to eleven display the same for the double normalised datasets. These figures are the estimated number N of molecules in the nucleus (see Sect. 9.7.2 for details and further remarks), the starting value q_0^* of the recovery curve, the fraction of bleached molecules f_{bl} and the intermediate fraction f_{int} as obtained by image analysis

		# data	Triple normalised				Double normalised			
Phase	Index	points	N	q_0^*	f_{bl}	f_{int}	N	q_0^*	f_{bl}	f_{int}
G1	1	390	11,702	0.074	0.623	0.027	13,303	0.152	0.603	0.026
G1	2	778	38,453	0.106	0.493	0.059	45,828	0.127	0.426	0.059
G1	3	778	2,059	0.093	0.613	0.035	2,334	0.171	0.596	0.034
G1	4	480	784	0.106	0.585	0.039	1,017	0.251	0.552	0.038
G1	5	778	3,727	0.062	0.707	0.015	5,447	0.295	0.663	0.015
G1	6	778	778	0.082	0.577	0.031	973	0.226	0.556	0.030
G1	7	778	2,208	0.076	0.625	0.023	2,675	0.206	0.612	0.022
G1	8	778	1,157	0.076	0.679	0.017	1,583	0.273	0.655	0.017
G1	9	778	3,877	0.051	0.663	0.013	5,411	0.258	0.622	0.013
G1	10	778	446	0.077	0.631	0.026	632	0.286	0.592	0.027
early S	1	778	3,155	0.072	0.599	0.030	3,672	0.158	0.573	0.030
early S	2	778	1,709	0.031	0.506	0.017	1,989	0.095	0.468	0.016
early S	3	778	858	0.040	0.520	0.018	925	0.095	0.515	0.016
early S	4	778	1,688	0.085	0.667	0.029	1,907	0.163	0.650	0.029
early S	5	778	1,647	0.069	0.554	0.031	1,776	0.114	0.543	0.031
early S	6	778	3,038	0.048	0.591	0.019	3,436	0.119	0.569	0.018
early S	7	778	2,105	0.040	0.487	0.020	2,391	0.107	0.464	0.020
early S	8	778	2,673	0.024	0.638	0.013	3,259	0.139	0.597	0.013
early S	9	774	962	0.028	0.611	0.012	1,083	0.095	0.585	0.012
early S	10	778	26,892	0.037	0.550	0.017	30,425	0.099	0.522	0.016
early S	11	778	33,914	0.050	0.622	0.020	39,914	0.136	0.586	0.019
early S	12	778	723	0.035	0.516	0.018	834	0.117	0.493	0.018
early S	13	778	1,016	0.056	0.530	0.028	1,181	0.145	0.510	0.028
early S	14	778	180	0.058	0.577	0.027	203	0.123	0.554	0.027
early S	15	778	4,038	0.054	0.642	0.057	4,920	0.173	0.608	0.057
early S	16	778	1,584	0.043	0.495	0.021	1,739	0.098	0.481	0.020
early S	17	778	11,799	0.055	0.606	0.007	13,392	0.121	0.578	0.007
early S	18	778	17,827	0.055	0.569	0.024	20,431	0.133	0.545	0.023
early S	19	778	233	0.014	0.628	0.011	283	0.133	0.593	0.011
early S	20	778	267	0.073	0.607	0.029	477	0.365	0.524	0.026
early S	21	754	2,128	0.037	0.627	0.024	3,470	0.294	0.548	0.024
early S	22	778	6,443	0.040	0.656	0.017	7,931	0.136	0.602	0.017
early S	23	778	1,883	0.101	0.632	0.041	2,135	0.131	0.578	0.041
early S	24	778	1,076	0.080	0.532	0.029	1,429	0.259	0.513	0.026
early S	25	778	901	0.079	0.570	0.026	1,148	0.238	0.553	0.025
early S	26	778	1,105	0.085	0.575	0.031	1,376	0.232	0.560	0.031

(continued)

Table C.1 (continued)

Phase	Index	# data points	Triple normalised N	q_0^*	f_{bl}	f_{int}	Double normalised N	q_0^*	f_{bl}	f_{int}
late S	1	779	1,790	0.062	0.427	0.032	1,875	0.095	0.425	0.032
late S	2	779	402	0.052	0.464	0.028	436	0.104	0.456	0.028
late S	3	779	1,243	0.064	0.536	0.035	1,708	0.260	0.502	0.035
late S	4	779	476	0.033	0.446	0.024	599	0.174	0.427	0.024
late S	5	779	1,034	0.119	0.574	0.054	1,223	0.238	0.559	0.054
late S	6	779	2,097	0.089	0.591	0.033	2,701	0.247	0.564	0.033
late S	7	779	1,650	0.054	0.479	0.035	1,995	0.183	0.462	0.035
late S	8	779	1,111	0.097	0.491	0.050	1,139	0.097	0.479	0.050
late S	9	779	967	0.121	0.459	0.052	1,277	0.311	0.452	0.052
late S	10	779	543	0.075	0.568	0.028	745	0.276	0.537	0.028
late S	11	777	911	0.073	0.554	0.093	1,164	0.229	0.527	0.093

Table C.2 Model choice by means of BIC as defined in Eq. (9.22) on p. 345 for different modifications of the least squares estimation in Sect. 9.7.3: First, the datasets may be either triple normalised or double normalised as described in Sects. 9.6.1 and 9.6.2, respectively. Second, the intermediate fraction f_{int}, introduced in Sect. 9.4, may be either set to zero or equal to the experimentally obtained values from Table C.1. Third, the initial value q_0^* of the recovery curves may either be kept fixed or treated as a free parameter. This table displays the number of mobility classes chosen by the BIC

		$f_{int} = 0$				$f_{int} > 0$			
		Triple		Double		Triple		Double	
Phase	Index	fixed	free	fixed	free	fixed	free	fixed	free
G1	1	2	2	2	2	2	2	2	2
G1	2	3	3	3	3	3	3	3	3
G1	3	2	2	2	2	2	2	2	2
G1	4	2	2	2	2	2	2	2	2
G1	5	3	3	3	3	3	3	3	3
G1	6	3	2	3	2	3	2	2	2
G1	7	3	3	3	3	3	3	3	3
G1	8	4	2	3	3	3	3	3	3
G1	9	2	2	2	2	2	2	2	2
G1	10	2	2	2	2	2	2	2	2
early S	1	3	3	3	3	3	3	3	3
early S	2	2	2	3	2	3	2	3	2
early S	3	2	2	2	2	2	2	2	2
early S	4	3	3	3	4	3	3	4	3
early S	5	3	3	4	4	3	3	3	3

(continued)

Table C.2 (continued)

		$f_{\text{int}} = 0$				$f_{\text{int}} > 0$			
		Triple		Double		Triple		Double	
Phase	Index	fixed	free	fixed	free	fixed	free	fixed	free
early S	6	3	3	3	3	3	3	3	3
early S	7	3	3	3	3	3	3	3	3
early S	8	3	2	3	2	3	2	3	2
early S	9	2	2	2	2	2	2	2	2
early S	10	3	3	3	3	3	3	3	3
early S	11	3	3	3	3	3	4	3	3
early S	12	3	3	4	3	3	3	3	3
early S	13	3	3	3	3	3	3	3	3
early S	14	3	2	4	2	3	2	3	2
early S	15	2	2	2	2	2	2	2	2
early S	16	3	2	3	2	3	2	3	3
early S	17	3	3	3	3	3	4	3	3
early S	18	3	3	3	3	3	3	3	3
early S	19	3	3	3	2	3	3	3	2
early S	20	4	3	4	3	4	3	3	3
early S	21	3	3	3	3	4	4	4	3
early S	22	3	2	3	2	4	3	3	4
early S	23	3	3	3	3	4	3	3	3
early S	24	3	3	4	3	3	4	3	3
early S	25	3	3	3	3	4	3	3	4
early S	26	3	3	3	3	3	3	3	3
late S	1	3	3	3	3	3	3	3	4
late S	2	3	3	3	3	3	3	3	3
late S	3	3	3	3	3	3	4	3	3
late S	4	4	3	3	3	3	3	3	3
late S	5	3	3	3	4	3	3	3	4
late S	6	3	4	3	3	3	4	3	3
late S	7	3	3	4	3	3	3	3	4
late S	8	3	3	4	3	3	3	3	3
late S	9	3	3	3	4	3	3	3	4
late S	10	3	3	3	3	3	4	3	3
late S	11	3	3	3	3	3	3	3	3

The following presents a deterministic description for diffusion-coupled FRAP in case of M mobility classes. The proceeding in the derivation of this model is analogous to that in Sect. 9.5 and hence not shown here. The derivation of a corresponding diffusion approximation is straightforward as well along the lines of Chap. 4.

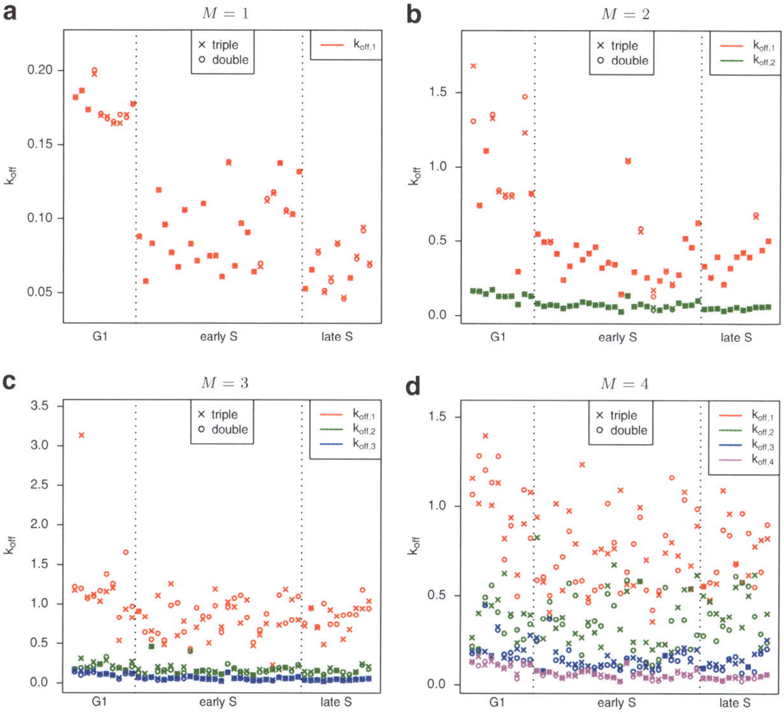

Fig. C.4 Least squares estimates for $k_{\mathrm{off},i}$ with $f_{\mathrm{int}} = 0$ and fixed starting value q_0^*. The underlying datasets are triple normalised (according estimates are marked with a *cross*) or double normalised (estimates are represented by a *circle*). The figures display the estimates for the parameters $k_{\mathrm{off},1}, \ldots, k_{\mathrm{off},M}$ in the deterministic kinetic model with $M = 1, \ldots, 4$ mobility classes. In each plot, the distinct time series are ordered according to their phase and index as in Table C.1, and the respective results are presented from the left to the right

Let $u_{\mathrm{bl}}^{\mathrm{free}}$ denote the fraction of unbleached free molecules in the bleached section of the nucleus and $u_{\mathrm{unbl}}^{\mathrm{free}}$ the fraction of unbleached free molecules in the unbleached section. Furthermore, for $i = 1, \ldots, M$, define $u_{\mathrm{bl}}^{\mathrm{bound},i}$ as the fraction of unbleached type i-bound molecules in the bleached section and $u_{\mathrm{unbl}}^{\mathrm{bound},i}$ as the fraction of unbleached type i-bound molecules in the unbleached section. These variables are non-negative and sum up to one. Given suitable initial values, their dynamics can be described by the set of ODEs

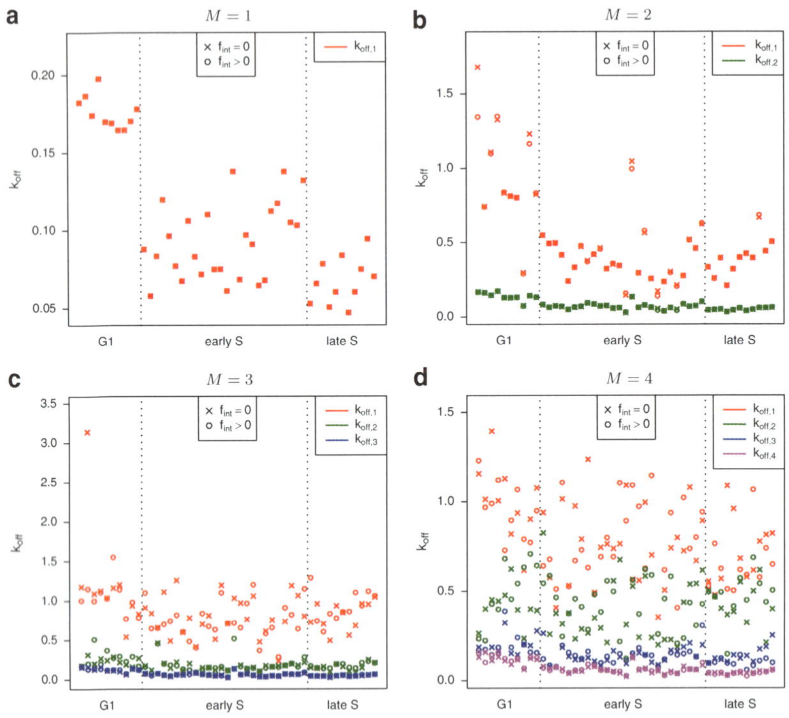

Fig. C.5 Least squares estimates for $k_{\mathrm{off},i}$, based on triple normalised datasets with fixed starting value q_0^*. The intermediate fraction f_{int} is either set to zero (according estimates are marked with a *cross*) or equal to the values from Table C.1 (estimates are represented by a *circle*). The figures display the estimates for the parameters $k_{\mathrm{off},1}, \ldots, k_{\mathrm{off},M}$ in the deterministic kinetic model with $M = 1, \ldots, 4$ mobility classes. In each plot, the distinct time series are ordered according to their phase and index as in Table C.1, and the respective results are presented from the left to the right

$$\frac{du_{\mathrm{bl}}^{\mathrm{free}}}{dt} = -\, u_{\mathrm{bl}}^{\mathrm{free}} \sum_{i=1}^{M} k_{\mathrm{on},i} + \sum_{i=1}^{M} k_{\mathrm{off},i} u_{\mathrm{bl}}^{\mathrm{bound},i} + k_{\mathrm{diff}} \left(f_{\mathrm{bl}} u_{\mathrm{unbl}}^{\mathrm{free}} - (1 - f_{\mathrm{bl}}) u_{\mathrm{bl}}^{\mathrm{free}} \right)$$

$$\frac{du_{\mathrm{unbl}}^{\mathrm{free}}}{dt} = -u_{\mathrm{unbl}}^{\mathrm{free}} \sum_{i=1}^{M} k_{\mathrm{on},i} + \sum_{i=1}^{M} k_{\mathrm{off},i} u_{\mathrm{unbl}}^{\mathrm{bound},i} - k_{\mathrm{diff}} \left(f_{\mathrm{bl}} u_{\mathrm{unbl}}^{\mathrm{free}} - (1 - f_{\mathrm{bl}}) u_{\mathrm{bl}}^{\mathrm{free}} \right)$$

$$\frac{du_{\mathrm{bl}}^{\mathrm{bound},i}}{dt} = k_{\mathrm{on},i} u_{\mathrm{bl}}^{\mathrm{free}} - k_{\mathrm{off},i} u_{\mathrm{bl}}^{\mathrm{bound},i}$$

$$\frac{du_{\mathrm{unbl}}^{\mathrm{bound},i}}{dt} = k_{\mathrm{on},i} u_{\mathrm{unbl}}^{\mathrm{free}} - k_{\mathrm{off},i} u_{\mathrm{unbl}}^{\mathrm{bound},i},$$

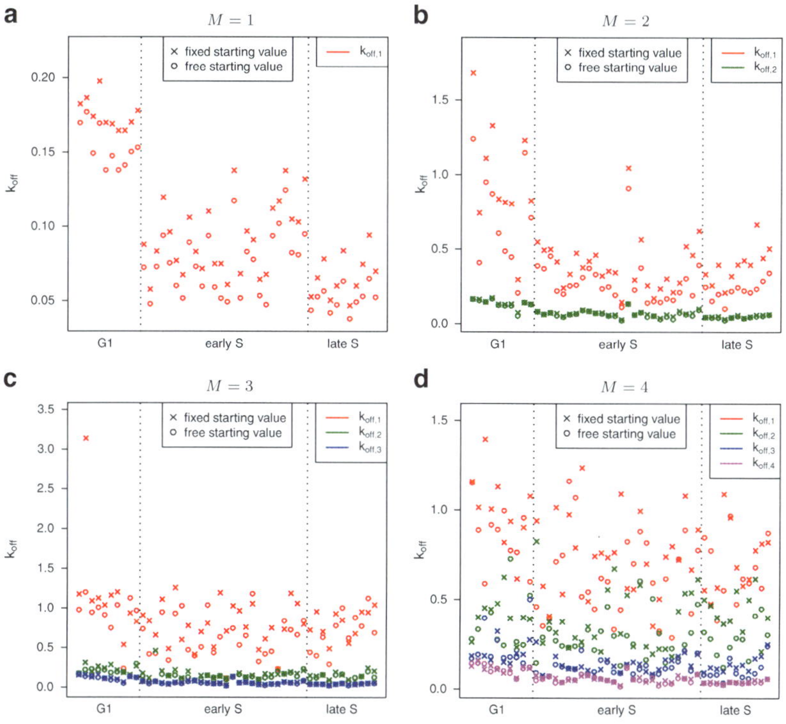

Fig. C.6 Least squares estimates for $k_{\mathrm{off},i}$, based on triple normalised datasets with $f_{\mathrm{int}} = 0$. The starting value q_0^* of the recovery curve is either kept fixed (according estimates are marked with a *cross*) or free, i.e. determined by the optimisation procedure (estimates are represented by a *circle*). The figures display the estimates for the parameters $k_{\mathrm{off},1}, \ldots, k_{\mathrm{off},M}$ in the deterministic kinetic model with $M = 1, \ldots, 4$ mobility classes. In each plot, the distinct time series are ordered according to their phase and index as in Table C.1, and the respective results are presented from the left to the right

where $i = 1, \ldots, M$. The observed variable is the fluorescence intensity

$$q = \frac{u_{\mathrm{bl}}^{\mathrm{free}} + \sum_{i=1}^{M} u_{\mathrm{bl}}^{\mathrm{bound},i}}{f_{\mathrm{bl}}} .$$

The process is fully described by, for example, the state vector

$$\left(q, u_{\mathrm{bl}}^{\mathrm{free}}, u_{\mathrm{unbl}}^{\mathrm{free}}, u_{\mathrm{bl}}^{\mathrm{bound},1}, \ldots, u_{\mathrm{bl}}^{\mathrm{bound},M-1}, u_{\mathrm{unbl}}^{\mathrm{bound},1}, \ldots, u_{\mathrm{unbl}}^{\mathrm{bound},M-1}\right)'.$$

Index

C. Fuchs, *Inference for Diffusion Processes: With Applications in Life Sciences*,
DOI 10.1007/978-3-642-25969-2, © Springer-Verlag Berlin Heidelberg 2013